12th International Workshop on the Physics of Excited Nucleons (NSTAR 2019)

EPJ Web of Conferences Volume 241 (2020)

Beijing, China
19 - 24 May 2019

Editors:

Reinhard Beck
Annika Thiel
Ulrike Thoma
Yannick Wunderlich

ISBN: 978-1-7138-1919-6

Printed from e-media with permission by:

Curran Associates, Inc.
57 Morehouse Lane
Red Hook, NY 12571

Some format issues inherent in the e-media version may also appear in this print version.

For additional information, please contact EDP Sciences – Web of Conferences
at the address below.

EDP Sciences – Web of Conferences
17, Avenue du Hoggar
Parc d'Activité de Courtabœuf
BP 112
F-91944 Les Ulis Cedex A
France

Phone: +33 (0) 1 69 18 75 75
Fax: +33 (0) 1 69 28 84 91

contact-edps@webofconferences.org

Additional copies of this publication are available from:

Curran Associates, Inc.
57 Morehouse Lane
Red Hook, NY 12571 USA
Phone: 845-758-0400
Fax: 845-758-2633
Email: curran@proceedings.com
Web: www.proceedings.com

TABLE OF CONTENTS

Foreword

The *International Workshop on the Physics of Excited Nucleons, NSTAR2019*, the 18th workshop in the esteemed cycle of NSTAR conferences, was held at the University of Bonn, Germany, from June 10-14, 2019. It was the latest of a series of successful conferences at the Rensselear Polytechnic Institute (1988), Florida State University (1994 and 2005), Jefferson Lab (1995, 2000 and 2011), INT Seattle (1996), George Washington University (1997), ECT* Trento (1998), Mainz (2001), Pittsburgh (2002), the LPSC Grenoble (2004), University of Bonn (2007), Institute of High Energy Physics Beijing (2009), Valencia University (2013), Osaka University (2015) and the University of South Carolina (2017).

The goal of *NSTAR2019* was to bring together experts in all areas of physics relevant to baryon spectroscopy, in experiment, phenomenology and theory. Latest results were presented in 20 plenary talks and 59 parallel contributions. All the proceedings, which we received before the deadline are collected in this volume. The workshop was attended by 114 scientists from 41 universities and laboratories from 16 countries around the world.

Exciting new high-precision data were shown from facilities in Asia, Europe and the US, e.g. BESIII, ELSA, GSI, JLab, MAMI and Spring-8. Large-acceptance detectors provide complete angular distributions in many reaction channels. Over the last years, a special focus has been placed on the measurement of single- and double polarization observables. Many new polarization measurements, as well as their impact on the determination of the nucleon excitation spectrum, have been shown and discussed during the meeting.

On the theory side, new results were presented on *ab initio* calculations of the nucleon resonance spectrum on the basis of Quantum Chromodynamics, either directly in terms of quarks and gluons through lattice gauge theory or via Dyson-Schwinger approaches. Chiral effective-field theories were discussed as well, where resonances are described in terms of nucleons and mesons as effective degrees of freedom. Contributions on quark potential models, capable of shedding light on the nucleon resonance spectrum in more phenomenological terms, were also presented. Methods for the solution of inverse scattering problems, i.e. amplitude analysis and partial-wave analysis, which are very important for hadron physics, were discussed in some detail. Through them, an interface between such models and experiment is established which enables a well-defined comparison between calculations and measurements.

We thank all participants for their valuable contributions and the lively discussions which made the workshop the success that it was. The conference was made possible through the existing financial support of our institutional partners.

We are grateful for the advice of the International Advisory Committee and the invaluable help of the Organizing Committee. We would like to extend special thanks to the students Vitalij Adam, Achim Bruchertseifer, Luca Gottardi, Julian Günther, Kai Habermann, Daria Henrichs, Lena Herrmann, Jonas Kohlen, Philipp Krönert, Jutta Kurth, Philipp Mahlberg, Lisa Richter and Jan Schultes who helped to run the meeting smoothly and to the co-workers of the "Helmholtz-Institut für Strahlen- und Kernphysik", Andrea Kehr, Barbara Kraus and Dr. Marcus Grüner who handled the numerous administrative details. We thank Birgit Grötsch and Dr. Philipp Hoffmeister for their expert help in computer services. Further thanks go to Armin Leiendecker and Udo Lütz for the technical help. For the preparation of the proceedings Isabelle Houlbert's and Elodie Guillaud's help was indispensible. The cooperation with the EDP Sciences Production Office was always a pleasure.

After the fruitful workshop at Bonn University we are looking forward to the next meeting which is planned to be held at Venice (Italy) in 2021.

Bonn, September 2020

Reinhard Beck
Annika Thiel
Ulrike Thoma
Yannick Wunderlich

The Editors

edp sciences

EPJ Web of Conferences

Statement of Peer review

In submitting conference proceedings to EPJ Web of Conferences, I certify to the Publisher that I adhere to the **Policy on Publishing Integrity** of the journal in order to safeguard good scientific practice in publishing.

1. All articles have been subjected to peer review administered by the proceedings editors.
2. Reviews have been conducted by expert referees, who have been requested to provide unbiased and constructive comments aimed, whenever possible, at improving the work.
3. Proceedings editors have taken all reasonable steps to ensure the quality of the materials they publish and their decision to accept or reject a paper for publication has been based only on the merits of the work and the relevance to the journal.

Title, date and place of the conference NSTAR 2019, 10.-14.06.2019, Bonn

Proceedings editor(s): Yannick Wunderlich, Annika Thiel

Date and editor's signature 09.09.20 A. Thiel, Y. Wunderlich

17, avenue du Hoggar - PA de Courtabœuf – BP 112 - 91944 Les Ulis Cedex A (France)
Tél. : 33 (0)1 69 18 75 75 - Fax : 33(0)1 69 07 45 17 – www.edpsciences.org

http://doi.org/10.1051/epjconf/202024101001

Recent results from the CBELSA/TAPS experiment at ELSA

Farah Afzal for the CBELSA/TAPS collaboration[1,*]

[1]Helmholtz-Institut für Strahlen- und Kernphysik, Universität Bonn, Germany

Abstract. In order to gain a better understanding of the dynamics inside the nucleon and of the non-perturbative regime of QCD, the nucleon excitation spectra and the properties of nucleon resonances are investigated. An essential experimental tool to achieve this goal is the study of different photoproduction reactions. Partial wave analyses are performed in order to obtain information about the contributing resonances. A complete experiment is needed to extract the underlying amplitudes unambiguously, which requires the measurement of carefully chosen single and double polarization observables in addition to the unpolarized cross section. The CBELSA/TAPS experiment in Bonn offers the possibility to measure several polarization observables using a linearly or circularly polarized photon beam and with a longitudinally or transversely polarized target. This contribution gives an overview of recently measured polarization observables in different final states. The impact of the new data is discussed.

1 Introduction

Comparisons of the nucleon or delta excitation spectra to either quark model predictions [1] or latest lattice QCD calculations [2] reveal that the dynamics inside the nucleons are not well understood in the non-perturbative regime of QCD. In particular, more states are predicted than there have been experimentally observed so far. This feature is known as the *missing resonances* problem. One possible explanation for this could be that most of the states, which were found in the past, were observed in πN scattering experiments. In case some resonances may not couple strongly to the πN, they could have been missed. Therefore, it is crucial to probe the excitation spectra using different production mechanisms such as the photoproduction of mesons. Several facilities around the world have been studying photoproduction reactions like the CLAS collaboration at JLab [3–5], the A2 collaboration at MAMI [6], the LEPS collaboration at SPring-8 [7] and the CBELSA/TAPS collaboration at ELSA [8–14].

The determination of the resonance parameters from the data is a complicated task since the resonances are broad and overlap strongly. Thus, partial wave analyses (PWA) are necessary to disentangle the contributing resonances and their parameters. Different PWA approaches exist e.g. the BnGa [15], the JüBo [16], the SAID [17] or MAID [18] PWA. In order to obtain a unique solution, it is not sufficient to only measure the total unpolarized cross section as it is only sensitive to the squared moduli of the partial wave amplitudes. Using a polarized photon beam, a polarized target or by measuring the recoil nucleon polarization, it is possible to measure different polarization observables. Table 1 shows all the possible combinations of single and double polarization observables that can be measured in the photoproduction of single pseudoscalar mesons. These polarization observables have a sensitivity to interference terms between different partial waves. Therefore, the polarization observables enhance the possibility to detect contributions from small partial waves, via interference of the small waves with large dominant partial waves.

For more than a decade the main focus of the CBELSA/TAPS experiment in Bonn has been dedicated to acquiring polarization observables for different final states. This contribution reports on the latest measured data and their impact.

2 Experimental setup

The ELectron Stretcher Accelerator (ELSA) [19] provides unpolarized or longitudinally polarized electrons with an energy of up to 3.2 GeV. The electron beam is delivered either to the BGO-OD [20] or to the CBELSA/TAPS experimental area (see Figure 1). The electrons impinge on a radiator located inside the goniometer tank and produce a photon beam via bremsstrahlung. The scattered electrons are deflected according to their momenta in the magnetic field of the tagging system and detected by scintillation fibers and bars.

A thin diamond crystal is used as radiator in order to have coherent bremsstrahlung. The diamond crystal lattice can be positioned precisely with respect to the electron beam, which allows to control the coherent edge position [21]. A maximal polarization degree of 65% is achieved for a coherent edge position of $E_\gamma = 950$ MeV and of 35% for $E_\gamma = 1850$ MeV. Longitudinally polarized electrons in combination with an amorphous radiator are utilized to get

*e-mail: afzal@hiskp.uni-bonn.de

Table 1. Table lists all possible polarization observables that can be obtained by using a polarized photon beam, a polarized target by measuring the recoil nucleon polarization degree. Some observables can be measured in two different ways, e.g. P can be measured as a single polarization observable or as a double polarization observable with linearly polarized photon beam and a transversely polarized target. All information are taken from [22].

photon beam		**target**			**recoil nucleon**			**target and recoil**			
		x	y	z	-	-	-	x	z	x	z
		-	-	-	x'	y'	z'	x'	x'	z'	z'
-	σ_0	-	T	-	-	P	-	$T_{x'}$	$L_{x'}$	$T_{z'}$	$L_{z'}$
linear	$-\Sigma$	H	$(-P)$	G	$O_{x'}$	$(-T)$	$O_{z'}$	$(-L_{z'})$	$(T_{z'})$	$(L_{x'})$	$(-T_{x'})$
circular		F	-	$-E$	$C_{x'}$	-	$C_{z'}$	-	-	-	-

circularly polarized photons. Here, a maximal polarization degree of around 65% could be reached for the maximal beam photon energy that is possible to tag for an ELSA energy of $E_0 = 2400$ MeV.

The photons impinge a target which is located in the center of the Crystal Barrel (CB) calorimeter [23]. As target material an unpolarized liquid hydrogen is used to perform measurements on free protons. A longitudinally or transversely polarizable frozen-spin butanol or deuterized butanol target is available for the measurement of polarization observables with polarized protons or neutrons, respectively [24]. A maximal polarization degree of $p_T = 84\%$ and high relaxation times of around 2000 h were reported for the transversely polarized butanol target of most recent beamtimes, which has been possible due to a collaborative effort between the target group of the CBELSA/TAPS collaboration in Bonn and the target group of the A2 collaboration in Mainz [25]. In addition, a carbon foam target is used to study the unpolarized background contribution of the butanol target. Furthermore, in-medium modification of mesons are investigated by utilizing different solid targets.

The CB calorimeter consists of 1320 CsI(Tl) crystals that cover a large polar angular range from 30° to 156° and the full azimuthal angular range. Each crystal is now read out by two avalanche photodiodes instead of the previously used PIN photodiodes. The new readout electronics allows to use the CB in the first level trigger which is crucial for a high trigger efficiency of complete neutral final states, e.g. $n\pi^0$. The upgrade was successfully completed and new data were taken since December 2017.

In forward direction, the MiniTAPS calorimeter [26] is located, which is made of 216 BaF_2 crystals and readout by photomultipliers in order to gain a fast trigger signal.

A cylindrically formed scintillation fiber detector [27] is located around the target. Together with scintillation plates, that are present in front of all the crystals that fulfill $\theta < 30°$, the identification of charged particles is possible.

2.1 Determination of polarization observables

Since the CBELSA/TAPS experiment is equipped with two electromagnetic calorimeters, it is ideally suited for detecting photons in the final state. Thus, this experiment focuses on reconstructing neutral mesons that decay into two or more photons. Figure 2 shows a typical two photon invariant mass spectrum. It is possible to obtain high statistics data sets with very low background contributions for single meson final states like the $p\pi^0$ and $p\eta$ final states. In addition, final states like $p\omega$ or $p\eta'$ can be selected as well. Furthermore, multi-meson final states like $p\pi^0\pi^0$ or $p\pi^0\eta$ have been successfully selected.

As already mentioned, the CBELSA/TAPS experiment provides the possibility to measure with a polarized photon beam and with a polarized target. Thus, the CBELSA/TAPS setup can be used to measure the polarization observables of the *beam-target* category (see Table 1). The polarized cross section $\frac{d\sigma}{d\Omega}_{\text{pol}}$ for a single pseudoscalar meson reads [22]

$$\begin{aligned}\frac{d\sigma}{d\Omega}_{\text{pol}} = \frac{d\sigma}{d\Omega}_0\Big[1 - p_\gamma^{\text{lin}}\mathbf{\Sigma}\cos(2\varphi) \\ + p_x\left(p_\gamma^{\text{lin}}\mathbf{H}\sin(2\varphi) + p_\gamma^{\text{circ}}\mathbf{F}\right) \quad (1)\\ - p_y\left(p_\gamma^{\text{lin}}\mathbf{P}\cos(2\varphi) - \mathbf{T}\right) \quad (2)\\ - p_z\left(-p_\gamma^{\text{lin}}\mathbf{G}\sin(2\varphi) + p_\gamma^{\text{circ}}\mathbf{E}\right)\Big], \quad (3)\end{aligned}$$

where p_γ^{lin} and p_γ^{circ} are the degree of linearly and circularly polarized photons, p_x, p_y and p_z give the target polarization degree in transverse and longitudinal direction and $\frac{d\sigma}{d\Omega}_0$ is the unpolarized differential cross section.

For the measurement of the beam asymmetry Σ data were taken with a linearly polarized photon beam and a liquid hydrogen target in 2013. Here, the coherent edge settings were chosen at high beam photon energies of $E_\gamma = 1750$ MeV and $E_\gamma = 1850$ MeV, where high precision data is scarce for many final states.
Using a polarized frozen-spin butanol target instead of the hydrogen target, allows to measure the double polarization observable H. In addition, the polarization observables T and P, whereby P is measured here as a double polarization observable (see Table 1), can be accessed simultaneously as well. Data were taken with this configuration before and after the CBELSA/TAPS upgrade using different coherent edge settings (from $E_\gamma = 950$ MeV to $E_\gamma = 1300$ MeV) [10, 11]. Furthermore, the double polarization observable G was measured in a separate beam time using a longitudinally polarized butanol target in combination with linearly polarized photons [12, 13]. Besides, the

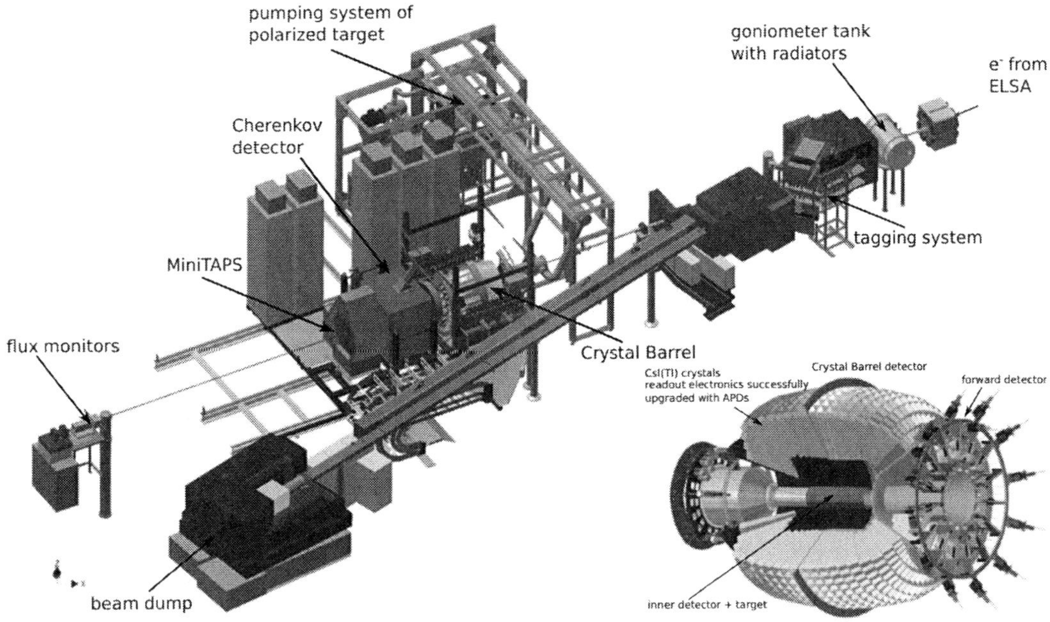

Figure 1. Overview of the CBELSA/TPAS experimental area. The electron beam from ELSA (upper right corner) impinges a thin radiator inside the goniometer tank and a photon beam is produced via bremsstrahlung. The photon energy can be determined with the tagging system that deflects the scattered electrons according to their momenta and detects them via scintillation fibers and bars. The photons hit a target at the center of the Crystal Barrel calorimeter, which is shown in large in the right bottom corner. It is complemented by the MiniTAPS detector in forward direction. The inner detector, which consists of scintillation fibers, and the scintillation plates, that are mounted in front of the forward crystals, are utilized for the identification of charged particles.

helicity asymmetry E can be accessed by using circularly polarized photons and a longitudinally polarized butanol target [8, 9]. It is defined as [22]

$$E = \frac{\sigma_{1/2} - \sigma_{3/2}}{\sigma_{1/2} + \sigma_{3/2}}. \quad (4)$$

Here, $\sigma_{1/2}$ and $\sigma_{3/2}$ are the spin dependent cross sections for antiparallel or parallel spin configuration of the incoming beam photon and target, respectively.

A complication due to the used butanol target is that it does not only contain contributions from the polarizable free hydrogen nuclei but also from the unpolarized bound carbon and oxygen nuclei. The carbon background was studied in detail using additional measurements with a carbon foam target. This allowed the determination of the so-called dilution factor, which gives the amount of polarizable hydrogen nuclei of the selected data. It is given by

$$d = 1 - s_c \frac{N_C}{N_B}, \quad (5)$$

where N_C and N_B are the count rates of the selected carbon and butanol data, respectively and s_C is a scaling factor that is needed to account mainly for the differences in the photon fluxes during the measurements performed with the carbon and the butanol target. References [9, 11, 13] demonstrate in detail how the dilution factor can be determined from the data.

2.2 Recent results

The CBELSA/TAPS collaboration has measured several polarization observables for the photoproduction reactions $\gamma p \rightarrow p\pi^0$ [8–13], $\gamma p \rightarrow p\eta$ [14], $\gamma p \rightarrow p\eta'$ [28], $\gamma p \rightarrow p\pi^0\pi^0$ [29, 30], $\gamma p \rightarrow p\pi^0\eta$ [31] and $\gamma p \rightarrow p\omega$ [32]. Some of the recently obtained results will be discussed in the following.

2.2.1 *Single meson photoproduction: $\gamma p \rightarrow p\pi^0$*

The $p\pi^0$ channel is the most intensely studied photoproduction final state due to its large cross section. It has the largest database for the unpolarized cross section and polarization observables available now, to which the CBELSA/TAPS collaboration contributed significantly.

Figure 3 depicts preliminary results for the beam asymmetry Σ as a function of $\cos\theta_{\pi^0}$ covering a large energy range from $E_\gamma = 1131$ MeV to $E_\gamma = 1779$ MeV. The new data is in good agreement with previously measured data of the GRAAL, CLAS and LEPS collaborations. A Legendre moment analysis revealed sensitivity of this

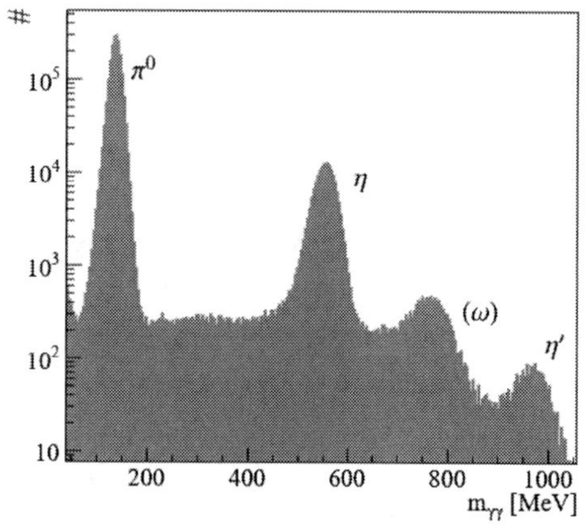

Figure 2. A typical two photon invariant mass is plotted. Clear peaks are visible for the π^0, η, (ω) and η' mesons. The background is very low for the $p\pi^0$ and $p\eta$ final states. The peak of the ω meson appears here due to a lost photon since it can decay to $\omega \to \pi^0\gamma$ [33].

and the CLAS data up to the G-waves [34]. This high sensitivity can only be achieved through the high precision of the data and more importantly due to the full angular coverage of the data. Latest fits from the BnGa and JüBo PWA describe the data well. Within the BnGa PWA evidence was found for the $\Delta(2200)\frac{7}{2}^-(G_{37})$ resonance which can not be the parity partner of the $\Delta(1950)\frac{7}{2}^+(F_{37})$ resonance due to its much higher mass [35].

The helicity asymmetry E was determined for a large energy range from E_γ = 600 MeV to E_γ =2300 MeV and used to determine the spin dependent total cross sections $\sigma_{1/2}$ and $\sigma_{3/2}$ [8, 9], which are shown in Figure 4. It is remarkable that already in the second resonance region (E_γ = 660 MeV to E_γ = 900 MeV) discrepancies exist between data and PWA predictions. These differences could be traced back to the $N(1535)\frac{1}{2}^-(S_{11})$ resonance contribution, which was underestimated by the MAID and SAID-SN11 PWA. Instead both PWAs overestimated the $N(1520)\frac{3}{2}^-(D_{13})$ contribution to the $p\pi^0$ final state. The third resonance region (E_γ = 900 MeV to E_γ = 1200 MeV) is also not well described by the different PWA predictions: while the MAID, SAID-SN11 and BnGa-2011-02 PWA underestimate the $\sigma_{1/2}$ contribution, the JüBo group overestimates it. The description of the data was significantly improved by new fits of the PWA groups.

The results for the polarization observables T, P and H are shown in Figure 5 [10, 11]. The data are in good agreement with previously existing data and exceed them in terms of statistics, angular and energy coverage. Furthermore, the observable H has been measured for the first time. The data show a relatively good agreement also to different PWA predictions.
After the CB upgrade finished, new data were taken for T, P and H, which are depicted in Figures 6 and 7. The new preliminary data shows a very good agreement to the already published CBELSA/TAPS data. In addition, an overall improvement of the statistical precision could be achieved by a factor of approximately two. Furthermore, the existing database for the polarization observables P and H could be extended to E_γ = 1300 MeV.

2.2.2 Single meson photoproduction: $\gamma p \to p\eta$

The database of the $p\eta$ final state is not well established in comparison to the $p\pi^0$ final state since the $p\eta$ cross section is much smaller. Therefore, only a scarce energy and angular coverage exists with larger statistical error bars. Nevertheless, it is prudent to study the $p\eta$ final state since the η has an isospin of $I = 0$ and can only couple to N^* resonances and therefore provides an isospin filter.

The polarization observables E, G, T, P and H have been determined by the CBELSA/TAPS collaboration for the $p\eta$ final state as well. Figure 8 shows the results for the double polarization observable G. This new data provides much needed constraints to the $p\eta$ photoproduction amplitudes since none of the PWAs could predict the angular dependence of G. A new fit of the BnGa PWA including this new data resulted in the precise determination of $N^* \to p\eta$ branching ratios [14]. Here, a remarkable result is the new determined branching ratio of 0.33 ± 0.04 for the $N(1650)\frac{1}{2}^-(S_{11})$ resonance [14]. This branching ratio was listed as 0.14 − 0.22 [33] in the past and thus, was significantly smaller than the branching ratio for the $N(1535)\frac{1}{2}^-(S_{11})$ with $0.42^{+0.13}_{-0.12}$ [33], which was topic of intense discussions [36].

The beam asymmetry Σ was determined for the $p\eta$ final state as well, providing a precise data-set with a large energy and full angular coverage. A Legendre moment analysis was performed here as well, which showed evidence for the $p\eta'$-cusp in the E_{0+} multipole of $p\eta$ [37, 38]. Thus, the unprecedented precision in the new beam asymmetry data puts a reinforced requirement on the PWA groups to implement all relevant singularities correctly into their amplitude parametrizations.

2.2.3 Double meson photoproduction: $\gamma p \to p\pi^0\pi^0$

Multi-meson photoproduction final states, like the $p\pi^0\pi^0$ final state, play an important role at high center-of-mass energies, where cascading decays can be observed. Figure 9 shows a Dalitz plot for the $p\pi^0\pi^0$ final state. The visible horizontal and vertical bands indicate the intermediate resonances ($\Delta(1232)\frac{3}{2}^+$, $N(1520)\frac{3}{2}^-$ and $N(1680)\frac{5}{2}^+$) over which the cascading decays can occur. In addition, studying the $p\pi^0\pi^0$ final state has the advantage of having fewer non-resonant contributions in comparison to the double charged pion channel.

Within the CBELSA/TAPS collaboration the $p\pi^0\pi^0$ final state has been analyzed utilizing the full three

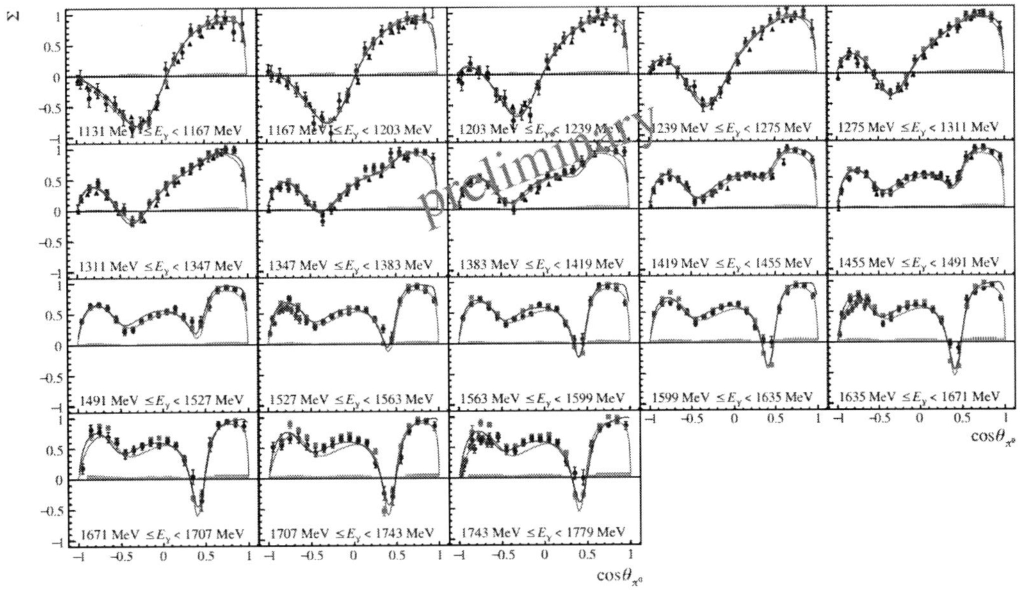

Figure 3. The preliminary results for the beam asymmetry Σ (blue points) are shown as a function of $\cos\theta_{\pi^0}$ for a beam photon energy range from $E_\gamma = 1131$ MeV to $E_\gamma = 1779$ MeV for the $p\pi^0$ final state. The data are compared to data from the CLAS collaboration (green squares) [5], from the GRAAL collaboration (black triangles) [39] and from the LEPS collaboration (red points) [7]. In addition the data are compared to the latest PWA solutions of BnGa [40] and JüBo-2017 [16]. The systematic error is represented by the gray area.

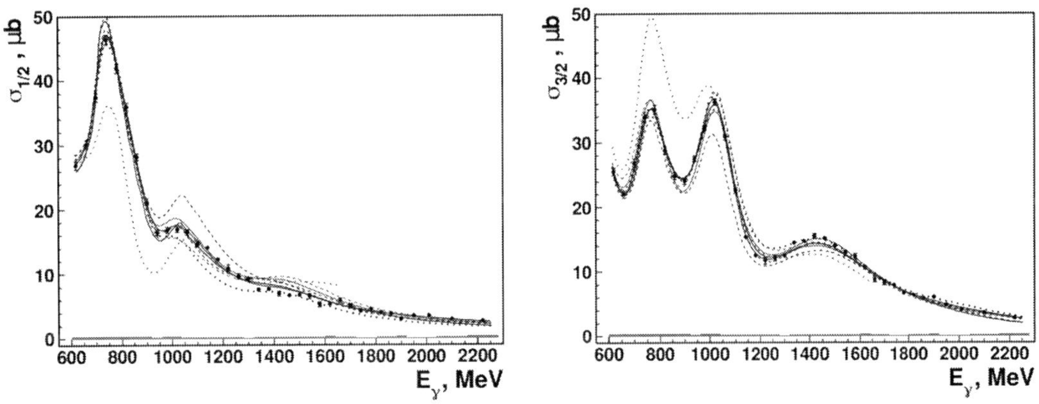

Figure 4. The spin dependent cross sections $\sigma_{1/2}$ (left) and $\sigma_{3/2}$ (right), which were extracted from the helicity asymmetry E measurement, are plotted for the $p\pi^0$ final state (black points) together with different PWA solutions (solutions including this data: BnGa-2014-02 (black solid) [15], BnGa-2014-01 (black dashed) [41], SAID (PD03) [41] (red solid, curve calculated used solution PD03 for E and CM12 for the cross section), JüBo-2016-01 (blue solid) [41], and not including this data: BnGa-2011-02 (black dotted) [42], SAID CM12 (red dashed) [17], JüBo-2013-01 (blue dashed) [43], MAID2007 (green dotted) [18]). The systematic error is represented by the gray area. Taken from [9].

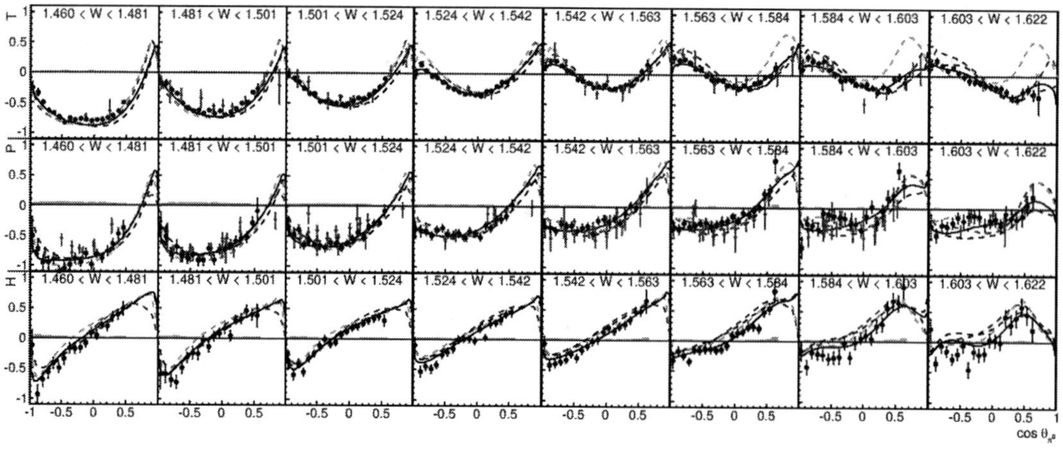

Figure 5. The results for the polarization observables T, P and H are shown for selected energy bins for the $p\pi^0$ final state (black points). The data is compared to previous measurements (red points) [44] and to different PWA predictions (dashed curves: BnGa-2011 (black) [42], MAID (green) [18], and SAID CM12 (blue) [17]) and PWA fit (BnGa-2014-02 (blak solid line) [15, 41]). Taken from [10].

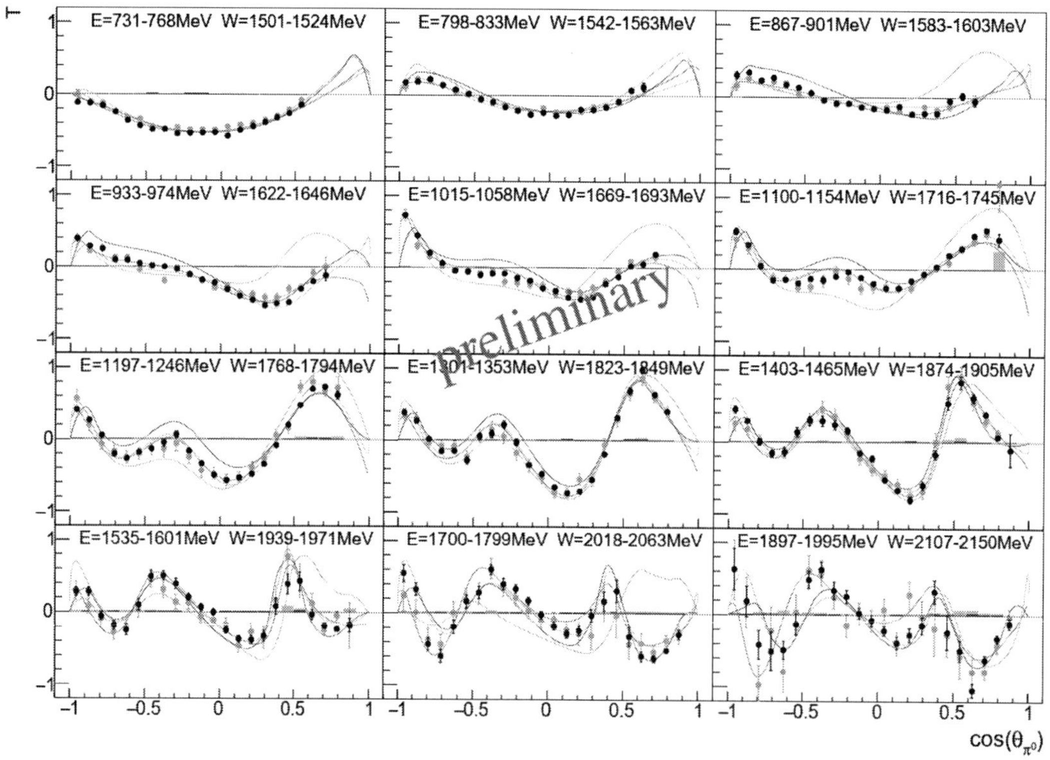

Figure 6. Preliminary results are shown for the polarization observable T, P for the $p\pi^0$ final state. This new data (black points), which was taken after the CB upgrade from December 2017 to October 2018, is compared to already published data of the CBELSA/TAPS collaboration [10, 11]. In addition, different PWA curves are plotted as well (JüBo-2017 (orange line) [16], BnGa-2014-02 (red line) [15, 41], SAID-CM12 (blue line) [17] and MAID-2007 (green line) [18]).

http://doi.org/10.1051/epjconf/202024101001

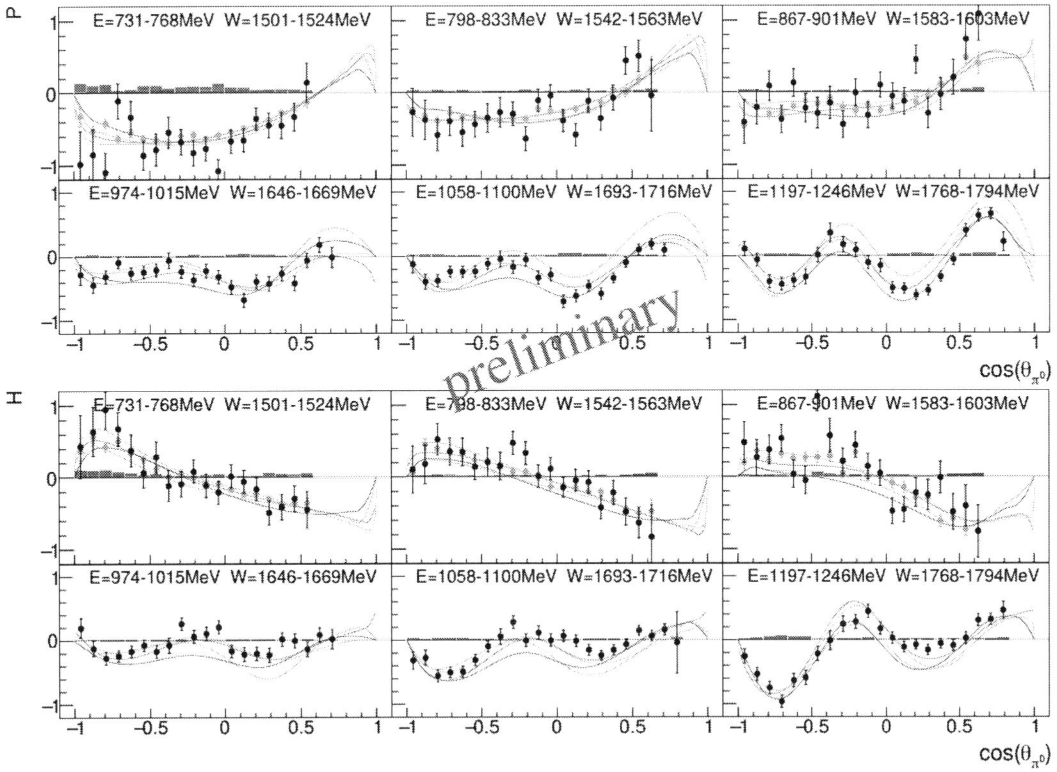

Figure 7. Preliminary results are shown for the polarization observables P and H for the $p\pi^0$ final state. This new data (black points), which was taken after the CB upgrade from December 2017 to October 2018, is compared to already published data of the CBELSA/TAPS collaboration [10, 11]. In addition, different PWA curves are plotted as well (JüBo-2017 (orange line) [16], BnGa-2014-02 (red line) [15, 41], SAID-CM12 (blue line) [17] and MAID-2007 (green line) [18]).

body kinematics. Thus, more polarization observables than the ones listed in Table 1 become accessible [45]. The results for these observables are shown in references [29, 30]. Using this new data in their latest PWA, the BnGa group was able to determine the branching ratios of N^* and Δ^* resonances into $N\pi, \Delta\pi, N(1440)\pi, N(1520)\pi, N(1535)\pi, N(1680)\pi$ or $N\sigma$ [29, 46]. The extracted branching ratios show that, depending on the structure of the baryon wave function, some resonances prefer to decay into the ground state nucleons or deltas, while others decay with a significant branching ratio into excited nucleon or delta states [29, 30].

2.3 Impact of polarization observables

Using the determined double polarization observables G, E, T, P and H of the $p\pi^0$ final state by the CBELSA/TAPS collaboration, stronger constraints could be imposed on the photoproduction multipoles of the $p\pi^0$ final state by the BnGa, JüBo and SAID groups. As an example the real and imaginary parts of the E_{0+}, E_{1+}, M_{1-} and M_{2-} multipoles are shown in Figure 10. It becomes apparent that the fit error bars of the BnGa PWA decrease by a factor of 2.25 due to the inclusion of these new polarization measurements [10]. In addition, the deviation between the different PWA groups was investigated by taking the variance of all three PWA (BnGa, JüBo and SAID) summed over all $p\pi^0$ photoproduction multipoles up to $L = 4$ [15]. Here, a clear reduction of the variance was observed due to inclusion of the new polarization data. Thus, the multipoles of the different PWA are slowly converging towards one solution.

Based on the latest polarization data from several collaborations, e.g. A2, CLAS, GRAAL, and CBELSA/TAPS, several resonances have been upgraded regarding their star-rating in the PDG [47].

2.4 Summary and Outlook

The CBELSA/TAPS collaboration has measured almost all of the polarization observables of the beam-target category for single and multi-meson final states with high precision, covering a large region in phase-space. The results were used by several PWA groups to constrain their solutions. A comparison of the multipoles shows that the new polarization data helps to reach the goal of

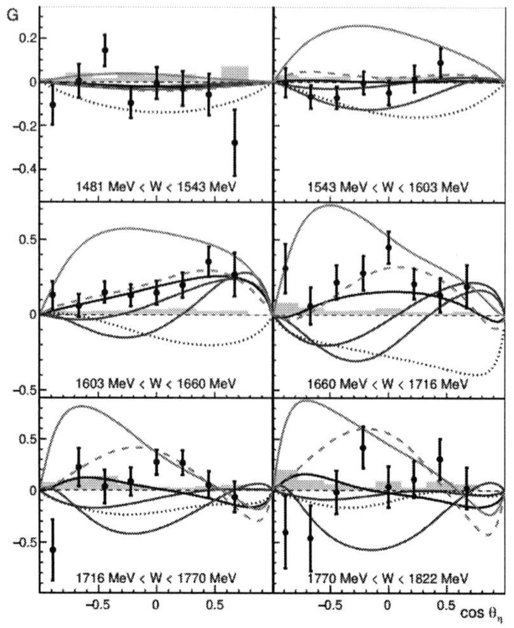

Figure 8. The double polarization observable G is shown for the $p\eta$ final state (black points) together with different PWA solutions (latest fit of BnGa (black line) [40], BnGa2011-02 (red line) [42], ηMAID2018 (green line) [48], SAID-GE09 (dark blue dotted) [49], JüBo-2015 (light blue dashed line) [50] and JüBo-2015-3 (light blue solid line) [51]). Taken from [14].

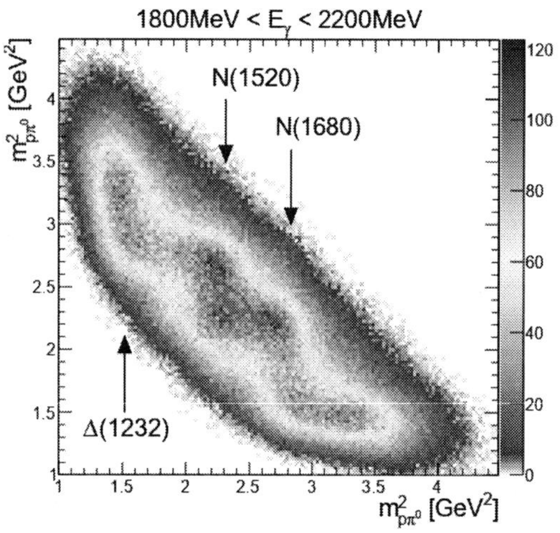

Figure 9. A Dalitz plot is depicted for the reaction $\gamma p \to p\pi^0\pi^0$ for an energy range of 1800 MeV < E_γ < 2200 MeV. The horizontal and vertical bands indicate the intermediate states.

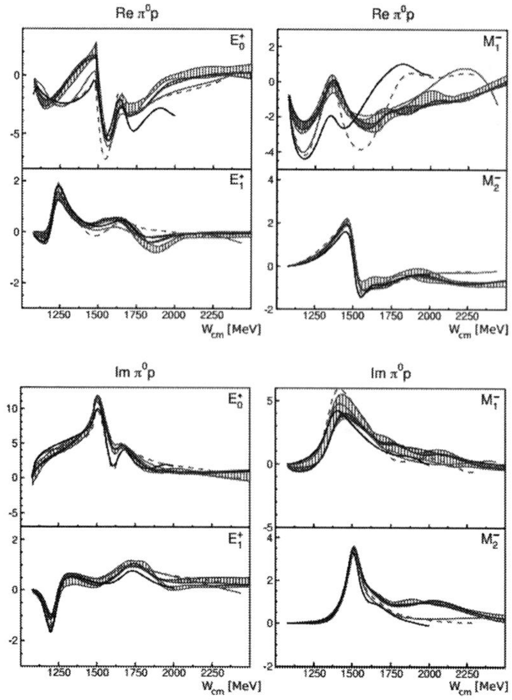

Figure 10. the real and imaginary parts of the E_{0+}, E_{1+}, M_{1-} and M_{2-} multipoles of the $p\pi^0$ final state are depicted. The red areas represent a variety of different fits of the BnGa2011-01 and BnGa2011-02 PWA [42], that do not include the new polarization data, while the blue areas represent the variety of solutions when including the new polarization data (BnGa-2014-02 PWA solution [15, 41]). In addition, the MAID (black line) [18], SAID-CM12 (solid green line) [17], SAID-SN11 (green dashed line) [52] and the JüBo-2015 (magenta line) [50] PWA fit solutions are shown as well. Taken from [11].

finding one unique solution. The results have now entered the PDG and the star-rating of several resonances was improved.

The CBELSA/TAPS experiment was successfully upgraded with a new readout electronics for the CB detector. The new data, that has been taken after the upgrade, show promising results with improved statistical precision. The finished detector upgrade allows to measure polarization observables for final states that involve neutral particles exclusively, with a high trigger efficiency. These final states, e.g. $n\pi^0$, are in particular important since there is only very few data available for these final states.

References

[1] U. Loring, B. C. Metsch and H. R. Petry, Eur. Phys. J. A **10**, 395 (2001).

[2] R. G. Edwards, J. J. Dudek, D. G. Richards and S. J. Wallace, Phys. Rev. D **84**, 074508 (2011).

[3] V. Burkert, these proceedings.

[4] S. Strauch, these proceedings.
[5] M. Dugger et al. [CLAS Collaboration], Phys. Rev. C **88**, (2013).
[6] S. Costanza, these proceedings.
[7] M. Sumihama et al. [LEPS Collaboration], Phy. Lett. B **657**, 32 (2007).
[8] M. Gottschall et al. [CBELSA/TAPS Collaboration], Phys. Rev. Lett. **112**, 012003 (2014).
[9] M. Gottschall et al. [CBELSA/TAPS Collaboration], arXiv:1904.12560v1 [nucl-ex] (2019).
[10] J. Hartmann et al. [CBELSA/TAPS Collaboration], Phys. Rev. Lett. **113**, 062001 (2014).
[11] J. Hartmann et al. [CBELSA/TAPS Collaboration], Phys. Lett. B **748**, 212 (2015).
[12] A. Thiel et al. [CBELSA/TAPS Collaboration], Phys. Rev. Lett. **109**, 102001 (2012).
[13] A. Thiel et al. [CBELSA/TAPS Collaboration], Eur. Phys. J. A **53**, 8 (2017).
[14] J. Müller et al. [CBELSA/TAPS Collaboration], Phys. Lett. B **803**, 135323 (2020).
[15] A. V. Anisovich et al., Eur. Phys. J. A **52**, 284 (2016).
[16] D. Rönchen, M. Döring and U.-G. Meißner, Eur. Phys. J. A **54** 110 (2018).
[17] R.L. Workman, et al., Phys. Rev. C **86**, 015202 (2012).
[18] D. Drechsel, S.S. Kamalov, L. Tiator, Eur. Phys. J. A **34**, 69 (2007).
[19] W. Hillert, Eur. Phys. J. A **28S1**, 139 (2006).
[20] T. Jude, these proceedings.
[21] D. Elsner et al., Eur. Phys. J. A **39**, 373 (2009).
[22] A. M. Sandorfi et al., J. Phys. G **38**, 053001 (2011).
[23] E. Aker et al., Nucl. Instr. Meth. A **321**, 69-108 (1992).
[24] Ch. Bradtke, H. Dutz, H. Peschel et al., Nucl. Instr. Meth. A **436**, 430-442 (1999).
[25] S. Runkel, these proceedings.
[26] R. Novotny, IEEE Trans. Nucl. Sci. NS **38**, 379-385 (1991).
[27] G. Suft et al., Nucl. Instr. Meth. A **531**, 416-424 (2005).
[28] F. N. Afzal [CBELSA/TAPS Collaboration], EPJ Web Conf. **73**, 04005 (2014).
[29] T. Seifen et al. [CBELSA/TAPS Collaboration], "Polarization observables in double neutral pion photoproduction", in preparation.
[30] T. Seifen [CBELSA/TAPS Collaboration], these proceedings.
[31] E. Gutz et al. [CBELSA/TAPS Collaboration], Eur. Phys. J. A **50**, 74 (2014).
[32] H. Eberhardt et. al. [CBELSA/TAPS Collaboration], Phys. Lett. B **750**, 453-458 (2015).
[33] C. Patrignani et al. (Particle Data Group), Chin. Phys. C **40**, 100001 (2016) and 2017 update.
[34] Y. Wunderlich et al., Eur. Phys. J. A **53**, no. 5, 86 (2017).
[35] A. V. Anisovich et. al., Phys. Lett. B **766**, 357-361 (2017).
[36] B. Krusche et al., Prog. Part. Nucl. Phys. **51**, 399 (2003).
[37] Y. Wunderlich, these proceedings.
[38] F. Afzal et al. [CBELSA/TAPS Collaboration], "Observation of the $p\eta'$ cusp in the new precise beam asymmetry Σ data for $\gamma p \to p\eta$", submitted to Phys. Rev. Lett.
[39] O. Bartalini et al. [GRAAL Collaboration], Eur. Phys. J. A **26**, 399 (2005).
[40] A. Anisovich *et al.*, Phys. Lett. B **785** 626-630 (2018).
[41] V. Sokhoyan et al. [CBELSA/TAPS Collaboration], Eur.Phys. J. A **51**, 95 (2015).
[42] A. V. Anisovich et al., Eur. Phys. J. A **48**, 15 (2012).
[43] D. Rönchen et al., Eur. Phys. J. A **49**, 44 (2013).
[44] W.J. Briscoe, D. Schott, I.I. Strakovsky, and R.L. Workman, GWU analysis center, gwdac.phys.gwu.edu/
[45] W. Roberts and T. Oed, Phys. Rev. C **71**, 055201 (2005).
[46] A. Thiel et al., Phys. Rev. Lett. **114**, 091803 (2015).
[47] M. Tanabashi et al. (Particle Data Group), Phys. Rev. D **98**, 030001 (2018) and 2019 update.
[48] L. Tiator et al., Eur. Phys. J. A **54**, 210 (2018).
[49] SAID: http://gwdac.phys.gwu.edu/
[50] D. Rönchen *et al.*, Eur. Phys. J. A **51** 70 (2015).
[51] I. Senderovich et al., Phys. Lett. B **775**, 64 (2016).
[52] R.L. Workman, et al., Phys. Rev. C **85**, 025201 (2012).

η' beam asymmetry at threshold using the BGO-OD experiment

Stefan Alef[1,*], *Patrick* Bauer[1], *Reinhard* Beck[2], *Alessandro* Braghieri[13], *Philip* Cole[3], *Rachele* Di Salvo[4], *Daniel* Elsner[1], *Alessia* Fantini[4,5], *Oliver* Freyermuth[1], *Francesco* Ghio[6,7], *Anatoly* Gridnev[8], *Daniel* Hammann[1], *Jürgen* Hannappel[1], *Thomas* Jude[1], *Katrin* Kohl[1], *Nikolay* Kozlenko[8], *Alexander* Lapik[9], *Paolo* Levi Sandri[10], *Valery* Lisin[9], *Giuseppe* Mandaglio[11,12], *Roberto* Messi[4,5], *Dario* Moricciani[4], *Vladimir* Nedorezov[9], *Dmitry* Novinsky[8], *Paolo* Pedroni[13], *Andrei* Polonski[9], *Björn-Eric* Reitz[1], *Mariia* Romaniuk[4], *Georg* Scheluchin[1], *Hartmut* Schmieden[1], *Victorin* Sumachev[8], *Viacheslav* Tarakanov[8], and *Christian* Tillmanns[1]

[1]Rheinische Friedrich-Willhelms-Universität Bonn, Physikalisches Institut, Nußallee 12, 53115 Bonn, Germany
[2]Helmholtz-Institut fuer Strahlen- und Kernphysik, Universitaet Bonn, Nussallee 1-16, D-53115 Bonn Germany
[3]Lamar University, Department of Physics, Beaumont, Texas, 77710, USA
[4]INFN Roma "Tor Vergata", Rome, Italy
[5]Università di Roma "Tor Vergata", Via della Ricerca Scientifica 1, 00133 Rome, Italy
[6]INFN sezione di Roma La Sapienza, P.le Aldo Moro 2, 00185 Rome, Italy
[7]Istituto Superiore di Sanita, Viale Regina Elena 299, 00161 Rome, Italy
[8]Petersburg Nuclear Physics Institute, Gatchina, Leningrad District, 188300, Russia
[9]Russian Academy of Sciences Institute for Nuclear Research, prospekt 60-letiya Oktyabrya 7a, Moscow 117312, Russia
[10]INFN - Laboratori Nazionali di Frascati, Via E. Fermi 40, 00044 Frascati, Italy
[11]INFN sezione Catania, 95129 Catania, Italy
[12]Universita degli Studi di Messina, Via Consolato del Mare 41, 98121 Messina, Italy
[13] INFN sezione di Pavia, Via Agostino Bassi, 6 - 27100 Pavia, Italy

Abstract.
The unexpected nodal structure of the beam asymmetry recently reported by the GRAAL collaboration in η' photoproduction very close to threshold could be explained by a previously unobserved narrow resonance.
The BGO-OD experiment is ideally suited to verify this measurement via the detection of forward going charged particles which in the threshold region of interest allows the identification of the reaction $\gamma p \to \eta' p$ solely based on the proton going in the forward direction. This yields unprecedented statistics if, in the missing mass analysis of the η' meson, the background can be sufficiently well controlled. Preliminary results using a linearly polarised photon beam are shown. The reaction $\gamma p \to \eta' p$ was identified in the BGO forward spectrometer, with simulated data used to seperate signal and background.

1 Introduction

As can be seen in figure 1 the η' beam asymmetry has already been measured by the GRAAL experiment [1]. There an unexpected nodal structure was observed in the $\Theta_{\eta'}^{CMS}$ (polar angle of the η' in the centre of mass frame) dependance of the beam asymmetry that vanishes rapidly with higher beam energies.
The polarised cross section for the reaction can be written as the unpolarised cross section including an additional term as shown in equation 1.

$$\left(\frac{d\sigma}{d\Omega}\right)^{\text{pol}} = \left(\frac{d\sigma}{d\Omega}\right)^{\text{unpol}} \left(1 - P_\gamma\left(E_\gamma\right)\Sigma\left(E_\gamma, \Theta_{\eta'}^{CMS}\right)\cos\left(2\phi_{\eta'}\right)\right) \quad (1)$$

$P_\gamma\left(E_\gamma\right)$ is the photon beam polarisation, $\Sigma\left(E_\gamma, \Theta_{\eta'}^{CMS}\right)$ the beam asymmetry and $\phi_{\eta'}$ the azimuthal angle of the η'. An expansion of $\Sigma\left(E_\gamma, \Theta_{\eta'}^{CMS}\right)$ in partial waves can be done, leading to the expression shown in equation 2-4 [2]. This is valid for a truncation at $l_{max} = 3$ (F-waves).

$$\Sigma = \left(1 - x^2\right)\sum_{k=0}^{4} a_k x^k \quad (2)$$

$$x = \cos(\Theta) \quad (3)$$

$$a_1 = a_1^{SF} + a_1^{PD} + a_1^{DF} \quad (4)$$

For the lowest order (k=0) this results in a $\sin^2(\Theta)$ modulation of Σ which is not observed in the data. k=1 yields the observed $\sin^2(\Theta) \cdot \cos(\Theta)$ modulation. The corresponding coefficient a_1 consists of interference terms between S-F, P-D and D-F-wave.
The observed behaviour can be explained by a narrow resonance close to threshold. This is further supported by results from the Bonn-Gatchina partial wave analysis group which show that including a narrow resonance ($D_{13}(1900)$) with a mass of $M_{\eta' p} = 1900 \pm 1$ MeV and a width of $\Gamma_{\eta' p} < 3$ MeV close to threshold much more accuratly describes the experimental data [3], see figure 2. EtaMAID can describe the observed behaviour as well us-

*e-mail: alef@physik.uni-bonn.de

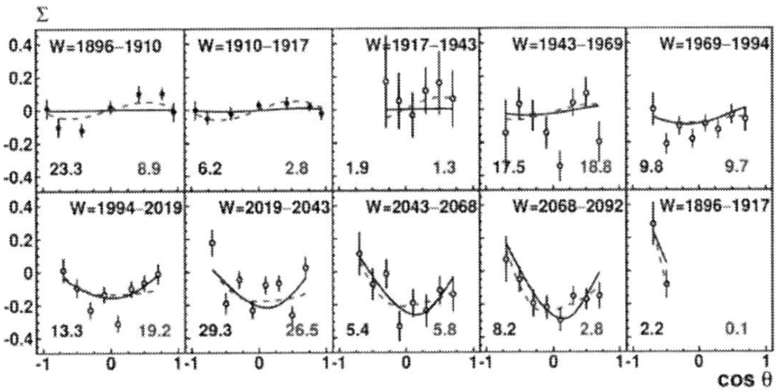

Figure 2. Partial wave analysis results from the Bonn-Gatchina partial wave analysis group. The red dotted lines show the fits with a narrow resonance close to the η' threshold included whereas the solid black lines show the same fits without the resonance included [3].

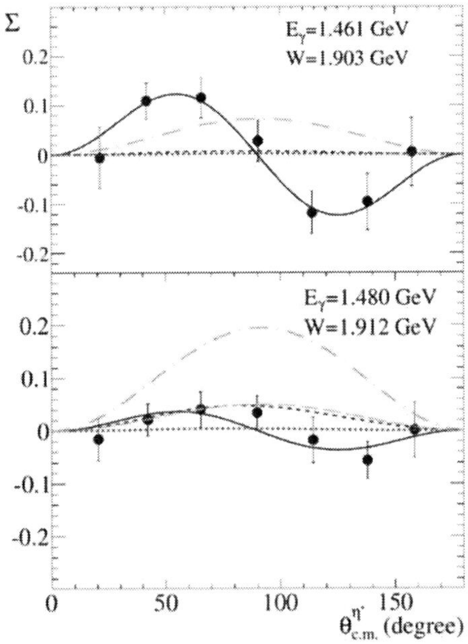

Figure 1. η' beam asymmetry results from the GRAAL experiment. The colored lines are different theoretical calculations, references found in [1]. Figure taken from [1].

ing a narrow S_{11} (1900) resonance [2].
Experimentally the determination of the beam asymmetry is done by extracting the normalised η' yields for two perpendicular polarisation planes (+ and -), substracting the two and dividing out the sum to remove detector inefficiencies. This is shown in equation 5.

$$A = \frac{N_+ - N_-}{N_+ + N_-} = P_\gamma\left(E_\gamma\right)\Sigma\left(E_\gamma, \Theta^{CMS}_{\eta'}\right)\cos(2\phi_{\eta'}) \quad (5)$$

From this the beam asymmetry can be calculated by applying a $\cos(2\phi_{\eta'})$ fit to the $\phi_{\eta'}$ distribution of A in different $\cos\left(\Theta^{CMS}_{\eta'}\right)$ bins.

2 The BGO-OD experiment at ELSA

The BGO-OD experiment is located at the ELSA [4] (Electron-Stretcher-Accelerator) facility at the physics institute of the University of Bonn. ELSA is a three stage accelerator which can boost electrons up to 3.2 GeV. The electrons are produced via a thermionic gun, from which they are first accelerated using a linear accelerator. Subsequently they enter a booster synchroton, where they are accelerated up to 1.6 GeV. Several runs of the synchroton are used to fill the stretcher ring, where the electrons reach their final energy of up to 3.2 GeV. From there they are fed to one of the two hadron physics experiments (BGO-OD and Crystal Barrel) or into an area for detector tests.
At BGO-OD (see figure 3) the electron beam enters the goniometer tank at which point bremsstrahlung is produced via a thin copper or diamond radiator. The copper radiator is employed to produce an unpolarised photon beam whereas the diamond radiator is used to create a linearly polarised photon beam through coherent bremsstrahlung. Using a magnet and a hodoscope (the tagging system) the post-bremsstrahlung electron's energy can be determined. The produced real photon beam passes two collimators and enters the central part of the setup. It consists of the target filled with liquid hydrogen surrounded by two cylindrical Multi-Wire-Proportional-Chambers for charged particle tracking, a plastic scintillator barrel for particle ID and the BGO ball, consisting of 480 $Bi_4Ge_3O_{12}$ crystals. The ball covers nearly 4π in acceptance and is ideally suited for the detection of neutral particles. In polar angle it covers $22° - 155°$.
The central detectors are complimented by the forward spectrometer which is well suited for the identification and momentum determination of charged particles. It consists of tracking detectors in front and behind a large open dipole magnet and time-of-flight walls at the end. The forward spectrometer covers polar angles up to 12° which leaves an acceptance gap between the central detectors and the forward spectrometer. This is filled by the SciRi (Scintillating-Ring) detector.
Two detectors are located at the very end of the experiment for flux measurement.

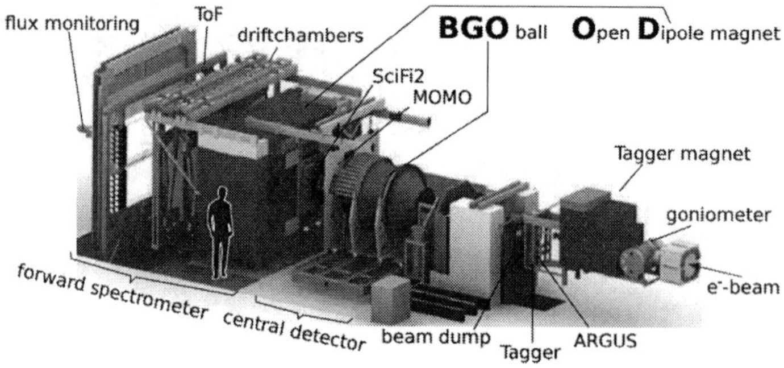

Figure 3. The BGO-OD setup.

3 Analysis of $\gamma p \rightarrow \eta' p$

Due to the Lorentz boost, protons will be boosted in a forward direction close to threshold, thus entering the forward spectrometer, where their momentum and direction can be determined. No other particles are reconstructed in this analysis, resulting in high statistics at the expense of a significant background.
After identifying the proton, the mising mass, m_{miss}, to the proton can be calculated as shown in 6.

$$m^2_{\text{miss}} = \left(p_\gamma + p_{\text{target}} - p_p\right)^2 \tag{6}$$

p_γ is the 4-momentum of the incoming photon, p_{target} the 4-momentum of the target proton and p_p the momentum of the forward going proton.
The resulting missing mass spectrum is shown in figure 4. On top of a continuous background, the peak of the η' is clearly visible alongside peaks from others mesons. The spikes around 600 MeV in the spectrum are binning artifacts from the tagging system.
For the determination of the beam asymmetry the data is subsequently binned as shown in table 1.

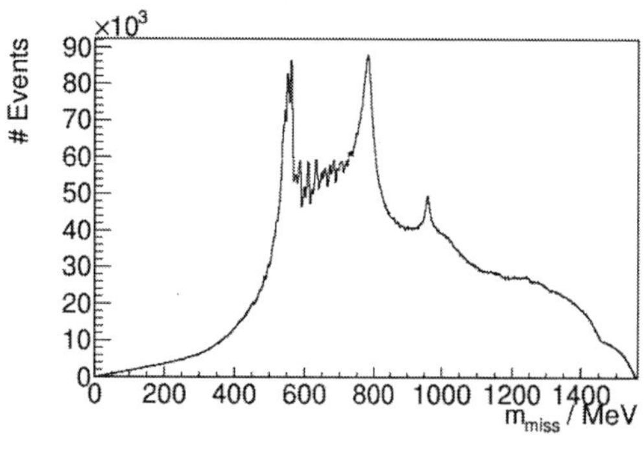

Figure 4. Spectrum of the missing mass to the forward going proton.

Table 1. Binning for the beam asymmetry determination

Parameter	Minimum	Maximum	Bins
E_γ	1446 MeV	1466 MeV	1
$\cos\left(\Theta^{CMS}_{\eta'}\right)$	-1.0	1.0	7
$\Phi_{\eta'}$	$-180°$	180°	7

4 η' number extraction

In figure 4 the η' peak is clearly visible above background. This is not the case anymore if the data is split up into the aforementioned bins (see table 1) as can be seen in figure 5.
The missing mass spectra of several simulated reactions are fit simultaneously to the missing mass spectrum of the real data, which makes it possible to determine the contributions of the different reactions to the total spectrum [1].
To ensure that there is an η' contribution in the real data, the fits have been performed with and without the η' included. Figure 5 shows two example bins to demonstrate that this channel is needed for a proper fit.

5 Preliminary beam asymmetry results

Having extracted the η' yields from the different bins, the beam asymmetry can be calculated.
Ideally the polarisation is equal for both polarisation planes, however the polarisation degree varies slightly with orientation and time, so a weighted mean, $P_\gamma\left(E_\gamma, \pm\right)$, is determined for each orientation and divided out, see equation 7-9.

$$\frac{N_+ - N_-}{N_+ + N_-} = P_\gamma\left(E_\gamma\right)\Sigma\left(E_\gamma, \Theta^{CMS}_{\eta'}\right)\cos(2\phi_{\eta'}) \tag{7}$$

$$\tilde{A} = \frac{\tilde{N}_+ - \tilde{N}_-}{N_+ + N_-} = \Sigma\left(E_\gamma, \Theta^{CMS}_{\eta'}\right)\cos(2\phi_{\eta'}) \tag{8}$$

$$\tilde{N}_\pm = \frac{N_\pm}{P_\gamma\left(E_\gamma, \pm\right)} \tag{9}$$

By fitting a $\cos(2\phi_{\eta'})$ to $\tilde{A}$ the beam asymmetry can be extracted. The particle numbers are carefully flux normalised.
Figure 6 shows the fits to the $\phi_{\eta'}$ distribution of $\tilde{A}$ in the

[1]The fits were performed using the RooFit library of Root [5].

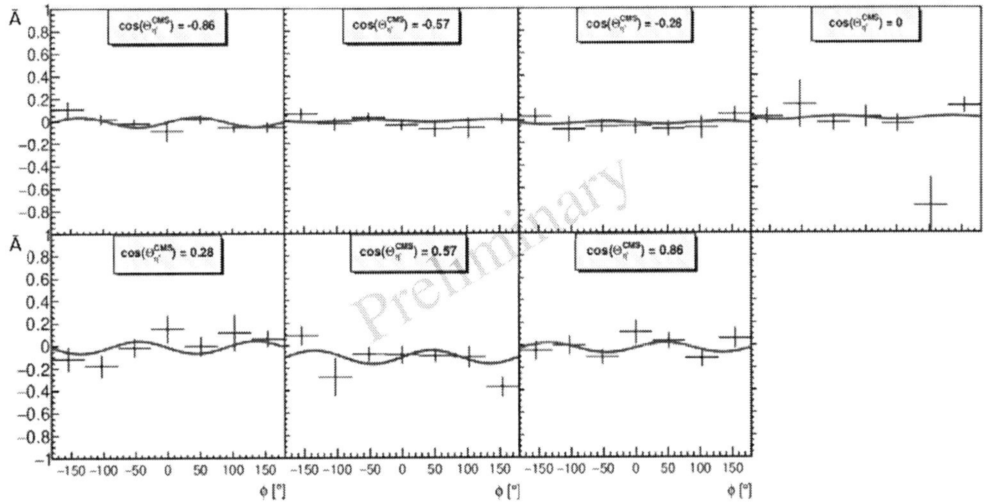

Figure 6. Fits to the $\phi_{\eta'}$ distribution of $\tilde{A}$ in the different $\cos\left(\Theta_{\eta'}^{CMS}\right)$ bins. Only statistical errors shown.

different $\cos\left(\Theta_{\eta'}^{CMS}\right)$ bins. These fits are preliminary and represent a snapshot of the current status of this work. The resulting beam asymmetry is shown in figure 7.

6 Conclusion

It has been shown, that the reaction $\gamma p \to \eta' p$ can be identified at the BGO-OD experiment using only proton identification in the forward spectrometer and fitting the resulting missing mass spectrum.
This technique yields high statistics, but results in a large amount of background. Despite this, the data can be well decribed in the chosen energy range close to threshold with several simulated reaction channels. These fits are challenging due to the similar nature of signal and background and work is still ongoing.
Furthermore systematic checks on the signal/background extraction will be performed in the future.
A very preliminary first look at the resulting beam asymmetry is also given.

This work was supported by the Deutsche Forschungsgemeinschaft project numbers 388979758 and 405882627. Our Russian collaborators thank the Russian Scientific Foundation (grant RSF number 19-42-04132) for financial support.

References

[1] P. Levi Sandri et. al., First measurement of the Σ Beam Asymmetry in η' Photoproduction off the Proton near Threshold, arXiv:1407.6991v2 [nucl-ex], 4 (2015)
[2] L. Tiator et. al., Eta and Etaprime Photoproduction on the Nucleon with the Isobar Model EtaMAID2018, arXiv:1807.04525v2 [nucl-th], 21 (2018)
[3] A. Anisovich et. al., Proton–η' interactions at threshold, arXiv:1803.06814v1 [nucl-ex], 2 (2018)
[4] W. Hillert, Eur. Phys. J. A28, s01, (2006)
[5] W. Verkerke et. al., The RooFit toolkit for data modeling, arXiv:physics/0306116v1 [physics.data-an] (2003)

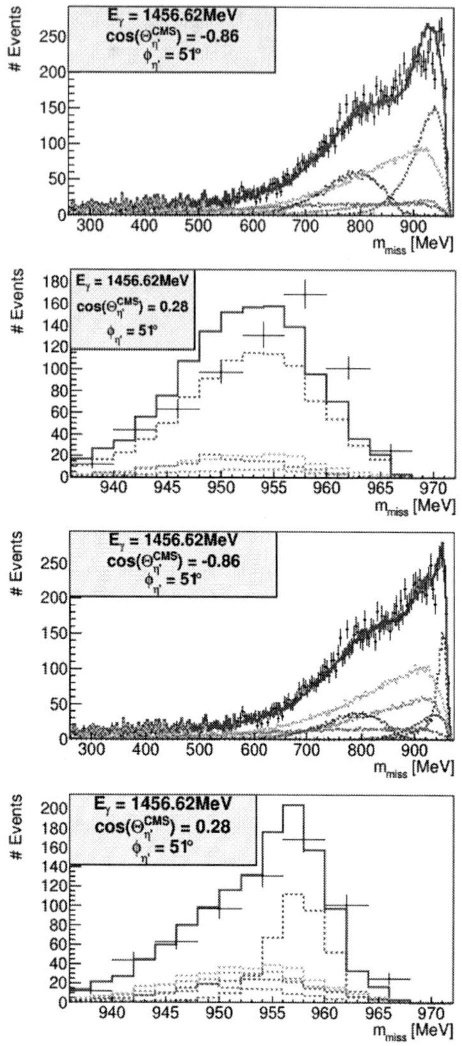

Figure 5. Fits in two different bins without the η' (top two plots) and the same bins with the η' included (bottom two plots). The black data points are the real data, the total fit is shown in solid red, the $\eta' p$ contribution in dotted red, $2\pi p$ in purple, $3\pi p$ in light blue, $\eta\pi^0 p$ in green, $\eta\pi^0\pi^0 p$ in dark purple, ωp in dark blue and $\pi^0 p$ in brown. Note that ωp and $\pi^0 p$ only contribute in the most backward $\cos\left(\Theta_{\eta'}^{CMS}\right)$ bins.

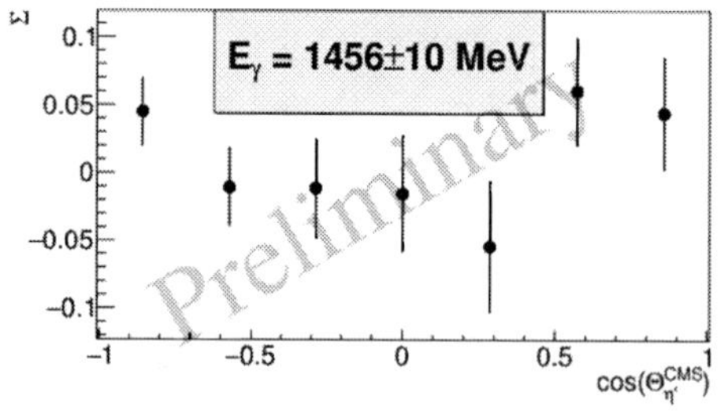

Figure 7. Beam asymmetry results. The errors are the errors of the $\cos(2\phi_{\eta'})$ fit.

Study of baryonic resonances and ρ meson production in the reaction $pp \rightarrow pp\pi^+\pi^-$ at 3.5 GeV with HADES

Amel Belounnas[1,3,*]
for the HADES collaboration

[1]Institut de Physique Nucléaire d'Orsay, CNRS-IN2P3, Univ. Paris Sud, Univ. Paris-Saclay, 91406 Orsay Cedex, France.

Abstract. The High Acceptance Di-Electron Spectrometer (HADES) installed at GSI in Darmstadt and designed to investigate dielectron production in heavy-ion collisions in the range of kinetic beam energies 1-3 A GeV is also an excellent detector for charged hadron detection. We discuss the analysis of the $pp \rightarrow pp\pi^+\pi^-$ channel at an incident energy of 3.5 GeV, using results from pp→ppπ^0, pp→pnπ^+ and $pp \rightarrow pK\Lambda$ measured at the same energy by HADES. A consistent description of the differential cross sections as a function of the invariant mass and the production angles is achieved by a simple resonance model taking into account the single and the double excitation of baryonic resonances with masses up to 1.9 GeV. The contributions can be quantified and confronted to the inputs of transport models or to the predictions of dedicated models for nucleon-nucleon reactions. In addition, using specific kinematical cuts, the total production cross section and the angular distribution of the ρ meson have been also extracted. As both baryon resonances and the ρ meson are important sources of di-electrons, these results will provide valuable constraints for the interpretation of the dielectron spectra measured by the HADES collaboration.

1 Introduction

The High Acceptance DiEelectron Spectrometer (HADES) [1] installed at GSI in Darmstadt was built to investigate dielectron production in order to study the hot and dense hadronic medium produced in heavy-ion collisions in the 1-2 AGeV energy range. Proton- or pion-nucleus reactions are also investigated to probe cold nuclear matter. Moreover, the experimental programme also comprises elementary reactions (pp, quasi-free np, and πp) to study more selectively the different dilepton sources (see P. Salabura's report at this conference [2] for an overview). In particular, baryonic resonances are important sources of dileptons through two mechanisms: their Dalitz decays (e.g. $\Delta/N^\star \rightarrow Ne^+e^-$) and the mesonic decays with subsequent dilepton decay ($\pi^0 \rightarrow \gamma e^+e^-$, $\eta \rightarrow \gamma e^+e^-$, $\omega/\rho \rightarrow e^+e^-$). According to Vector Meson Dominance [3], electromagnetic transitions between hadrons are mediated by vector mesons, which strongly connects the Dalitz decay of baryons to their coupling to vector mesons. The ρ meson is expected to play a big role, over a broad kinematic region, due to its large width and strong coupling to baryons. The possibility to measure simultaneously with the HADES detector meson and di-electron production is therefore a great advantage. In this way, constraints on reaction mechanisms, and in particular on the role of intermediary baryon resonances, obtained by the study of the meson production channels can indeed be used for the interpretation of the dielectron production.

This motivated previous studies of the HADES collaboration in the 1π, η, $K\Lambda$ and $K\Sigma$ production channels in pp reactions from 1.25 to 3.5 GeV [4–7]. The 2π production channel offers the additional possibility to study the double baryon excitation, and, at high enough energies, ρ meson production. A previous study of the 2π production by the HADES collaboration in the $pn \rightarrow pn\,\pi^+\pi^-$ channel at 1.25 GeV [8] showed strong contributions of both the double $\Delta(1232)$ and of the N(1440) excitation. This contribution focuses on the analysis of the $pp \rightarrow pp\,\pi^+\pi^-$ reaction at 3.5 GeV. At such high energies, one can also expect the exctation of higher resonances. Therefore, the reaction is well suited to study the single and double excitation of many baryon resonances the production of the ρ meson, either directly or via the decay of baryon resonances.

2 Previous HADES studies

HADES (High Acceptance Di-Electron Spectrometer) was designed for the detection of e^+e^- pairs, but it is also an excellent detector for charged hadron detection. It consists of six identical sectors covering polar angles between 18° and 85° with respect to the beam axis. A hadron blind RICH detector is used for the identification of e^+ and e^-. The momentum vectors of produced particles are reconstructed by means of the four drift chambers (MDC), two placed before and two placed behind the magnetic field region provided by six coils of a super-conducting toroid. In the detector configuration used for this experiment, time-of-flight information was given by two time-of-flight walls

*e-mail: ramstein@ipno.in2p3.fr

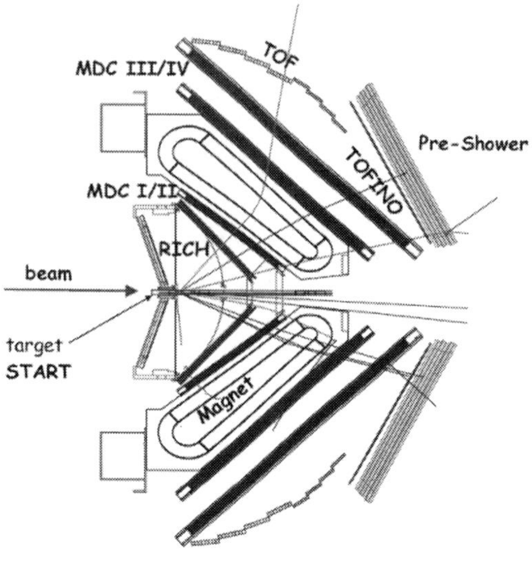

Figure 1. Schematic layout of the HADES detector in the configuration used for this experiment.

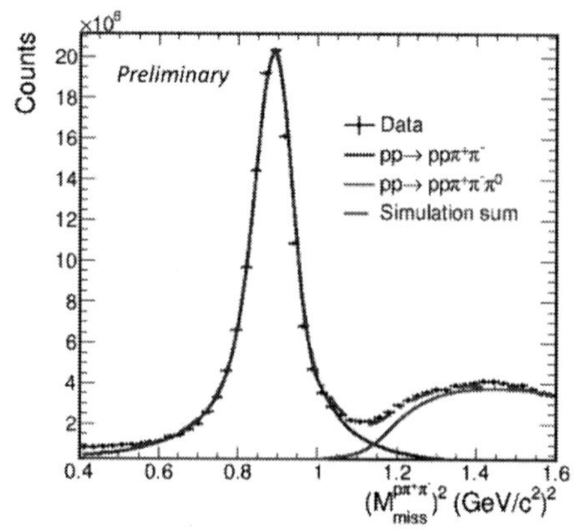

Figure 2. Squared missing mass distribution in the $pp \to pp\pi^+\pi^-$X reaction compared to simulations for the two pion production (blue) and three pion production (green). The total is shown as a red curve.

based on plastic scintillators covering polar angles, larger (TOF) and smaller (TOFINO) than 45°, respectively. For the analysis of the $pp \to pp\pi^+\pi^-$ reaction, the strategy is to select events with 3 prongs (one proton and two oppositely charged pions) and to apply a missing mass selection. In a first step, events with at least three tracks not correlated to RICH rings, one with a negative curvature, two with a positive curvature are selected. Particle identification of pions and protons is based on the correlation between momentum and time-of-flight. However, as no START detectors could be used in this experiment, only the information about the differences between time-of-flights was measured. To solve this problem, the negative track is used as a reference and is identified as a negative pion. Then, for each possible identification of positive tracks, *i.e.* π^+ or p, the time-of-flight differences with respect to the negative pion calculated using the reconstructed track length and the momentum are compared to the measured ones. The best combination is selected, by means of a χ^2 test taking into account errors on the time measurements.

The distribution of missing mass of the reaction $pp \to p\pi^+\pi^-X$ ($M_{miss}^{p\pi^+\pi^-}$) displays a clear peak corresponding to a missing proton (fig. 2). The structure at larger missing masses is mostly due to events with both a missing nucleon and a missing pion. This is corroborated by results of simulations for the reactions $pp \to pp\pi^+\pi^-$ and $pp \to pp\pi^+\pi^-\pi^0$, which, as shown in fig. 2 reproduces fairly well both structures. The missing mass distribution is weakly sensitive to the differential distributions in the model, which were taken assuming production according to the available phase space. But, it is important to note that the width of the proton peak in the missing mass distribution is very well reproduced by the simulation. This confirms that the modeling of the detector resolution adopted in the simulation and based on a parametrization of full scale GEANT simulations (see below) gives very realistic results. One can however notice an excess yield in the low missing mass tail, which is due to a small contamination of misidentified particles. To take into account both sources of background, the selection of the signal was performed in 25 bins of $p\pi^-$ invariant masses ($M_{p\pi^-}$) and 20 bins of cosine of the angle of the $p\pi^-$ pair in the center-of-mass ($\cos\theta_{p\pi^-}^{CM}$). For each bin, the three pion contribution was first subtracted based on the simulation. Then, the resulting missing mass distribution was fitted using the sum of two gaussians, representing the signal and of one polynomial for the background (fig. 4). This allowed to calculate the probability of an event to be a signal event, as a function of $M_{miss}^{p\pi^+\pi^-}$, $M_{p\pi^-}$, $\cos\theta_{p\pi^-}^{CM}$. The background subtracted differential distributions are then obtained by applying this probability as a weight to each event. In order to determine the geometrical acceptance and the reconstruction efficiency for each particle, detector hits produced by π^+, π^- and protons taking into account the experimental effects (detector geometry, energy loss, detector response,...) were produced using GEANT simulations and processed through the reconstruction and analysis code as for the real events. Acceptance and efficiency matrices were calculated as a function of the momentum, the polar and azimuthal angles in the laboratory [9]. The normalization of the experimental yield is obtained using the analysis of events produced in elastic scattering, as described in [10]. The final differential distributions are obtained after background subtraction, efficiency corrections and normalization as displayed in fig. 3. The error bars include statistical and uncorrelated systematic errors due to efficiency corrections and background subtraction. The differential cross sections are also affected by a global error of 7%, mainly due to the normalization uncertainty.

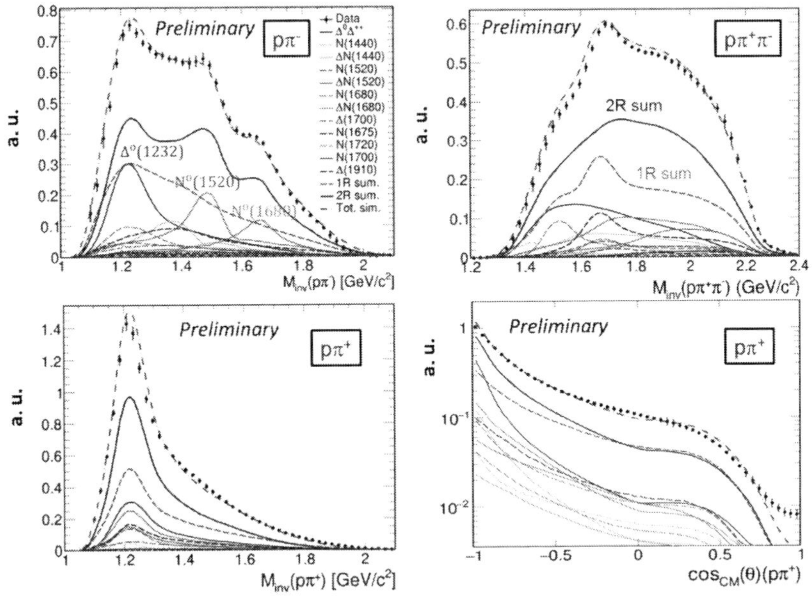

Figure 3. $p\pi^-$(top left), $p\pi^+\pi^-$(top right), $p\pi^+$ (bottom left) invariant masses and distribution of the center-of-mass angle of the $p\pi^+$ system. Data (black dots) are compared to the results of the fit with the resonance model (dashed red). The total double (blue) and single (dashed green) contributions are shown together with the most important individual contributions (see color code on the picture).

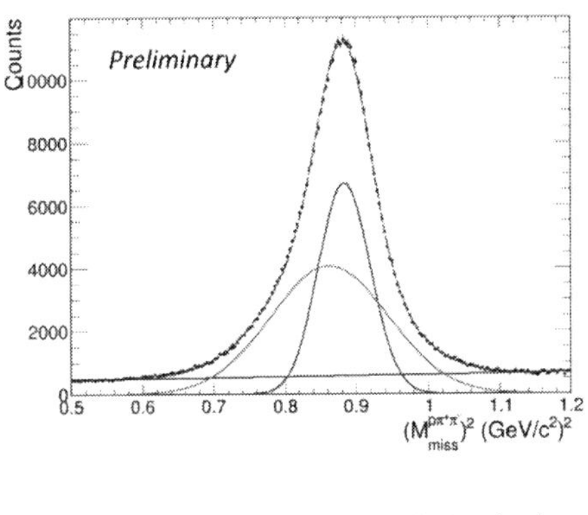

Figure 4. Squared missing mass distribution in the $pp \to pp\pi^+\pi^-X$ reaction after subtraction of the three pion production contribution. The red curve represents the result of a fit with the sum of two Gaussians (violet and green curves) and of a polynomial (blue curve).

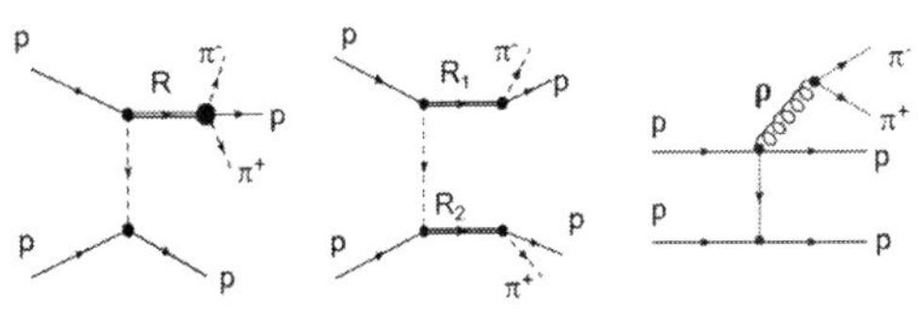

Figure 5. Diagrams of the three processes taken into account in the simulation of the $pp \to pp\pi^+\pi^-$ reaction. Left: Excitation of one positively charged resonance followed by decay into $p\pi^+\pi^-$. Center: Excitation of one doubled charged and one neutral resonance followed by their decay into $p\pi^-$ and $p\pi^+$ respectively. Right: direct ρ meson production.

3 HADES resonance model

To describe our data, we developed a model for the $pp \to pp\pi^+\pi^-$ reaction based on three different processes: the excitation of one resonance, with subsequent decay into a proton and two pions ($R \to p\pi^+\pi^-$) (fig. 5 Left), the excitation of two resonances decaying into a proton and a pion ($R_1, R_2 \to p\pi$) (fig. 5 Center) and the direct ρ production followed by its decay ($\rho \to \pi^+\pi^-$) (fig. 5 Right).

Useful constraints on the single resonance excitation in the pp reaction at 3.5 GeV are available thanks to previous data analyses of the HADES collaboration of respectively $NN\pi$ [10] and $pK\Lambda$ [11]. In [10], a model based on an incoherent sum of resonant contributions decaying in the πN channel has been used. Using a fit of the measured differential cross sections in the $pp\pi^0$ and $pn\pi^+$ channels, the cross sections for the excitation of various baryon resonances (Δ(1232), N(1440), N(1535), Δ(1700)) could be established (see table 1). However, for other baryons, due to the overlap with neighboring states, only upper limits for the cross sections could be extracted, as shown in table 1. The dependence of differential cross section on the four-momentum transfer, which, for fixed masses of the reaction products, is fully correlated to the angular distri-

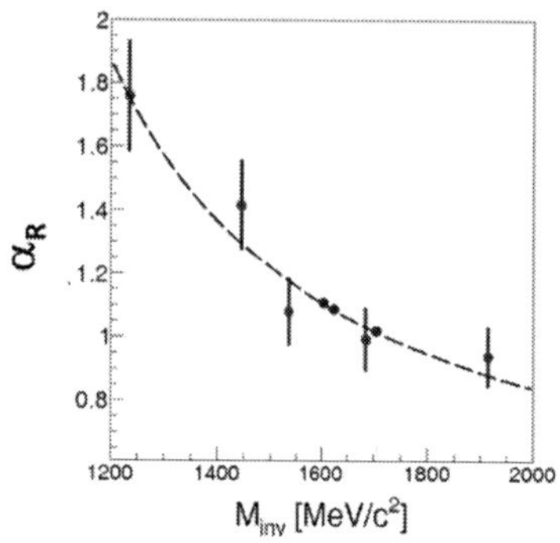

Figure 6. Dependency of the constant α_R used for the parametrization of the four-momentum transfer dependence of the cross section (Eq. 1) on the resonance mass M^R_{pole} obtained from fits to the data of 1π production analysis [10].

bution, was parametrized as

$$\frac{d\sigma}{dt} \sim t^{-\alpha_R}, \tag{1}$$

where t is the four momentum transfer $t = (p - p_R)^2$ at the resonance excitation vertex, calculated between the four-momentum vectors of the outgoing resonance (p_R) and the incoming nucleon (p). The parameter α_R could be fitted to the data and showed a decrease as a function of the mass of the resonance (see fig. 6).

Resonance	$\Gamma_{N\pi\pi}/\Gamma_{tot}(\%)$	σ_R [mb]
$N(1440)$	65	1.50 ± 0.37
$N(1520)$	55	1.8 ± 0.3
$N(1535)$	46	0.15 ± 0.05
$\Delta(1600)$	15	$< 0.24 \pm 0.10$
$\Delta(1620)$	25	$< 0.10 \pm 0.03$
$N(1650)$	45	$< 0.81 \pm 0.13$
$N(1675)$	45	$< 1.65 \pm 0.27$
$N(1680)$	65	$< 0.9 \pm 0.15$
$N(1720)$	20	$< 4.4 \pm 0.7$
$\Delta(1700)$	15	$< 0.45 \pm 0.16$
$\Delta(1905)$	15	$< 0.85 \pm 0.53$
$\Delta(1910)$	25	$< 0.38 \pm 0.16$

Table 1. Cross sections in units of mb for excitation of a positively charged resonance extracted from the one pion analysis [10].

Resonance	$\Gamma_{K\Lambda}/\Gamma_{tot}(\%)$	$\sigma(N^* \to pK\Lambda)[\mu b]$
$N^*(1650)$	7 ± 4	$8.6 \pm 0.6 \pm 2.1$
$N^*(1710)$	15 ± 10	$11.7 \pm 1.0 \pm 2.8$
$N^*(1720)$	4 ± 1	$2.4 \pm 1.3 \pm 0.4$
$N^*(1875)$	4 ± 2	$1.5 \pm 1.3 \pm 0.4$
$N^*(1880)$	2 ± 1	$14.9 \pm 0.2 \pm 3.6$

Table 2. Cross sections in units of μb for the single positively charged resonances extracted from $pp \to pK\Lambda$ analysis (third column) [11] . The ratio of $K\Lambda$ to total branching ratios derived from values in [12] is indicated in the second column.

In [11], a Partial Wave Analysis developed by the Bonn-Gatchina group was conducted for several data sets measured in the reaction $pp \to pK\Lambda$ at various energies. In this way, the amplitudes for excitation of five heavy resonances ($N^*(1650), N^*(1710), N^*(1720), N^*(1875), N^*(1880)$) in the $pp \to pK\Lambda$ channel could be determined (see table 2).

Both results were used as starting inputs in our model for the single resonance contributions in the $pp\pi^+\pi^-$ channel. The decays into the different $N\pi\pi$ channels were taken into account, with $N\rho$ branching ratio from the preliminary results of a Bonn-Gatchina PWA analysis [13] and the total, $\Delta\pi$ and Nσ branching ratios consistent with [12] within errors. For the double resonance excitation, no experimental information is existing in our energy range. We took the branching ratios to Nπ from [14], as in [10] and for the angular distributions, we adapted the parametrization Eq.(1) to take into account the excitation of baryons R1 and R2 at the two vertices:

$$\frac{d\sigma}{dt} \sim t^{-0.5(\alpha_{R1}+\alpha_{R2})} \tag{2}$$

where α_{R1} and α_{R2} are the parameters determined for the single excitation of the two baryons. The direct ρ contribution was generated according to uniform phase space.

Following the approach of [10], the simulations of the resonance production and decay were performed with PLUTO++ [15], a simulation framework for heavy ion and hadronic reactions. The decay channels were added incoherently and filtered with the HADES acceptance matrices deduced from GEANT simulations, as mentioned above. The cross sections for the different contributions were determined from a fit of the four invariant mass distributions $M_{inv}(p,\pi^+)$, $M_{inv}(p,\pi^-)$, $M_{inv}(p,\pi^+,\pi^-)$ and $M_{inv}(\pi^+,\pi^-)$, based on a χ^2 minimization. Fig. 3 displays some examples of comparison of experimental distributions measured in the HADES acceptance with the fit results. Many different iterations have been performed, in order to check the robustness of the fitting procedure and to reduce the number of parameters. It can be observed that the cocktail describing the data results from several significant contributions, corresponding either to the single or to the double baryon excitation. For the double resonance excitation, which in the pp$\pi^+\pi^-$ exit channel, implies the excitation of a double charged Δ , preliminary results show that only the $\Delta(1232)^{++}$ is found to have a significant contribution. The associated neutral baryon is mostly a $\Delta(1232)^0$, but significant contributions from N(1440), N(1520), N(1535) and N(1680) are also found. The $\Delta(1232)^{++}$ is excited very copiously, either in association with another resonance or produced in the decay of a heavier baryon. The cross sections for the single resonance excitation are found to be compatible with the previous analyses [10, 11]. In addition, the new analysis allows to determine the cross sections for the single excitation of some resonances, for which only upper limits could be provided previously. The results are also being compared to predictions of transport models. Most of the used models include double resonance excitation, but the cross sections are very scattered, even for the $\Delta(1232)^{++}\Delta(1232)^{0}$. Our data can therefore be used as constraints for the double resonance excitation, which is particularly important in view of the upcoming HADES and CBM experiments which will be performed at higher energies at the FAIR facility. In order to test the validity of adding incoherently

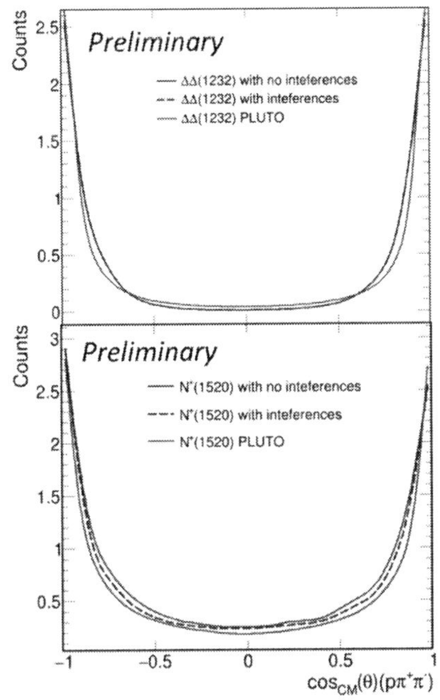

Figure 7. Comparison of the results of a Lagrangian model with (blue) and without (red) interferences between the different graphs. The simulations with the resonance model are shown in green. Top: Distribution of the $(p\pi^{+})$ centre of mass angular distribution for the double $\Delta^{++}(1232)$ contribution. Bottom: Distribution of the $(p\pi^{+}\pi^{-})$ centre of mass angular distribution for excitation of the $N^{+}(1520)$ decaying into $p\pi^{+}\pi^{-}$.

the cross sections of the different contributions in the resonance model, we developed a test Lagrangian model [16] for the main contributions, i.e. the double $\Delta(1232)$ and the N(1520) excitations. The values of coupling constants and cut-offs were taken as in [17]. The amplitudes of all contributing graphs, i.e. the direct, exchange, antisymmetrization graphs and for the N(1520) contribution, the decay into the various $\pi\pi N$ channels , i.e. $\Delta(1232)\pi$, ρN, $\sigma\pi$ was calculated. To evaluate the effect of the interferences, the calculation of the cross section from the squared sum of amplitudes was compared to an "incoherent" estimate deduced from the sum of squared amplitudes It was found that the effect of interferences was below 10% in all phase space regions (see fig. 7). In addition, it was interesting to observe that the angular distribution in the Lagrangian model is quite similar to the one due the parametrization of the four-momentum transfer dependence in our simple resonance model (see Eq. 1 and Eq. 2).

Dedicated models for the $pp \rightarrow pp\pi^{+}\pi^{-}$ [17, 18] can also be confronted to our data. The model of [17] is a Lagrangian model, based on meson exchange (π, σ, ρ, ω,...). It was developed for nucleon-nucleon reactions at incident energies below 2.2 GeV, and considers $\Delta(1232)^{++}\Delta(1232)^{0}$ as the only contributing double resonance excitation. In addition, it uses branching ratios from [14], which are in general, much larger than in [13]. The model of [18] uses experimental pion-nucleon amplitudes in a pion exchange approach and also considers one baryon exchange graphs. It is characterized by a strong contribution of double $\Delta(1232)$ excitation, hence producing a quite steep angular distribution, but also by a rather large yield coming from heavy resonances decaying to $p\pi^{+}\pi^{-}$.

4 ρ contribution

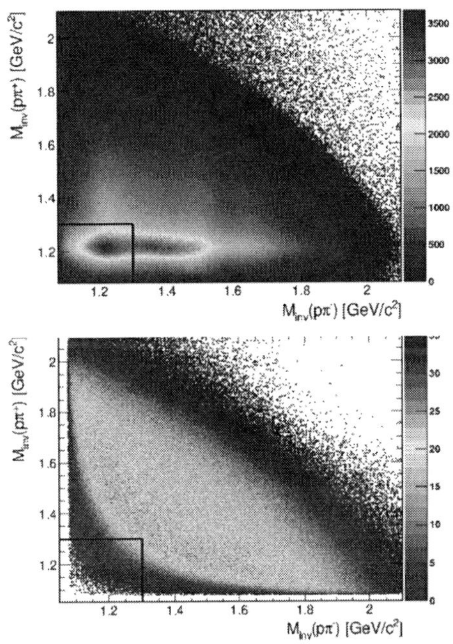

Figure 8. Correlation between $p\pi^{+}$ invariant mass and $p\pi^{-}$ invariant mass for data (top) and for simulation of the $pp \rightarrow pp\rho$ reaction (bottom). The black lines shows the first rejection ($M_{inv}(p\pi^{+}) < 1.3$ GeV/c^2 and $M_{inv}(p\pi^{-}) < 1.3$ GeV/c^2) applied to enhance the ρ signal.

The $\pi^{+}\pi^{-}$ channel is well suited to investigate the production of the ρ meson. However, the very large contribution from baryon resonances needs to be suppressed. To do so, cuts on the $p\pi$ invariant masses are first ap-

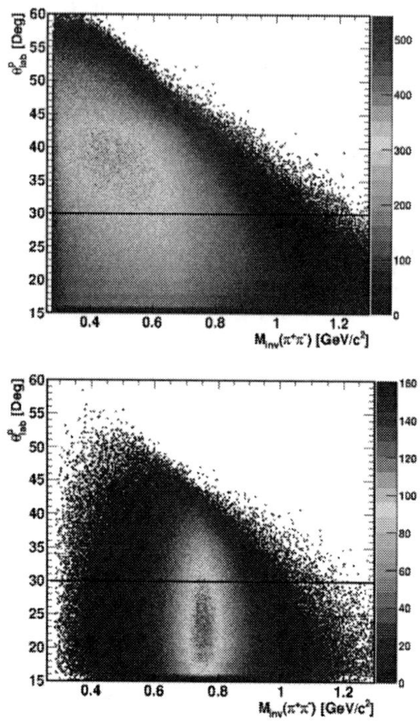

Figure 9. Correlation between the proton lab angle and the $\pi^+\pi^-$ invariant mass for data (top) and for simulation of the $pp \to pp\rho$ reaction (bottom). The black line shows the second rejection ($\theta_p > 30°$) applied to enhance the ρ signal.

plied (fig. 8), which suppresses very efficiently the contribution of the Δ(1232) resonance, while affecting very weakly the ρ production. Then, an additional threshold on the proton angle is applied to suppress protons emitted backwards in the center-of-mass which still mostly come from baryon resonance production (fig. 9). After both rejection criteria, a structure in the $\pi^+\pi^-$ invariant mass distribution appears around the ρ meson mass (see fig. ??). The production of the ρ meson via baryon resonance decay is included in our simulation with branching ratios from the recent Bonn-Gatchina PWA. The corresponding ρ meson spectral function is significantly distorted from a Breit-Wigner, due to the coupling to the light baryons (e.g. N(1520)) which favors ρ mesons in the low mass tail. The observed excess with respect to the resonant contribution can be considered as due to the direct ρ production (see fig. 5 Right). Its yield can be extracted in bins of $\pi^+\pi^-$ angle. Preliminary results show a significant enhancement for forward/backward emission, which is consistent with the dominance of nucleon currents in the $pp \to pp\rho$ reaction, as also observed by DISTO [20]. Despite the model dependency, this analysis gives information on the ρ production mechanism and the role of baryon resonances in it. To enable a more direct comparison to data, the usual method for extracting the ρ production cross section based

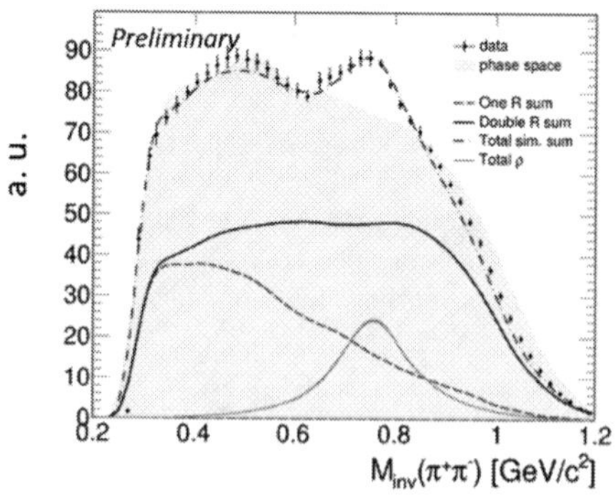

Figure 10. $\pi^+\pi^-$ invariant mass distribution after applying the kinematical rejections $M_{inv}(p\pi^+) < 1.3$ GeV/c^2, $M_{inv}(p\pi^-) < 1.3$ GeV/c^2 and $\theta_{lab}(p) < 30°$, compared to phase space (yellow area). The blue (respectively dashed green) curves represent the double (respectively single) resonance contribution from the resonance model (see text) and the red curve is the direct ρ contribution. The dashed red curve shows the total [19].

on a simple Breit-Wigner fit was also applied, although it suppresses the contribution due to the light baryons. The analysis is in progress and results on the cross section of the ρ meson production are expected soon.

5 Conclusion

These preliminary results show the interest of the $pp \to pp\pi^+\pi^-$ reaction to constrain in a quantitative way the baryon resonance contribution and the ρ meson production, which are important sources of e^+e^- pairs. The two contributions are strongly coupled due to the possible decay of baryon resonances in the ρN channel. In addition, according to the Vector Meson Dominance, it is expected that the electromagnetic coupling to baryons is fully mediated by the vector mesons. These data provide important information on nucleon-nucleon reaction mechanisms in general and constitute in particular important constraints for the models used for dielectron production, which will be essential in the context of the future experiments at FAIR with the HADES and CBM set-ups, at similar incident energy per nucleon than the present experiment.

6 Acknowledgements

SIP JUC Cracow, Cracow (Poland), National Science Center, 2016/23/P/ST2/040 POLONEZ, 2017/25/N/ST2/00580, 2017/26/M/ST2/00600; TU Darmstadt, Darmstadt (Germany), VH-NG-823, DFG GRK 2128, DFG CRC-TR 211, BMBF:05P18RDFC1; Goethe-University, Frankfurt (Germany) and TU Darmstadt, Darmstadt (Germany), ExtreMe Matter Institute EMMI at GSI Darmstadt; TU München, Garching (Germany), MLL München, DFG EClust 153, GSI TMLRG1316F, BmBF 05P15WOFCA, SFB 1258, DFG

FAB898/2-2; NRNU MEPhI Moscow, Moscow (Russia), in framework of Russian Academic Excellence Project 02.a03.21.0005, Ministry of Science and Education of the Russian Federation 3.3380.2017/4.6; JLU Giessen, Giessen (Germany), BMBF:05P12RGGHM; IPN Orsay, Orsay Cedex (France), CNRS/IN2P3; NPI CAS, Rez, Rez (Czech Republic), MSMT LM2015049, OP VVV CZ.02.1.01/0.0/0.0/16 013/0001677, LTT17003.

References

[1] G. Agakichiev et al. (HADES), Eur. Phys. J. **A41**, 243 (2009), 0902.3478
[2] P. Salabura *et al.* (HADES), *Exploring time like transitions in pp, π^-p and heavy ion reactions with HADES*, NSTAR2019
[3] J.J. Sakurai, Phys. Rev. Lett. **22**, 981 (1969)
[4] G. Agakishiev et al. (HADES), Eur. Phys. J. **A48**, 74 (2012), 1203.1333
[5] G. Agakishiev et al. (HADES), Eur. Phys. J. **51**, 137 (2015)
[6] G. Agakishiev et al., Eur. Phys. J. **A50**, 82 (2014), 1403.3054
[7] G. Agakishiev et al. (HADES), Phys. Lett. **B742**, 242 (2015), 1410.8188
[8] G. Agakishiev et al., Phys. Lett. **B750**, 184 (2015), 1503.04013
[9] A. Belounnas (HADES), PoS **Hadron2017**, 213 (2018)
[10] G.. Agakishiev et al. (HADES), Eur. Phys. J. **A50**, 82 (2014), 1403.3054
[11] R. Münzer et al., Phys. Lett. **B785**, 574 (2018), 1703.01978
[12] M. Tanabashi et al. (Particle Data Group), Phys. Rev. D **98**, 030001 (2018)
[13] V.V. Sarantsev, private communication
[14] K. Nakamura et al. (Particle Data Group), J. Phys. G **37**, 075021 (2010)
[15] I. Fröhlich et al., PoS **ACAT2007**, 076 (2007), 0708.2382
[16] Van de Wiele, J., private communication
[17] X. Cao, B.S. Zou, H.S. Xu, Phys.Rev. **C81**, 065201 (2010), 1004.0140
[18] A.P. Jerusalimov, A.V. Belyaev, V.P. Ladygin, A.K. Kurilkin, A.Yu. Troyan, Yu.A. Troyan, Eur. Phys. J. **A51**, 83 (2015)
[19] A. Belounnas (HADES), EPJ Web Conf. **199**, 02012 (2019)
[20] F. Balestra et al. (DISTO), Phys. Rev. Lett. **89**, 092001 (2002), hep-ex/0209004

N* Experiments and what they tell us about Strong QCD Physics

V. D. Burkert

December 16, 2019

Abstract I give an overview on experimental studies of the spectrum and the structure of the excited states of the nucleon and what we can learn about their internal structure. One focus is on the efforts to obtain a more complete picture of the light-quark baryon excitation spectrum employing electromagnetic beams that will allow us to draw some conclusions on the symmetries underlying the spectrum. For the higher mass excitations, the full employment of coupled channel approaches is essential when searching for new excited states in the large amounts of data already accumulated in different channels involving a variety of polarization observables. The other focus is on the study of transition form factors and helicity amplitudes and their dependences on Q^2, especially on some of the more prominent resonances, especially $\Delta(1232)\frac{3}{2}^+$, $N(1440)\frac{1}{2}^+$, and negative parity states $N(1535)\frac{1}{2}^-$, and $N(1675)\frac{5}{2}^-$. These were obtained in pion and eta electroproduction experiments off proton targets and have already led to further insights in the active degrees-of-freedom as a function of the distance scale involved.

Keywords light-quark baryon excitation, electroexcitation of nucleon resonances, quark core, meson-baryon contributions

PACS 12.39.Ki, 13.40.Gp, 13.40.Hq, 14.20.Gk

V.D. Burkert
Thomas Jefferson National Accelerator Facility
Newport News, Virginia 23606, USA
E-mail: burkert@jlab.org

1 Excited baryon states in the history of the Universe

For this talk the organizers asked me to address what we learn about strong QCD (sQCD), i.e. QCD in the domain where perturbative methods fail in describing nucleon resonances transitions. Talking about a similar topic from the theory perspective, Nathan Isgur said this in the concluding talk at $N*2000$: "*I am convinced that completing this chapter in the history of science will be one of the most interesting an fruitful areas of physics for at least the next thirty years.*" We are now 19 years into this 30 years prediction, and the physics of N*'s continues to go strong, while many related issues remain to be explored.

As we are trying to make progress in the complex world of physical sciences, we should not lose sight of what physics is all about: understanding the origin and the history of our universe, and the laws underlying the observations. In this meeting we also address how excited states of the nucleon fit in to our understanding of the forces and the dynamics of matter in the history of the universe. On the internet we find beautiful representations of the phases through which the universe evolved from the Big Bang (BB) to our times as shown in Fig. 1. There are some marked events that have been of particular significance during the early phases of the its history, such as the quark-gluon plasma of non-interacting colored quarks and gluons, and the forming of protons and neutrons. During this transition dramatic events occur - chiral symmetry is broken, quarks acquire mass dynamically, baryon resonances occur abundantly, and colored quarks and gluons are confined. This crossover process is governed by the excited hadrons, this is schematically shown in the generic QCD phase diagram in Fig. 1. In this process strong

http://doi.org/10.1051/epjconf/202024101004

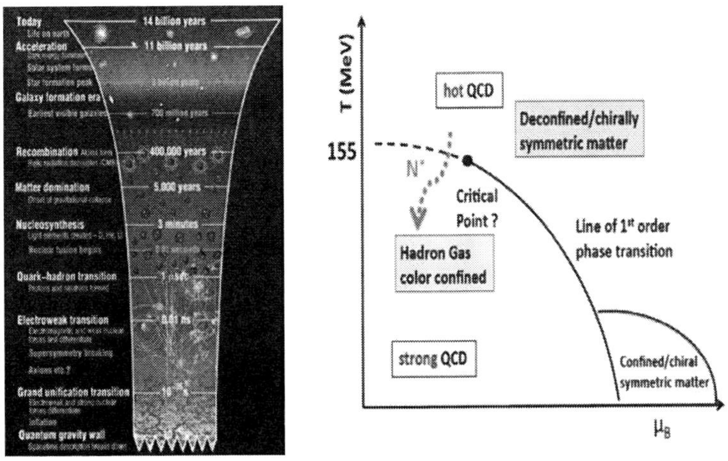

Fig. 1 Left panel: The evolution of the Universe. The line denoted as *Quark-hadron transition*, is where protons and neutrons are formed. Existing electron accelerators as CEBAF, ELSA, and MAMI, and colliders as BES III have sufficient energy reach to access this region and study processes in isolation that occurred during this transition in the microsecond old universe and resulted in the freeze out of baryons. Right panel: A generic phase diagram for the transition from the de-confined quark-gluon state to the confined hadron state.

QCD (sQCD) is born as the process describing the interaction of colored quarks and gluons. These are the phenomena that we are exploring with electron and hadron accelerators - the full discovery of the baryon (and meson) spectrum, the role of chiral symmetry breaking and the generation of dynamical quark mass in confinement. While we can not recreate the exact condition in the laboratory, with existing accelerators we can explore these processes in isolation. With electron machines and high energy photon beams in the few GeV energy range we search for undiscovered excitations of nucleons and other baryons.

As the Universe expands and cools down the coupling of quarks to the gluon field becomes stronger and quarks become more massive and form excited states in abundance. This eventually gives way to the forming stable nucleons. In the heavy-quark sector It has been demonstrated [1,2] that the entire complement of excited states as predicted in the quark model [3] is needed to be included in these calculation to explain what is observed in "hot QCD" lattice calculations. Including only resonances from the Review of Particle Physics (RPP) is insufficient to explain the computations within hot QCD. Similar projections have been made in the light-quark sector that we will discuss in the second part of this report. The close relationship of the baryon resonance spectrum and the evolution of the early universe makes the experimental search for the "missing resonances" an even more compelling experimental program.

2 Search for missing baryon states

Vigorous spectroscopy programs are currently underway at various particle accelerators in the quest for undiscovered excited mesons and baryons. Experiments at electron machines such as CEBAF at JLAB in the US, ELSA at Bonn University in Germany, and MAMI at the Johannes Gutenberg University at Mainz in Germany, focus on the s-channel excitation of protons and neutrons to N^* and Δ^* states.

The excited states of the nucleon have been studied experimentally since the 1950's [4]. They contributed to the discovery of the quark model in 1964 by Gell-Mann and Zweig [5,6], and were critical for the discovery of "color" degrees of freedom as first introduced by Greenberg [7]. The 3-quark quark structure of baryons resulted in the prediction of a wealth of excited states with underlying spin-flavor and orbital symmetry of $SU(6) \otimes O(3)$. The predictions led to a broad experimental effort to search for these states. Of the many states predicted in the quark model, only a fraction have been observed, even today. Searches for the "missing" states and detailed studies of the resonance structure are now mostly carried out using electromagnetic probes and have been a major focus of hadron physics for the past two decades [8]. A broad experimental effort has been underway for the past two decades, with measurements of exclusive meson photoproduction and electroproduction reactions, including many polarization observables. Precision data and the development of multi-channel partial wave analysis procedures have

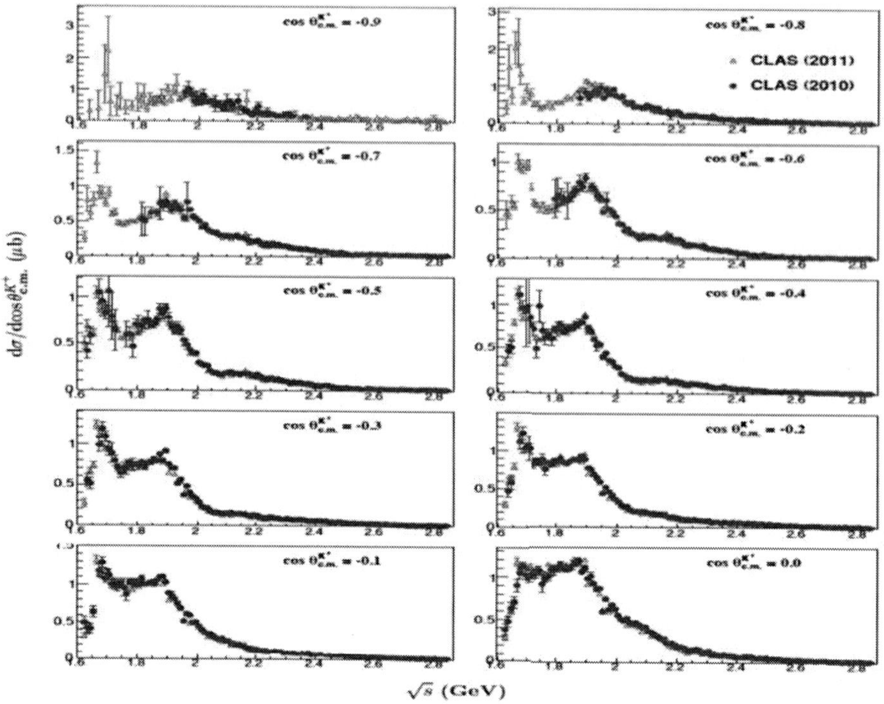

Fig. 2 Invariant mass dependence of the $\gamma p \to K^+\Lambda$ differential cross section in the backward polar angle range. There are 3 structures visible that may indicate resonance excitations, at 1.7, 1.9, and 2.2 GeV. The blue full circles are based on the topology $K^+p\pi^-$, the red open triangles are based on topology K^+p or $K^+\pi^-$, which extended coverage towards lower W at backward angles and allows better access to the resonant structure near threshold.

resulted in the discovery of a series of excited states of the nucleon in the mass range of 1.9 to 2.2 GeV, others have been upgraded in their status of likelihood of existence as entered in the bi-annual Reviews of Particle Physics [9–12]. We will discuss some of these states in the following sections.

Accounting for the complete excitation spectrum of the nucleon (protons and neutrons) and understanding the effective degrees of freedom is perhaps the most important and certainly the most challenging task of hadron physics. The experimental N* program currently focusses on the search for new excited states in the mass range above 2 GeV using energy-tagged photon beams in the few GeV range, and the study of the internal structure of prominent resonances in meson electroproduction.

A quantitative description of baryon spectroscopy and the structure of excited nucleons must eventually involve solving QCD for a complex strongly interacting multi-particle system. Recent advances in Lattice QCD led to predictions of the nucleon spectrum in QCD with dynamical quarks [13], albeit with still large pion masses of 396 MeV. Lattice prediction can therefore only be taken as indicative of the quantum numbers of excited states and not of the masses of specific states. In parallel, the development of dynamical coupled channel models is being pursued with new vigor. The EBAC group at JLab as well as others have shown [14] that dynamical effects can result in significant mass shifts of the excited states. As a particularly striking result, a very large shift was found for the Roper resonance pole mass to 1365 MeV downward from its bare mass of 1736 MeV. This result has clarified the longstanding puzzle of the incorrect mass ordering of $N(1440)\frac{1}{2}^+$ and $N(1535)\frac{1}{2}^-$ resonances in the constituent quark model. Developments on the phenomenological side go hand

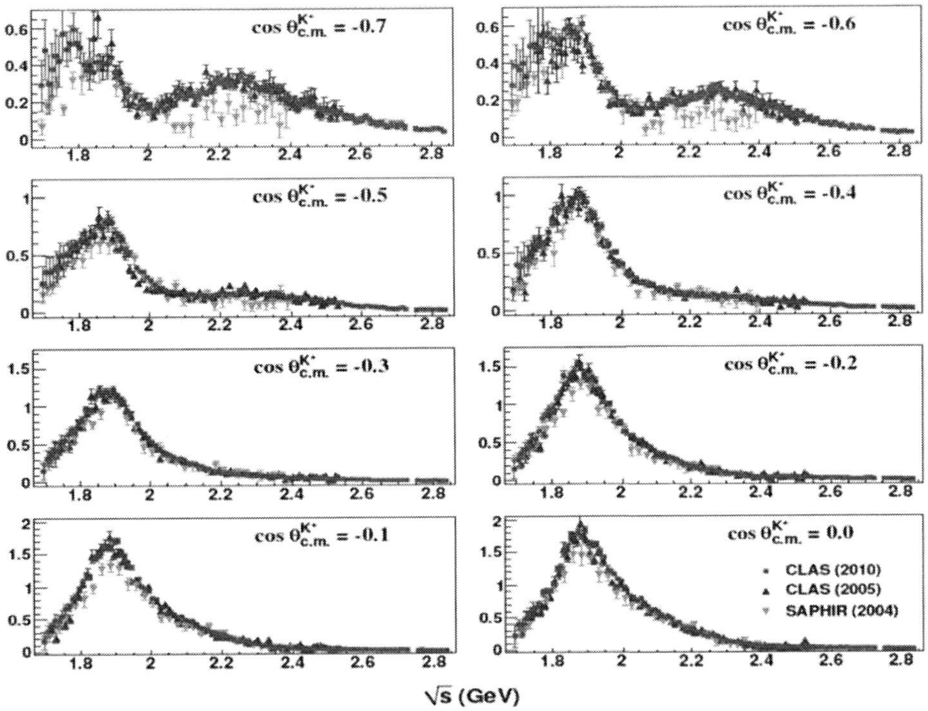

Fig. 3 Invariant mass dependence of the $\gamma p \to K^+\Sigma^0$ differential cross section in the backward polar angle range.

in hand with a world-wide experimental effort to produce high precision data in many different channel as a basis for a determination of the light-quark baryon resonance spectrum. On the example of experimental results from CLAS, the strong impact of precise meson photoproduction data is discussed. Several reviews have recently been published on this and related subjects [15–19] where many details can be found.

3 Establishing the N* Spectrum

The complex structure of the light-quark (u & d quarks) baryon excitation spectrum complicates the experimental search for individual states. As a result of the strong interaction, resonances are wide, often 200 MeV to 300 MeV, and are difficult to uniquely identify when only differential cross sections are measured. Most of the excited nucleon states listed in the Review of Particle Physics (RPP) prior to 2012 have been observed in elastic pion scattering $\pi N \to \pi N$. However there are important limitations in the sensitivity to the higher mass nucleon states that may have very small $\Gamma_{\pi N}$ decay widths, and the extraction of resonance contributions then becomes exceedingly difficult in this channel.

Estimates for alternative decay channels have been made in quark model calculations[20] for various channels. This has led to a major experimental effort at Jefferson Lab, ELSA, GRAAL, and MAMI to chart differential cross sections and polarization observables for a variety of meson photoproduction channels. At JLab with CLAS, many different final states have been measured with high precision on the proton [21–37], many of them are now employed in single- and in multi-channel analyses [38]. Recently, the first measurements of open strangeness processes on neutron targets have been published [39,40].

State N((mass)J^P	PDG 2010	PDG 2018
N(1710)1/2$^+$	***	****
N(1880)1/2$^+$		***
N(2100)1/2$^+$	*	***
N(1895)1/2$^-$		****
N(1900)3/2$^+$	**	****
N(1875)3/2$^-$		***
N(2120)3/2$^-$		***
N(2060)5/2$^-$		***
Δ(1600)3/2$^+$	***	****
Δ(1900)1/2$^-$	**	***
Δ(2200)7/2$^-$	*	***

Fig. 4 Evidence for 11 N and Δ states in RPP 2010 compared with RPP 2018 [12].

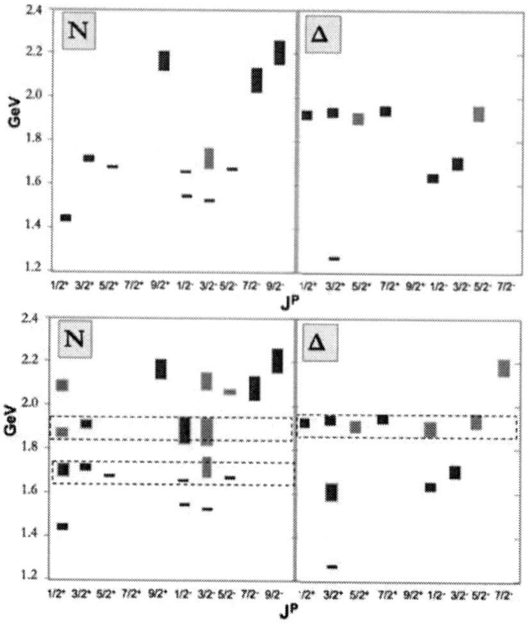

Fig. 5 Top panel: Nucleon and Δ resonance spectrum below 2.2 GeV in RPP 2010 [9]. Bottom panel: Nucleon and Δ resonance spectrum below 2.2 GeV in RPP 2018 [12]. The green frames highlight the new states and states with improved start ratings compared to 2010. The light brown color indicate 3* states, the dark color indicates 4* states. The dashed frames indicate apparent mass degeneracy of states with masses near 1.7 GeV and 1.9 GeV and different spin and parity. The Bonn-Gatchina analysis includes all the $K^+\Lambda$ and $K^+\Sigma^0$ cross section and polarization data, as well as pion photoproduction data.

3.1 New states from open strangeness photoproduction

Here one focus has been on measurements of $\gamma p \to K^+\Lambda$ and $\gamma p \to K^+\Sigma$. Using a linearly polarized photon beam several polarization observables can be measured by analyzing the parity violating decay of the recoil $\Lambda \to p\pi^-$. It is well known that the energy-dependence of a partial-wave amplitude for one particular channel is influenced by other reaction channels due to unitarity constraints. To fully describe the energy-dependence of a production amplitude other reaction channels must be included in a coupled-channel approach. Such analyses have been developed by the Bonn-Gatchina group[41], at JLab[42], at Jülich[43] and other groups. More recent measurements of single and double polarization data in the same channel [33] are shown in Fig. 8.

The data sets with the highest impact on resonance amplitudes in the mass range above 1.7 GeV have been kaon-hyperon production using a spin-polarized photon beam and the polarization of the Λ or Σ^0 is also measured by analyzing the parity violating decay $\Lambda \to p\pi^-$. The high precision cross section and polarization data [29,30,32,35,36] provide nearly full polar angle coverage and span the $K^+\Lambda$ invariant mass range from threshold to 2.9 GeV, hence covering the full nucleon resonance domain where new states could be discovered.

The backward angle $K^+\Lambda$ data in Fig.2 show clear resonance-like structures at 1.7 GeV and 1.9 GeV that are particularly prominent and well-separated from other structures at backward angles, while at more forward angles (not shown) t-channel processes become prominent and dominate the cross section. The broad enhancement at 2.2 GeV may also indicate resonant behavior although it is less visible at more central angles with larger background contributions. The $K^+\Sigma$ channel also indicates significant resonant behavior as seen in Fig. 3. The peak structure at 1.9 GeV is present at all angles with a maximum strength near 90 degrees, consistent with the behavior of a $J^P = \frac{3}{2}^+$ p-wave. Other structures near 2.2 to 2.3 GeV are also visible. Still, only a full partial wave analysis can determine the underlying resonances, their masses and spin-parity. The task is somewhat easier for the $K\Lambda$ channel, as the iso-scalar nature of the Λ selects isospin-$\frac{1}{2}$ states to contribute to the $K\Lambda$ final state, while both isospin-$\frac{1}{2}$ and isospin-$\frac{3}{2}$ states can contribute to the $K\Sigma$ final state.

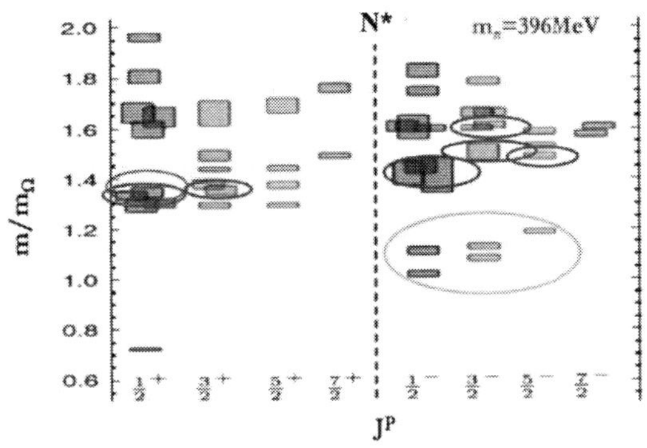

Fig. 6 Nucleon resonance spectrum below 2.2 GeV from LQCD [45]. The new discovered Nucleon states from RPP 2018 [12] are indicated by the red and blue ellipses with their spin-parity assignments. The observed masses deviate from the predicted ones as the lattice calculation used 396MeV pion mass, and no multi channel coupling is included. The green ellipses are well-known states that come out in the LQCD work with higher masses.

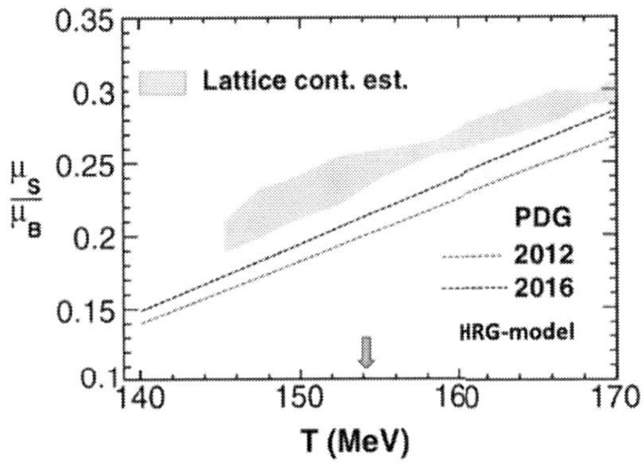

Fig. 7 The ratio of baryon chemical potential of strangeness versus all baryons for the RPP 2014 and RPP 2016. The hashed grey area show the LQCD calculation in "hot QCD". The straight lines are calculations within a hadron gas model. The 2016 line, which includes more N/Δ baryon states, moves closer to the LQCD area. Note that only 3* and 4* states are included. If the newly discovered states in RPP 2018 (seven states that are now at 3* or 4* status) were included this line would be moving even closer to the LQCD area.

These cross section data together with the Λ and Σ recoil polarization and polarization transfer data to the Λ and Σ had strong impact on the discovery of several new nucleon states [44]. They also provided new evidence for several candidate states that had been observed previously but lacked confirmation as shown in Fig. 4. It is interesting to observe that four of the observed nucleon states have nearly degenerate masses near 1.9 GeV, as seen in Fig. 5. Similarly, the new Δ state appears to complete a mass degenerate multiplet near 1.9 GeV as well. There is no obvious mechanism for this apparent degeneracy. Nonetheless, all new states may be accommodated within the symmetric constituent quark model based on $SU(6) \otimes O(3)$ symmetry group as far as quantum numbers are concerned. As discussed in section 1 for the case of the Roper resonance $N(1440)\frac{1}{2}^+$, the masses of all pure quark model states need to be corrected for dynamical coupled channel effects to compare them with observed resonances. The same applies to the Lattice QCD predictions [45] for the nucleon and Delta spectrum.

Coming back to the evolution of the early universe, we may check what is the expected effect of these newly discovered states on the evolution of the universe near the cross over temperature $T_c \approx 155$ MeV? Figure 7 shows the ratio of the baryon chemical potential of strangeness carrying baryons over all baryons versus temperature near the crossover temperature between de-confined and confined and conditions. The graph clearly shows the significant impact of the recently included new baryons by the PDG in the 2016 edition of the RPP in comparison to the 2012 edition. Furthermore, adding new states found since 2016 and now included in the 2018 edition of RPP will bring the HRG model closer to the Lattice QCD calculation [46], demonstrating the strong impact excited nucleon states have in the transition from the phase of free colored quarks and gluons to quarks and gluons confined in protons and neutrons. [1]

3.2 Vectormeson photoproduction

Double-polarization measurements of single pion photoproduction data [34] have contributed significantly to the discovery of the high-mass $\Delta(2200)7/2^-$ state [47], although the state couples just at the 3.5% level to Nπ Nevertheless in the mass range above 2.0 GeV resonances tend to decouple from simple final states like $N\pi$, $N\eta$, and $K\Lambda$, the search for undiscovered high-mass states requires to consider more complex final states with multi-mesons $N\pi\pi$ or vector mesons, such as $N\omega$, $N\phi$, and $K^*\Sigma$. The study of such final states adds significant complexity as many more amplitudes can contribute to these photoproduction processes compared to single pseudo-scalar meson production. As is the case for $N\eta$ production, the $N\omega$ channel is selective to isospin $\frac{1}{2}$ nucleon states only. The CLAS collaboration has collected a tremendous amount of data in the $p\omega$ channel [27,28], including single and double

[1] Increasing the number of excited baryons will lower μ_B at a given temperature, and hence increase the ratio of $\frac{\mu_S}{\mu_B}$

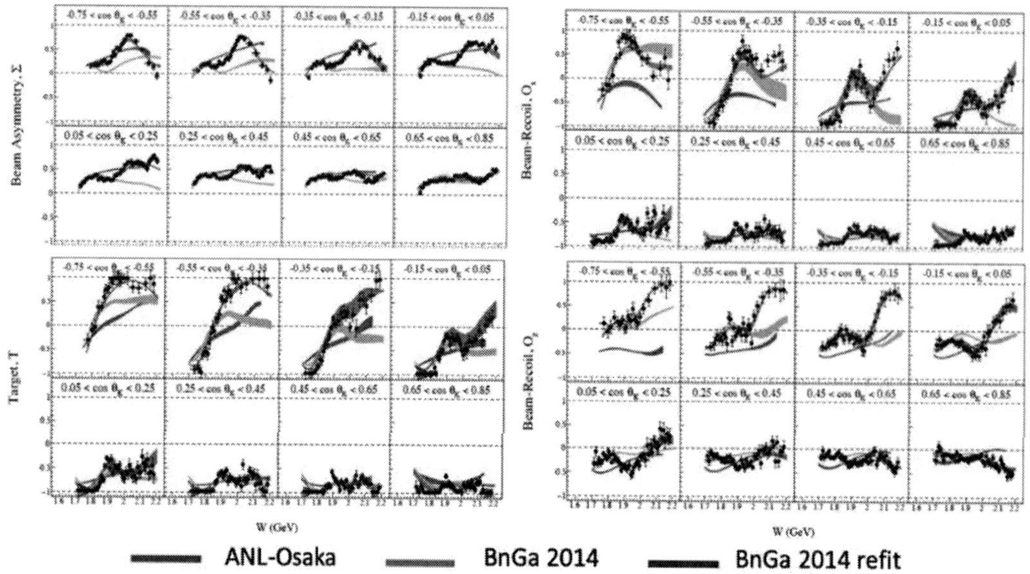

Fig. 8 Beam and Target polarization asymmetries (left panel), and beam-recoil polarization asymmetries (right panel) measured by the CLAS collaboration in the $\gamma p \to K^+\Lambda$ channel. The projections from earlier analyses. shown in the red and green bands, show large discrepancies with the data in the higher W range. A refit by the BnGa group (blue band) shows that the discrepancies do not require new excited states but could be accommodated by just adjusting some parameters.

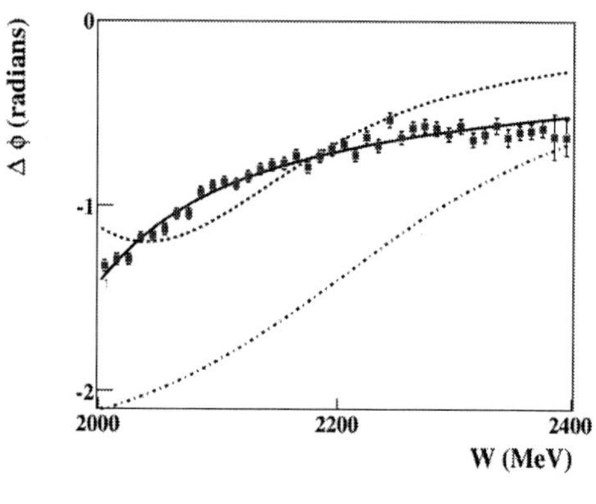

Fig. 9 Phase motion of the partial wave fit to the $\gamma p \to p\omega$ differential cross section and spin density matrix elements. 3 resonant states, the subthreshold resonance $N(1680)\frac{5}{2}^+$, $N(2190)\frac{7}{2}^-$, and the missing $N(2000)\frac{5}{2}^+$ are needed to fit the data (solid line). Fits without $N(2000)\frac{5}{2}^+$ (dashed-dotted line), or without $N(1680)\frac{5}{2}^+$ (dashed line) cannot reproduce the data.

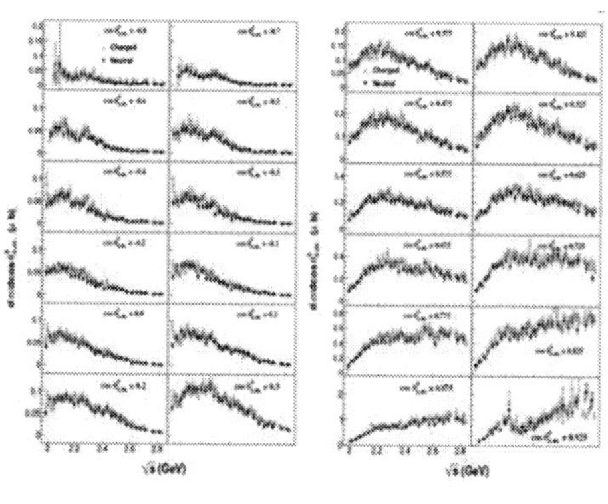

Fig. 10 Differential cross sections in a nearly full angular range for $\gamma p \to p\phi$ production.

polarization measurements [48–52]. The CLAS collaboration performed a single channel event-based analysis, whose results are shown in Fig. 9, and provided further evidence for the $N(2000)\frac{5}{2}^+$. Also a large amount of $p\phi$ [53,54] final states on differential cross sections and spin-density matrix elements have been published, although they have not systematically included in the more complex multi-channel analyses. Photoproduction of ϕ mesons is also considered a potential source of new excited nucleon states in the mass range above 2 GeV. Differential cross sections and spin-density matrix elements have been measured for $\gamma p \to p\phi$ in a mass range

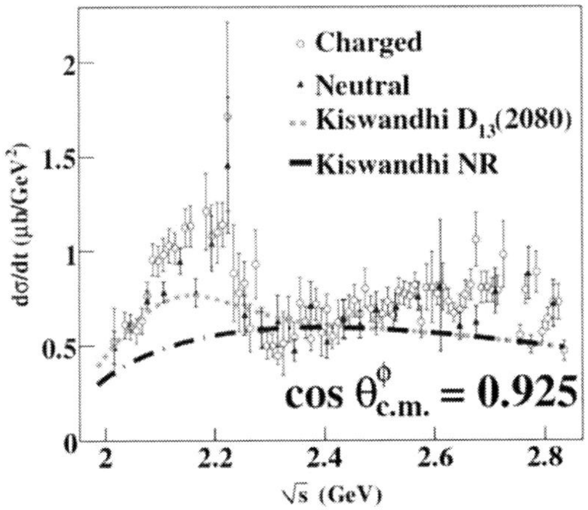

Fig. 11 Differential cross sections of $\gamma p \to p\phi$ production for the most forward angle bin. The two curves refer to fits without (dashed) and with (dotted) a known resonance at 2.08 GeV included.

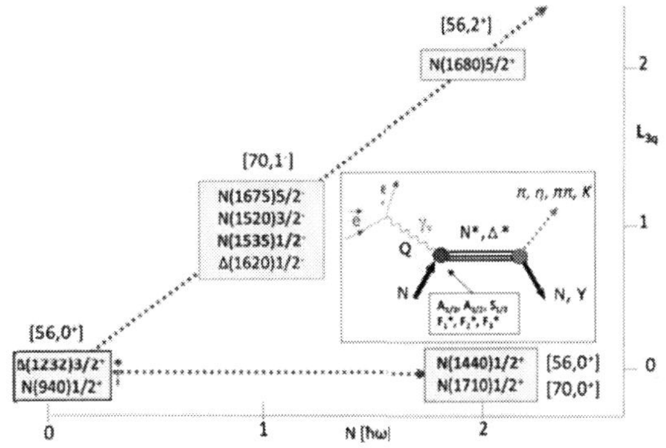

Fig. 12 Schematics of $SU(6) \otimes O(3)$ supermultiplets with some of the excited states that have been explored in $ep \to e'\pi^+ n$,$ep \to e'p\pi^\circ$, $ep \to e'p\eta$, and $ep \to e'p\pi^+\pi^-$ electroproduction experiments. The inset shows the helicity amplitudes and electromagnetic multipoles often used to describe the data. Only the states highlighted in red are discussed here.

up to nearly 3 GeV. In Fig. 10 structures are seen near 2.2 GeV in the forward most angle bins and at very backward angles for both decay channels $\phi \to K^+K^-$ and $\phi \to K_l^0 K_s^0$, and with the exception of the smallest forward angle bin the structures are more prominent at backward angles. Only a multi-channel partial wave analysis will be able to pull out any significant resonance strength. Fig. 11 shows the differential cross section $d\sigma/dt$ of the most forward angle bin. A broad structure at 2.2 GeV is present, but does not show the typical Breit-Wigner behavior of a single resonance. It also does not fit the data in a larger angle range, which indicates that contributions other than genuine resonances may be significant. The forward and backward angle structures may also hint at the presence of dynamical effects possibly due to molecular contributions such as diquark-anti-triquark contributions [55], the strangeness equivalent to the recently observed hidden charm P_c^+ states. Another process that has promise in the search for new excited baryon states, including those with isospin-$\frac{3}{2}$ is $\gamma p \to K^*\Sigma$ [56]. In distinction to the vector mesons discussed above, diffractive processes do not play a role in this channel, which then may allow better direct access to s-channel resonance production.

4 Structure of excited nucleons

Meson photoproduction has become an essential tool in the search for new excited baryons. The exploration of the internal structure of excited states and the effective degrees of freedom contributing to s-channel resonance excitation requires the use of electron beams, where the virtuality (Q^2) of the exchanged photon can be varied to probe the spatial structure. Electroproduction of final states with pseudoscalar mesons (e.g. $N\pi$, $p\eta$, $K\Lambda$) have been employed with CLAS, leading to new insights into the scale dependence of effective degrees of freedom, e.g. meson-baryon, constituent quark, and dressed quark contributions. Several excited states, shown in Fig. 12 assigned to their primary $SU(6) \otimes O(3)$ supermultiplets have been studied. The $N\Delta(1232)\frac{3}{2}^+$ transition is now well measured in a large range of Q^2 [57–59]. Results on the magnetic transition form factor G_{Mn} and on the quadrupole transition ratios are shown in Fig. 13 and Fig. 14. Two of the prominent higher mass states, the Roper resonance $N(1440)\frac{1}{2}^+$ and $N(1535)\frac{1}{2}^-$ are shown in Fig. 15 and in Fig. 17, respectively, as representative examples [60,61,63] of a wide program at JLab [64–72]. For these three states advanced relativistic quark model calculations [73] and QCD calculations from DSE [74] and Light Cone sum rule [75] are available, for the first time employing QCD-based modeling of the excitation of the quark core.

There is agreement with the data at $Q^2 > 1.5$ GeV^2 for the latter two states, while the meson-baryon contributions for the $\Delta(1232)$ are more extended, and agreement with the quark based calculations is reached at $Q^2 > 4$ GeV^2. The calculations deviate significantly from the data at lower Q^2, which indicates significant non quark core effects. For the Roper resonance such contributions have been described successfully in dynamical meson-baryon models [76] and in effective field

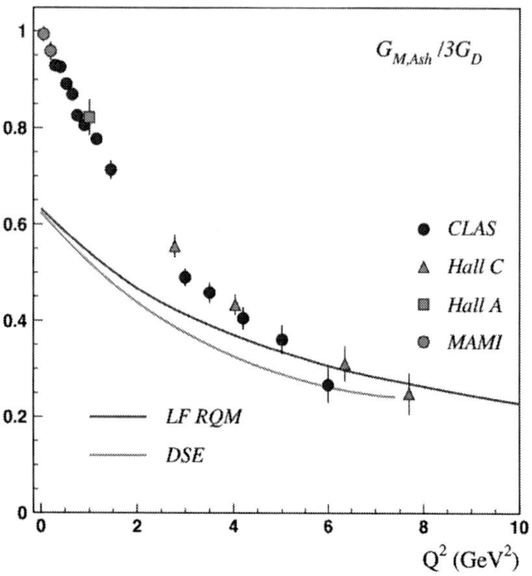

Fig. 13 (Color online) The magnetic NΔ transition form factor normalized to the dipole form.factor compared with the LF RQM with running quark mass, and with DSE/QCD. Both are close to the data at high Q^2. At $Q^2 < 3\text{GeV}^2$ meson-baryon contributions are significant.

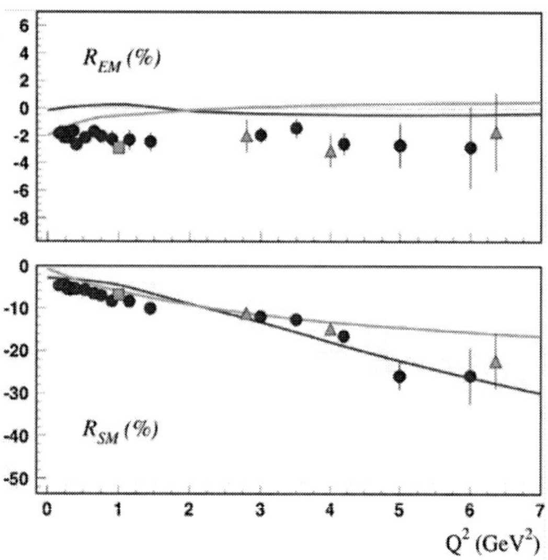

Fig. 14 (Color online) The ratios R_{EM} and R_{SM} for the $N\Delta(1232)$ transition. The solid red curves are the LF RQM predictions, the green curves are from the DSE approach. References to data as in Fig. 13.

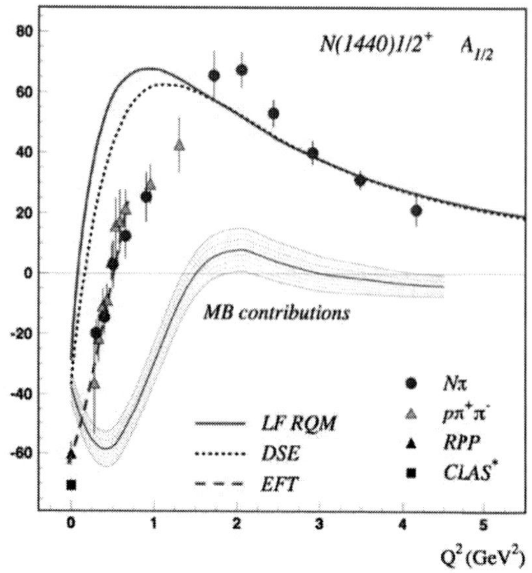

Fig. 15 The transverse helicity amplitudes $A_{1/2}$ for the Roper resonance $N(1440)\frac{1}{2}^+$. Data are from CLAS compared to LF RQM with running quark masses (solid line), and with projections from the DSE/QCD approach (dotted line). The shaded band indicates non 3-quark contributions inferred from the difference of the LF RQM curve and the CLAS data. The red dashed line is the EFT calculation that describes the data at small Q^2.

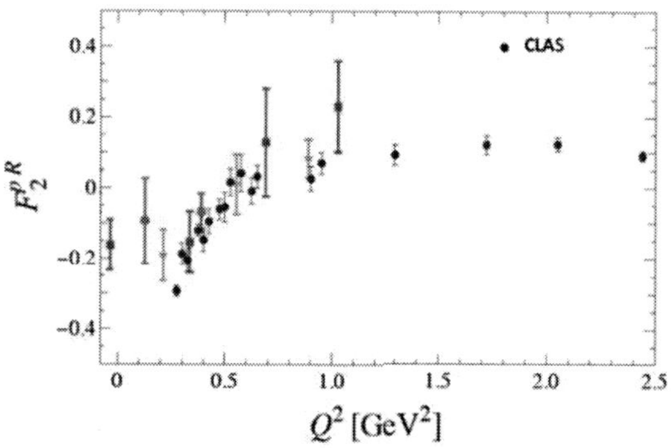

Fig. 16 The $pN^+(1440)1/2^+$ transition amplitude $F_2(Q^2)$ from LQCD [62] compared to CLAS results.

theory [77]. Calculations on the Lattice for the N-Roper transition amplitudes have been carried out with dynamical quarks [62]. The results agree with the data in the range $Q^2 < 1.0$ GeV^2, where data and calculations overlap Fig. 16.

Knowledge of the helicity amplitudes in a large Q^2 allows for the determination of the transition charge densities on the light cone in transverse impact parameter space (b_x, b_y) [78]. Figure 18 shows the comparison of $N(1440)\frac{1}{2}^+$ and $N(1535)\frac{1}{2}^-$. There are clear differences in the charge transition densities between the two

 http://doi.org/10.1051/epjconf/202024101004

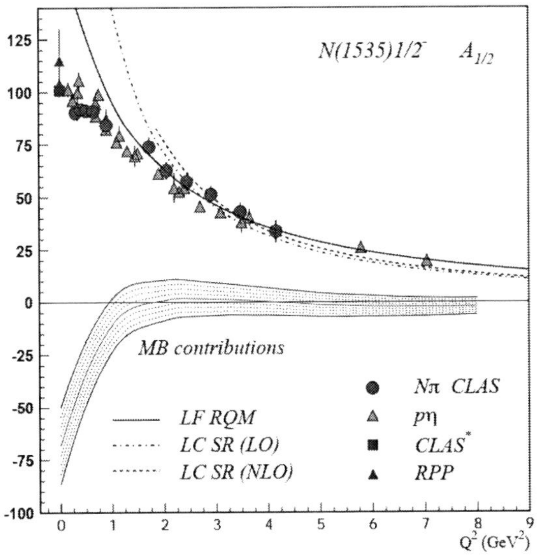

Fig. 17 The transverse amplitude $A_{1/2}$ for the $N(1535)\frac{1}{2}^-$ resonance compared to LF RQM calculations (solid line) and QCD computation within the LC Sum Rule approach.

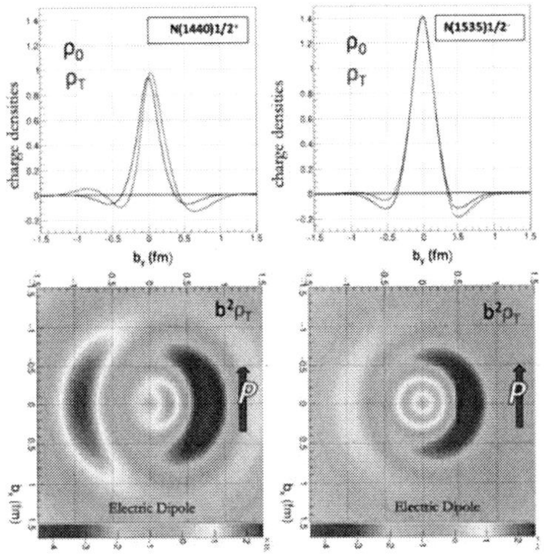

Fig. 18 Left panels: $N(1440)$, top: projection of charge densities on b_y, bottom: transition charge densities when the proton is spin polarized along b_x. Right panels: same for $N(1535)$. Note that the densities are scaled with b^2 to emphasize the outer wings. Color code:negative charge is blue, positive charge is red. Note that all scales are the same.

states. The Roper state has a softer positive core and a wider negative outer cloud than $N(1535)$ and develops a larger shift in b_y when the proton is polarized along the b_x axis. New electroproduction data on the Roper and several higher mass states have been obtained in the 2-pion channel, specifically in $ep \to e'p\pi^+\pi^-$ [65].

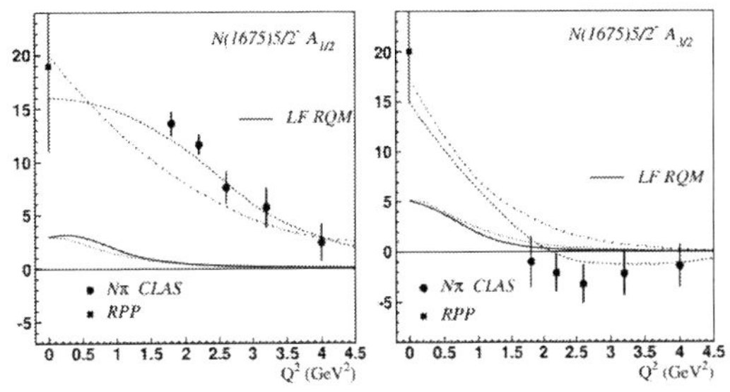

Fig. 19 Helicity amplitude $A_{1/2}$ (left) and $A_{3/2}$ (right) for $N^+(1675)5/2^-$ off proton target.

4.1 The $N(1675)5/2^-$ state - revealing the meson-baryon contributions

In previous discussions we have assumed that meson-baryon degrees of freedom provide significant strength to the resonance excitation in the low Q^2 domain where quark based approaches LF RQM, DSE/QCD, and LCSR calculations fail to reproduce the transition amplitudes quantitatively. Our conclusion rests, in part, with this assumption. But, how can we be certain of the validity of this assumption?

The $N(1675)5/2^-$ resonance allows testing this assumption, quantitatively. Figure 19 shows our current knowledge of the transverse helicity amplitudes $A_{1/2}(Q^2)$ and $A_{1/2}(Q^2)$, for proton target compared to RQM [79] and hypercentral CQM [80] calculations. The specific quark transition for a $J^P = 5/2^-$ state belonging to the $SU(6) \otimes O(3)] = [70, 1^-]$ supermultiplet configuration, in non-relativistic approximation prohibits the transition from the proton in a single quark transition. This suppression, known as the Moorhouse selection rule [81], is valid for the transverse transition amplitudes $A_{1/2}$ and $A_{3/2}$ at all Q^2. It should be noted that this selection rule does apply to the transition from protons but not from neutrons. Modern quark models, that go beyond single quark transitions, confirm quantitatively the suppression resulting in very small transition amplitudes from protons but large ones from neutrons. The measured helicity amplitudes off the protons are almost exclusively due to meson-baryon contributions as the dynamical coupled channel (DCC) calculation indicates (dashed line). The close correlation of the DCC calculation and the measured data for the case when quark contributions are nearly absent, supports the phenomenological description of the helicity amplitudes in terms of a 3-quark core that dominate at high Q^2 and meson-baryon contributions that can make important contributions at lower Q^2.

5 Conclusions and Outlook

Over the past five years eight baryon states in the mass range from 1.85 to 2.15 GeV have been either discovered or evidence for the existence of states has been significantly strengthened. To a large degree this is the result of adding very precise photoproduction data in open strangeness channels to the data base that is included in multi-channel partial wave analyses, especially the Bonn-Gatchina PWA. The possibility to measure polarization observables in these processes has been critical. In the mass range above 2 GeV more complex processes such as vector mesons or $\Delta\pi$ may have sensitivity to states with higher masses but require more complex analyses techniques to be brought to bear. Precision data in such channels have been available for a few years but remain to be fully incorporated in multi-channel partial wave analyses processes. The light-quark baryon spectrum is likely also populated with hybrid excitations [13] where the gluonic admixtures to the wave function are dominating the excitation. These states appear with the same quantum numbers as ordinary quark excitations, and can only be isolated from ordinary states due to the Q^2 dependence of their helicity amplitudes [82], which is expected to be quite different from ordinary quark excitations. This requires new electroproduction data especially at low Q^2 [83] with different final states and at masses above 2 GeV.

Despite the very significant progress made in recent years to further establish the light-quark baryon spectrum and explore the internal structure of excited states, much remains to be done. A vast amount of precision data already collected needs to be included in the multi-channel analysis frameworks, and polarization data are still to be analyzed.

There are approved proposals to study resonance excitations at much higher Q^2 and with higher precision at Jefferson Lab with CLAS12 [84,85], which may reveal the transition to the bare quark core contributions at short distances.

A new avenue of experimental research has recently been opened up with the data-based extraction of the first mechanical property of the proton - its internal pressure distribution [86]. Mechanical properties of resonance transitions have recently been explored for the $N(1535)3/2^- \rightarrow N(938)$ gravitational transition form factors calculations [87]. In order to access these new gravitational form factors experimentally, the nucleon to resonance transition GPDs have to be studied. The framework for studying the $N \rightarrow N(1535)$ transition GPDs, which would enable experimental access to these mechanical properties, still remains to be developed.

I like to thank Inna Aznauryan and Viktor Mokeev for numerous discussions on the subjects discussed in this presentation.

6 Acknowledgment

This work was supported by the US Department of Energy under contract No. DE-AC05-06OR23177. This work was supported by the U.S. Department of Energy, Office of Science, Office of Nuclear Physics, under Contract No. DE-AC05-06OR23177.

References

1. A. Bazavov *et al.*, Phys. Rev. Lett. **113**, 072001 (2014) doi:10.1103/PhysRevLett.113.072001 [arXiv:1404.6511 [hep-lat]].
2. A. Bazavov *et al.*, Phys. Lett. B **737**, 210 (2014) doi:10.1016/j.physletb.2014.08.034 [arXiv:1404.4043 [hep-lat]].
3. S. Capstick and N. Isgur, Phys. Rev. D **34**, 2809 (1986) [AIP Conf. Proc. **132**, 267 (1985)]. doi:10.1103/PhysRevD.34.2809, 10.1063/1.35361
4. H. L. Anderson, E. Fermi, E. A. Long and D. E. Nagle (1952) Phys. Rev. **85**, 936 .
5. M. Gell-Mann, Phys. Lett. **8**, 214 (1964).
6. G. Zweig, CERN Reports, TH 401 and 412 (1964).
7. O. W. Greenberg, Phys. Rev. Lett. **13**, 598 (1964); arXiv:0803.0992 [physics.hist-ph].
8. V. D. Burkert and T. S. H. Lee Int. J. Mod. Phys. E **13**, 1035 (2004).
9. J. Beringer *et al.* [Particle Data Group], Phys. Rev. D **86**, 010001 (2012). doi:10.1103/PhysRevD.86.010001
10. K. A. Olive *et al.* [Particle Data Group], Chin. Phys. C **38**, 090001 (2014). doi:10.1088/1674-1137/38/9/090001
11. C. Patrignani *et al.* [Particle Data Group], Chin. Phys. C **40**, no. 10, 100001 (2016). doi:10.1088/1674-1137/40/10/100001
12. M. Tanabashi *et al.* [Particle Data Group], Phys. Rev. D **98**, no. 3, 030001 (2018). doi:10.1103/PhysRevD.98.030001
13. J. J. Dudek and R. G. Edwards, Phys. Rev. D **85**, 054016 (2012)
14. N. Suzuki et al., Phys. Rev. Lett. **104**, 042302 (2010) [arXiv:0909.1356 [nucl-th]].
15. E. Klempt and J. M. Richard, Rev. Mod. Phys. **82**, 1095 (2010)
16. L. Tiator, D. Drechsel, S. S. Kamalov and M. Vanderhaeghen, Eur. Phys. J. ST **198**, 141 (2011).
17. I. G. Aznauryan and V. D. Burkert, Prog. Part. Nucl. Phys. **67**, 1 (2012)
18. I. G. Aznauryan *et al.*, Int. J. Mod. Phys. E **22**, 1330015 (2013)
19. V. Crede and W. Roberts, Rept. Prog. Phys. **76**, 076301 (2013)
20. S. Capstick and W. Roberts, Phys. Rev. D **49**, 4570 (1994)
21. M. Dugger *et al.*, Phys. Rev. Lett. **96**, 062001 (2006) [Phys. Rev. Lett. **96**, 169905 (2006)]
22. M. Dugger *et al.* [CLAS], Phys. Rev. C **79**, 065206 (2009) doi:10.1103/PhysRevC.79.065206

23. P. T. Mattione *et al.* [CLAS], Phys. Rev. C **96**, no. 3, 035204 (2017) doi:10.1103/PhysRevC.96.035204
24. I. Senderovich *et al.* [CLAS], Phys. Lett. B **755** (2016) 64 doi:10.1016/j.physletb.2016.01.044
25. M. Williams *et al.* [CLAS Collaboration], Phys. Rev. C **80**, 045213 (2009)
26. P. Collins *et al.* [CLAS], Phys. Lett. B **771**, 213 (2017) doi:10.1016/j.physletb.2017.05.045
27. M. Williams *et al.* [CLAS], Phys. Rev. C **80**, 065209 (2009)
28. M. Williams *et al.* [CLAS], Phys. Rev. C **80**, 065208 (2009)
29. R. K. Bradford *et al.* [CLAS], Phys. Rev. C **75**, 035205 (2007)
30. R. Bradford *et al.* [CLAS], Phys. Rev. C **73**, 035202 (2006)
31. D. Ho *et al.* [CLAS], Phys. Rev. Lett. **118**, no. 24, 242002 (2017) doi:10.1103/PhysRevLett.118.242002
32. M. E. McCracken *et al.* [CLAS], Phys. Rev. C **81**, 025201 (2010)
33. C. A. Paterson *et al.* [CLAS], Phys. Rev. C **93**, no. 6, 065201 (2016) doi:10.1103/PhysRevC.93.065201
34. S. Strauch *et al.* [CLAS], Phys. Lett. B **750**, 53 (2015) doi:10.1016/j.physletb.2015.08.053
35. B. Dey *et al.* [CLAS], Phys. Rev. C **82**, 025202 (2010)
36. J. W. C. McNabb *et al.* [CLAS], Phys. Rev. C **69**, 042201 (2004)
37. E. Golovatch *et al.* [CLAS], Phys. Lett. B **788**, 371 (2019) doi:10.1016/j.physletb.2018.10.013
38. A. V. Anisovich *et al.*, Phys. Lett. B **772**, 247 (2017) doi:10.1016/j.physletb.2017.06.052
39. N. Compton *et al.* [CLAS], Phys. Rev. C **96**, no. 6, 065201 (2017) doi:10.1103/PhysRevC.96.065201
40. D. Ho *et al.* [CLAS], Phys. Rev. C **98**, no. 4, 045205 (2018) doi:10.1103/PhysRevC.98.045205
41. A. Anisovich, R. Beck, E. Klempt, V. Nikonov, A. Sarantsev and U. Thoma, Eur. Phys. J. A **48**, 15 (2012)
42. B. Julia-Diaz, T.-S. H. Lee, A. Matsuyama and T. Sato, Phys. Rev. C **76**, 065201 (2007)
43. D. Rönchen *et al.*, Eur. Phys. J. A **50**, no. 6, 101 (2014)
44. A. V. Anisovich *et al.*, Phys. Rev. Lett. **119**, no. 6, 062004 (2017) doi:10.1103/PhysRevLett.119.062004 [arXiv:1712.07549 [nucl-ex]].
45. R. G. Edwards, J. J. Dudek, D. G. Richards and S. J. Wallace, Phys. Rev. D **84**, 074508 (2011) doi:10.1103/PhysRevD.84.074508 [arXiv:1104.5152 [hep-ph]].
46. S. Chatterjee, D. Mishra, B. Mohanty and S. Samanta, Phys. Rev. C **96**, no. 5, 054907 (2017) doi:10.1103/PhysRevC.96.054907 [arXiv:1708.08152 [nucl-th]].
47. A. V. Anisovich *et al.*, Phys. Lett. B **766**, 357 (2017) doi:10.1016/j.physletb.2016.12.014
48. P. Collins *et al.* [CLAS], Phys. Lett. B **773**, 112 (2017) doi:10.1016/j.physletb.2017.08.015
49. Z. Akbar *et al.* [CLAS], Phys. Rev. C **96**, no. 6, 065209 (2017) doi:10.1103/PhysRevC.96.065209 [arXiv:1708.02608 [nucl-ex]].
50. P. Roy *et al.* [CLAS], Phys. Rev. C **97**, no. 5, 055202 (2018) doi:10.1103/PhysRevC.97.055202
51. P. Roy *et al.* [CLAS], Phys. Rev. Lett. **122**, no. 16, 162301 (2019) doi:10.1103/PhysRevLett.122.162301
52. A. V. Anisovich *et al.* [CLAS], Phys. Lett. B **771**, 142 (2017). doi:10.1016/j.physletb.2017.05.029
53. H. Seraydaryan *et al.* [CLAS Collaboration], Phys. Rev. C **89**, no. 5, 055206 (2014)
54. B. Dey *et al.* [CLAS Collaboration], Phys. Rev. C **89**, no. 5, 055208 (2014)
55. R. F. Lebed, Phys. Rev. D **92**, no. 11, 114030 (2015)
56. A. Sarantsev, talk at this workshop
57. K. Joo *et al.* [CLAS], Phys. Rev. Lett. **88**, 122001 (2002)
58. M. Ungaro *et al.* [CLAS], Phys. Rev. Lett. **97**, 112003 (2006)
59. V. V. Frolov *et al.*, Phys. Rev. Lett. **82**, 45 (1999)
60. I. G. Aznauryan *et al.* [CLAS], Phys. Rev. C **78**, 045209 (2008)
61. I. G. Aznauryan *et al.* [CLAS], Phys. Rev. C **80**, 055203 (2009)
62. H. W. Lin and S. D. Cohen, AIP Conf. Proc. **1432**, no. 1, 305 (2012) doi:10.1063/1.3701236 [arXiv:1108.2528 [hep-lat]].
63. V. D. Burkert and C. D. Roberts, Rev. Mod. Phys. **91**, no. 1, 011003 (2019) doi:10.1103/RevModPhys.91.011003 [arXiv:1710.02549 [nucl-ex]]
64. V. I. Mokeev *et al.* [CLAS], Phys. Rev. C **86**, 035203 (2012)
65. V. I. Mokeev *et al.*, Phys. Rev. C **93**, no. 2, 025206 (2016) doi:10.1103/PhysRevC.93.025206
66. G. V. Fedotov *et al.* [CLAS], Phys. Rev. C **98**, no. 2, 025203 (2018) doi:10.1103/PhysRevC.98.025203
67. E. L. Isupov *et al.* [CLAS], Phys. Rev. C **96**, no. 2, 025209 (2017) doi:10.1103/PhysRevC.96.025209
68. H. Denizli *et al.* [CLAS], Phys. Rev. C **76**, 015204 (2007)
69. C. S. Armstrong *et al.* [JLab E94014 Collaboration], Phys. Rev. D **60**, 052004 (1999)
70. H. Egiyan *et al.* [CLAS], Phys. Rev. C **73**, 025204 (2006)
71. K. Park *et al.* [CLAS], Phys. Rev. C **77**, 015208 (2008)
72. K. Park *et al.* [CLAS], Phys. Rev. C **91**, 045203 (2015)
73. I. G. Aznauryan and V. D. Burkert, Phys. Rev. C **92**, no. 3, 035211 (2015)
74. J. Segovia et al., Phys. Rev. Lett. **115**, no. 17, 171801 (2015)
75. I. V. Anikin, V. M. Braun and N. Offen, Phys. Rev. D **92**, no. 1, 014018 (2015)
76. I. T. Obukhovsky et al., Phys. Rev. D **84**, 014004 (2011)
77. T. Bauer, S. Scherer and L. Tiator, Phys. Rev. C **90**, no. 1, 015201 (2014)
78. L. Tiator and M. Vanderhaeghen, Phys. Lett. B **672**, 344 (2009)
79. I. G. Aznauryan and V. Burkert, Phys. Rev. C **95**, no. 6, 065207 (2017) doi:10.1103/PhysRevC.95.065207 [arXiv:1703.01751 [nucl-th]].
80. E. Santopinto and M. M. Giannini, Phys. Rev. C **86**, 065202 (2012) doi:10.1103/PhysRevC.86.065202 [arXiv:1506.01207 [nucl-th]].
81. R. G. Moorhouse, Phys. Rev. Lett. **16**, 772 (1966). doi:10.1103/PhysRevLett.16.772
82. Z. p. Li, V. Burkert and Z. j. Li, Phys. Rev. D **46**, 70 (1992).
83. L. Lanza, talk at this workshop, A. D' Angelo et al., Jefferson Lab Letter of Intent LOI12-15-004 (2015).
84. L.Elouadrhiri, R. Gothe, V. Mokeev, talks at this workshop.
85. V. D. Burkert, "Jefferson Lab at 12 GeV: The Science Program," Ann. Rev. Nucl. Part. Sci. **68**, 405 (2018). doi:10.1146/annurev-nucl-101917-021129
86. V. D. Burkert, L. Elouadrhiri and F. X. Girod, Nature **557**, no. 7705, 396 (2018). doi:10.1038/s41586-018-0060-z
87. U. Özdem and K. Azizi, arXiv:1912.06375 [hep-ph].

Double polarisation observable E and helicity dependent cross section for single π^0 photoproduction off proton and neutron

Susanna Costanza[1,2,*] and *Federico* Cividini[3] for the A2 Collaboration

[1]Dipartimento di Fisica, Università degli Studi di Pavia, I-27100 Pavia, Italy
[2]INFN, Sezione di Pavia, I-27100 Pavia, Italy
[3]Institut für Kernphysik, University of Mainz, D-55099 Mainz, Germany

Abstract. Photon-induced reactions, like meson photoproduction, allow to excite the nucleon, to have access to many different polarisation observables and are an essential tool to disentangle the role of the different electromagnetic multipoles due to the change of sign of some contributions and the presence of interference terms between different multipole amplitudes. In addition, the use of polarised beams and/or targets allow to access additional observables which are fundamental in order to accurately determine the nucleon resonance properties. The A2@MAMI collaboration is carrying out a broad and systematic study on this topics, both on the proton and the neutron. The experiments are performed at the tagged photon beam facility of the MAMI accelerator in Mainz, using circularly and linearly polarised photons on longitudinally polarised proton and deuteron targets, for energies ranging from the pion production threshold up to 1.6 GeV. Hadronic reaction products are then measured with the large acceptance Crystal Ball spectrometer, complemented by charged particle and vertex detectors for tracking and identification. An overview of the results obtained so far for the double polarisation observable E (circularly polarised photon beam on a longitudinally polarised target) on the single π^0 photoproduction off the proton and the neutron will be given. Furthermore, new results on the helicity-dependent total and differential cross sections on the deuteron will be presented.

1 Introduction

The internal structure of the nucleon, specifically its spin structure, has been a central issue for many theoretical models and experiments of nuclear and particle physics since many years. The study of the excited states of the nucleon is of particular interest because they are related to fundamental properties of the strong interaction, just like the excitation spectra of atoms reflect the properties of the electromagnetic interaction.
In this energy regime, Quantum Chromodynamics (QCD) is non-perturbative; therefore, only phenomenological quark models can provide reliable interpretations of the experimental results. Nevertheless, while these models are quite in agreement with the experimental findings in the low mass region of the spectrum, there are still a lot of higher lying states predicted by theories but not yet observed.
Experimentally, the "missing resonances" issue can be rooted in data analysis relying mostly on pion scattering, which will miss states weakly coupling to πN: if higher lying states couple mainly to decay channels involving heavier mesons or decay via intermediate excited states, they will be missed [1].
An experimental alternative aimed to a better understanding of the nucleon excitation spectra is represented by photon-induced reactions, like meson photoproduction: these reactions allow to excite the nucleon, to explore production states with multiple mesons, like $\pi\pi$, $\pi\eta$, ... and to investigate resonances which decay preferentially via intermediate excited states.
In addition, photoproduction reactions combined with the use of polarised beams and/or targets allow to have access to many different polarisation observables, which are needed to accurately determine the nucleon resonance properties.
For every fixed value of energy and angle (W, θ), it is possible to define a set of 16 polarisation observables: along with the unpolarised cross section σ, there are 3 single polarisation observables and 12 double polarisation observables, which are grouped in classes depending on the beam and target or recoil nucleon polarisation (table 1). By measuring 7 (8) properly chosen observables, it is possible to obtain a model-independent complete analysis [2, 3].

2 The A2 experiment at MAMI

Up to now, most efforts have been devoted to studying proton excitation but, since the electromagnetic excitations are isospin dependent, also measurements of meson-photoproduction off the neutron are required.
Thanks to the use of a polarised photon beam and polarised proton, deuteron and ^{3}He targets, the A2 MAMI collaboration is carrying out a broad experimental program, both on the proton and the neutron, in order to accurately inves-

*e-mail: susanna.costanza@pv.infn.it

http://doi.org/10.1051/epjconf/202024101005

Table 1. Polarisation observables.

Photon polarisation		Target polarisation			Recoil nucleon polarisation			Target and recoil polarisation			
		X	Y	$Z_{(beam)}$	X′	Y′	Z′	X′ X	X′ Z	Z′ X	Z′ Z
unpolarised	σ	-	T	-	-	P	-	$T_{X'}$	$L_{X'}$	$T_{Z'}$	$L_{Z'}$
linear	-Σ	H	(-P)	-G	$O_{X'}$	(-T)	$O_{Z'}$	$(-L_{Z'})$	$(T_{Z'})$	$(L_{X'})$	$(-T_{X'})$
circular	-	F	-	-E	$C_{X'}$	-	$C_{Z'}$	-	-	-	-

tigate the properties of baryonic resonances, through experiments devoted to the study of double-polarisation observables.

2.1 The detector

The experimental setup of the A2 collaboration is located at the MAMI tagged photon beam facility in Mainz (Germany). The MAMI accelerator provides a polarised electron beam with energy up to 1557 MeV, which hits a radiator foil, producing bremsstrahlung photons. The electrons are then deflected in the magnetic field (1.8 T) of the Glagow-Mainz magnetic spetrometer (energy resolution of ~2 MeV), while the photons are collimated and impinged on the target. Its cell is sitting inside the central detector system and surrounded by the cylindrical Particle Identification Detector (PID), made of 24 plastic scintillating strips used to distinguish the charged from the neutral particles. The PID is enclosed in two cylindrical Multi-Wire Proportional Chambers (MWPCs), used to identify and track the charged particles. The MWPCs are themselves contained in the Crystal Ball (CB) spectrometer: it's a large solid angle, highly segmented photon and hadron spectrometer consisting of 672 NaI(Tl) crystals.
The combined information from these detectors allow both precise energy and angular measurements, as well as particle identification in the azimuthal (ϕ) and polar (θ) angular regions from 0° to 360° and from 21° to 159°, respectively. Furthermore, particles flying in the forward direction, outside the acceptance region of CB, are detected by the TAPS spectrometer, an electromagnetic calorimeter comprised of 366 BaF_2 and 72 $PbWO_4$ crystals and covering the polar angular region from 1° to 20°.
A schematic view of the experimental setup is shown in Fig. 1.

2.2 The polarised beam and target

For the experiments described in the following, an elliptically polarised photon beam, with a linear and a circular polarisation component, was used.
Usually, linearly polarised photons are produced via coherent bremsstrahlung of unpolarised electrons impinging on a crystal radiator, like a diamand one. A circular polarisation of the photons, instead, is obtained when longitudinally polarised electrons hit any radiator. During data taking, longitudinally polarised electrons were used in combination with a diamond radiator, resulting in an elliptical polarisation of the photon beam.

Depending on the alignment of the diamond radiator with respect to the electron beam, it was possible to cover a large energy range by setting the coherent peak position at six different energies, from 350 MeV to 850 MeV (see Fig. 2). The linear polarisation degree was calculated according to [5].
Concerning the circular polarisation degree, it strongly depends on the polarisation degree of the electron beam, which was measured every day through a Mott measurement at the MAMI accelerator [6], and on the energy of the produced photons. The circular polarisation degree was calculated as a function of the photon beam energy ([7]) and it is shown in Fig. 3.
In the experiments described in the following sections, a frozen-spin butanol (C_4H_9OH) [9] and deuterated butanol (C_4OD_{10}) targets were used.
Concerning the butanol target, the 2 cm long target cell was located in a $^3He/^4He$ dilution refrigerator and the target nucleons were polarised via Dynamic Nuclear Polarisation (DNP) [10] to an initial polarisation degree up to 90%. During the data taking, a magnetic field of 0.68 T together with the 25 mK temperature provided by the dilution refrigerator ensured long relaxation times (up to 2000 h) before repolarisation was required. Since only the hydrogen nuclei of the butanol can be polarised, a carbon foam target was used in a dedicated beamtime to study background contributions from unpolarised carbon nuclei. In the case of the frozen-spin target using deuterated butanol, a starting value of target polarisation up to 70%, with a relaxation time of ~200 hours, was obtained, thanks to a new doping material [11]. Also in the case of gaseous ^3He an initial polarisation value up to 70% was obtained, via the metastability exchange optical pumping (MEOP [12]) method, with a total relaxation time of ~20 hours.

3 Helicity dependent cross section for single π^0 photoproduction

Fig. 4 shows the preliminary results of the helicity dependence of the total cross section $\Delta\sigma = (\sigma_a - \sigma_p)$ on the deuteron for the semi-exclusive channel $\gamma d \rightarrow \pi^0 X$. The A2 data (blue circles) are compared to the results of the GDH Collaboration (red circles) [13], showing good agreement. Since the GDH Collaboration collected data only in the energy region of the $\Delta(1232)$ resonance, it is therefore clear that the A2 results can provide a useful contribution in an energy region where no other data are available.

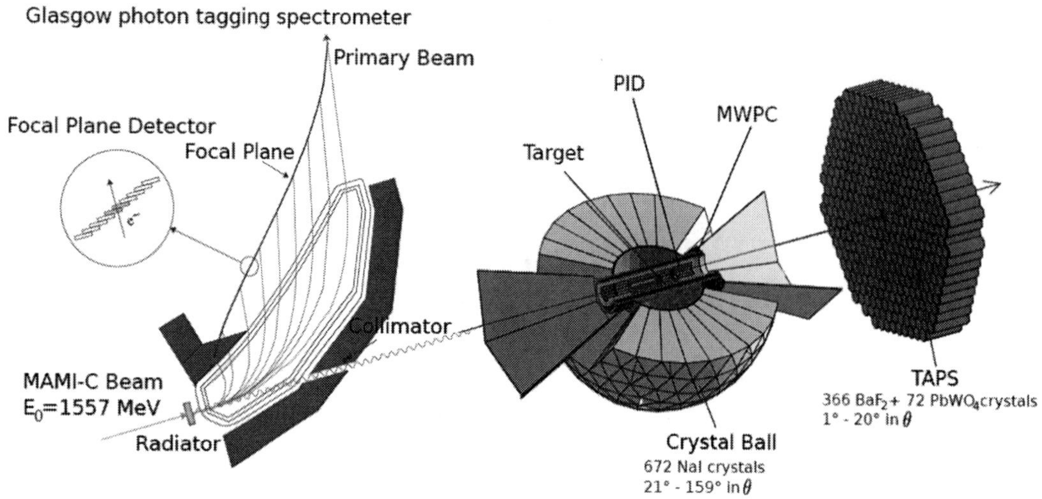

Figure 1. Schematic view of the A2 experimental setup.

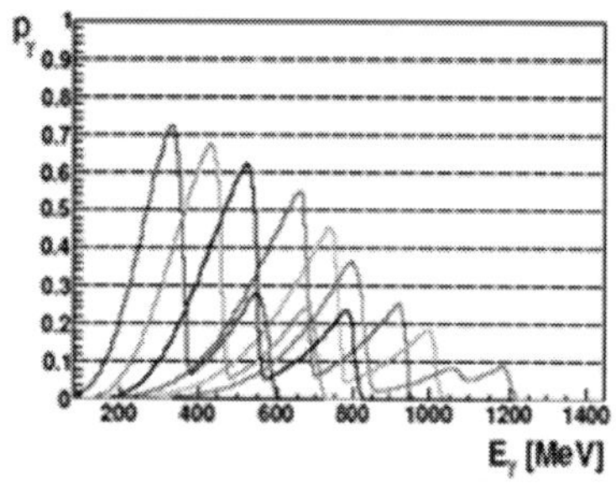

Figure 2. The linear polarisation degree for six different coherent peak settings, from 350 MeV to 850 MeV [4].

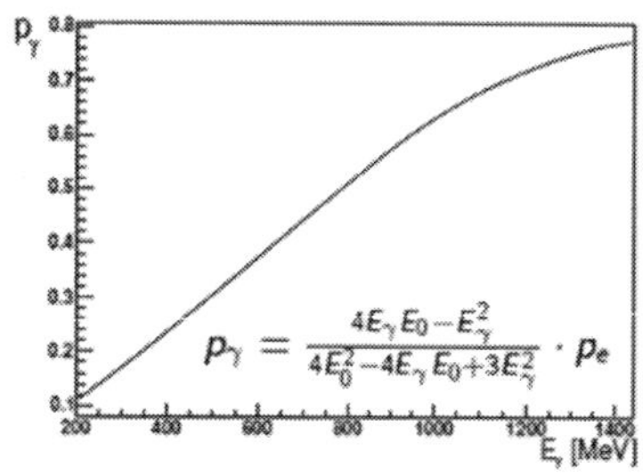

Figure 3. The calculated circular polarisation degree as a function of the photon beam energy [8].

The A2 experimental data are also compared to the predictions of the MAID multipole analysis (red dashed line) [14]: the discrepancy is due to the fact that the model is the sum of the free proton and free neutron contributions, hence it does not include the nuclear effects, which are not negligible in the deuteron target case. A better agreement is shown between the data and the MAID predictions corrected by A. Fix with a contribution based on [15], that takes into account the Fermi motion of the nucleons inside the deuteron (Impulse Approximation). In this case too, the predictions fail to describe the Δ-resonance region: in fact, in this region the contribution of the Final State Interaction (FSI), which is not taken into account by the model, is high. Instead, for $E_\gamma > 500$ MeV the experimental results follow the expectations of the MAID + IA model.
These new data can provide a strong constraint to all nucleon models that describe the $N-N$ and $\pi-N$ interactions. Ten energy bins of the preliminary results of the helicity dependent differential cross section for the partial channel $\gamma d \rightarrow \pi^0 X$ are shown in Fig. 5. As for the total cross section, the A2 data (blue circles) are compared to the predictions from MAID, without (red line) and with (green line) Impulse Approximation correction.

4 Double polarisation observable E

The experimental data collected with the frozen-spin butanol [9] and deuterated butanol targets were used to extract the double polarisation observable E in the photoproduction reactions $\gamma p \rightarrow \pi^0 p$ and $\gamma n \rightarrow \pi^0 n$ (the last one, only in case of the deuterated butanol target).
As shown in table 1, the double polarisation observable E requires both the polarisation of the beam and of the target. It's derived from the differential cross section for pseudoscalar meson photoproduction; since we used elliptically polarised photons in combination with a longitudinally polarised target, the cross section is given by:

$$\begin{aligned} \frac{d\sigma}{d\Omega}(\theta,\phi) = \frac{d\sigma}{d\Omega_0}(\theta)[1 - P_{lin}\Sigma\cos(2(\alpha-\phi)) \\ - P_z(-P_{lin}G\sin(2(\alpha-\phi)) + P_{circ}E]. \end{aligned} \quad (1)$$

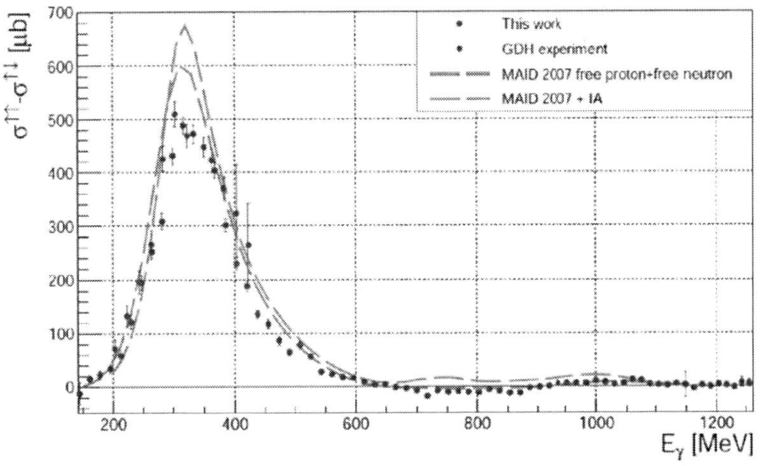

Figure 4. Helicity dependent total cross section for the semi-exclusive channel $\gamma d \to \pi^0 X$. The A2 results (blue circles) are compared to the results from the GDH Collaboration (red circles) [13], to the MAID model (red dashed line) [14] and to the MAID model + IA correction by A. Fix based on [15] (green dashed line).

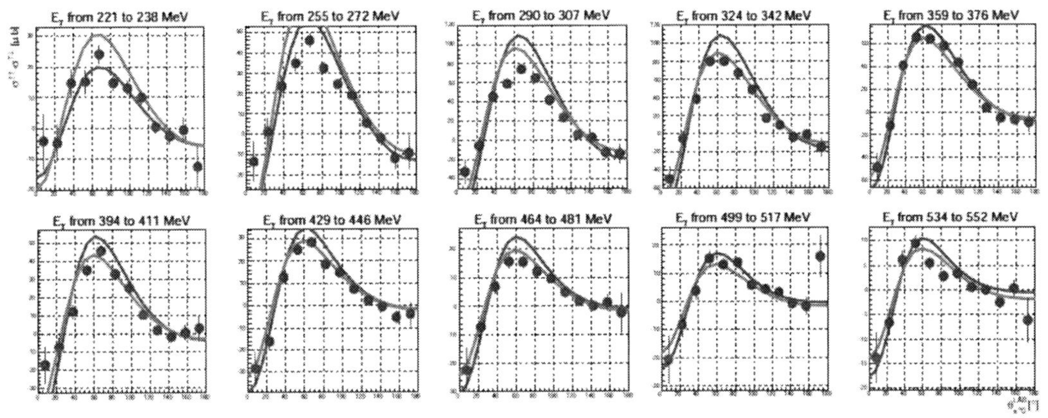

Figure 5. Helicity dependent differential cross section for the semi-exclusive channel $\gamma d \to \pi^0 X$. The A2 results (blue circles) are compared to the MAID model without (red line) [14] and with (green line) the Impulse Approximation correction by A. Fix based on [15].

We can extract the E observable by integrating eq. 1 over ϕ:

$$N_B\Big|_{\pm\alpha}^{\pm P_z}(\theta) = N_B(\theta) \cdot [1 - P_{circ}P_z E] \quad \rightarrow$$

$$E = \frac{\sigma^{1/2} - \sigma^{3/2}}{\sigma^{1/2} + \sigma^{3/2}} = \frac{N_B^{1/2} - N_B^{3/2}}{N_B^{1/2} + N_B^{3/2}} \cdot \frac{1}{d} \cdot \frac{1}{P_{circ}P_z}, \qquad (2)$$

where $N_B^{1/2}$ and $N_B^{3/2}$ are the helicity-dependent count rates, d is the dilution factor and $P_{circ}P_z$ are the degrees of target and beam polarisation.

In the case of our targets, the count rates accounts also for the contribution of both the polarisable free protons in hydrogen nuclei and the unpolarised bound nucleons in carbon and oxygen nuclei. We call dilution factor d ($d = N_{free}/(N_{free} + N_{bound})$) the amount of polarisable free protons in the data, which was determined in a dedicated data taking using a carbon foam target. d was obtained as $d = 1 - s \cdot N_C/N_B$, where the coplanarity distribution of the carbon data (N_C) is scaled to the one of the butanol data (N_B). Possible differences in acceptance, photon flux and target density during the carbon and butanol beamtimes are taken into account in the formula by s [16, 20].

4.1 Experimental results

The double polarisation observable E was extracted from the data according to Eq. 2 as a function of the beam energy E_γ for the entire angular range $-1 \leq \cos\theta_\pi \leq 1$.

Fig. 6 shows the results for the E observable in selected energy bins (from 1270 MeV to 1630 MeV) for the photoproduction channel $\gamma p \rightarrow \pi^0 p$, for free (green triangles) [18] and quasi-free (blue circles) protons, obtained with the frozen-spin butanol and deuterated butanol targets, respectively. The experimental points are also compared to the results of an independent analysis of a different data set collected by the A2 collaboration [19].
The E asymmetries on free and on quasi-free protons are in good agreement: this is due to the fact that the FSI affects both the numerator and the denominator of Eq. 2, hence it almost cancels out. Its global contribution on E is smaller than in the cross section case, where it plays an important role, especially in the Δ-resonance region, as shown in Figs. 4-5. This conclusion proves that the polarisation asymmetry can be used to access directly the internal structure of the nucleon, without the need of model-dependent corrections.
In Fig. 7, the A2 results for quasi-free protons (blue circles) are compared to different partial wave analyses solutions: the Bonn-Gatchina free proton model (red line) [21], the free proton model from SAID (black line) [22], the free proton model from MAID (magenta line) [14], MAID + IA (cyan line) and MAID + IA + FSI (green line) with the Final State Interaction contribution based on [15].
The agreement is mostly good, in particular with the BnGa2014-02 PWA solution. As compared to the data and to the other models, the MAID predictions overestimate σ_a and underestimates σ_p for energies up to $W = 1350$ MeV, while the situation is reversed for higher energies.
Similar results and their comparison with the PWA predictions are shown in Fig. 8 for the E observable for the photoproduction channel $\gamma n \rightarrow \pi^0 n$, as a function of $\cos\theta_{\pi^0}^{CM}$, for selected energy bins.
As for the proton target, the FSI effect on the E asymmetry is very marginal: thanks to this, the study of the observables which allow a better understanding of the nucleon structure is feasible even without the availability of free neutron targets.

5 Legendre analysis

The Legendre expansions provide a model-independent approach for the presentation of data for pion photoproduction reactions: in fact, it is possible to use a combination of multipoles and associated Legendre polynomials to describe the energy and angular dependence of the polarisation observables [23].
In the case of the E observable, $\hat{E}$ can be fitted using the following truncated partial wave analysis [24]:

$$\hat{E} = E(W,\theta) \cdot \frac{d\sigma}{d\Omega}(W,\theta) = \sum_{k=0}^{2l_{max}} A_k^E(W) P_k^0(\cos\theta) \quad (3)$$

where A_k^E are the Legendre coefficients for the polynomial functions truncated at l_{max} and $P_k^0(\cos\theta)$ are the associated Legendre polynomials of order zero. The selected value of l_{max} determines the sensitivity at different waves and multipoles.

Fig. 9 shows a sample of $\hat{E}$ for the $\pi^0 p$ channel, fitted with Legendre polynomials truncated at different l_{max} values.
In the case of the deuterated butanol data, due to the poor statistics, $l_{max} = 2$ would be the best trade off between fit efficacy and wave contributions. Nevertheless, in order to include a comparison with the free proton results (with much more statistics), the fit parameter results with $l_{max} = 4$ are shown in Fig. 10.
General features of the $A_k(W)$ coefficients are their marked decrease for large values of k and the sharp structures seen for each $A_k(W)$ coefficient in the region of the $\Delta(1232)3/2^+$. The contribution of this resonance is direct (without interference) only to the $A_0(W)$ and $A_2(W)$ coefficients; for the other coefficients, since there are no other resonances nearby, the presence of the peaked structure is due to the interference of the $\Delta(1232)3/2^+$ with other nonresonant partial waves. In particular, $A_1(W)$ should show the interference of the $\Delta(1232)3/2^+$ with the $S_{11}(1535)1/2^-$ and $D_{13}(1520)3/2^-$ states. $A_k(W)$ coefficients with $k > 2$ should reveal the $\Delta(1232)3/2^+$ interference with states with higher k. It is then clear that via such interference effects, it is possible to study the contributions from very high partial-wave amplitudes.
The fit with the Legendre polynomials has been performed for the $\pi^0 n$ channels too, and similar considerations can be drawn for the $A_k(W)$ coefficients obtained from the fit.

6 Conclusions

Useful tools for the understanding of the nucleon excitation spectra are double polarised pion photoproduction experiments, like the ones performed by the A2 Collaboration at MAMI with an elliptically polarised photon beam hitting on a longitudinally polarised frozen-spin butanol or a deuterated butanol target.
The experimental data provide new precise results on the double-polarisation measurements of the total and differential cross sections for the partial $\gamma N \rightarrow \pi^0 X$ channels on the proton and on the neutron. The results are compared to the existing model predictions and to the few available results. These new data are increasing the available statistics, in particular on the neutron, thus providing an important testing ground for the existing models.
Moreover, these data allow the extraction of the E asymmetry from quasi-free protons and quasi-free neutrons. The results are in good agreement with the existing data from the CBELSA/TAPS Collaboration and provide new measurements in an extended energy range.
Furthermore, it's possible to perform an alternative analysis with the A2 precise data: through the extraction of the coefficients of the Legendre polynomials we are able to reveal specific correlations and interferences between resonant states of definite parities.
These new, high-quality doubly-polarised pion-photoproduction data sets give a valuable input to the study of the nucleon structure and excitation spectra of protons and neutrons, by providing a contribution to the partial wave analysis models, especially for the neutron,

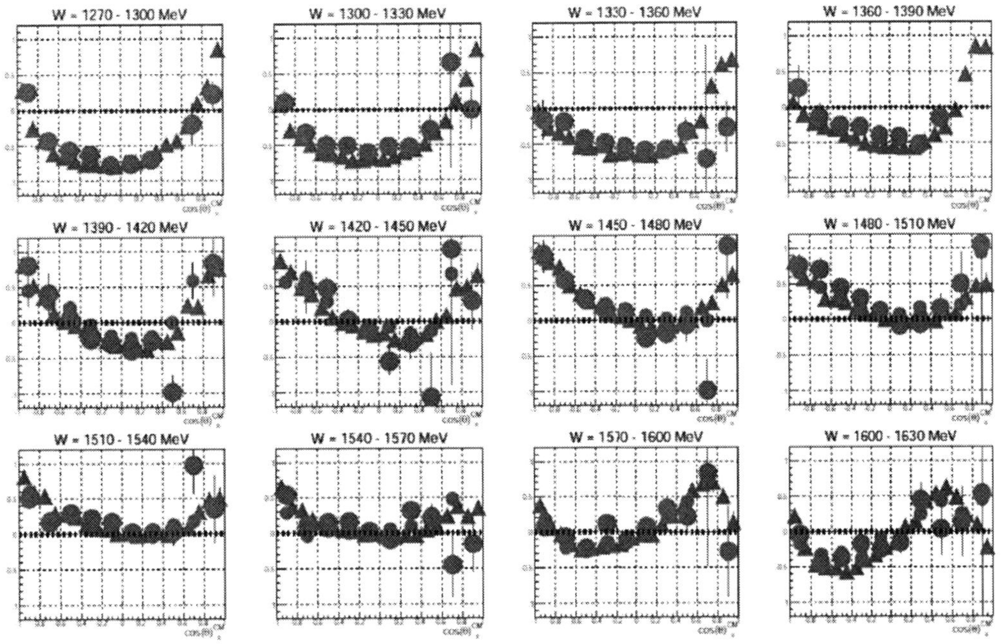

Figure 6. Results for the double polarisation observable E for the photoproduction channel $\gamma p \rightarrow \pi^0 p$, as a function of $\cos\theta_{\pi^0}^{CM}$ for selected 30 MeV energy bins, for free (green triangles) and quasi-free (blue circles) protons [20]. For $W_{CM} > 1390$ MeV, the data are also compared to the results in [19].

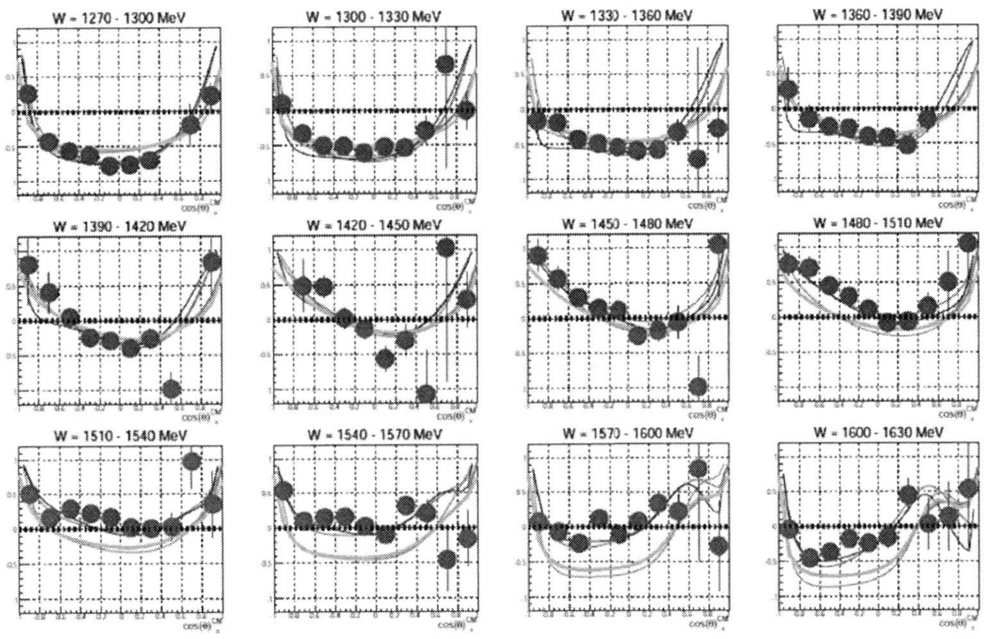

Figure 7. Results for the double polarisation observable E for the photoproduction channel $\gamma p \rightarrow \pi^0 p$, as a function of $\cos\theta_{\pi^0}^{CM}$ for selected 30 MeV energy bins, for quasi-free (blue circles) protons [20]. The results are compared to the BnGa2014-02 [21] (red line), SAID-CM12 [22] (black line), MAID [14] (magenta line), MAID + IA (cyan line), MAID + IA + FSI [15] (green line) partial wave solutions.

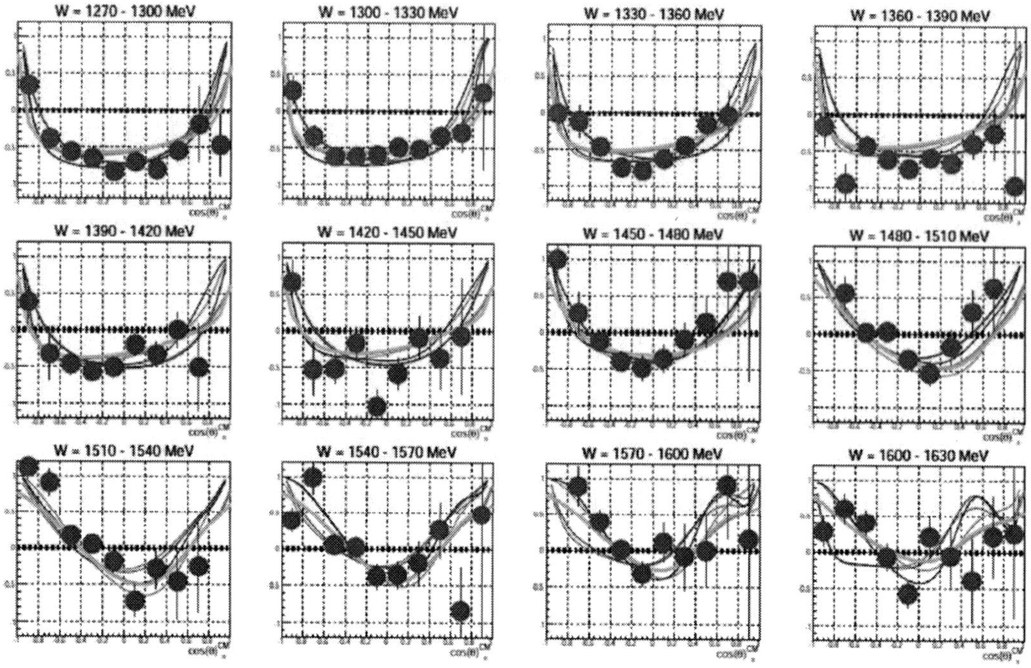

Figure 8. Results for the double polarisation observable E for the photoproduction channel $\gamma n \to \pi^0 n$, as a function of $\cos\theta_{\pi^0}^{CM}$ for selected 30 MeV energy bins, for quasi-free (blue circles) neutrons [20]. The results are compared to the Bonn-Gatchina free neutron model [21], without (dark red line) and with (red line) Fermi motion contribution (IA), SAID-CM12 [22] (black line), MAID [14] (magenta line), MAID + IA (cyan line), MAID + IA + FSI [15] (green line) partial wave solutions.

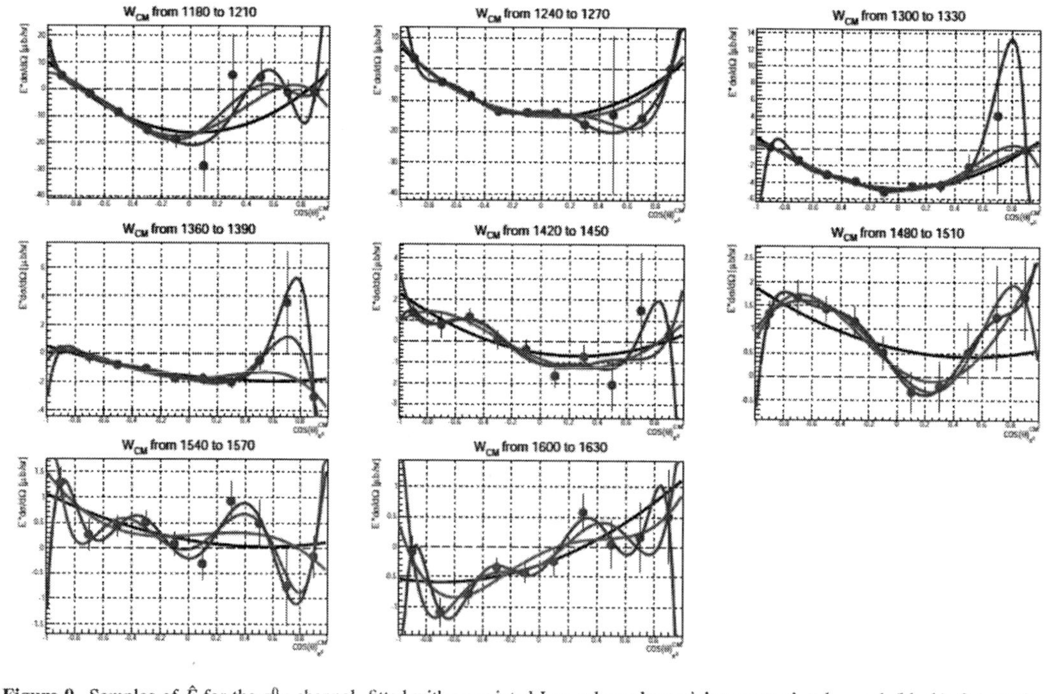

Figure 9. Samples of $\hat{E}$ for the $\pi^0 p$ channel, fitted with associated Legendre polynomials truncated at $l_{max} = 1$ (black), 2 (green), 3 (red) and 4 (blue) [20].

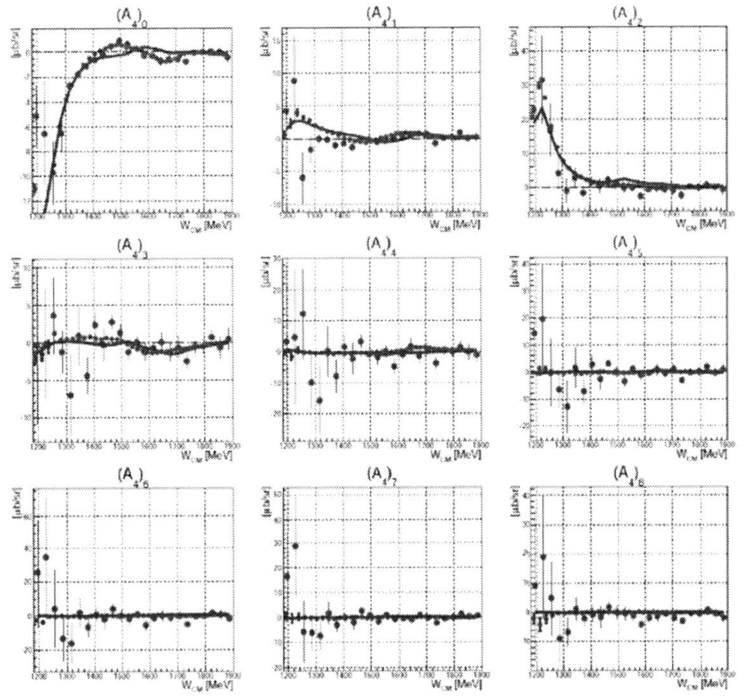

Figure 10. Legendre coefficients for $\hat{E}$ for the $\pi^0 p$ channel with $l_{max} = 4$ [20]. The results for the quasi free proton (blue circles) are compared to the ones for the free proton [18] (red circles).

and by allowing to constrain the multipole solutions of the different analyses.

References

[1] B. Krusche and S. Schadmand, arXiv:nucl-ex/0306023v1

[2] I. S. Barker, A. Donnachie, J. K. Storrow, Nucl. Phys. **B95**, 347 (1975)

[3] W.-T. Chiang and F. Tabakin, Phys. Rev. C **55**, 2054 (1997)

[4] K. Spieker, *Measurement of the double polarisation observables G and E in neutral and positive pion photoproduction off the proton*, Ph.D. thesis, Rheinische Friedrich-Wilhelms-Universität Bonn (2019)

[5] L. Livingston, *Polarization from Coherent Bremsstrahlung Enhancement*, CLAS NOTE 2011-020

[6] V. Tioukine, K. Aulenbacher and E. Riehn, Rev. Sci. Instrum. **82**, 033303 (2011)

[7] A.H. Olsen and L.C. Maximon, Phys. Rev. **114**, 887 (1959)

[8] F.N. Afzal et al., Proceedings of PSTP205, PoS(PSTP2015)036 (2016)

[9] D. Goertz et al., Proceedings of PSTP205, PoS(PSTP2015)009 (2016)

[10] D.G. Crabb and W. Meyer, Annu. Rev. Nucl. Part. Sci. **47**, 67 (1997)

[11] S. Goertx et al., Nucl. Instr. Meth. A **526**, 43 (2004)

[12] F. Colegrove et al., Phys. Rev. Lett. **132**, 2561 (1963)

[13] H. Dutz at al., Phys. Rev. Lett. **94**, 162001 (2005)

[14] D. Drechsel, S. Kamalov and L. Tiator, Eur. Phys. A **34**, 69 (2007)

[15] H. Arenhövel and A. Fix, Phys. Rev. C **72**, 064004 (2005)

[16] S. Costanza, Eur. Phys. J. A **142**, 01008 (2017)

[17] F. Cividini and S. Costanza, PoS(Hadron2017)063 (2018)

[18] F.N. Afzal, *Measurement of the beam and helicity asimmetries in the reactions $\gamma p \rightarrow \pi^0 p$ and $\gamma p \rightarrow \eta p$*, PhD thesis, Rheinischen Friedrich-Wilhelms-Universität Bonn (2019)

[19] M. Dieterle et al., Phys. Lett. B **770**, 523 (2017)

[20] F. Cividini, *Measurement of the helicity dependence for single π^0 photoproduction on the deuteron*, PhD thesis, Johannes Gutenberg-Universität Mainz (2019)

[21] E. Gutz et al., Eur. Phys. J. A **50**, 74 (2014)

[22] R.L. Workman, M.W. Paris, W.J. Briscoe, I.I. Strakovsky, Phys. Rev. C **86**, 015202 (2012)

[23] W.J. Briscoe et al., arXiv:1908.02730v1 [nucl-ex]

[24] Y. Wunderlich, F. Afzal, A. Thiel and R. Beck, Eur. Phys. J. A **53(5)**, 86 (2017)

http://doi.org/10.1051/epjconf/202024101006

Measurement of $\pi^0\pi^{+/-}$ photoproduction off the deuteron and dbutanol targets

Debdeep Ghosal[1,*]
on behalf of A2 Collaboration

[1]University of Basel (Basel, CH)

Abstract. Recent experiments using the Crystal Ball/TAPS setup at the MAMI accelerator in Mainz, Germany continue to study the properties and the excitation spectrum of the nucleon with meson photoproduction. Electromagnetic excitations of the proton and neutron are essential for understanding their isospin decomposition. The electromagnetic coupling of photons to protons is different than that of neutrons in certain states. Cross-section data alone is not sufficient for separating resonances, whereas polarization observables play a crucial role being essential in disentangling the contributing resonant and non-resonant amplitudes. Preliminary results of the polarization observable E of double π production measured with a polarized solid deuterated Butanol target are shown with comparison to predictions of recent analyses.

1 Introduction

Quark models describe the behavior of quarks in nucleons at medium energies. Meson photoproduction allows to investigate the excited states (resonances) of nucleons. Unfortunately, many states are overlapping and cannot be easily differentiated from each other and many have been predicted, but not yet detected [1]. The states that have not yet been observed could be missing because they have not been seen experimentally until now or do not exist at all. Most earlier experiments were performed with pion beams and some resonances might couple less strongly to pions and couple more strongly to rare channels and have circumvented detection. Most results still arise from experiments on the proton, which does not allow for as much information regarding the isospin structure of the electromagnetic transitions. Therefore, advances in experiments on the neutron can provide an integral piece to the understanding of the nucleon spectrum [2]. Unfortunately, it is not possible to perform measurements on the free neutron, but a deuterated butanol target has allowed for the possibility to study spin effects with quasi-free neutrons.

We have preliminary results for the contribution of different channels with proton targets. Final results need of course a more profound analysis including interference effects etc. From the outcomes of total absorption experiment [3], there was discussion whether there is a significant contribution from $D_{13} \rightarrow N\rho$, because due to the predicted in-medium modifications of the ρ this could modify the D_{13} shape. One of the motivations for the present experiments with proton and deuteron targets is the extraction of a much more precise $D_{13} \rightarrow N\rho$ branching ratio for those two targets.

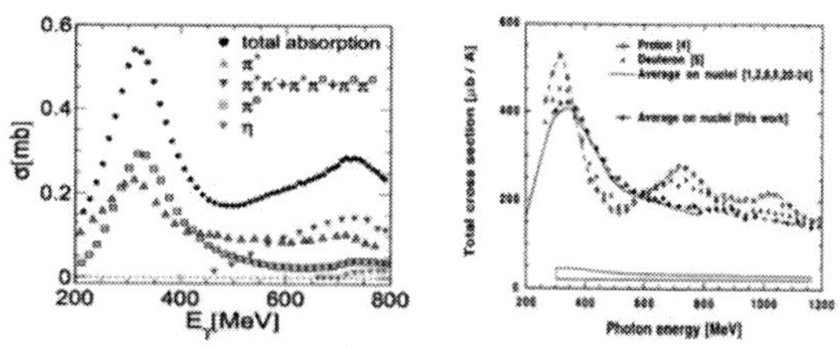

Figure 1: In the above left figure, much of the bump in the second resonance region is due to double pion production. The right figure above is also related to an old problem seen in total photoabsorption where the second resonance bump is completely suppressed for heavier nuclei [3].

For years, cross section data has been used to learn about the nucleon spectrum. However, cross section data alone is not enough to distinguish the broad overlapping resonances. Polarization observables can provide understanding of these overlapping resonances by discovering more information about the complex helicity amplitudes, which describe the interaction between photon beams and nucleons. These amplitudes are fully determined when a complete set of measurements is performed and give rise to the cross section, complemented by polarization observables including beam, target, and recoil asymmetries. Here the observable E (longitudinally polarized target and circularly polarized photon beam) will be discussed.

The relationship between the cross section and polarization observable E can be written in terms of the helicity of beam and target as follows:

*e-mail: debdeep.ghosal@unibas.ch

$$E_{version1} = \frac{\sigma_{1/2} - \sigma_{3/2}}{\sigma_{1/2} + \sigma_{3/2}} = \frac{\sigma_{diff}}{\sigma_{sum}} \quad (1)$$

$$E_{version2} = \frac{\sigma_{diff}}{2\sigma_{unpol.}} \quad (2)$$

where $\sigma_{1/2}$ is the cross section of events when beam and target polarizations are anti-parallel; $\sigma_{3/2}$ is the cross section of events with when they are parallel.

2 Experimental Setup

The measurements were performed at the MAMI-C accelerator in Mainz, Germany [4]. A pictorial diagram of the main experimental setup has been shown below. A longitudinally polarized electron beam of energy 1557 MeV and polarization degree of 80% is delivered into the A2 experimental Hall. Circularly polarized photons are produced from a radiator and energy tagged using the Glasgow-Mainz photon tagger [6] with energies between 470 and 1450 MeV. The targets consists of a deuterated butanol material (dButanol) cooled to a low temperature with the deuterons either transversally or longitudinally polarized up to 80%. The target is surrounded by a cylindrical particle identification detector (PID) made up of 24 plastic scintillator strips that each cover 15° in the azimuthal angle. The PID is then surrounded by a multi-wire proportional chamber (MWPC) and the MWPC is surrounded by the spherical Crystal Ball (CB) calorimeter [6]. The CB consists of 672 NaI(Tl) crystals and covers 20° to 160° in the polar angle. In the forward direction, a hexagonal two arm photon spectrometer (TAPS) built from 72 $PbWO_4$ (two innermost rings) and 366 BaF_2 crystals (remaining rings) is present. A veto wall present in front of TAPS is used for particle identification. The combination of the CB and TAPS provides an almost 4 π acceptance in the center of mass frame with a high angular and energy resolution.

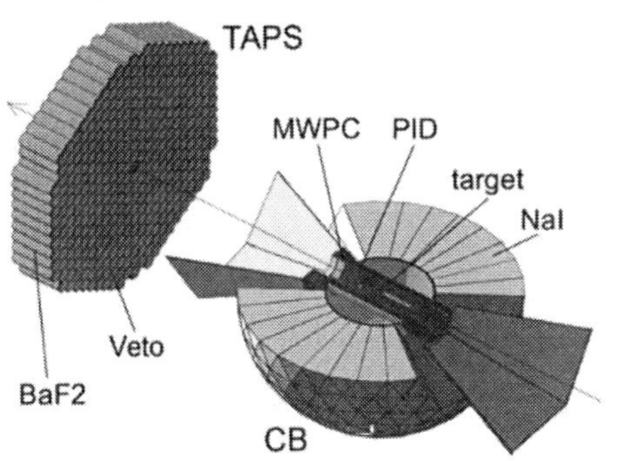

Figure 2: Schematic diagram of the experimental setup [5].

3 Analysis and Results

3.1 Event selection for the two amplitudes

Using information from the detectors, events are collected and then selected based on the number of charged or neutral hits. Neutral mesons are identified via a χ^2 test, which tries to find the best combination of photon clusters for the meson invariant mass. To eliminate accidental coincident tagger photons, coincidence time cuts are applied and random background subtraction was performed. To separate the background from the signal, kinematic cuts were applied for each W and $\cos\theta$ bin.

The analysis of $\gamma p(n) \rightarrow \pi^0\pi^+ n(n)$ requires the decay photons of the π^0 meson to be detected as well as the π^+ and the recoil neutron. Therefore, all events with 3 neutral clusters and 1 charged cluster are selected. A dE-E analysis was used to identify the charged pion [7]. The main source of background in this reaction comes from the following reaction that has the same final state: $\gamma p(n) \rightarrow \eta\pi^+ n$

The analysis of $\gamma n(p) \rightarrow \pi^0\pi^- p(p)$ requires the decay photons of the π^0 meson to be detected as well as the π^- and the recoil proton. Therefore, all events with 2 neutral clusters and 2 charged clusters are selected [7].

3.2 Polarization observable

A real photon can be either circularly or linearly polarised. Circularly polarised photons consists of two plane waves that have the same amplitude, but have a 90° difference in phase. For linearly polarised photons, this electric field vector is composed of only one plane wave. The target (such as proton or neutron) can be polarised in three different directions (x, y, z) as well as the recoil (proton or neutron) in x', y', z'. Thus, in total, $2 \times 3 \times 3 = 18$ double polarisation observables are theoretically possible. However, this number is reduced to only 12 observables since some combinations do not yield additional information. They can be divided into three groups as per the table below among which this paper is interested in E only.

Table 1: Double Polarization observables can be divided into three groups [8].

Beam-Target	Beam-Recoil	Target-Recoil
G, H, E, F	$O_x, O_z, C_x C_z$	T_x, T_z, L_x, L_z

Different versions were used to extract the beam asymmetry E. Version 1 refers to where a normalization with twice of the unpolarized cross section σ_0 is used, which was measured with a liquid deuterium target, and does not need to utilize carbon background subtraction since the background is canceled in the difference of the two helicity states ($\sigma_\Delta = \sigma_{1/2} - \sigma_{3/2}$). Version 2 refers to the normalization of the numerator using the sum of the two spin configurations ($\sigma_\Sigma = \sigma_{1/2} + \sigma_{3/2}$) [9]. Here, the background from unpolarized carbon and oxygen nuclei inside the target have to be subtracted, allowing for events only on polarized protons and neutrons.

Preliminary results regarding comparison plots of total cross sections in terms of E_γ(photon energy) and W(center of mass energy) for dButanol targets are shown in Fig 4 and Fig. 5. Here only one version(in terms of E-observable) to calculate the total cross sections have been presented both in terms of direct and carbon subtracted

Table 2: Overview of different versions used to extract E asymmetry

version	E
1	$\frac{\sigma_{1/2}-\sigma_{3/2}}{2\sigma_0}$
2	$\frac{\sigma_{1/2}-\sigma_{3/2}}{\sigma_{1/2}+\sigma_{3/2}}$

Table 3: Overview of different versions used to extract two helicity state cross section components; where, $\sigma_\Sigma = \sigma_{1/2} + \sigma_{3/2}$ and $\sigma_\Delta = \sigma_{1/2} - \sigma_{3/2}$

version	$\sigma_{1/2}$	$\sigma_{3/2}$
1	$\sigma_0(1+E)$	$\sigma_0(1-E)$
2	$\frac{\sigma_\Sigma+\sigma_\Delta}{2}$	$\frac{\sigma_\Sigma-\sigma_\Delta}{2}$

method. Before that, efficiency comparison of the two interested channels has been presented below.

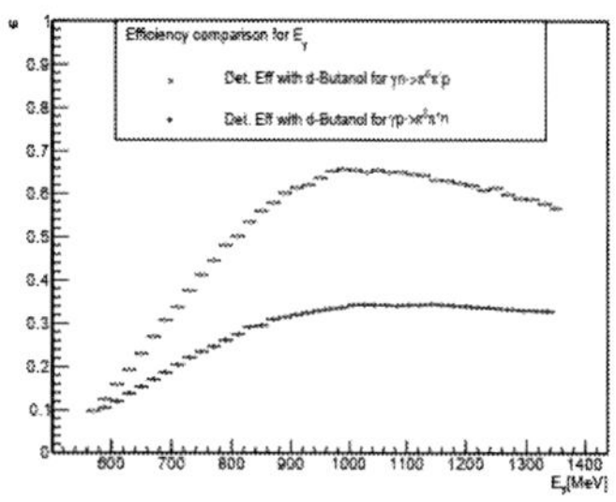

Figure 3: Detection efficiency comparison for the channels $\gamma p \to \pi^0\pi^+ n$ (red dots) and $\gamma n \to \pi^0\pi^- p$ (green dots) are shown above. It also indicates the higher efficiency for proton detection than that for the channel with neutron in final state.

3.3 E observable: Carbon subtraction method (Version-1)

For the determination of the carbon background dedicated beam-times were carried out, with the same experimental conditions as for the dButanol, but with a carbon target of identical density. For a proper background subtraction, the relative contributions of carbon and oxygen nuclei to the dButanol, also called dilution factor, has to be determined. This was done by comparing missing mass distributions of dButanol data, carbon data and deuterium data. If the individual data sets are properly normalized, the spectra of carbon and deuterium should add up to the dButanol. Missing mass distributions are especially suited for this comparison, as contributions from different nuclei are well separated, due to different Fermi momenta [7].

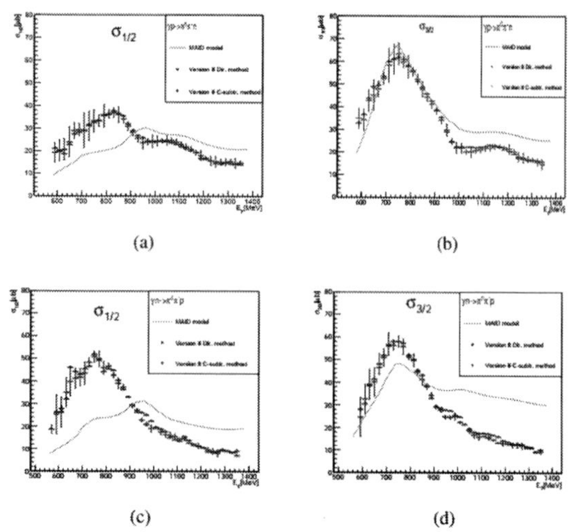

Figure 4: Total cross section comparison in terms of photon energy for $\gamma p \to \pi^0\pi^+ n$ (upper row) and $\gamma n \to \pi^0\pi^- p$ (bottom row). Different colored symbols in each case represent the cross sections determined with the direct and carbon subtracted method with both of the aforementioned versions, and the green line shape corresponds to that predicted by MAID model [10].

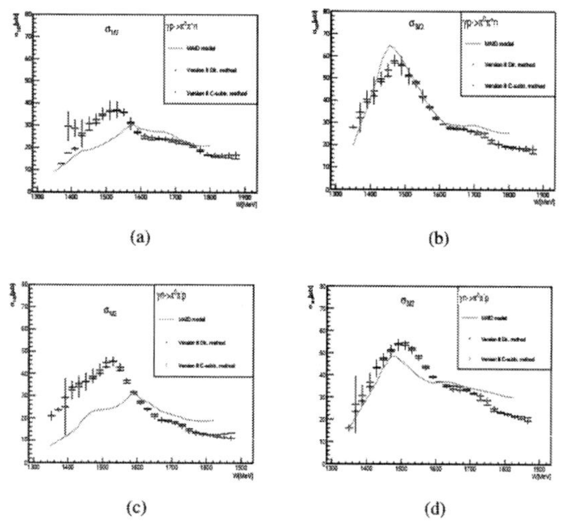

Figure 5: Total cross section comparison in terms of center of mass energy for $\gamma p \to \pi^0\pi^+ n$ (upper row) and $\gamma n \to \pi^0\pi^- p$ (bottom row). Different colored symbols in each case represent the cross sections determined with the direct and carbon subtracted method with both of the aforementioned versions, and the green line shape corresponds to that predicted by MAID model [10].

Preliminary results of E observable for $\gamma p \to \pi^0\pi^+ n$ and $\gamma n \to \pi^0\pi^- p$ channels both in terms of photon energy and center of mass energy are shown in Fig. 6.

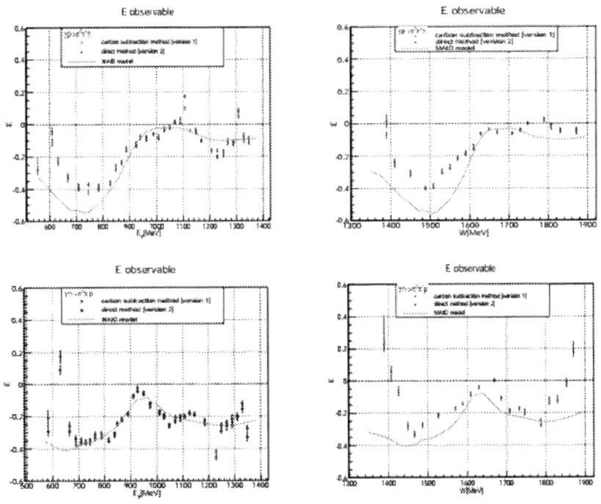

Figure 6: E-observable plots in terms of E_γ (photon energy) and W (center of mass energy) for $\gamma p \to \pi^0\pi^+ n$ (upper row) and for $\gamma n \to \pi^0\pi^- p$ (bottom row); Red dots are for the method with carbon subtraction and the Blue dots correspond to the direct method; and the green lineshape corresponds to that of MAID model prediction [11].

4 Conclusion and Outlook

For two of these interested channels, there is significant non-vanishing E-observable asymmetry for both in terms of photon energy and center of mass energy. Extracted results from the two different methods (version-1 and version-2) agree mostly with each other. The comparative study with the MAID model prediction is also quite satisfactory for the E asymmetry plots and fair for the helicity state cross section plots. Further investigation on background subtraction is under process for the final results. Data from further beamtimes with dButanol target is also being analyzed.

Acknowledgement

All of the results presented here have been obtained inside A2 collaboration at the Mainz MAMI accelerator. The support from Swiss National Fund (SNF) is kindly acknowledged.

References

[1] R.G. Edwards *et al.*, Phys. Rev. **D84**, 074508 (2011)
[2] https://jazz.physik.unibas.ch/site/talks/krusche_dnp08.pdf
[3] F. Zehr and B. Krusche *et al.*, Eur. Phys. J. **A48(7)**, 98 (2012)
[4] K. H. Kaiser *et al.*, NIM **A593**, 159 (2008)
[5] https://indico.cern.ch/event/739938/contributions/3411611/attachments/1859839/3056160/NSTAR_dghosal.pdf
[6] J. C. George *et al.*, Eur. Phys. J. **A37**, 129 (2008)
[7] https://edoc.unibas.ch/55107/1/thesis_kaeser_2017.pdf
[8] https://edoc.unibas.ch/39089/1/Lilian_Witthauer.pdf
[9] https://www.epj-conferences.org/articles/epjconf/pdf/2016/25/epjconf_meson2016_03009.pdf
[10] https://maid.kph.uni-mainz.de/twopion/
[11] https://maid.kph.uni-mainz.de/maid2007/polariz.html

Non-strange dibaryons studied in coherent double neutral-meson photoproduction on the deuteron

Takatsugu Ishikawa[1,*], *Hisako* Fujimura[1], *Hiroshi* Fukasawa[1], *Ryo* Hashimoto[1], *Qinghua* He[1], *Yuki* Honda[1], *Takahiro* Iwata[2], *Shun* Kaida[1], *Hiroki* Kanda[3], *Jirohta* Kasagi[1], *Atsushi* Kawano[4], *Shuzo* Kuwasaki[1], *Kazushige* Maeda[3], *Shin'ichi* Masumoto[5], *Manabu* Miyabe[1], *Fusashi* Miyahara[1], *Keiichi* Mochizuki[1], *Norihito* Muramatsu[1], *Akihiko* Nakamura[1], *Ken'ichi* Nawa[1], *Shoei* Ogushi[1], *Yasuyuki* Okada[1], *Ken'yu* Okamura[1], *Yoshihito* Onodera[1], *Kyoichiro* Ozawa[6], *Yasunobu* Sakamoto[4], *Hajime* Shimizu[1], *Hiroyuki* Sugai[1], *Koutaku* Suzuki[1], *Yasuhisa* Tajima[2], *Shin'ichiro* Takahashi[1], *Yusuke* Taniguchi[1], *Yusuke* Tsuchikawa[1], *Hirohito* Yamazaki[1], *Ryuji* Yamazaki[1], and *Hiroshi* Yoshida[2]

[1]Research Center for Electron Photon Science (ELPH), Tohoku University, Sendai 982-0826, Japan
[2]Department of Physics, Yamagata University, Yamagata 990-8560, Japan
[3]Department of Physics, Tohoku University, Sendai 980-8578, Japan
[4]Department of Information Science, Tohoku Gakuin University, Sendai 981-3193, Japan
[5]Department of Physics, University of Tokyo, Tokyo 113-0033, Japan
[6]Institute of Particle and Nuclear Studies, High Energy Accelerator Research Organization (KEK), Tsukuba 305-0801, Japan

Abstract. The $B = 2$ bound/resonance state (dibaryon) is an interesting object, which can be a molecule consisting of two baryons or a spatially compact hexaquark hadron object. The $\gamma d \to \pi^0\pi^0 d$ reaction has been experimentally investigated at incident energies ranging from 0.58 to 1.15 GeV to study non-strange dibaryons. The angular distributions of deuteron emission in the γd center-of-mass energy cannot be reproduced by quasi-free production of two neutral pions followed by deuteron coalescence. Additionally a 2.14-GeV peak is observed in the $\pi^0 d$ invariant mass distribution. These suggest a sequential process such as $\gamma d \to R_{IS} \to \pi^0 R_{IV} \to \pi^0\pi^0 d$ with an isoscalar dibaryon R_{IS} and an isovector dibaryon R_{IV}. Since the mass of the observed isoscalar dibaryons are close to the sum of the nucleon (N) and nucleon resonance (N^*) masses, an S-wave NN^* molecule may play a role as a doorway to a dibaryon.

1 Introduction

The structure of hadrons is one of the most important subjects to be studied in the non-perturbative domain of quantum chromodynamics. A $B = 2$ system (dibaryon) is an interesting object to study its basic configuration from a molecule-like state consisting of two baryons such as the deuteron to a spatially-compact hexaquark hadron state. Understanding dibaryons would not only give a clue to the solution of the current problem in hadron physics but also provide an insight into the nuclear equation of state and the interior of a neutron star [1].

The search for dibaryons has a long history [2]. Recent observations of the $d^*(2380)$ dibaryon [3, 4] with $M = 2.37$ GeV, $\Gamma = 0.07$ GeV, and $I(J^\pi) = 0(3^+)$ in the $pn \to \pi^0\pi^0 d$ reaction have opened the door to study non-strange dibaryons. The $d^*(2380)$ may be attributed to a spatially compact hexaquark hadron state and/or an isoscalar $\Delta\Delta$ quasi-bound state $\mathcal{D}_{03}$, which was predicted by Dyson and Xuong [5] as a member of the sextet non-strange dibaryons $\mathcal{D}_{IJ}$ with isospin I and spin J. Many experimental investigations of the sextet members have been made, and candidates for almost all the members are found: the deuteron $\mathcal{D}_{01}$, the $^1S_0$-NN virtual state $\mathcal{D}_{10}$, an isovector ΔN state $\mathcal{D}_{12}$ in the $\pi^+ d \to pp$ reaction [6–9], an isotensor ΔN state $\mathcal{D}_{21}$ in the quasi-free $pp \to \pi^+\pi^- pp$ reaction [10], and an $I = 3$ $\Delta\Delta$ dibaryon $\mathcal{D}_{30}$ in the $pp \to \pi^-\pi^-\pi^+\pi^+ pp$ reaction [11]. However, the dibaryonic interpretations of the experimental data for some members are still questionable.

From now on, we focus on $\mathcal{D}_{12}$ which is a key state of the sextet dibaryons. Experimentally, the evidence for $\mathcal{D}_{12}$ is given as the $^3P_2$ multipole strength at $M = 2.18$ GeV in $\pi^\pm d$ elastic scattering by a partial-wave analysis [7]. The corresponding $^1D_2$-pp amplitude also shows the same structure in the $\pi^+ d \to pp$ reaction [8]. The SAID group provides a pole for $\mathcal{D}_{12}$ [9] from a combined analysis of πd elastic scattering, the $\pi^+ d \to pp$ reaction, and pp elastic scattering. Hoshizaki shows the $\mathcal{D}_{12}$ state is required to explain the pp and πd phase parameters [12], and excludes the interpretation of the state as a cusp or a virtual state [13]. Platonova and Kukulin also claim that conventional meson-exchange models do not explain the $pp \to \pi^+ d$ reaction [14]. Recently, preliminary results for $\mathcal{D}_{12}$ candidates observed in the $\gamma d \to \pi^+\pi^- d$ reaction are reported [15, 16], showing a peak with $M = 2.1$–2.2 GeV and $\Gamma \simeq 0.1$ GeV in both the $\pi^\pm d$ invariant-mass distribu-

*e-mail: ishikawa@lns.tohoku.ac.jp

tions. The peak position is close to the sum of the N and Δ masses, and the width is much narrower than the Δ width. Although these considerations and observations seem to show the existence of $\mathcal{D}_{12}$, a possibility of quasi-free (QF) Δ production cannot be excluded in the interpretation of the experimental data.

There is no doubt that the $\pi^{\pm}d$ system has a resonance-like structure around 2.15 GeV. We study the $\gamma d \to \pi^0\pi^0 d$ reaction, aiming to find and observe $\mathcal{D}_{12}$ in the $\pi^0 d$ subsystem. Possible production mechanisms for the $\gamma d \to \pi^0\pi^0 d$ reaction are classified as follows:

1. dibaryon production: sequential $\pi^0\pi^0$ emission from the deuteron with intermediate isoscalar dibaryon $R_{\rm IS}$ and isovector dibaryon $R_{\rm IV}$ ($\gamma d \to R_{\rm IS} \to \pi^0 R_{\rm IV} \to \pi^0\pi^0 d$).

2. QF-$\pi\pi$ production: $\pi^0\pi^0$ is photoproduced on the QF participant nucleon N_p, followed by coalescence with the spectator nucleon N_s into a deuteron.

3. QF-π production: π^0 is photoproduced on N_p, followed by coalescence with N_s into $R_{\rm IV}$, finally $R_{\rm IV}$ decays into $\pi^0 d$.

4. direct-$\pi\pi$ production: $\pi^0\pi^0$ is directly photoproduced from the deuteron.

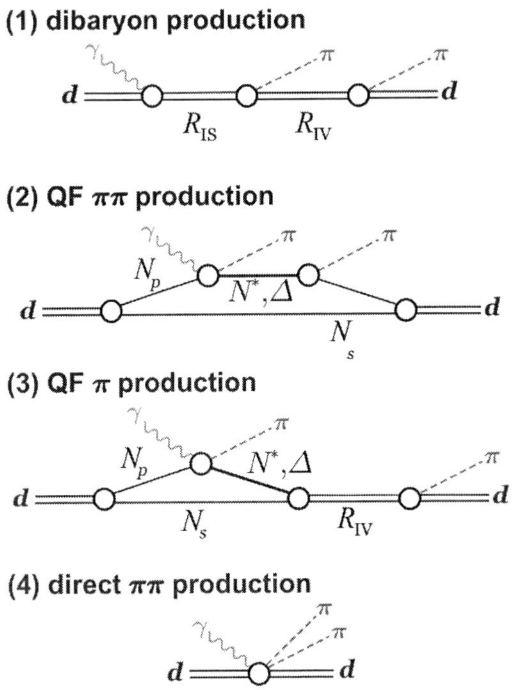

Figure 1. Possible mechanisms for the $\gamma d \to \pi^0\pi^0 d$ reaction. The $R_{\rm IS}$ and $R_{\rm IV}$ denote isoscalar and isovector dibaryons, respectively. The N_p and N_s stand for the participant and spectator nucleons in the quasi-free process, respectively.

In the QF-$\pi\pi$ production mechanism, coalescence into a deuteron requires that the two π^0s should be emitted so as to compensate for the momentum given to N_p. In this case, the angular distribution of deuteron emission in the γd center-of-mass (CM) frame shows strongly backward peaking. As for the QF-π production mechanism, the condition for coalescence into $R_{\rm IV}$ makes the distribution sideway peaking at the incident photon energy around 1 GeV. A completely-different rather-flat distribution is obtained in the dibaryon and direct-$\pi\pi$ production mechanisms. It should be noted that the angular distributions are mainly determined by the deuteron or $R_{\rm IV}$ masses and that the other detailed kinematic conditions do not affect the behaviors of the angular distributions. The most serious background of QF N^* or Δ excitation is kinematically separable from the dibaryon production mechanism.

2 Experiment

A series of meson photoproduction experiments [21] were carried out using the energy-tagged photon beams [22] at the Research Center for Electron Photon Science (ELPH), Tohoku University, Japan. The photon beam was produced by a bremsstrahlung process with a carbon fiber (radiator) with a diameter of 11 μm from the 0.93 and 1.20 GeV circulating electrons in a synchrotron called the STretcher Booster (STB) ring [23]. Each produced photon was energy-tagged by measuring the trajectory of the corresponding post-bremsstrahlung electron with a photon-tagging detector STB-Tagger II in the bending magnet just after the radiator location. The tagging energy of the photon beam ranged from 0.57 to 0.88 and from 0.74 to 1.15 GeV for 0.93- and 1.20-GeV circulating electrons, respectively. The target used in the experiments was liquid deuterium with a thickness of 45.9 mm. The incident photon energies for the two circulating electron energies give γd center of mass energies $W_{\gamma d}$s from 2.38 to 2.61 and from 2.51 to 2.80 GeV. The lowest photon energy for the 0.93-GeV circulating electrons corresponds to the centroid of the $d^*(2380)$ resonance.

All the final-state particles in the $\gamma d \to \pi^0\pi^0 d$ reaction were measured using an electromagnetic calorimeter (EMC) complex, FOREST [24]. FOREST consists of three different EMCs as shown in Fig. 2: the forward EMC consists of 192 pure CsI crystals, the central consists of 252 lead scintillating fiber modules, and the backward consists of 62 lead glass Cherenkov counters, respectively. A plastic scintillator hodoscope (PSH) is placed in front of each calorimeter to identify charged particles. The solid angle of FOREST is approximately 88% in total. The typical photon-tagging rates were 2.8 and 20 MHz, and the photon transmittance ratios (so-called tagging efficiency) were approximately 42% and 53% [22] for the 0.93 and 1.20 circulating electrons, respectively. The trigger condition of the data acquisition (DAQ) was made for detecting more than one final-state particles in coincidence with a photon-tagging signal. The details of the trigger condition are described elsewhere [24]. The average trigger rates were 1.1 and 1.7 kHz, and the average DAQ efficiency

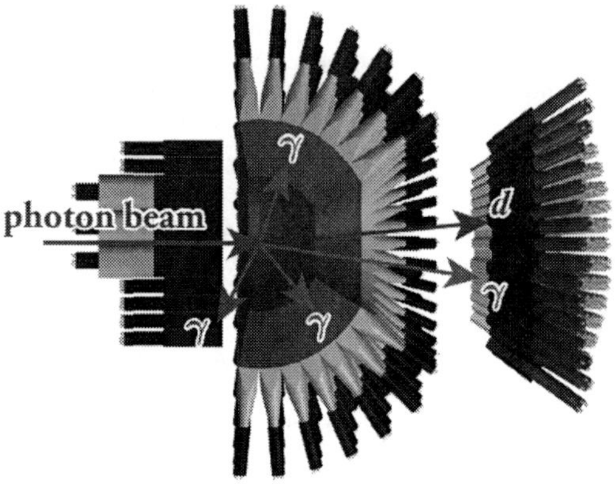

Figure 2. Cross sectional view of FOREST. It consists of three EMCs: the forward (right in the figure), central, and backward (left in the figure) EMCs are comprised of 192 pure CsI crystals, 252 lead scintillating-fiber modules, and 62 lead-glass Cherenkov counters respectively. The energy resolution of each calorimeter is approximately 3% (forward), 7% (central), and 5% (backward) for 1-GeV photons, respectively. A PSH (omitted in the figure) is placed in front of each EMC to identify charged particles.

were 85% and 79% for the 0.93 and 1.20 circulating electrons, respectively.

3 Total cross sections

The analysis of the $\gamma d \to \pi^0\pi^0 d$ reaction was made in the same way as in Refs. [25, 26]. We selected the events containing four neutral particles and a charged particle, and applied a kinematic fit with six constraints: four-momentum conservation, and every $\gamma\gamma$ invariant mass being the π^0 mass. Sideband-background subtraction was performed for accidental-coincidence events detected in STB-Tagger II and FOREST. Events in which the χ^2 probability was higher than 0.4 were selected to reduce those from background processes. The most competitive background process was the QF $\gamma p' \to \pi^0\pi^0 p$ reaction having 100 times higher cross section at maximum [27]. To remove the events from this background process, we made the lower limit of the χ^2 probability relatively higher as 0.4.

The total cross section of the $\gamma d \to \pi^0\pi^0 d$ reaction can be obtained from the equation

$$\sigma = \frac{N_{\pi^0\pi^0 d}}{N'_\gamma N_\tau \eta_{\rm acc} \left\{{\rm BR}(\pi^0 \to \gamma\gamma)\right\}^2}, \qquad (1)$$

which uses the number of events for the $\gamma d \to \pi^0\pi^0 d$ reaction $N_{\pi^0\pi^0 d}$, the effective number of incident photons N'_γ, the number of target deuterons per unit area $N_\tau = 0.237$ b^{-1}, the acceptance of the final state $\pi^0\pi^0 d \to \gamma\gamma\gamma\gamma d$ detection $\eta_{\rm acc}$, and the branching ratio of the neutral pion to the two-photon decay BR($\pi^0 \to \gamma\gamma$). The number of incident photons N_γ is determined by multiplying the number of photon-tagging signals after the counting-loss correction by the corresponding photon transmittance. The N'_γ is a product of multiplying N_γ and the DAQ efficiency. The acceptance of $\gamma\gamma\gamma\gamma d$ detection is estimated by a Monte-Carlo simulation based on Geant4 [28]. Here, the event generation is modified from pure phase-space generation to reproduce the following three measured distributions: the $\pi\pi$ invariant mass $M_{\pi\pi}$, the πd invariant mass $M_{\pi d}$, and the deuteron emission angle $\cos\theta_d$ in the γd-CM frame. Fig. 3 shows the total cross section σ as a function of the γd CM energy $W_{\gamma d}$. The systematic uncertainty of σ includes the uncertainty of event selection in the kinematic fit, that of acceptance owing to the uncertainties of the $M_{\pi\pi}$, $M_{\pi d}$, and $\cos\theta_d$ distributions in event generation of the simulation, that of detection efficiency of deuterons, and that of normalization resulting from the numbers of target deuterons and incident photons.

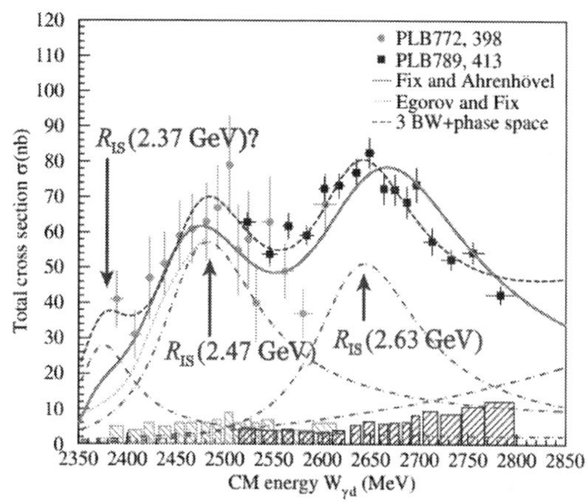

Figure 3. Total cross section σ as a function of $W_{\gamma d}$. The circles (cyan) and squares (blue) show σs obtained in Ref. [25] and in Ref. [26], respectively. The horizontal error of each data point corresponds to the coverage of the incident photon energy, and the vertical error shows the statistical error of σ. The solid and dotted curves (green) show theoretical calculations given in Ref. [19] and in Ref. [20] based on the QF-$\pi\pi$ production mechanism, respectively. The dashed curve (red) shows the fitted function (5) expressed by a sum of three Breit-Wigner peaks and phase-space contributions. Each contribution is shown in a dashed dotted curve (red). The lower hatched histograms show the systematic errors of σ.

The excitation function is not monotonically increasing but shows resonance-like behavior peaked at around 2.47 and 2.63 GeV. The two-peak structure is similar to the excitation function of the $\gamma N \to \pi^0\pi^0 N$ reaction with two peaks at the γN-CM energy $W_{\gamma N}$ of ~1.5 and ~1.7 GeV [17, 18], corresponding to the second- and third-resonance regions of the nucleon. A naive interpretation of this behavior may be a QF excitation of the nucleon in the deuteron, followed by coalescence into the deuteron. The solid curve in Fig. 3 shows a calculation performed by Fix and Ahrenhövel (FA) based on the QF-$\pi\pi$ production mechanism [19]. The calculation reproduces the data surprisingly well. A calculation performed by Egorov and Fix based [20] on the QF-$\pi\pi$ mechanism also gives a similar excitation function as shown in Fig. 3 (dotted curve). However, as discussed later in section 4.2, the kinematic condition for the obtained data completely differs from the QF-$\pi\pi$ and QF-π production mechanisms.

Although the measured cross sections are well-reproduced in the FA calculation, the discrepancy seems observed in the lowest incident photon energy region (~0.57 GeV). This discrepancy may be explained by excitation of the $d^*(2380)$ dibaryon resonance. The $d^*(2380)$ contribution is estimated by fitting the function

$$\sigma(W_{\gamma d}) = \sigma_{d^*}\mathrm{BW}(W_{\gamma d}) + \sigma_{\mathrm{FA}}(W_{\gamma d}) \tag{2}$$

to the data obtained for the 0.92 GeV circulating electrons, where $\mathrm{BW}(W_{\gamma d})$ denotes the relativistic Breit-Wigner function [29] corresponding to the expected $d^*(2380)$ contribution with a centroid of $M = 2.37$ GeV and a width of $\Gamma = 68$ MeV. The σ_{FA} stands for the total cross section in the FA calculation. The χ^2/dof of the fit is 10.1/15, and the obtained parameter is

$$\sigma_{d^*} = 18.4 \pm 9.2 \text{ nb}. \tag{3}$$

The $d^*(2380)$ contribution is not statistically significant, and the upper limit of the total cross section is found to be 34 nb at $W_{\gamma d} = 2.37$ GeV (90% confidence level).

4 Differential cross sections

The differential cross sections, $d\sigma/dM_{\pi\pi}$, $d\sigma/dM_{\pi d}$, and $d\sigma/d\Omega_d$, are obtained for each group of photon-tagging channels divided into four groups. All the differential cross sections obtained are plotted in Fig. 2 in Ref. [26]. Here, we focus on the $M_{\pi d}$ and $d\sigma/d\Omega_d$ distributions for the highest photon-tagging group. The $M_{\pi d}$ distribution is investigated to determine the mass and width of a peak near 2.15 GeV. The differential cross section as a function of the deuteron emission angle $d\sigma/d\Omega_d$ is shown to study the reaction mechanism for the $\gamma d \to \pi^0\pi^0 d$ reaction.

4.1 $M_{\pi d}$ distribution

The $M_{\pi d}$ distribution shows two peaks for the higher incident photon energies. The centroid of the low-mass peak is ~ 2.15 GeV independently of the incident energy. However, that of the high-mass peak decreases with a decrease in the incident energy, and finally the two peaks are merged into a bump. The high-mass peak reflects the appearance of the 2.15-GeV peak in the $M_{\pi d}$ distribution between the other pion and deuteron (reflection). This interpretation becomes obvious by looking at the correlation between the two $M_{\pi d}$s (see Fig. 3(b) in Ref. [26]). The $M_{\pi d}$ distribution of the FA calculation shown in Fig. 4 (dotted) also shows a similar distribution having two peaks, which are caused by QF $\pi^0\Delta$ production.

We have analyzed the $M_{\pi d}$ distribution in detail. We consider the sequential (dibaryon and QF-π production

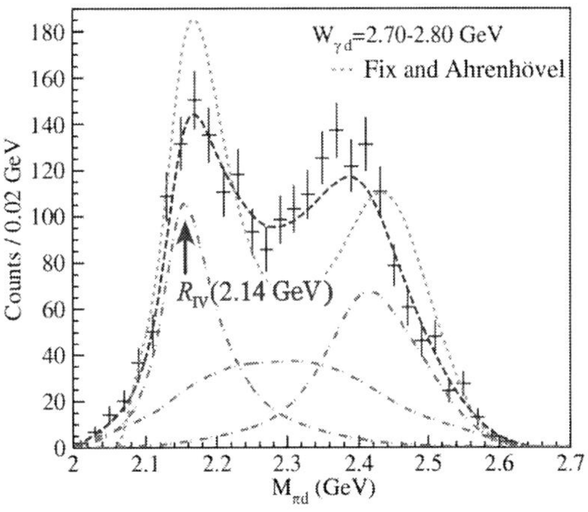

Figure 4. Typical πd invariant mass $M_{\pi d}$ distribution. The dashed curves (red) show the fitted functions, expressed as a sum of a Breit-Wigner (BW) peak, its reflection, and phase-space contributions, to the data. Contributions of a BW peak and its reflection are summed up at an amplitude level. Each contribution is plotted in a dashed dotted curve (magenta). The dotted curve (green) correspond to the FA calculations with arbitrary normalization.

mechanisms) and non-sequential (QF-$\pi\pi$ and direct-$\pi\pi$ production meachisms) processes of two-π^0 emission. Here, the contribution from the latter process is assumed to be proportional to the phase space. The mass and width are determined by fitting a function, expressed as a sum of a Breit-Wigner (BW) peak, its reflection, and phase-space contributions, to the $M_{\pi d}$ data. The function is given by convolution of a Gaussian with an experimental mass resolution of $\sigma_M = 0.011$ GeV, and

$$N(m_1) = \int_{m_2} \left(\alpha \left|A^L_{M,\Gamma}(m_1) + A^L_{M,\Gamma}(m_2)\right|^2 + C\right) \times V_{\mathrm{PS}}(m_1, m_2)\, dm_2, \tag{4}$$

where $V_{\mathrm{PS}}(m_1, m_2)$ expresses the phase-space contribution, $A^L_{M,\Gamma}(m) = \left(m^2 - M^2 + iM\Gamma\right)^{-1}$ represents the BW amplitude with mass M and width Γ. The acceptance is taken into account depending on $M_{\pi\pi}$, $M_{\pi d}$, and $\cos\theta_d$ to estimate $V_{\mathrm{PS}}(m_1, m_2)$. The parameters obtained are $M = 2.14\pm0.01$ GeV and $\Gamma = 0.09\pm0.01$ GeV. The mass is slightly lower than the sum of the N and Δ masses (~ 2.17 GeV), and the width is narrower than that of Δ (~ 0.12 GeV) [30]. It should be noted that the FA calculation does not reproduce the $M_{\pi d}$ distribution, either. The centroid of the second peak is higher at higher incident energy.

4.2 Angular distribution of deuteron emission

A great difference between the FA calculation based on the QF-$\pi\pi$ mechanism and the experimental data appears in the angular distribution of deuteron emission $d\sigma/d\Omega_d$ in Fig. 5. The experimentally obtained $d\sigma/d\Omega_d$ shows a gradually-increasing behavior with decrease of $\cos\theta_d$. While the FA calculation based on the QF-$\pi\pi$ production mechanism provides a strongly backward-peaking behavior. The QF-π production mechanism, giving a sideway peak in $d\sigma/d\Omega_d$ at high incident energies in Fig. 5 (dotted curve in magenta), is also rejected. A sideway peak corresponding to the QF-π production mechanism is not observed in the experimental distribution. Since neither backward nor sideway peaks are not observed in the $d\sigma/d\Omega_d$ distribution, the sequential and non-sequential processes in section 4 one-to-one correspond to the dibaryon and direct-$\pi\pi$ production mechanisms, respectively.

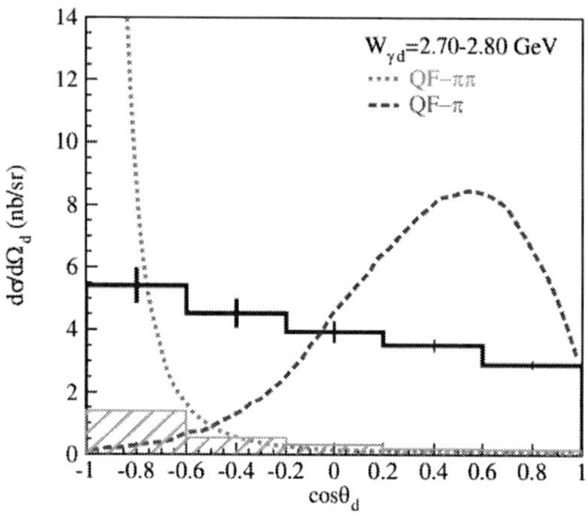

Figure 5. Typical differential cross section $d\sigma/d\Omega_d$. The dotted (green) and dashed (magenta) curves show $d\sigma/d\Omega_d$ as a function of $\cos\theta_d$ for the QF-$\pi\pi$ and QF-π mechanisms, respectively, where their yields are normalized so that the total cross section should be the same. The lower hatched histogram shows the systematic errors of $d\sigma/d\Omega_d$.

We conclude that the peak at 2.14 GeV in the $M_{\pi d}$ spectrum is attributed to a dibaryon state, which can be generated in neither the QF-$\pi\pi$ production mechanism nor the QF-π production mechanism. This conclusion gives the dominant sequential process $\gamma d \to R_{\rm IS} \to \pi^0 R_{\rm IV} \to \pi^0\pi^0 d$, and motivates the following interpretation for the resonance-like structure in Fig. 3: a dibaryon state can be formed from a deuteron and a photon; the generated state may be a loosely-coupled molecular state; it plays a role as a doorway to a more complicated dibaryon state. Of particular importance is the fact that there is no spectator nucleon in the observed reaction. We fit a function expressed by a sum of three BW peaks and phase-space contributions to the data. The function is given by

$$\sigma(W_{\gamma d}) = \sigma_{\rm PS}(W_{\gamma d})\left\{1 + \sum_{i=0}^{2} \alpha_i L_{M_i,\Gamma_i}(W_{\gamma d})\right\}, \quad (5)$$

where $\sigma_{\rm PS}(W_{\gamma d})$ denotes σ for the phase-space contribution, $L_{M,\Gamma}(W_{\gamma d})$ represents a BW function with the centroid of M and width of Γ. Here, three BW peaks are summed up at a cross section level. The $M_0 = 2.37$ GeV and $\Gamma_0 = 0.07$ GeV are fixed to the values for $d^*(2380)$ [4], and the absolute values for $\sigma_{\rm PS}(W_{\gamma d})$ are determined to fit the phase-space component in each $M_{\pi d}$ spectrum (see Fig. 4). The fitted function and each contribution to it are also plotted in Fig. 3 (dashed-and-dotted curve in magenta). The obtained parameters for the two peaks are $(M_1, \Gamma_1) = (2.47, 0.12)$ GeV and $(M_2, \Gamma_2) = (2.63, 0.13)$ GeV. Isoscalar dibaryons appear not only at 2.37 GeV but also at 2.47 and 2.63 GeV. It should be noted that each observed peak may be comprised of several overlapping resonances and that the WASA-at-COSY data covering masses up to 2.56 GeV do not show any signature of the 2.47-GeV dibaryon.

5 Spin-parity

Information on the spin-parity of the dibaryon states has been deduced from angular distributions of π^0s obtained for the events from $M_{\pi d} = 2.05$–2.25 GeV in Fig. 4. We define π_1 and π_2 as follows: π_2 is one of the two pions giving $2.05 \le M_{\pi d} < 2.25$ GeV, and π_1 is the other pion. Here, we assume π_1 is emitted first in the dominant dibaryon production mechanism, leaving the $\pi_2 d$ system with $2.05 \le M_{\pi d} < 2.25$ GeV, and π_2 is emitted subsequently. We combine the data of the two highest-energy groups, which covers $W_{\gamma d}$ from 2.66 to 2.80 GeV, to analyze angular distributions. Thus, we expect the reaction sequence $\gamma d \to R_{\rm IS}(2.63\ {\rm GeV}) \to \pi_1 R_{\rm IV}(2.14\ {\rm GeV}) \to \pi_1\pi_2 d$ together with the direct $\pi_1\pi_2$ production mechanism. Fig. 6(a) shows the deduced π_1 angular distribution after the acceptance correction is applied in the γd-CM frame with the z axis taken along the incident photon direction. The experimental distribution is well-reproduced by a sum of two terms, constant and proportional to $\cos\theta$. This $\cos\theta$ dependence can be understood as a result of interference between π_1-emission amplitudes with different parities. The π_2 angular distribution in the rest frame of the $\pi_2 d$ system (π_1-π_2 angular correlation) after the acceptance correction is applied is shown in Fig. 6(b), where the z axis is defined to be opposite to the direction of π_1 emission. Unlike Fig. 6(a), the distribution shows almost 90° symmetry. This implies that the 2.14-GeV resonance is made of a single J^π state or mixed states with the same parity (no interference between different parities). The FA calculation (provided in the same kinematic condition) completely fails to reproduce both the angular distributions, due to the difference in the underlying reaction mechanism as discussed previously (QF-$\pi\pi$ production mechanism adopted in the FA calculation is not observed in the experiment). A sharp peak at 0° in Fig. 6(a) is the reflection of the backward peak in $d\sigma/d\Omega_d$. In Fig. 6(b), the distribution takes an upward-convex shape being opposite to the experiment.

We calculate the π_1 and π_2 angular distributions for the reaction sequence $\gamma d \to R_1 \to \pi_1 R_2 \to \pi_1\pi_2 d$ using the density matrix (statistical tensor) formalism [31], where spins of R_1 and R_2 are denoted by J_1 and J_2, respectively. The formalism incorporates the exchange symmetry between π_1 and π_2. A possible contribution from

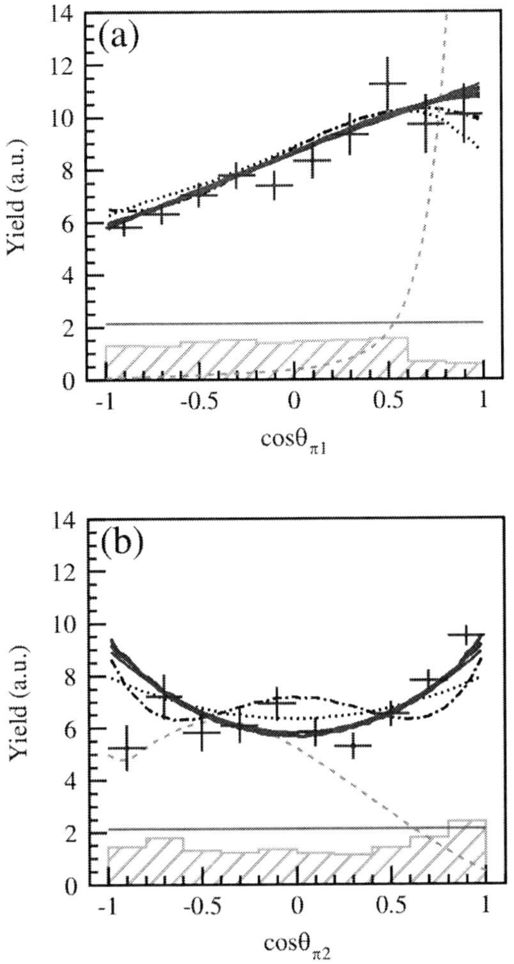

Figure 6. Acceptance-corrected angular distributions for π_1 in the γd-CM frame (z axis: the photon beam direction) (a), and for π_2 in the $\pi_2 d$ rest frame (z axis: the opposite direction to π_1) (b). Events with $M_{\pi_2 d}$ = 2.05–2.25 GeV and $W_{\gamma d}$ = 2.66–2.80 GeV are selected. The lower hatched histograms (cyan) show the corresponding systematic errors. The dashed curves (green) show the corresponding distributions in the FA calculations. The angular distributions are plotted with a shaded band (red) for $J_2^\pi = 1^+$, 2^+, and 3^-, and with dotted and dash-dotted curves (black) for $J_2^\pi = 1^-$ and 2^-, respectively. The solid horizontal lines (magenta) show the phase-space contributions.

non-sequential direct $\pi^0\pi^0$ production mechanism is assumed to be proportional to the phase-space contribution, of which the fraction is determined from the $M_{\pi d}$ spectrum in Fig. 4 for each photon-tagging group. Since the π_2 angular distribution is almost symmetric with respect to $\cos\theta_{\pi_2} = 0$, the interference effect may be small. Thus, the contributions from the sequential dibaryon production and non-sequential direct $\pi^0\pi^0$ production mechanisms are summed up incoherently. A set of the amplitudes of sequential processes $A_{\Lambda\Lambda}$ is determined for all the $\Lambda = (L_0, J_1, L_1, J_2, L_2)$ combinations to reproduce the measured π_1 and π_2 angular distributions simultaneously (20 data points), where L_0, L_1, and $L_2 \leq 2$ denote angular momenta carried by the incident photon, π_1 emission, and π_2 emission, respectively. An amplitude for a mixed state is given by $A_{\Lambda\Lambda'} = (A_{\Lambda\Lambda}A_{\Lambda'\Lambda'})^{1/2}$. The S-wave NN^* molecular states are assumed to play a role as a doorway to R_1 since all the observed isoscalar dibaryons are close to the sum of NN^* masses as discussed in section 6. Here, the considered N^*s are $D_{15}(1675)$, $F_{15}(1680)$, and $P_{13}(1720)$, which give dominant contributions in the $\gamma N \to \pi^0\pi^0 N$ reaction in the relevant energy region [18]. Hence, J_1^πs under consideration are 1^+, $2^\pm$, and $3^\pm$. Additionally, R_2 is assumed to be a single resonance, namely $J_2 = J_2'$ and $L_2 = L_2'$.

In Fig. 6, also shown are the angular distributions calculated for $J_2^\pi = 1^\pm$, $2^\pm$, and 3^+. The $J_2^\pi = 0^\pm$ assignments are already excluded because of an isotropic distribution for π_2 emission. The assignments of $J_2^\pi = 1^+$, 2^+, and 3^- show almost the same quality to reproduce the angular distributions ($29 < \chi^2 < 32$). We reject the $J_2^\pi = 1^-$ and 2^- assignments, giving worse distributions (χ^2 = 47 and 55), respectively, with a confidence level of higher than 99.7% (3σ), and leave the possibility of $J_2^\pi = 1^+$, 2^+ and 3^-. Regarding J_1^π, major components are 1^+ (~70%) and 2^- (~20%) for the case of $J_2^\pi = 1^+$ and 2^+, while they are distributed widely to $J_1^\pi = 1^+$, $2^\pm$, and 3^+ for $J_2^\pi = 3^-$. The $J_2^\pi = 2^+$ assignment not only coincides with the energy dependence of the $^3P_2$-πd amplitude, but also supports the existence of the predicted $\mathcal{D}_{12}$ state at 2.14 GeV. The $J_2^\pi = 3^-$ assignment is consistent with the energy dependence of the $^3D_3$-πd amplitude (about a half strength of $^3P_2$). There is no experimental sign for a 1^+ state, although two isovector 0^- and 2^- states at 2.2 GeV have been reported recently [32]. Most probably, R_{IS}(2.63 GeV) is a mixing state between $J_1^\pi = 1^+$ (~70%) and 2^- (~20%), and $J_2^\pi = 2^+$ for R_{IV}(2.14 GeV).

6 Discussion

A rather-flat angular distribution of deuteron emission completely differs from the QF-$\pi\pi$ and QF-π production mechanisms. In addition, the $M_{\pi d}$ distributions shows a peak at 2.14 GeV. The present work suggests a sequential process $\gamma d \to R_{IS} \to \pi^0 R_{IV} \to \pi^0\pi^0 d$ is dominant with two 2.47- and 2.63-GeV R_{IS} and a 2.14-GeV R_{IV}. Fig. 7 shows the deuteron hadronic excited spectrum. The three R_{IS}s are photo-excited from the deuteron with masses of 2.37, 2.47, and 2.63 GeV. The spin-parity of R_{IS}(2.37 GeV) is 3^+, and that of R_{IS}(2.63 GeV) is found to be most probably a mixing of 1^+ and 2^-. The R_{IS}(2.63 GeV) decays into R_{IV}(2.14 GeV) by emitting π^0 with angular momenta of 0 and 1, and the R_{IV}(2.14 GeV) decays into the deuteron by emitting π^0 with an angular momentum of 1.

The sum of the nucleon N and $N(1440)P_{11}$ nucleon resonance masses (the $NN(1440)P_{11}$ mass) is very close to the R_{IS}(2.37 GeV) mass. The masses of $NN(1535)S_{11}$ and $NN(1520)D_{13}$ masses are close to the

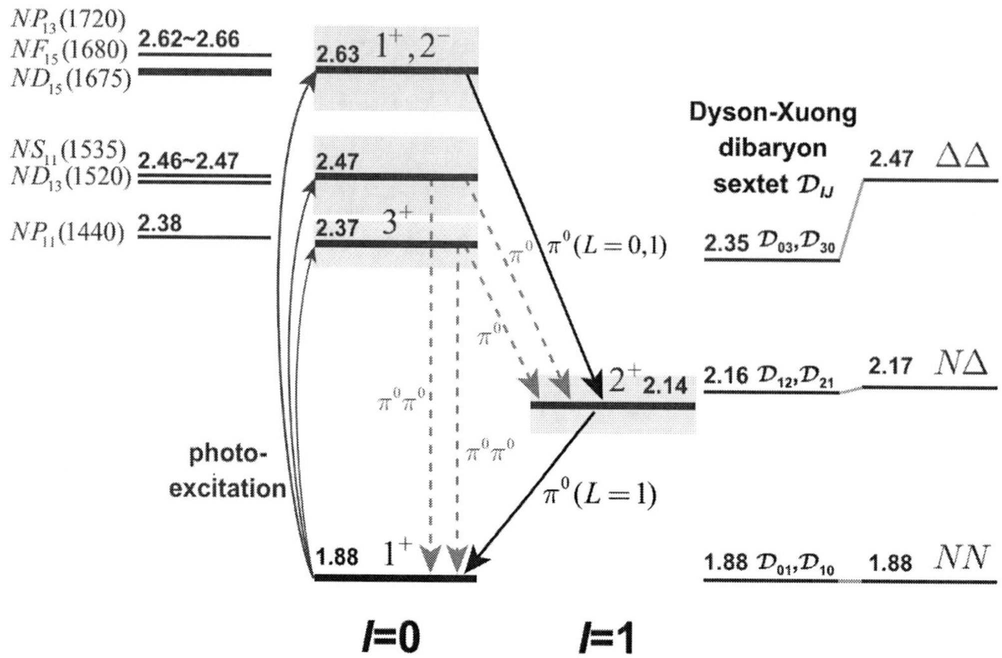

Figure 7. Deuteron hadronic excitation spectrum. The three R_{IS}s are photo-excited from the deuteron with masses of 2.37, 2.47, and 2.63 GeV. The spin-parity of R_{IS}(2.37 GeV) is 3^+, and that of R_{IS}(2.63 GeV) is found to be a mixing of 1^+ and 2^-. The R_{IS}(2.63 GeV) decays into R_{IV}(2.14 GeV) by emitting π^0 with angular momenta of 0 and 1, and the R_{IV}(2.14 GeV) decays into the deuteron by emitting π^0 with an angular momentum of 1. The left-side levels show the sum of the nucleon and nucleon resonance masses, and the right-side levels show the masses of the dibaryon sextet member predicted by Dyson and Xuong [5], and the sum of the two baryon masses.

R_{IS}(2.47 GeV) mass. The $NN(1675)D_{15}$, $NN(1680)F_{15}$, and $NN(1720)P_{13}$ masses are close to the R_{IS}(2.63 GeV) mass. Thus, we consider the possibility that the S-wave NN^* molecular states may play a role as a doorway to the three R_{IS} states.

The R_{IS}(2.37 GeV), or $d^*(2380)$, has been considered as a mixing state between a spatially compact hexaquark hadron state and an isoscalar $\Delta\Delta$ quasi-bound state. The R_{IV} would be also attributed to be the $N\Delta$ quasi-bound state since its mass is lower than the $N\Delta$ mass, and its width is narrower than that of Δ. The dibaryon sextet picture predicted by Dyson and Xuong [5] is now established owing to strong evidence for the existence of the 2.14-GeV isovector dibaryon in the $\pi^0 d$ channel. Additionally, our work shows evidence of the 2.47 and 2.63-GeV isoscalar dibaryons in the $\pi^0\pi^0 d$ channel. These findings would give a base to explore dibaryon states lying at higher masses.

To study dibaryon states, it is advantageous to find two neutral mesons and the deuteron in the final state. The $\gamma d \to \pi^0 \eta d$ reaction also satisfy this condition. The angular distribution of deuteron emission in this reaction also shows rather flat, suggesting dibaryon production mechanism seems dominant. Since the available phase space is very limited in this reaction at incident energies below 1.2 GeV, it is difficult to determine the mass, width, and spin-parity of the intermediate dibaryon states using the current data. The maximum energy for circulating electrons in the synchrotron at ELPH has increased after the 2011 earthquake, allowing us to study the $\gamma d \to \pi^0 \eta d$ reaction in more detail.

7 Summary

The total and differential cross sections have been measured for the $\gamma d \to \pi^0\pi^0 d$ reaction at $E_\gamma = 0.75$–1.15 GeV. The measured angular distribution of deuteron emission is rather flat, suggesting that a sequential process $\gamma d \to R_{IS} \to \pi^0 R_{IV} \to \pi^0\pi^0 d$ is dominant. The total cross section σ as a function of $W_{\gamma d}$ shows isoscalar dibaryons R_{IS} with masses of 2.47 and 2.63 GeV. The σ for R_{IS} with a mass of 2.37 GeV corresponding to the $d^*(2380)$ dibaryon observed by the WAST-at-COSY collaboration is found to be 18.4 ± 9.2 nb at the central mass. The $\pi^0 d$ invariant-mass distributions show an isovector dibaryon R_{IV} with a mass of 2.14 GeV. Most probably, R_{IS}(2.63 GeV) is interpreted as a mixing state between $J_1^\pi = 1^+$ (~70%) and 2^- (~20%), and $J_2^\pi = 2^+$ for R_{IV}(2.14 GeV). The details of the analysis and discussion can be found elsewhere [26].

Acknowledgments

The authors express their gratitude to the ELPH accelerator staff for stable operation of the accelerators in the FOREST experiments. They acknowledge Mr. Kazue Matsuda, Mr. Ken'ichi Nanbu, and Mr. Ikuro Nagasawa for their technical assistance in the FOREST experiments. They also thank Prof. Alexander I. Fix for the theoretical calculations of the cross sections and fruitful discussion. This work was supported in part by JSPS KAKENHI Grants Nos. 17340063, 19002003, 24244022, 26400287, 16H02188, 19H01902, and 19H05141.

References

[1] A. Akmal, V.R. Pandharipande, D.G. Ravenhall, Phys. Rev. C **58**, 1804 (1998).
[2] H. Clement, Prog. Part. Nucl. Phys. **93**, 195 (2017).
[3] M. Bashkanov *et al.* (CELSIUS/WASA collaboration), Phys. Rev. Lett. **102**, 052301 (2009).
[4] P. Adlarson *et al.* (WASA-at-COSY collaboration), Phys. Rev. Lett. **106**, 242302 (2011).
[5] F.J. Dyson, N.-H. Xuong, Phys. Rev. Lett. **13**, 815 (1964).
[6] B.S. Neganov, L.B. Parfenov, J. Exp. Theor. Phys. **7**, 528 (1958).
[7] R.A. Arndt, I.I. Strakovsky, R.L. Workman, Phys. Rev C. **50**, 1796 (1994).
[8] R.A. Arndt, I.I. Strakovsky, R.L. Workman, D.V. Bugg, Phys. Rev C. **48**, 1926 (1993).
[9] C.H. Oh, R.A. Arndt, I.I. Strakovsky, R.L. Workman, Phys. Rev C. **56**, 635 (1997).
[10] P. Adlarson *et al.* (WASA-at-COSY collaboration), Phys. Rev. Lett. **121**, 052001 (2018),
[11] P. Adlarson *et al.* (WASA-at-COSY collaboration), Phys. Lett. B **762**, 455 (2016).
[12] N. Hoshizaki, Phys. Rev C. **45**, R1424 (1992); Prog. Theor. Phys. **89**, 251 (1992); Prog. Theor. Phys. **89**, 563 (1992).
[13] N. Hoshizaki, Prog. Theor. Phys. **89**, 569 (1992).
[14] M.N. Platonova, V.I. Kukulin, Phys. Rev. C **94**, 054039 (2016).
[15] R.A. Schumacher, talk at YITP workshop "Meson in Nucleus 2016" (MIN16) (2016); MIN16 website: ⟨`http://www2.yukawa.kyoto-u.ac.jp/~min2016/`⟩.
[16] H. Kanda, talk at ELPH workshop "Meson Production and Meson-Baryon Interaction" (MPMBI) (2015); MPMBI website: ⟨`http://www.lns.tohoku.ac.jp/workshop/c013/`⟩.
[17] Y. Assafiri, et al. (GRAAL collaboration), Phys. Rev. Lett. **90**, 222001 (2003); J. Ajaka, et al. (GRAAL collaboration), Phys. Lett. B **651**, 108 (2007).
[18] M. Dieterle, et al. (A2 collaboration), Eur. Phys. J. A **51**, 142 (2015).
[19] A. Fix, H. Arenhövel, Euro. Phys. J. A **25**, 115 (2005).
[20] M. Egorov, A. Fix, Nucl. Phys. A 933 (2015) 104.
[21] T. Ishikawa *et al.*, JPS Conf. Proc. **10**, 031001 (2016); PoS (Hadron 2013) **095** (2013); Few Body Sys. **54**, 1047 (2013).
[22] T. Ishikawa *et al.*, Nucl. Instrum. Meth. A **622**, 1 (2010); **811**, 124 (2016); Y. Matsumura *et al.*, Nucl. Instrum. Meth. A **902**, 103 (2018); Y. Obara *et al.*, Nucl. Instrum. Meth. A **922**, 108 (2019).
[23] F. Hinode, et al., in: Proceedings of 2005 Particle Accelerator Conference (2005) 2458.
[24] T. Ishikawa, et al., Nucl. Instrum. Meth. A **832**, 108 (2016).
[25] T. Ishikawa, et al., Phys. Lett. B **772**, 398 (2017).
[26] T. Ishikawa, et al., Phys. Lett. B **789**, 413 (2019).
[27] B. Krusche *et al.*, Eur. Phys. J. A **324**, 309 (1999).
[28] S. Agostinelli *et al.*, Nucl. Instrum. Meth. A **506**, 250 (2003);
J. Allison *et al.*, IEEE Trans. on Nucl. Sci. **53**, 270 (2006);
Geant4 website ⟨http://geant4.cern.ch/⟩.
[29] A.R. Bohm, Y. Sato, Physical Review D 71 (2005) 085018.
[30] C. Patrignani *et al.* (Particle Data Group), Chin. Phys. C **40**, 100001 (2016).
[31] L.C. Biedenharn, M.E. Rose, Rev. Mod. Phys. **25**, 729 (1953).
[32] V. Komarov, D. Tsirkov, et al., Phys. Rev. C **93**, 065206 (2016).

Strangeness Photoproduction at the BGO-OD experiment

Thomas Jude[1,*], *Stefan* Alef[1], *Patrick* Bauer[1], *Reinhard* Beck[2], *Alessandro* Braghieri[13], *Philip* Cole[3], *Rachele* Di Salvo[4], *Daniel* Elsner[1], *Alessia* Fantini[4,5], *Oliver* Freyermuth[1], *Francesco* Ghio[6,7], *Anatoly* Gridnev[8], *Daniel* Hammann[1], *Jürgen* Hannappel[1], *Katrin* Kohl[1], *Nikolay* Kozlenko[8], *Alexander* Lapik[9], *Paolo* Levi Sandri[10], *Valery* Lisin[9], *Giuseppe* Mandaglio[11,12], *Roberto* Messi[4,5], *Dario* Moricciani[4], *Vladimir* Nedorezov[9], *Dmitry* Novinsky[8], *Paolo* Pedroni[13], *Andrei* Polonski[9], *Björn-Eric* Reitz[1], *Mariia* Romaniuk[4], *Georg* Scheluchin[1], *Hartmut* Schmieden[1], *Victorin* Sumachev[8], *Viacheslav* Tarakanov[8], and *Christian* Tillmanns[1]

[1]Rheinische Friedrich-Willhelms-Universität Bonn, Physikalisches Institut, Nußallee 12, 53115 Bonn, Germany
[2]Helmholtz-Institut fuer Strahlen- und Kernphysik, Universitaet Bonn, Nussallee 1-16, D-53115 Bonn Germany
[3]Lamar University, Department of Physics, Beaumont, Texas, 77710, USA
[4]INFN Roma "Tor Vergata", Rome, Italy
[5]Università di Roma "Tor Vergata", Via della Ricerca Scientifica 1, 00133 Rome, Italy
[6]INFN sezione di Roma La Sapienza, P.le Aldo Moro 2, 00185 Rome, Italy
[7]Istituto Superiore di Sanita, Viale Regina Elena 299, 00161 Rome, Italy
[8]Petersburg Nuclear Physics Institute, Gatchina, Leningrad District, 188300, Russia
[9]Russian Academy of Sciences Institute for Nuclear Research, prospekt 60-letiya Oktyabrya 7a, Moscow 117312, Russia
[10]INFN - Laboratori Nazionali di Frascati, Via E. Fermi 40, 00044 Frascati, Italy
[11]INFN sezione Catania, 95129 Catania, Italy
[12]Universita degli Studi di Messina, Via Consolato del Mare 41, 98121 Messina, Italy
[13] INFN sezione di Pavia, Via Agostino Bassi, 6 - 27100 Pavia, Italy

Abstract.
The BGO-OD experiment at the ELSA accelerator facility uses an energy tagged bremsstrahlung photon beam to investigate the excitation structure of the nucleon. The setup consists of a highly segmented BGO calorimeter surrounding the target, with a particle tracking magnetic spectrometer at forward angles.
BGO-OD is ideal for investigating low momentum transfer processes due to the acceptance and high momentum resolution at forward angles. In particular, this enables the investigation of strangeness photoproduction where t-channel exchange mechanisms play an important role. This also allows access to low momentum exchange kinematics where extended, molecular structure may manifest in reaction mechanisms.
First key results at low t indicate a cusp-like structure in $K^+\Sigma^0$ photoproduction at $W = 1900\,\mathrm{MeV}$, line shapes and differential cross sections for $K^+\Lambda(1405)\rightarrow K^+\Sigma^0\pi^0$, and a peak structure in $K^0_S\Sigma^0$ photoproduction. The peak in the $K^0_S\Sigma^0$ channel appears consistent with meson-baryon generated states, where equivalent models have been used to describe the P_C pentaquark candidates in the heavy charmed quark sector.

1 Introduction

Hadron spectroscopy has been used for many years to understand the interactions of the constituents and the relevant degrees of freedom. Despite the wealth of data available using both pion and photon beams and a multitude of final states, there are still many resonances predicted by constituent quark models (CQMs) that have not been observed experimentally (see for example Ref. [1]). This is in spite of major advances in partial wave analysis working in tandem with experimental research (see for example Ref. [2]). Even some lowest lying states are difficult to reconcile within a CQM. The pattern of the mass and parity of the $N(1440)1/2^+$ and $N(1535)1/2^-$ for example are hard to describe assuming valence quarks in a mutually generated potential. In the strange sector, the mass ordering of the $\Lambda(1405)$ compared to the $N^*(1535)$ (despite the $\Lambda(1405)$ being a *uds* singlet state) and the mass difference to the $\Lambda(1520)$ spin orbit partner is hard to understand within a CQM framework.

Since the conception of the quark model, there have been descriptions of hadrons with more than three constituent quarks [3–5], and models explicitly including light mesons interacting as elementary objects have had markedly improved success in describing nucleon excitation spectra [6–8]. Such interactions may give rise to meson rescattering effects near production thresholds, and molecular like systems.

The BGO-OD experiment at the ELSA facility, Bonn, Germany, is a unique setup that is ideal to study molecular like interactions in strangeness photoproduction which may manifest in the reaction mechanisms. This proceedings is as follows: section 2 discusses parallels between

*e-mail: jude@physik.uni-bonn.de

the heavy charm and lighter strange quark sector, where there maybe equivalent evidence of molecular-like systems. The BGO-OD experiment and performance is described in sections 3 and 4, and the first preliminary results in associated strangeness photoproduction are shown in section 5.

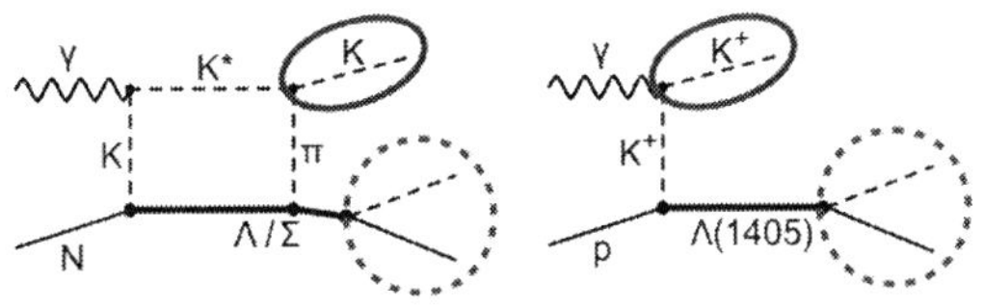

Figure 1. Example of kinematics ideally suited for BGO-OD.

2 Exotic hadron structure - parallels in strange and charmed quark sectors?

In the heavy, charmed quark sector, the pentaquark candidates, $P_C(4450)$ and $P_C(4380)$ were one of the first indications of baryons of more than three constituent quarks [9]. It is still an open question to whether these are five quark bound systems, or a molecular like system to some extent, with models successfully describing them as meson-baryon dynamically generated states (see for example, Ref. [10]). In the meson sector, the $X(3872)$ close to the $D^0\bar{D}^{0*}$ threshold has also been described as a molecular state [11].

In the lighter strange sector, models including meson-baryon interactions have had improved success. The $\Lambda(1405)$, lying close to the $\bar{K}N$ threshold appears to be dynamically generated to some extent [12–14], which is supported by Lattice QCD calculations [15]. A cusp-like structure observed in the $K^0_S\Sigma^+$ differential cross section at the K^* thresholds [16] has been successfully described by the destructive interference between two vector meson-baryon dynamically generated resonances [17–19]. There are remarkable comparisons to the heavier charmed pentaquarks described by an equivalent model [10]. Both share quantum numbers, $J^P = 3/2^-$, equivalent pion exchange transitions and close to two and three body thresholds.

To study systems where molecular like interactions may manifest, it is vital to access kinematic regions where the residual baryon remains at low momentum. Using a real, energy tagged photon beam, this corresponds to very forward meson angles. BGO-OD is ideal for studying such systems in associated strangeness photoproduction. The experiment comprises a forward spectrometer for charged particle identification at forward angles, coupled with a central calorimeter ideal to identify hyperon decays. Fig. 1 depicts two examples where the kinematics suit this setup. In in both scenarios, the K^+ travels forwards, leaving either a residual K^*Y system (left) or $\Lambda(1405)$ (right) at low momentum to decay almost isotropically. With the ability to reconstruct complicated mixed charge final states, BGO-OD is complementary to other facilities such as CBELSA-TAPS, Crystal Ball and CLAS.

3 The BGO-OD experiment at ELSA

The ELectron Stretcher Accelerator (ELSA) [20], shown in Fig. 2 is situated at the University of Bonn's Physikalisches Institut. ELSA is comprised of three parts. A LINAC delivers a 20 MeV electron beam to a booster synchrotron which accelerates this pulsed beam up to 1.6 MeV. This enters the stretcher ring, whereupon it is accelerated up to 3.2 GeV and accumulated to provide a 100% duty factor continuous wave beam. The beam is subsequently delivered to one of three experimental areas: a beam line for detector tests (bottom right), the CBELSA/TAPS experiment [21–23] or the BGO-OD experiment (both top left).

The BGO-OD experiment, shown in Fig. 3, is comprised of two main components. A central region, consisting of a calorimeter and charged particle tracking detectors, and a forward spectrometer for charged particle identification and momentum reconstruction.

The electron beam from ELSA impinges upon a thin radiator (bottom right of Fig. 3 within the goniometer tank) producing bremsstrahlung photons. A metal or crystal radiator can be used to produce either unpolarised (incoherent) or polarised (coherent) bremsstrahlung photons respectively. Photon energies are determined by momentum analysing the post-bremsstrahlung electrons in the Photon Tagger, over a range of 10% to 90% of incoming electron beam energy.

The photon beam is subsequently collimated and incident upon the target at the centre of the BGO Rugby Ball. Liquid hydrogen, liquid deuterium and carbon targets have previously been used. Fig. 4 shows a sliced view of the central calorimeter region. A barrel of 32 plastic scintillators surrounds the target to veto between charged and neutral particles and to provide charged particle identification via $\Delta E - E$ techniques. Two MultiWire Proportional Chambers (MWPCs) surround this for accurate determination of charged particle trajectories, reaction and decay vertices. The BGO Rugby Ball, consisting of 480 BGO crystals, covers the polar angle range of $25° - 155°$. Each crystal is 24 cm long, corresponding to 21 radiation lengths. Separate time readout per crystal and a resolution of approximately $2 - 3$ ns enable clean identification of neutral meson decays.

The small intermediate range between the BGO Rugby Ball and the Forward Spectrometer is covered by the Scintillating Ring detector (SciRi). SciRi is comprised of three rings of 32 scintillating plastic elements each to measure directions of charged particles.

The Forward Spectrometer is a combination of tracking detectors, open dipole magnet and Time of Flight Walls (ToF Walls). The Scintillating fibre detectors, MOMO and SciFi, track particles from the target before their trajectories are curved in the magnetic field of the open dipole magnet, with an integrated field of $\int Bdl \approx 0.71$ Tm. The acceptance of the forward spectrometer covers polar angles of 12° and 8° in the horizontal and vertical directions respectively.

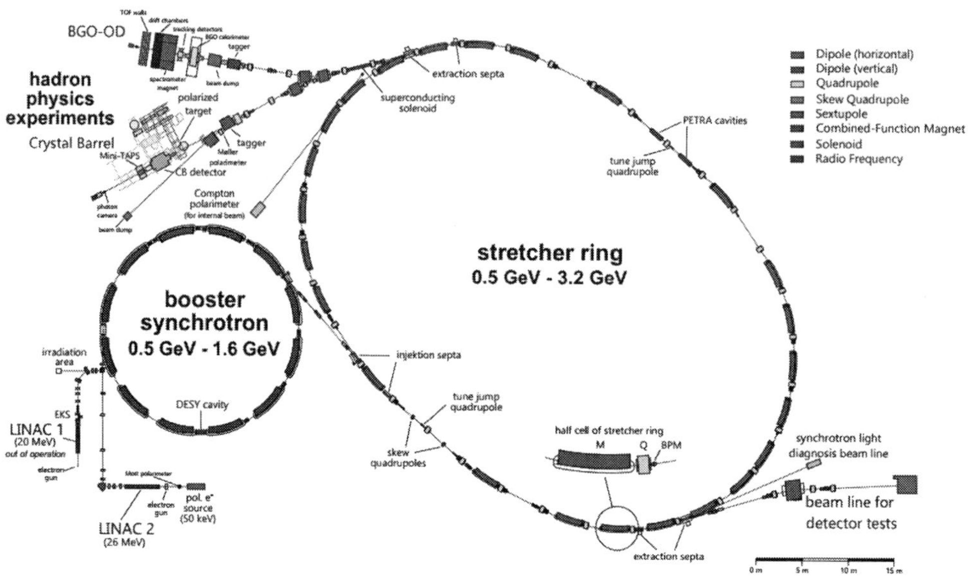

Figure 2. Overview of the ELSA accelerator facility. Figure adapted from Ref. [24].

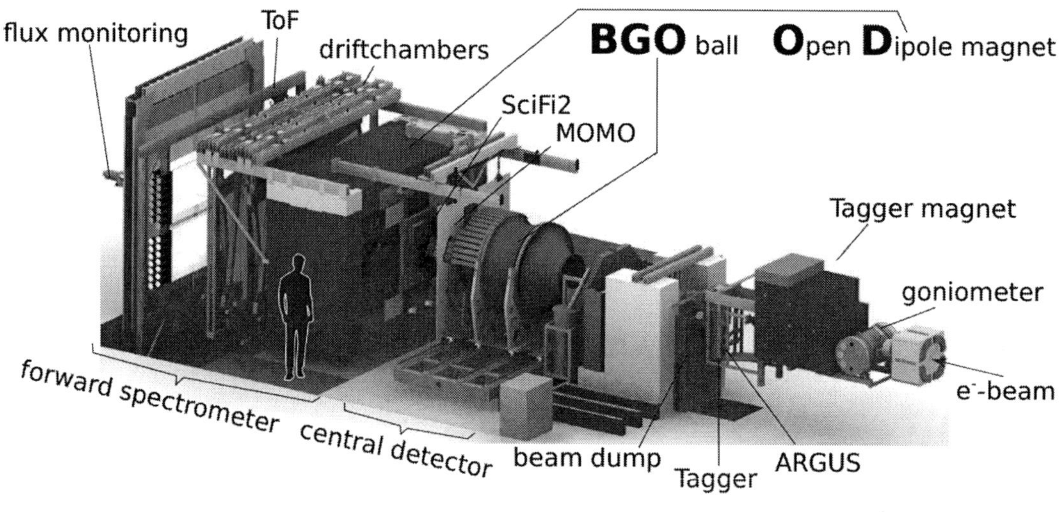

Figure 3. The BGO-OD experiment

Eight double-sided drift chambers trace particle trajectories downstream from the magnetic field. Each drift chamber is orientated to measure either horizontal, vertical or $\pm 9°$ from horizontal information (named X,Y,U and V respectively). Using accurate three dimensional maps of the magnetic field (both measured and simulated), particle trajectories and momenta are determined using combinations of all tracking detectors, with a resolution of $\delta p/p \approx 3\%$. Three ToF Walls measure the time of incident particles relative to the reaction start time. Particle β is determined after accounting for the trajectory length and energy loss en route.

Two photon monitoring detectors, GIM and Flumo are located behind the TOF walls to measure absolute photon flux.

For a detailed description of the BGO-OD setup and running conditions see Ref. [25].

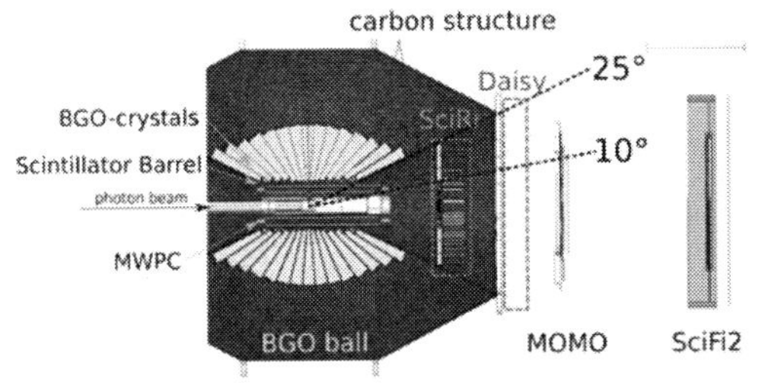

Figure 4. Slice view of the central detector consisting of the segmented BGO Rugby Ball, cylindrical inner scintillator barrel and MWPC, and the SciRi scintillator ring detector in forward direction. Also depicted are MOMO and SciFi, the front tracking fibre detectors of the forward spectrometer and the MRPC detector under construction.

4 BGO-OD performance

Fig. 5 shows the invariant mass of two photons identified in the BGO Rugby Ball after selecting a recoiling missing mass consistent with that of a proton. A clean identification of π^0 and η mesons can be made

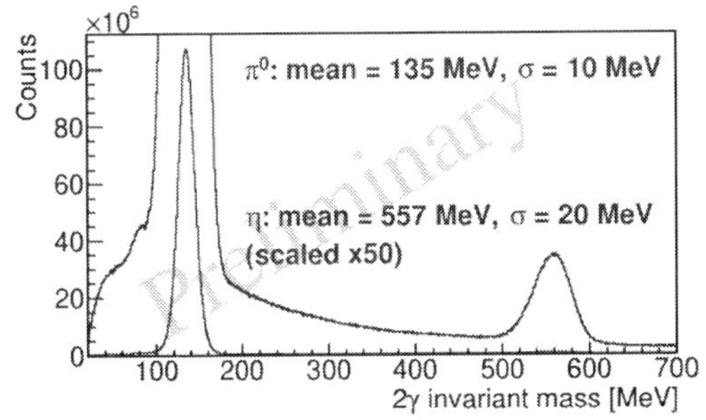

Figure 5. Neutral meson identification using the BGO Rugby Ball. The two photon invariant mass spectrum has peaks corresponding to the accepted π^0 and η masses. Approximate mean and sigma values given inset.

Numerous "bench mark" cross section measurements have been made to ensure the accuracy of the simulated geometry, trigger conditions, photon beam flux normalisation and event reconstruction (including kinematic fitting algorithms). One example is shown in Fig. 6, where the differential cross section for $\gamma p \rightarrow \omega p$ was determined with a limited data set. The data are consistent with previous measurements. This analysis employed the unique mixed charge identification available at BGO-OD, where the decay $\omega \rightarrow \pi^0\pi^+\pi^-$ was identified.

Fig. 7 demonstrates particle and momentum determination using the forward spectrometer. Particle β versus momentum yields characteristic loci for particle species (top). After selecting forward going protons, and with no additional selection criteria, reaction channels can be easily identified. The bottom plot shows the missing mass re-

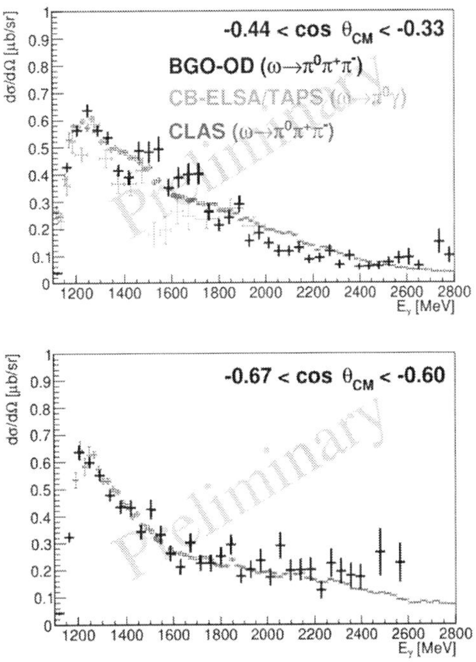

Figure 6. Preliminary $\gamma p \rightarrow \omega p$ differential cross sections identified via the mixed charge decay, $\omega \rightarrow \pi^+\pi^-\pi^0$ (black points). For comparison, data from Dietz *et al.* [26] and Williams *et al.* [27] are shown in blue and magenta respectively. Analysis by G. Scheluchin [28].

coiling from the forward proton, where peaks corresponding to single meson photoproduction are evident. The smooth distribution under the peaks originates from multiple mesons in the final state.

5 Preliminary results in strangeness photoproduction

The results shown below are from the first stage of an extensive strangeness photoproduction programme. Channels with both neutral and charged kaons are investigated, using both liquid hydrogen and deuterium targets. The forward spectrometer for charged particle identification and calorimeter for neutral meson and hyperon decays yield unique kinematics complementary to existing facilities.

5.1 Neutral kaon photoproduction

BGO-OD is able to identify K_S^0 via both the neutral and charged pion decays. The model of Ramos and Oset [29], which described the cusp in $\gamma p \rightarrow K_S^0\Sigma^+$ as the destructive interference of two dynamically generated resonances predicts the same states will constructively interfere in $\gamma n \rightarrow K_S^0\Sigma^0$, resulting in a peak in the cross section. BGO-OD is ideally suited to search for this "smoking gun" signal.

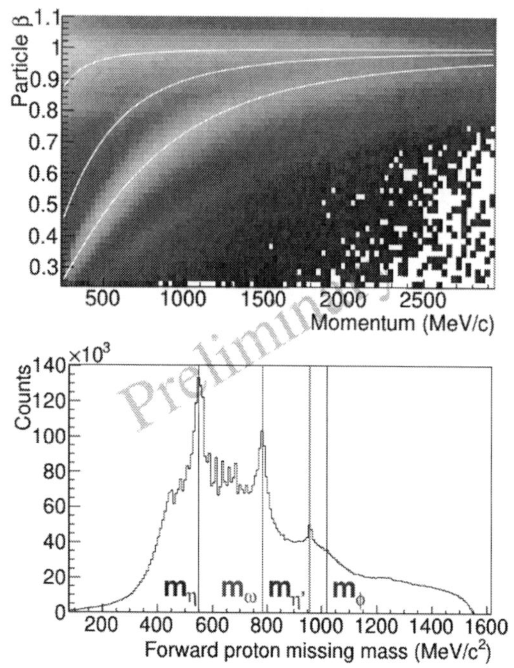

Figure 7. Particle identification in the forward spectrometer. Top: β versus momentum. Characteristic loci for π^+, protons and K^+ are indicated by white lines. Bottom: Mass recoiling from forward particles. Peaks from single meson photoproduction indicated.

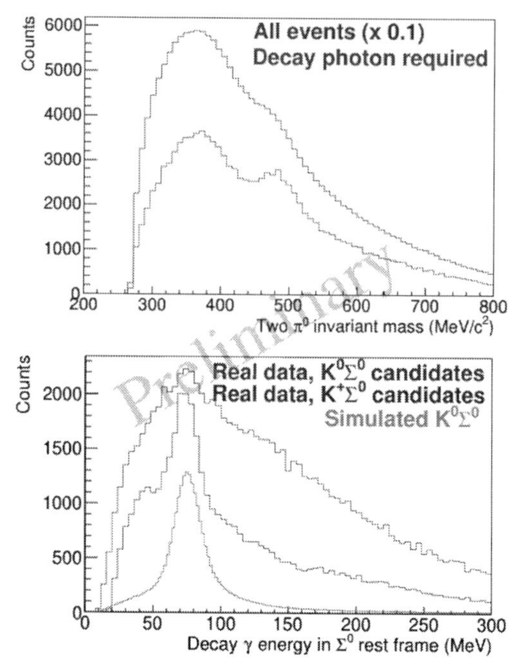

Figure 8. Top: Identification of $\gamma n \rightarrow K^0_S\Sigma^0$ via $K^0_S \rightarrow 2\pi^0$, with and without the $\Sigma^0 \rightarrow \gamma\Lambda$ decay additionally identified (red and blue lines respectively). Bottom: Energy of Σ^0 decay photon candidates in the Σ^0 rest frame for different scenarios labelled inset. Analysis by K. Kohl [30].

The reaction $\gamma p \rightarrow K^0_S\Sigma^0$ is identified via $K^0_S \rightarrow 2\pi^0$ in the BGO Rugby Ball and selecting a missing mass consistent with Σ^0. A peak at the K^0_S invariant mass can be seen in Fig. 8(top), above a broad distribution of background channels. This is dominantly uncorrelated three pion photoproduction yielding the same final state. To improve the signal to background ratio, the photon from the decay $\Sigma^0 \rightarrow \Lambda\gamma$ was additionally identified in the BGO Rugby Ball. After the photon momentum is boosted into the rest frame of the Σ^0, the fixed energy is close to the $\Sigma^0 - \Lambda$ mass difference of 77 MeV. This is shown in Fig. 8 (bottom), for simulated data (where the peak is immediately evident) and real data, where a small peak resides over a large background. To demonstrate the veracity of this technique, the channel $K^+\Sigma^0$ was used, where there is a much cleaner signal. The peak is consistent with the other data. After selecting events around this peak, the invariant mass of the K^0_S is more prominent above background.

To remove $\gamma p \rightarrow K^0_S\Sigma^+$ events from the yield, an identical analysis was performed using liquid hydrogen target data. The kinematics were smeared according to the momentum distribution of nucleons in deuterium, and the signal subtracted from the K^0_S invariant mass plot.

The selection of the $\Sigma^0 \rightarrow \gamma\Lambda$ decay removed almost all $\gamma n \rightarrow K^0_S\Lambda$ events. The remaining background was subtracted by scaling the known cross section [31] according to the measured efficiency and photon flux. This was approximately 16 % and 2 % of the yield at photon beam energies 1250 MeV and 1750 MeV respectively.

Fig. 9 shows an example of a preliminary differential cross section for $K^0_S\Sigma^0$ photoproduction. This shows a peak structure at approximately 1700-1800 MeV photon beam energy, reducing dramatically to the point where the signal is negligible above 2 GeV. The peak structure predicted by Oset and Ramos [29], arising from the coherent interference of the vector-meson dynamically generated resonances appears to be consistent. It should be noted that the model prediction is for the integrated cross section which has been scaled by a factor of five to approximately match the data. This is only to demonstrate that the shape of the structure appears consistent, but with no constraints upon the absolute scale. For further details see Ref. [32].

A complimentary analysis to identify the decay $K^0_S \rightarrow \pi^+\pi^-$ is underway [33], where it is anticipated that more forward angles can be reached. This analysis uses the excellent spatial resolution afforded by the MWPC to identify the K^0_S decay vertex outside of the target fiducial volume ($c\tau \approx 2.7$ cm).

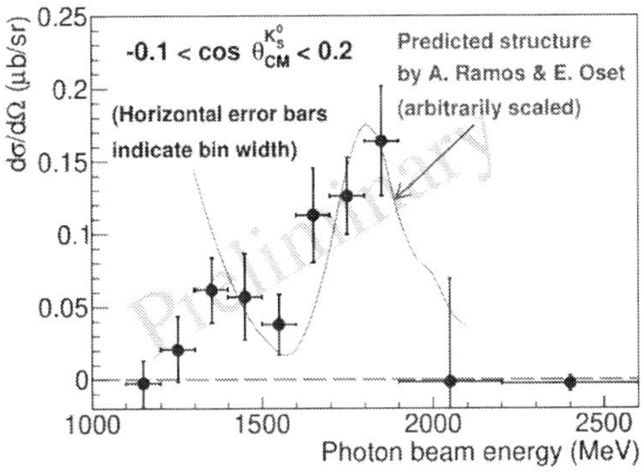

Figure 9. Preliminary $\gamma p \to K_S^0\Sigma^0$ differential cross section. The arbitrarily scaled peaked structure predicted by Ramos and Oset [29] is overlayed (see text for details). Analysis by K. Kohl [30].

5.2 $K^+(\Lambda(1405) \to \pi^0\Sigma^0)$ photoproduction - Lineshapes and differential cross sections

The unique BGO-OD setup allows identification of K^+Y states at extremely forward regions and at very low momentum transfer. Fig. 10 shows the missing mass from forward K^+ for all events where the K^+ momentum is below 1 GeV/c. Peaks corresponding to hyperon masses are immediately visible above a smooth background up to approximately 1900 MeV/c^2. The approximate fit used simulated data to describe the signal, and background arising from misidentified π^+ and positrons from pair production from the beam. This kinematic region covers many Λ and Σ^* states where the evidence of existence is poor or many parameters not accurately determined [34].

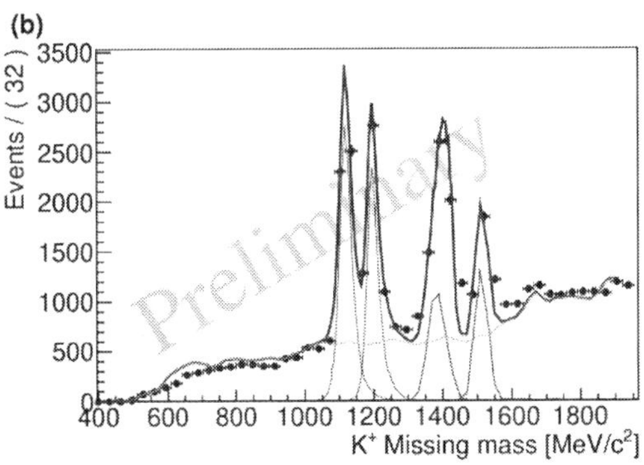

Figure 10. Missing mass recoiling from forward K^+. The spectrum is fitted with real data describing background from e^+ and π^+ (cyan line) and peaks corresponding to $K^+\Lambda$, $K^+\Sigma^0$, $K^+\Sigma^0(1385)/\Lambda(1405)$ (nearly mass degenerate) and $K^+\Lambda(1502)$ (red lines).

Different hyperon states where the masses overlap can be disentangled via identifying their decays in the BGO Rugby Ball. Fig. 11 shows the same spectra as in Fig. 10 on the x-axis, with an additional axis showing the missing mass from a combined $K^+\pi^0$ system. This has been used to separate the nearly mass degenerate $\Sigma^0(1385)$ and $\Lambda(1405)$.

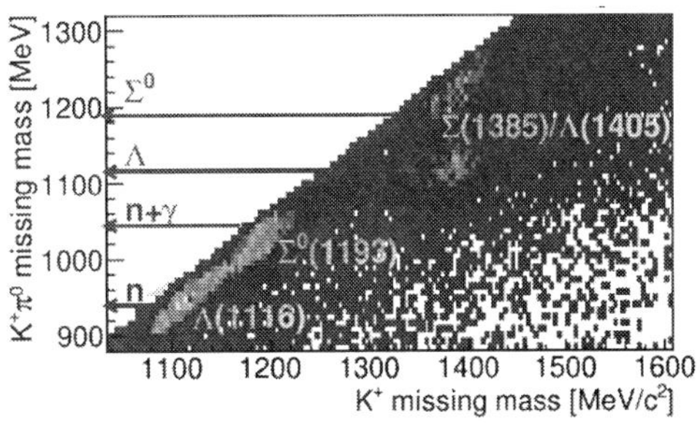

Figure 11. Missing mass from $K^+\pi^0$ system versus K^+ missing mass. Yellow labels indicate hyperon mass peaks and red arrows indicated expected missing mass from the $K^+\pi^0$ system.

BGO-OD is ideal to investigate the $\gamma p \to K^+(\Lambda(1405) \to \Sigma^0\pi^0)$ line shape ($\Lambda(1405)$ invariant mass) and differential cross section. In addition to providing high statistics data over kinematic regions accessible to other facilities, the kinematics can be extended to very forward, low momentum transfer regions using the forward spectrometer.

The $\gamma p \to K^+(\Lambda(1405) \to \Sigma^0\pi^0)$ signal has been determined by fitting simultaneously to the two dimensions in Fig. 11, and line shapes and differential cross sections measured [28]. Two distinct analyses were made. The *Full topology Analysis* has a nearly 4π angular acceptance, where every particle in the final state $\pi^0\gamma\pi^-p$ was identified and a kinematic fit employed. The *Forward Spectrometer Analysis* uses the high momentum resolution at forward angles to reconstruct each event from a forward K^+ and π^0 decay only.

Fig.12 shows the line shape of the $\Lambda(1405) \to \pi^0\Sigma^0$ decay for the *Full Topology Analysis*. The statistics are comparable with the previous data shown and provide a crucial constraint given that BGO-OD is ideally suited to identify the neutral $\Lambda(1405)$ decay. Line shapes at very forward angles using the *Forward Spectrometer Analysis* also show consistent results. The differential cross section for both analyses is shown in Fig. 13, where extremely forward angles can be accessed for the first time. This is also demonstrated in Fig. 14, where the data are converted into a differential cross section with respect to the square of the momentum exchange, $-t$. See Ref. [35] for further details.

5.3 Forward $K^+\Lambda$ and $K^+\Sigma^0$ photoproduction

Ground state, $K^+\Lambda$ and $K^+\Sigma^0$ photoproduction is virtually unconstrained at forward angles. The lack of data and discrepancies in the existing world data prevents accurate descriptions by isobar models and partial wave analyses,

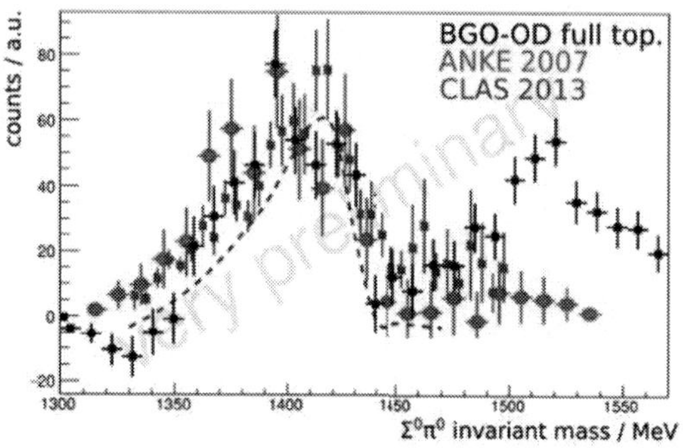

Figure 12. $\Lambda(1405) \rightarrow \pi^0\Sigma^0$ line shape (black points), compared to previous data from Zychor *et al.* [36] and Moriya *et al.* [37] the dashed line is the prediction of Nacher *et al.* [38] using a chiral unitary model where the $\Lambda(1405)$ is dynamically generated via K^-p interactions. Analysis by G. Scheluchin [28].

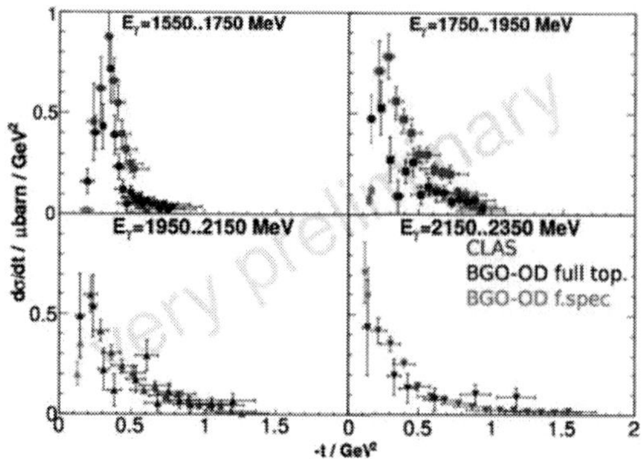

Figure 14. The same data shown in Fig. 13, but plotted vs $-t$ (black and magenta points show the full topology and forward K^+ analysis respectively). Blue points are previous data of Moriya *et al.* [37]. Analysis by G. Scheluchin [28].

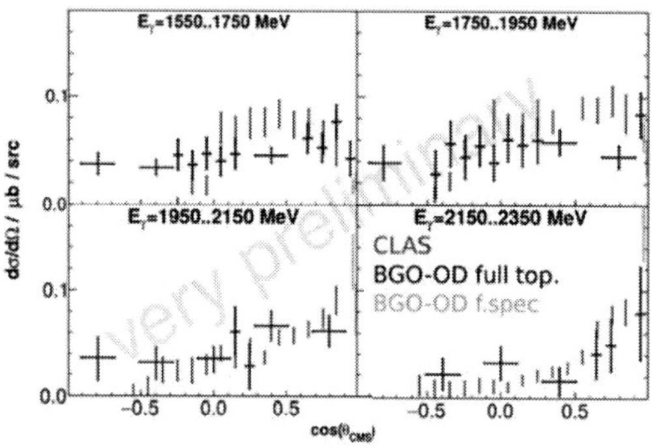

Figure 13. Preliminary $\gamma p \rightarrow K^+(\Lambda(1405) \rightarrow \pi^0\Sigma^0)$ differential cross section using a full topology and forward K^+ analysis (black and green points respectively). Blue points are previous data of Moriya *et al.* [37]. Analysis by G. Scheluchin [28].

where many solutions have large deviations (see for example, Ref. [39]). Data at this very low momentum exchange region is also crucial in constraining the elementary reaction processes in hypernuclei electroproduction, where the required low Q^2 is comparable to the photoproduction process. The excellent forward acceptance and high momentum resolution allows BGO-OD to resolve these issues.

Additional selection criteria were used to enhance the signal above background for $K^+\Lambda$ and $K^+\Sigma^0$ photoproduction. For $K^+\Lambda$ photoproduction, a diagonal selection cut was applied to the two dimensional plot in Fig. 11 to select the $\Lambda \rightarrow \pi^0 n$ decay. For $K^+\Sigma^0$ photoproduction, the Σ^0 decay photon was identified in the same manner as described in section 5.1. Fig. 15 shows the missing mass for forward K^+ (corresponding to $\cos\theta^{K^+}_{CM} > 0.9$) for two different photon beam intervals. The two selection criteria enhance the required channel. The non-K^+ background indicated by the cyan line in the figure mostly originates from positrons from pair production in the beam. This distribution was determined using an equivalent analysis for negatively charged particles, where the electron distribution was identical. The largest contribution of hadronic reaction background was determined as $\gamma p \rightarrow \pi^0\Delta^+$, where a π^+ from the Δ^+ decay was misidentified as a K^+. Simulated data were used to determine systematic effects of this upon the extracted signal yields.

Fig. 16 shows the $\gamma p \rightarrow K^+\Lambda$ differential cross section for $\cos\theta^{K^+}_{CM} > 0.9$. The statistical error is significantly smaller than previous datasets, allowing for the first time an accurate description at small angles, low momentum transfer regions where t-channel mechanisms dominate the reaction process. The high statistics allow a discrimination between previous datasets which exhibit discrepancies. It should be noted that the CLAS data are at the slightly more backward angle interval of $0.85 < \cos\theta^{K^+}_{CM} < 0.95$. The dominant systematic errors shown on the abscissa are the flux normalisation and beam spot position (4% and 5% respectively) up to approximately 1260 MeV photon beam energies. At higher energies the fitting to the signal and background is the dominant error (approximately 12%).

BGO-OD is able to measure this forward region in fine polar angle intervals. Fig. 17 shows four examples of the angular distribution of the differential cross section, enabling accurate extrapolations to the minimum possible momentum exchange at the most forward angle.

The recoil polarisation, P_Λ was accessed via the self-analysing weak decay of the Λ, given in Eq. 1, where $N_{\uparrow/\downarrow}$ are the π^0 yield above or below the reaction plane in the Λ rest frame and the weak decay parameter, $\alpha = 0.642$ [34].

$$P_\Lambda = \frac{2}{\alpha}\frac{N_\uparrow - N_\downarrow}{N_\uparrow + N_\downarrow} \qquad (1)$$

Fig. 18 shows the recoil polarisation for $\cos\theta^{K^+}_{CM} > 0.9$. The BGO-OD data are consistent with the previous data shown.

The differential cross section for $\gamma p \rightarrow K^+\Sigma^0$ for $\cos\theta_{CM} > 0.9$ is shown in Fig. 19. As for $K^+\Lambda$, the statistics are of unprecedented accuracy at this forward angle.

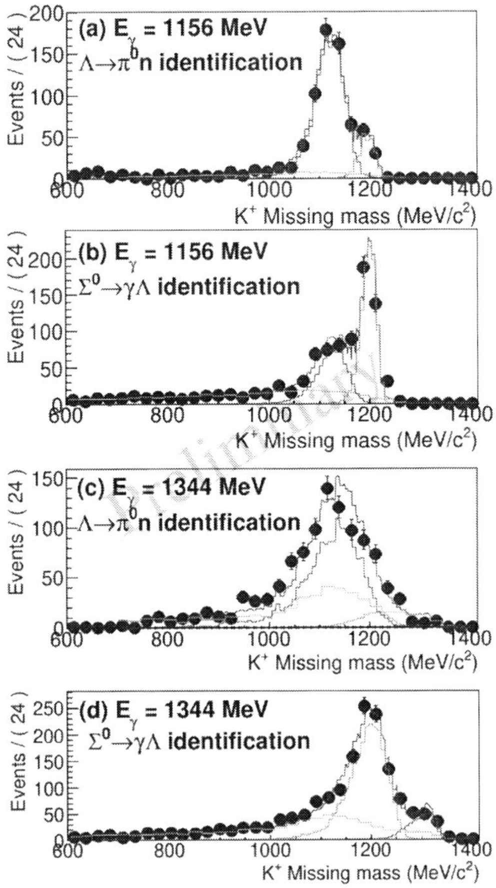

Figure 15. Missing mass recoiling from forward K^+ candidates (black points) for two different photon beam energies and selection criteria to either accentuate the $K^+\Lambda$ or $K^+\Sigma^0$ signal. The spectra are fitted with e^+ background (described in the text), simulated $K^+\Lambda$, $K^+\Sigma^0$ and $K^+\Sigma^0(1385)$ (cyan, red, green and purple lines respectively). The blue line is the summed fit.

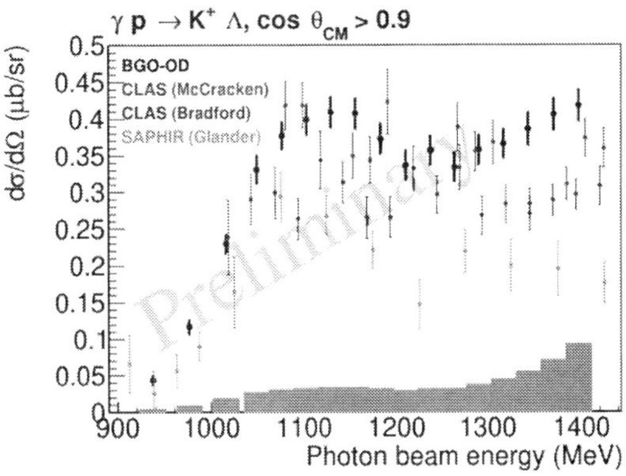

Figure 16. $\gamma p \to K^+\Lambda$ differential cross section for $\cos\theta_{CM} > 0.9$ (Black points). Systematic errors are the filled grey bars on the abscissa. Previous data of McCracken [40], Bradford [41] and Glander [42] shown in red, blue and green respectively. The data of McCracken and Bradford are at the more backward interval, $0.85 < \cos\theta_{CM} < 0.95$.

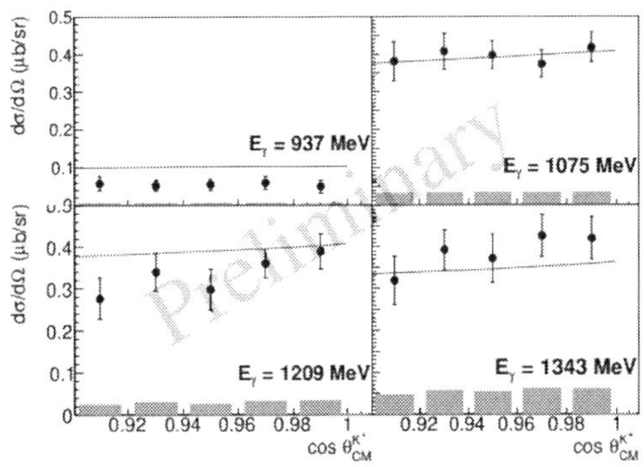

Figure 17. $\gamma p \to K^+\Lambda$ differential cross section for four example photon beam energies (labelled inset) versus $\cos^{K^+}_{CM}$. Systematic errors are the filled grey bars on the abscissa. The red line is the Bonn-Gatchina PWA solution [2].

The data suggest a "cusp-like" structure at a photon beam energy of approximately 1470 MeV where there is a drop in the cross section of approximately 25%. Previous data sets supported this however with insufficient statistics for a clear statement.

Similarly to $K^+\Lambda$ photoproduction, the data were divided into fine, 0.02 $\cos^{K^+}_{CM}$ intervals, two examples of which are shown in Fig. 20. The "cusp-like" structure appears more pronounced at the most forward, lowest momentum exchange region. This appears consistent with the expected behaviour of a molecular like system or sub-threshold effects playing important roles at very low momentum exchange regions. This is close to multiple thresholds of open and hidden strangeness, notably $K^+\Lambda(1405)$, $f_0 p$ (where the f_0 has been previously described as a $K\bar{K}$ molecule) and $\eta' p$, however mechanisms responsible are still under discussion.

6 Conclusions

The BGO-OD experiment is an ideal setup to investigate associated strangeness photoproduction at very forward angles and low momentum transfer where exotic behaviour may manifest. The first key results vindicate this argument, with a cusp in the $K^+\Sigma^0$ photoproduction cross section observed at very forward angles, and a peak in $K^0\Sigma^0$ photoproduction cross section at the K^* thresholds. The $\Lambda(1405) \to \pi^0\Sigma^0$ line shape and differential cross section is also measured over a broad kinematic region, with

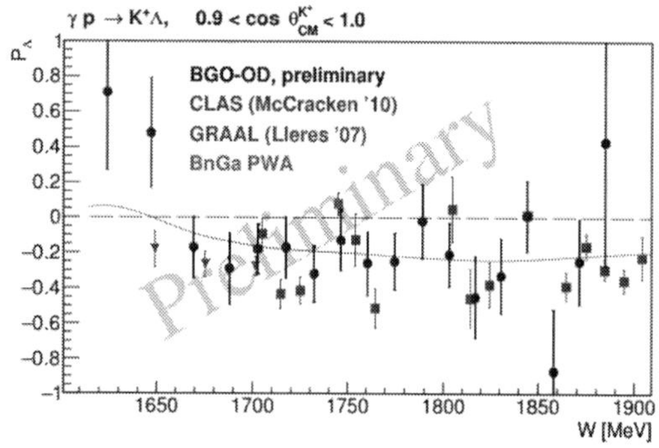

Figure 18. $\gamma p \rightarrow K^+\Lambda$ recoil polarisation for $\cos^{K^+}_{CM} > 0.9$ (black points). Previous data shown from McCracken *et al.* [40], Lleres *et al.* [43] and the Bonn-Gatchina PWA solution [44] shown in red triangles, purple triangles and magenta line respectively.

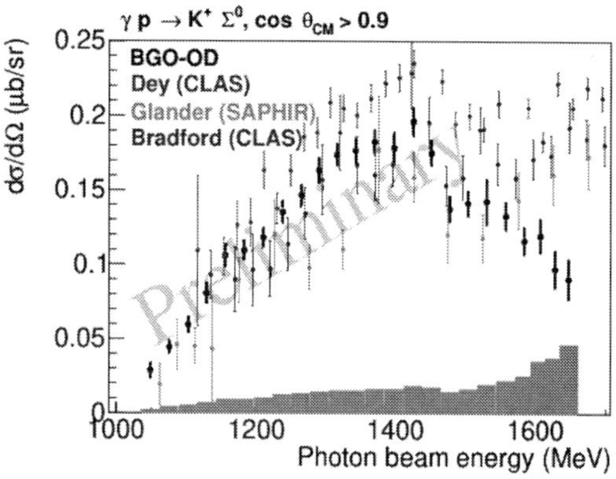

Figure 19. $\gamma p \rightarrow K^+\Sigma^0$ differential cross section for $\cos\theta_{CM} > 0.9$ (Black points). Systematic errors are the filled grey bars on the abscissa. Previous data of Dey [45], Glander [42] and Bradford [41] shown in red, blue and green respectively. The data of Dey and Bradford are at the more backward interval, $0.85 < \cos\theta_{CM} < 0.95$.

high statistics and extending to very low momentum transfer regions.

Acknowledgements

It is a pleasure to thank the ELSA staff for a reliable and stable operation of the accelerator and the technical staff of the contributing institutions for essential help in the realisation and maintenance of the apparatus.

This work was supported by the Deutsche Forschungsgemeinschaft project numbers 388979758 and 405882627. Our Russian collaborators thank the Russian Scientific Foundation (grant RSF number 19-42-04132) for financial support.

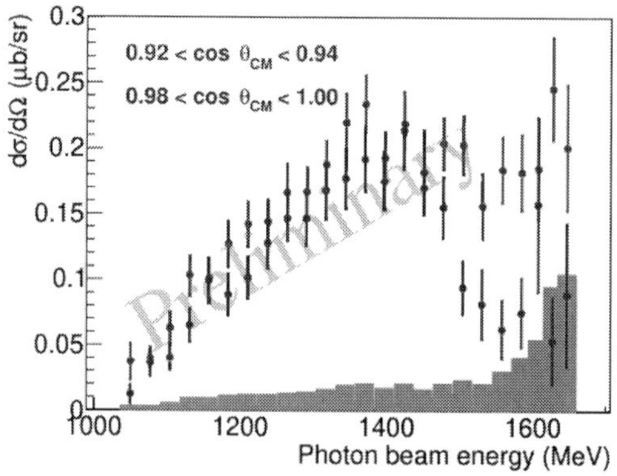

Figure 20. $\gamma p \rightarrow K^+\Sigma^0$ differential cross section for $0.92 < \cos\theta_{CM} < 0.94$ and $\cos\theta_{CM} > 0.98$ (red and blue points respectively). Systematic errors for the more backward angle interval are the filled grey bars on the abscissa.

References

[1] U. Löring, B.C. Metsch, H.R. Petry, Eur. Phys. J. A **10**, 395 (2001)
[2] A.V. Anisovich et al., Eur. Phys. J. A **50**, 129 (2014)
[3] G.M. M, Phys. Rev. Lett. **8**, 214 (1964)
[4] J.R. L., Phys. Rev. D **15**, 267 (1977)
[5] D. Strottmann, Phys. Rev. D p. 748 (1979)
[6] L.Y. Glozman, D.O. Riska, Physics Reports **268**, 263 (1996)
[7] C. Garcia-Recio, M.F.M. Lutz, J. Nieves, Phys. Lett. B **582**, 49 (2004)
[8] M.F.M. Lutz, E.E. Kolomeitsev, Phys. Lett. B **585**, 243 (2004)
[9] R. Aaij et al., Phys. Rev. Lett. **115**, 072001 (2015)
[10] J.J. Wu, R. Molina, E. Oset, B.S. Zou, Phys. Rev. Lett. **105**, 232001 (2010)
[11] N.A. Törnqvist, Phys. Lett. B **590**, 209 (2004)
[12] D. Jido, J.A. Oller, E. Oset, A. Ramos, U.G. Meissner, Nucl. Phys. A **725**, 181 (2003)
[13] R. Molina, M. Döring, Phys. Rev. D **94**, 056010, 079901 (2016)
[14] D. Jido, J.A. Oller, E. Oset, A. Ramos, U.G. Meissner, Nucl. Phys. A **725**, 181 (2003)
[15] J.M.M. Hall et al., **114**, 132002 (2015)
[16] R. Ewald et al., Phys. Lett. B **713**, 180 (2012)
[17] P. Gonzalez, E. Oset, J. Vijande, Phys. Rev. C **79**, 025209 (2009)
[18] S. Sarkar et al., Eur. Phys. J. A **44**, 431 (2010)
[19] E. Oset, A. Ramos, Eur. Phys. J. A **44**, 445 (2010)
[20] W. Hillert, Eur. Phys. J. A **28**, 139 (2006)
[21] E. Aker et al., Nucl. Instrum. Meth. A **321**, 69 (1992)
[22] A.R. Gabler et al., Nucl. Instrum. Meth. A **346**, 168 (1994)
[23] R. Novotny, IEEE Trans. Nucl. Sci. **38**, 379 (1991)
[24] W. Hillert et al., EPJ Web Conf. **134**, 05002 (2017)

[25] S. Alef et al., To be sumitted to Eur. Phys. J. A (2019)
[26] F. Dietz et al., Eur. Phys. J. A **51**, 6 (2015)
[27] M. Williams et al., Phys. Rev. C **80**, 065208 (2009)
[28] G. Scheluchin, Ph.D. thesis, Rheinische Friedrich-Wilhelms-Universität Bonn (2019)
[29] A. Ramos, E. Oset, Phys. Lett. B **727**, 287 (2013)
[30] K. Kohl, Ph.D. thesis, Rheinische Friedrich-Wilhelms-Universität Bonn (in preparation) (2019)
[31] N. Compton et al., Phys. Rev. C **96**, 065201 (2017)
[32] K. Kohl, in *The 12th International Workshop on the Physics of Excited Nucleons,* (2019), EPJ Web of Conferences
[33] B.E. Reitz, Ph.D. thesis, Rheinische Friedrich-Wilhelms-Universität Bonn (in preparation) (2019)
[34] M. Tanabashi, K. Hagiwara, K. Hikasa, K. Nakamura, Y. Sumino, F. Takahashi, J. Tanaka et al. (Particle Data Group), Phys. Rev. D **98**, 030001 (2018)
[35] G. Scheluchin, in *The 12th International Workshop on the Physics of Excited Nucleons* (2019), EPJ Web of Conferences
[36] I. Zychor et al., Phys. Lett. B **660** (2008)
[37] K. Moriya, R.A. Schumacher et al., Phys. Lett. B **87**, 035206 (2013)
[38] J.C. Nacher et al., Phys. Lett. B **455**, 55 (1999)
[39] D. Skoupil, P. Bydžovský, Phys. Rev. C **97**, 025202 (2018)
[40] M.E. McCracken et al., Phys. Rev. C **81**, 025201 (2010)
[41] R. Bradford et al., Phys. Rev. C **81**, 035202 (2006)
[42] K. Glander et al., Eur. Phys. J. A **19**, 251 (2004)
[43] A. Lleres et al., Eur. Phys. J. A **31**, 79 (2007)
[44] A. Anisovich, V. Kleber, E. Klempt, V. Nikonov, A. Sarantsev, U. Thoma, Eur. Phys. J. A **34**, 243 (2007)
[45] B. Dey et al., Phys. Rev. C **82**, 025202 (2010)

http://doi.org/10.1051/epjconf/202024101009

$K^0_S\Sigma^0$ photoproduction at the BGO-OD experiment

Katrin Kohl[1,*] *Stefan* Alef[1], *Patrick* Bauer[1], *Reinhard* Beck[2], *Alessandro* Braghieri[13], *Philip* Cole[3], *Rachele* Di Salvo[4], *Daniel* Elsner[1], *Alessia* Fantini[4,5], *Oliver* Freyermuth[1], *Francesco* Ghio[6,7], *Anatoly* Gridnev[8], *Daniel* Hammann[1], *Jürgen* Hannappel[1], *Thomas* Jude[1], *Nikolay* Kozlenko[8], *Alexander* Lapik[9], *Paolo* Levi Sandri[10], *Valery* Lisin[9], *Giuseppe* Mandaglio[11,12], *Roberto* Messi[4,5], *Dario* Moricciani[4], *Vladimir* Nedorezov[9], *Dmitry* Novinsky[8], *Paolo* Pedroni[13], *Andrei* Polonski[9], *Björn-Eric* Reitz[1], *Mariia* Romaniuk[4], *Georg* Scheluchin[1], *Hartmut* Schmieden[1], *Victorin* Sumachev[8], *Viacheslav* Tarakanov[8], and *Christian* Tillmanns[1]

[1]Rheinische Friedrich-Willhelms-Universität Bonn, Physikalisches Institut, Nußallee 12, 53115 Bonn, Germany
[2]Helmholtz-Institut fuer Strahlen- und Kernphysik, Universitaet Bonn, Nussallee 1-16, D-53115 Bonn Germany
[3]Lamar University, Department of Physics, Beaumont, Texas, 77710, USA
[4]INFN Roma Tor Vergata, Rome, Italy
[5]Università di Roma "Tor Vergata", Via della Ricerca Scientifica 1, 00133 Rome, Italy
[6]INFN sezione di Roma La Sapienza, P.le Aldo Moro 2, 00185 Rome, Italy
[7]Istituto Superiore di Sanita, Viale Regina Elena 299, 00161 Rome, Italy
[8]Petersburg Nuclear Physics Institute, Gatchina, Leningrad District, 188300, Russia
[9]Russian Academy of Sciences Institute for Nuclear Research, prospekt 60-letiya Oktyabrya 7a, Moscow 117312, Russia
[10]INFN - Laboratori Nazionali di Frascati, Via E. Fermi 40, 00044 Frascati, Italy
[11]INFN sezione Catania, 95129 Catania, Italy
[12]Universita degli Studi di Messina, Via Consolato del Mare 41, 98121 Messina, Italy
[13] INFN sezione di Pavia, Via Agostino Bassi, 6 - 27100 Pavia, Italy

Abstract.
The BGO-OD experiment at the ELSA accelerator facility uses an energy tagged bremsstrahlung photon beam to investigate the excitation structure of the nucleon via meson photoproduction.
The setup with a BGO calorimeter surrounding the target and an open dipole spectrometer covering the forward region is ideally suited for investigating low momentum transfer processes, in particular in strangeness photoproduction.
The associated photoproduction of K^0_S and hyperons is essential to understand the role of K^* exchange mechanisms. A cusp-like structure observed in the $\gamma p \to K^0_S\Sigma^+$ reaction at the K^* threshold is described by models including dynamically generated resonances from vector meson-baryon interactions. Such interactions are predicted to give a peak like structure in $K^0_S\Sigma^0$ photoproduction off the neutron. A very preliminary cross section is determined and compared to the prediction, the results appear to support the model.

1 Introduction

The cusp-like structure observed in the $\gamma p \to K^0\Sigma^+$ differential cross section by Ewald et al. [1] is not described in previous isobar and PWA models, for example SAID [2] and K-MAID [3]. It could be described however using an equivalent model to the one that predicted the charmonium-pentaquark candidate [4] observed by the LHCb collaboration [5] in 2015 which includes dynamically generated resonances from vector meson-baryon interactions [6]. Below K^* threshold, an intermediate $K^*\Lambda$ or $K^*\Sigma$ state feeds the cross section by conversion to the final state via pion exchange. Above threshold the K^* can be produced freely. Whereas this produces a cusp in $K^0\Sigma^+$ photoproduction due to destructive interference, constructive interference is predicted to give an enhancement in $K^0\Sigma^0$ photoproduction. Fig. 1 shows both cases for two different parameter sets. The upper plot includes the experimental data from the CBELSA/TAPS collaboration [1]. A measurement of the $\gamma n \to K^0\Sigma^0$ cross section and an observation of the predicted enhancement would strongly support this theory and help with the understanding of pentaquark states.

*e-mail: kohl@physik.uni-bonn.de

2 BGO-OD experiment

The BGO-OD experiment at the ELSA accelerator facility in Bonn is ideally suited to measure strangeness photoproduction, especially in the very forward region. ELSA provides an electron beam up to 3.2 GeV. Via bremsstrahlung a real photon beam is produced. By measuring the momentum of the electron, the energy of the photon can be calculated.

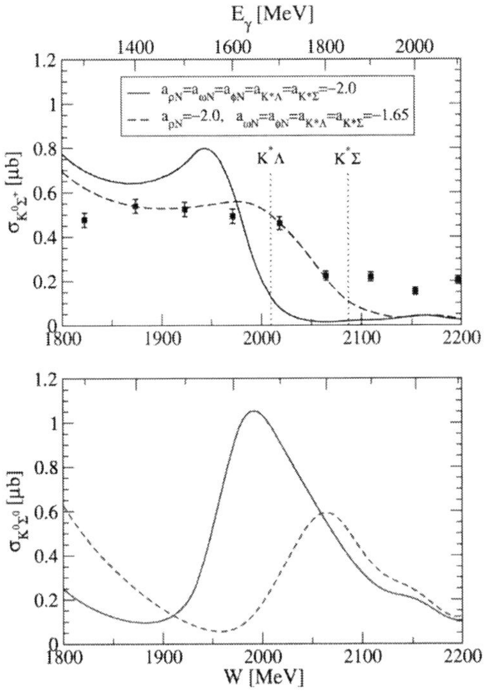

Figure 1. A prediction of the cross section of $\gamma p \rightarrow K^0\Sigma^+$ (upper plot) and $\gamma n \rightarrow K^0\Sigma^0$ (lower plot) is shown as solid line. The dotted line parameters were fitted to the data (black points [1]). Figure taken from Ref. [6].

The photon beam impinges on a fixed target, usually liquid hydrogen or deuterium. This is surrounded by a Multi-Wire-Proportional-Chamber for tracking, a scintillator barrel to provide charge information and an electromagnetic calorimeter made from BGO. This central detector is ideally suited for photon detection, but also able to detect other charged or neutral particles.

In the forward direction two tracking detectors (MOMO and SciFi) in front and eight driftchambers behind an open dipole magnet, together with time of flight measurement, allow excellent particle identification. The intermediate range is covered by SciRi, a Scintillating Ring detector, which provides directions of charged particles. An overview is given in figure 2, for detailed information on the performance of the BGO-OD experiment see Ref. [7].

3 Reconstruction

To determine a cross section of $\gamma n \rightarrow K^0\Sigma^0$ the decay channel $K^0\Sigma^0 \rightarrow \pi^0\pi^0\Lambda\gamma \rightarrow 5\gamma\Lambda$ is identified, by requesting exactly five neutral and any number of charged particle. The Λ is not reconstructed. The five photons are identified in the BGO-Ball and combined to two π^0 with one remaining photon. All combinations are allowed where no photon is used twice and the invariant masses of the π^0 are within ± 25 MeV of the nominal π^0 mass. If the remaining photon is from the Σ^0 decay (for details see section 3.1), the pair of π^0 are kept and the masses are set to the nominal π^0 mass. A cut on the missing mass to $2\pi^0$ is applied at ± 50 MeV around the Σ^0 mass. A plot of the invariant mass of $2\pi^0$ with these selections is shown in figure 3. A signal peak at around the K^0 mass is visible, but superimposed with a lot of background.

Using simulated data, the dominant background channels proved to be:

- $\gamma N \rightarrow 3\pi N$
- $\gamma N \rightarrow \eta N$
- $\gamma N \rightarrow \eta\pi N$
- $\gamma n \rightarrow K^0\Lambda$
- $\gamma p \rightarrow K^0\Sigma^+$

The last two are the most crucial ones, since they contribute to the signal peak. To remove part of this background the entire proton contamination is subtracted (see

Figure 2. Overview of the BGO-OD experiment.

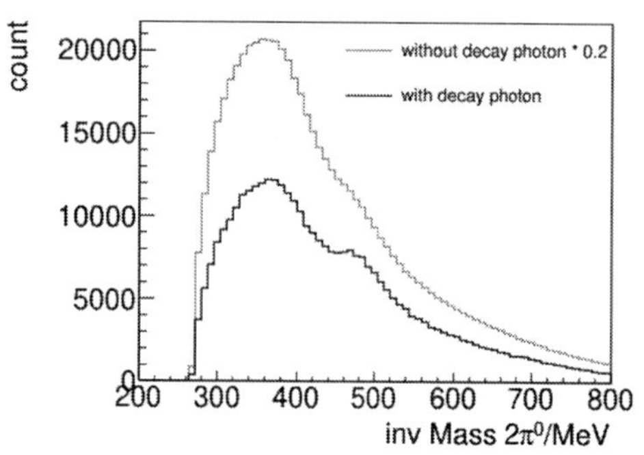

Figure 3. $2\pi^0$ invariant mass spectrum before and after selecting Σ^0 decay photon.

section 3.2). Finally the Root tool Roofit[8] is used to determine the number of signal events from the $2\pi^0$ invariant mass spectrum (section 3.3).

3.1 Tagging Σ^0 decay photon

To suppress background, the decay of the Σ^0 to a Λ and a photon was identified. In the rest frame of the Σ^0, the photon always has a constant energy of approximately the mass difference between Σ^0 and Λ of 75 MeV. Figure 4 shows the energy spectrum of a fifth photon in the BGO-calorimeter that was not used for the two π^0 reconstruction.A small peak is visible over a large background. Also shown is simulated $K^0\Sigma^0$ data, where the peak is immediately visible. To prove this technique the channel $K^+\Sigma^0$ was selected, making use of the much cleaner signal. It is evident, that in all three cases the peak in photon energy is at the same position.

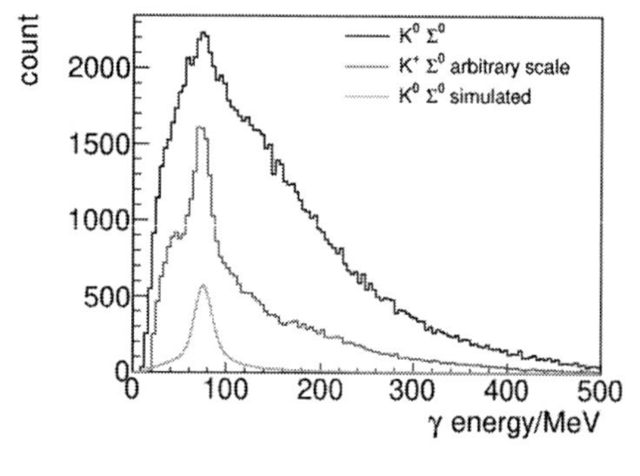

Figure 4. Energy spectrum of the Σ^0 decay photon for real and simulated data. To prove the technique is working correctly the channel $K^+\Sigma^0$ is studied.

By selecting events where this photon has an energy between 50 MeV and 105 MeV, the signal to background ratio is improved as shown in figure 3 which shows the $2\pi^0$ invariant mass spectrum. This does not only improve the signal to background ratio in general, but it also strongly suppresses the background channel $\gamma n \to K^0\Lambda$ which contributes to the signal peak, but is undesired. Only about 10% of this background channel survives, which will later be subtracted (see section 4).

3.2 Proton contamination

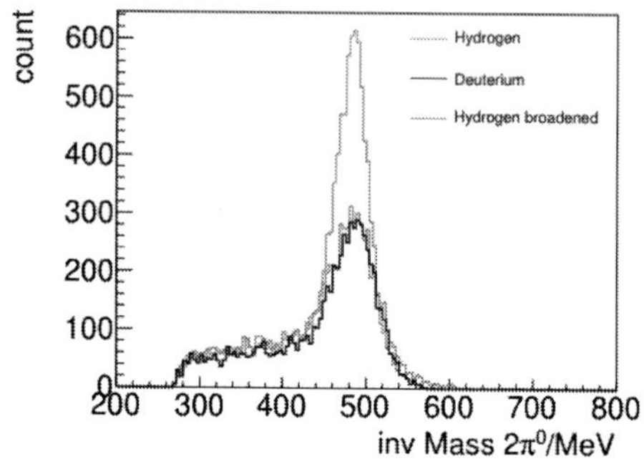

Figure 5. $2\pi^0$ invariant mass spectrum for simulated $\gamma p \to K^0\Sigma^+$ on hydrogen and deuterium target, as well as the hydrogen data set broadened according to the momentum distribution of nucleons in deuterium.

The liquid deuterium target consists of neutrons and protons. Since the investigated reaction only happens on the neutron, all reactions happening on the proton are background that have to be subtracted. This is done by doing the same analysis as before with a hydrogen target data set, which is broadened according to the momentum distribution of nucleons in deuterium. To get the correct amount of proton reactions in the deuterium data set, the hydrogen data set is scaled with the ratio of the photon flux.

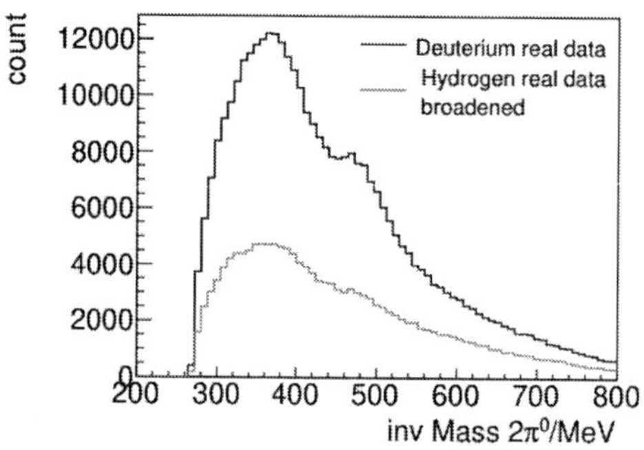

Figure 6. $2\pi^0$ invariant mass spectrum for deuterium data and broadened and scaled hydrogen data. The hydrogen data make up about half of the deuterium data.

Figure 5 shows the $2\pi^0$ invariant mass spectrum for simulated $\gamma p \to K^0\Sigma^+$ on hydrogen and deuterium targets, as well as the hydrogen data set broadened according

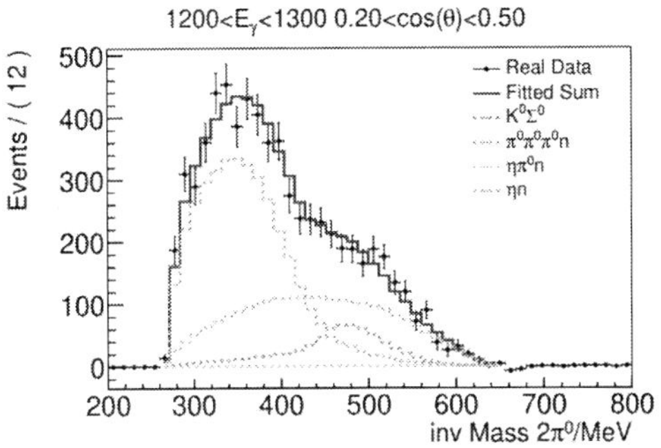

Figure 7. Example fit to $2\pi^0$ invariant mass spectrum. Simulated contributions to the fit labeled inset.

to the momentum distribution of nucleons in deuterium, after a cut on the missing mass to the $2\pi^0$ system, selecting the Σ^+. The broadening of the missing mass causes a decrease in statistic after this cut. Deuterium and broadened hydrogen data agree very well.

Figure 6 shows the $2\pi^0$ invariant mass spectrum for real data for deuterium and scaled and broadened hydrogen target data. The hydrogen target data set accounts for about half of the deuterium data set which could be expected considering the dominant background channels are uncorrelated π^0 and η production. What remains after subtraction are pure neutron channels, therefore also the background channel $\gamma n \rightarrow K^0\Sigma^+$ does not contribute to the signal peak anymore.

3.3 Fitting to data

The remaining background can be removed using the Root tool Roofit [8]. Phase space generated distributions of background channels and signal are fitted to the data. The amplitudes of the contributions are varied until the sum agrees to the data. Fig. 7 shows an example. Only points larger zero are taken into account. The spectrum is well described over the entire range and signal and background can be accurately separated. The tool allows to extract the yield of the single channels, the corresponding error bars include statistics and fit quality.

4 Differential cross section measurement

The extracted number of K^0 events is plotted as a function of beam energy in four bins in $\cos(\theta_{CM})$ in figure 8, where θ_{CM} is the center of mass polar angle. This is not the desired cross section yet, since there is still some contribution of $\gamma n \rightarrow K^0\Lambda$. To remove this contribution, the known cross section [9] is scaled with the respective reconstruction efficiency and also shown in figure 8. Within the error bars the calculated $K^0\Lambda$ contribution is in agreement with being smaller/equal than the measured yield as it is expected. After subtracting the $\gamma n \rightarrow K^0\Lambda$ contribution, a preliminary differential cross section of $\gamma n \rightarrow K^0\Sigma^0$ is shown in figure 9. Also included in this figure is the prediction by Oset and Ramos [6] in an arbitrary scale. It should be noted that this is the integrated cross section, not the plotted differential, scaled by a factor $\frac{1}{3}$ or $\frac{1}{2}$, depending on the angular region. From the integration of the solid angle a factor 2 is expected to the total cross section, which is consistent with the used scaling factors. The general shape and position of the predicted enhancement seem to agree to these preliminary results.

5 Conclusion

The results shown in figure 9 appear to support the model behind the prediction made for the cross section $\gamma n \rightarrow K^0\Sigma^0$ [6]. The general shape and position of the enhancement agree very well. More over, the enhancement appears to get stronger for more forward directions, as did the drop in the corresponding proton channel $\gamma p \rightarrow K^0\Sigma^+$ [1]. At this point it would be interesting to compare the results to predictions for the differential cross sections. Further plans are expanding the acceptance region to more forward angles, for example with the charged K^0 decay in the forward spectrometer.

This work was supported by the Deutsche Forschungsgemeinschaft project numbers 388979758 and 405882627. Our Russian collaborators thank the Russian Scientific Foundation (grant RSF number 19-42-04132) for financial support.

References

[1] R. Ewald et al., Phys. Lett. B **713**, 180 (2012)
[2] http://gwdac.phys.gwu.edu (27.8.2019)
[3] https://maid.kph.uni-mainz.de (27.8.2019)
[4] J.J. Wu, R. Molina, E. Oset, B.S. Zou, Phys. Rev. Lett. **105**, 232001 (2010)
[5] R. Aaij et al., Phys. Rev. Lett. **115**, 072001 (2015)
[6] A. Ramos and E. Oset, Phys. Lett. B **727**, 287 (2013)
[7] S. Alef et al, To be submitted to Eur. Phys. J. A (2019)
[8] W. Verkerke and D. P. Kirkby, eConf C **0303241** (2003) MOLT007 [physics/0306116].
[9] N. Compton et al Phys. Rev. C **96**, 065201 (2017)

http://doi.org/10.1051/epjconf/202024101009

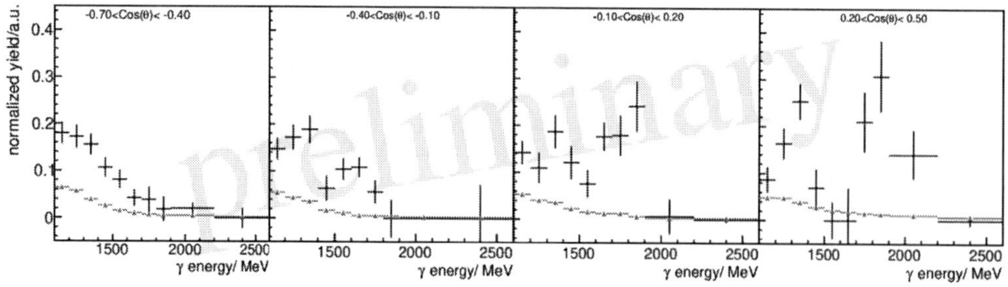

Figure 8. K^0 yield extracted from Roofit[8] (black points). Remaining $K^0\Lambda$ contribution, calculated from [9] (brown triangles). The error bars on the x-axis represent the bin width.

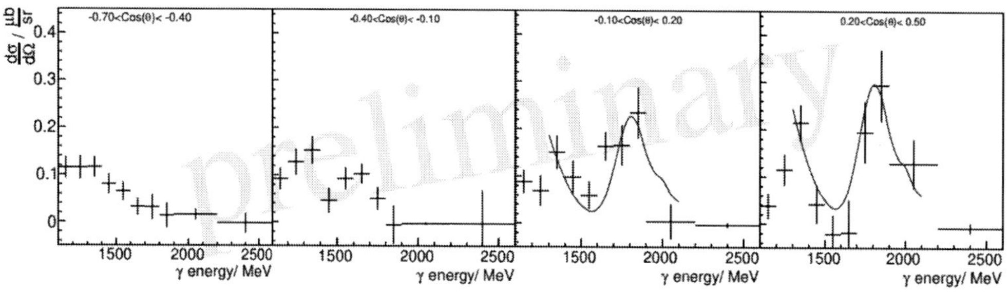

Figure 9. $\gamma n \rightarrow K^0\Sigma^0$ differential cross section (black points). Predicted enhancement by Oset and Ramos[6] in an arbitrary scale (red line). The error bars on the x-axis represent the bin width.

Recent results of charged pion and kaon photoproduction on the proton at SPring-8/LEPS

H. Kohri[1,2,*] for LEPS Collaboration

[1]Research Center for Nuclear Physics, Osaka University, Ibaraki, Osaka 567-0047, Japan
[2]Nagoya University, Furocho, Chikusa, Nagoya 464-8602, Japan

Abstract. We carry out photoproduction experiments by using linearly polarized photon beams with energies of 1.5-3 GeV at SPring-8/LEPS. One of our physics motivations is to understand reaction mechanisms of hadron photoproduction. We study (1) $\vec{\gamma}p \to \pi^-\Delta^{++}$, (2) $\vec{\gamma}p \to \pi^+n$, (3) $\vec{\gamma}p \to K^+\Lambda(K^+\Sigma^0)$, and (4) $\vec{\gamma}p \to \pi^+\Delta^0$ reactions which produce (1) $u\bar{u}$, (2) $d\bar{d}$, (3) $s\bar{s}$, and (4) $d\bar{d}$ quark-antiquark pairs in the final state. Differential cross sections and photon beam asymmetries have been measured at forward meson angles of 0.6< $\cos\theta_{cm}$ <1. Precise comparison between the $\pi^-\Delta^{++}$ and $\pi^+\Delta^0$ reactions plays an important role in the study of nucleon resonances. It enables to distinguish between N^* and Δ^* by taking the cross section ratios of the two reactions. We have measured preliminary cross section ratios $d\sigma(\pi^+\Delta^0)/d\sigma(\pi^-\Delta^{++})$. The ratios are found to be close to 1/3, which is expected from the exchange of isospin=1 in the t channel, for 0.966< $\cos\theta_{cm}$ <1 and all the photon energies. The ratios become larger than 1/3 and close to 1 for larger angles and higher energy regions. The larger ratios might be due to Δ resonances, giving the ratio 4/3, excited in the intermediate state.

1 Introduction

In quark models, there are more nucleon resonances than those experimentally observed so far [1, 2]. Since nucleon resonances have relatively wide widths and are overlapping each other, rich physics observables with a wide angular and energy range are needed to search for new resonances. One of our physics motivations is to understand reaction mechanisms of hadron photoproduction. Nucleon resonances excited in the intermediate states of the reactions are very important to understand the reaction mechanisms.

We constructed a beamline for studying hadron photoproduction at a few GeV at SPring-8 in 2000. Linearly or circularly polarized photon beams were produced by backward Compton scattering of deep-UV laser lights with a wavelength of 257 nm and 8 GeV electron beams [3]. Recoil electrons were tagged to determine energies of the photon beams (1.5-3 GeV). The photon polarization was 88% at 3 GeV and 28% at 1.5 GeV. The photon beams were incident on a liquid hydrogen target with an effective length of 16 cm. Charged particles produced at forward angles were momentum-analyzed by the LEPS spectrometer [4]. Particle identification was performed by the mass calculated from the momentum and Time-Of-Flight. The LEPS spectrometer was optimized to detect a K^+K^- pair from a ϕ-meson at forward angles [5–11]. We also extensively studied kaon photoproduction reactions [12–19]. We newly installed a plastic scintillation counter to veto e^+e^- produced by the pair creation and no aerogel cherenkov counter was used in order to detect high momentum charged pions [20].

*e-mail: kohri@rcnp.osaka-u.ac.jp

2 $u\bar{u}$, $d\bar{d}$, and $s\bar{s}$ productions

We precisely studied the (1) $\vec{\gamma}p \to \pi^-\Delta^{++}$ [21], (2) $\vec{\gamma}p \to \pi^+n$ [22], and (3) $\vec{\gamma}p \to K^+\Lambda(K^+\Sigma^0)$ [23] reactions. These reactions produce (1) $u\bar{u}$, (2) $d\bar{d}$, and (3) $s\bar{s}$ quark-antiquark pairs in the final states. Comparison among these reactions is of special importance to obtain a unified understanding of hadron photoproduction. Figure 1 shows the differential cross sections for the π^+n reaction [22].

The cross sections decrease monotonically as the photon beam energy increases for 0.6< $\cos\theta_{cm}$ <0.9. The energy dependence is different for 0.9< $\cos\theta_{cm}$ <1 and a small bump structure is observed for E_γ=1.5-2.2 GeV(W=1.9-2.2 GeV). The structure was also observed by DESY and was inferred to be due to nucleon resonances [24, 25]. In the nucleon resonance energies, CLAS measured the differential cross sections for wide angular regions and photon energies from 0.73 to 2.88 GeV [26]. Since the LEPS spectrometer is complementary to the CLAS spectrometer, the cross section data by LEPS are the first precise data for forward angles.

We also measured the differential cross sections for the $\gamma p \to \pi^-\Delta^{++}$ [21] and $K^+\Lambda(K^+\Sigma^0)$ [23] reactions. All the reactions have forward-peaking cross sections, which suggests the dominance of t-channel reaction mechanisms. The cross sections for the $\pi^-\Delta^{++}$ reaction by LEPS are the first high-statistics data. In the present data for the $K^+\Lambda(K^+\Sigma^0)$ reaction, the photon energy region was extended from 1.5-2.4 GeV [13] to 1.5-3 GeV [23].

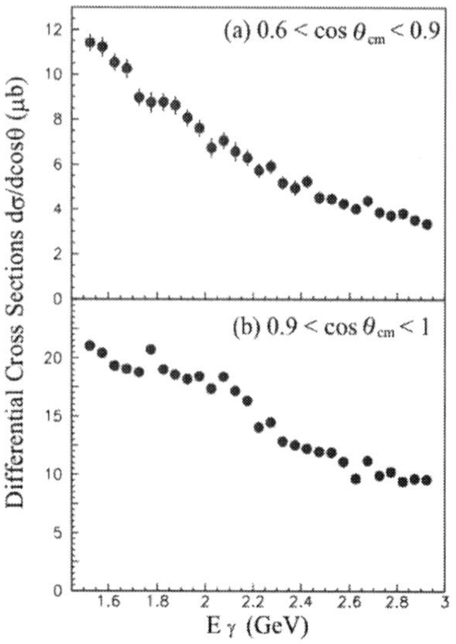

Figure 1. Differential cross sections for the $\gamma p \rightarrow \pi^+ n$ reaction for (a) 0.6< $\cos\theta_{cm}$ <0.9 and (b) 0.9< $\cos\theta_{cm}$ <1 [22].

The photon beam asymmetries obtained by using linearly polarized photon beams are sensitive to the reaction mechanisms. Figure 2 shows the asymmetries for the $\pi^+ n$ reaction for 0.6< $\cos\theta_{cm}$ <0.9 and 0.9< $\cos\theta_{cm}$ <1 [22]. The asymmetries are positive in this kinematical region, which is mostly explained by ρ-meson exchange in the t channel [27]. The asymmetries increase as the π^+ angle becomes smaller.

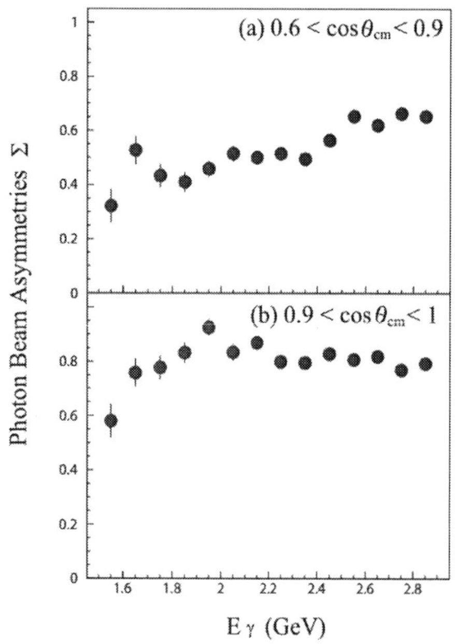

Figure 2. Photon beam asymmetries for the $\vec{\gamma} p \rightarrow \pi^+ n$ reaction for (a) 0.6< $\cos\theta_{cm}$ <0.9 and (b) 0.9< $\cos\theta_{cm}$ <1 [22].

The asymmetries for the $K^+\Lambda$ and $K^+\Sigma^0$ reactions [23] are also positive and smaller than those for the $\pi^+ n$ reaction. The positive asymmetries are mostly explained by K^*-exchange in the t channel [27]. Since the asymmetries gradually increase as the photon energy increases, the contribution from the K^*-exchange is inferred to become larger with increasing photon energy.

Figure 3 shows the asymmetries for the $\pi^-\Delta^{++}$ reaction [21]. It is interesting that only the $\pi^-\Delta^{++}$ reaction is found to have negative asymmetries in most of the present kinematical region. The negative asymmetries are explained by the dominance of π-exchange in the t channel [21]. In other pseudoscalar meson photoproduction reactions, such as $\vec{\gamma} p \rightarrow \pi^0 p$ [28, 29] and ηp [30, 31], positive asymmetries were observed at forward meson angles in this energy region. Pure $u\bar{u}$ quark-antiquark pair production might have different reaction mechanisms from the $d\bar{d}$ and $s\bar{s}$ productions. Negative asymmetries for the $\pi^-\Delta^{++}$ reaction are expected to be a key to obtain unified understanding of hadron photoproduction.

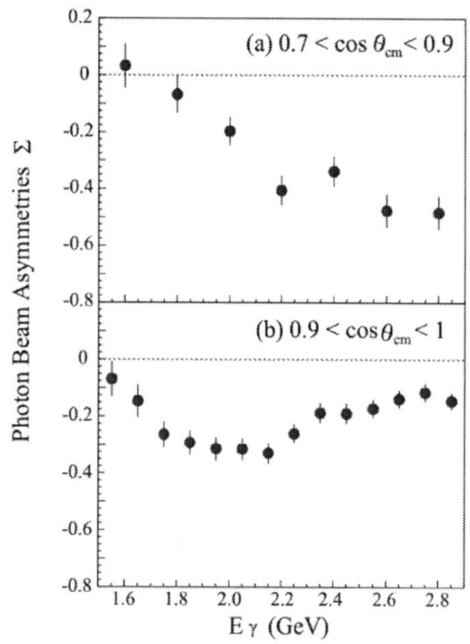

Figure 3. Photon beam asymmetries for the $\vec{\gamma} p \rightarrow \pi^-\Delta^{++}$ reaction for (a) 0.7< $\cos\theta_{cm}$ <0.9 and (b) 0.9< $\cos\theta_{cm}$ <1 [21].

3 Comparison between the $\gamma p \rightarrow \pi^-\Delta^{++}$ ($u\bar{u}$ production) and $\pi^+\Delta^0$ ($d\bar{d}$ production) reactions

The $\gamma p \rightarrow \pi^+\Delta^0$ reaction, which produces a $d\bar{d}$ pair in the final state, has an interesting relation to the $\pi^-\Delta^{++}$ reaction producing a $u\bar{u}$ pair. Clebsch-Gordan coefficients predict cross section ratios of the two reactions in the case of the excitation of isospin 1/2(N^*) and 3/2(Δ^*). If a nucleon resonance is excited in the intermediate state, the cross section ratio $d\sigma(\pi^+\Delta^0)/d\sigma(\pi^-\Delta^{++})$ becomes 1/3. In case a Δ resonance is excited, the cross section ratio becomes 4/3. In the energy regions relating to nucleon resonances, SAPHIR measured the total cross sections for both the reactions for photon energies from the threshold to 2.6 GeV [32]. Although SAPHIR also measured the differential cross sections for the $\pi^-\Delta^{++}$ reaction, those for the $\pi^+\Delta^0$ reaction were not measured because of low statistics.

Figure 4 shows the missing mass of the $\gamma p \to \pi^+ X$ and $\gamma p \to \pi^- X$ reactions. Since the LEPS spectrometer [4] has the same acceptance for positive and negative charged particles, precise comparison between the $\pi^+ X$ and $\pi^- X$ reactions is possible.

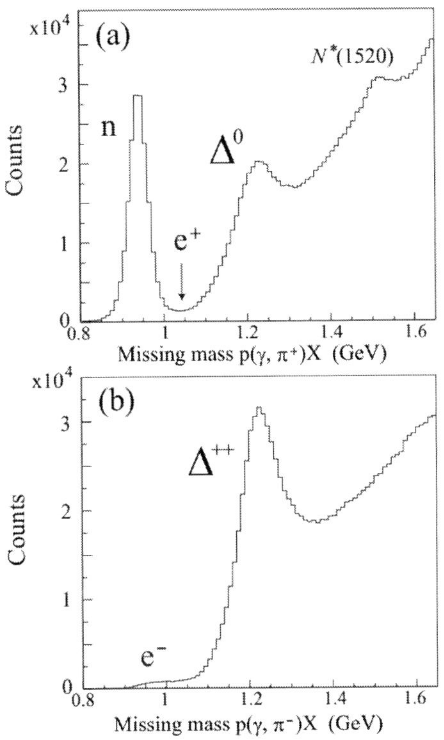

Figure 4. Missing mass of the (a) $\gamma p \to \pi^+ X$ and (b) $\gamma p \to \pi^- X$ reactions for 0.7< $\cos\theta_{cm}$ <1 and E_γ=1.5-2.95 GeV.

In the $\pi^+ X$ reaction, the neutron peak and $\Delta^0(1232)$ bump are clearly observed. The events between the peak and bump are due to positron mis-identification in the π^+ sample. The positron events contaminate the π^+ sample at very forward angles only. The yield of the $\pi^+\Delta^0$ reaction was obtained by fitting to the missing mass with a relativistic Breit-Wigner shape [33, 34] for the Δ^0 bump. The background under the bump was fitted with curves for 2π, 3π, ρ, and positron events which were reproduced by GEANT simulations [35]. The shapes of the curves were fixed and the heights of the curves were free parameters to fit to the background. The number of the $\pi^+\Delta^0$ events was about 190 k.

In the $\pi^- X$ reaction, the $\Delta^{++}(1232)$ bump is dominantly observed. The events around 1.0 GeV are due to mis-identification of electrons in the π^- sample. The electron events exist at very forward angles only. The yield of the $\pi^-\Delta^{++}$ reaction was obtained by fitting to the missing mass [21] with a similar method to the $\pi^+\Delta^0$ reaction. The number of the $\pi^-\Delta^{++}$ events was about 400 k which was roughly 2 times larger than that of the $\pi^+\Delta^0$ events.

Figure 5 shows preliminary cross section ratios $d\sigma(\pi^+\Delta^0)/d\sigma(\pi^-\Delta^{++})$ as a function of the photon beam energy. It is interesting that the ratios are found to be close to 1/3 for 0.966< $\cos\theta_{cm}$ <1 for all the energies. The ratio of 1/3 is expected from the exchange of isospin=1 in the t channel. The ρ-meson and π-meson are good candidates for the exchanged particle. This result means that the Clebsch-Gordan coefficients work excellently and our knowledge on the t-channel hadron photoproduction is correct in this kinematical region. The cross section ratios become larger than 1/3 for 0.9< $\cos\theta_{cm}$ <0.966 and E_γ >2.2 GeV. The ratios become close to 1 for 0.7< $\cos\theta_{cm}$ <0.9 and E_γ >2.4 GeV. One explanation for the ratios close to 1 might be a contribution from Δ resonances. If the Δ resonances are excited in the s channel, the ratios become 4/3. Another explanation is the exchange of isospin=2 exotic meson, which is expected to give a ratio of 2, in the t channel. Interference between the ρ-meson and π-meson exchanges might cause larger ratios. Advanced theoretical calculations introducing nucleon resonances and exotic meson exchanges are needed to understand the present results.

4 Summary

We have precisely studied photoproduction reactions which produce $u\bar{u}$, $d\bar{d}$, and $s\bar{s}$ quark-antiquark pairs for the first time. The differential cross sections for the $\gamma p \to \pi^-\Delta^{++}$, $\pi^+ n$, and $K^+\Lambda(K^+\Sigma^0)$ reactions are forward-peaking, which suggests the dominance of t-channel reaction mechanisms. The $\pi^+ n$ and $K^+\Lambda(K^+\Sigma^0)$ reactions have positive photon beam asymmetries obtained by using linearly polarized photon beams. The $\pi^-\Delta^{++}$ reaction is found to have negative asymmetries which are different from other pseudoscalar meson photoproduction reactions. We obtained preliminary differential cross sections for the $\gamma p \to \pi^+\Delta^0$ reaction producing a $d\bar{d}$ quark-antiquark pair. Precise comparison among these various quark-antiquark pair productions plays an important role in obtaining unified understanding of reaction mechanisms including nucleon and Δ resonances excited in the intermediate state.

We are developing a polarized HD target [36–39] for near future experiments at SPring-8/LEPS. The highest polarization for the proton target was 44±1% and the longest relaxation time was 8±2 months. Skills for the transportation of the polarized HD target from Osaka University to SPring-8 are needed before starting physics experiments. We are also developing a large acceptance spectrometer for charged particles at SPring-8/LEPS2. Introducing the polarized HD target to these experiments is

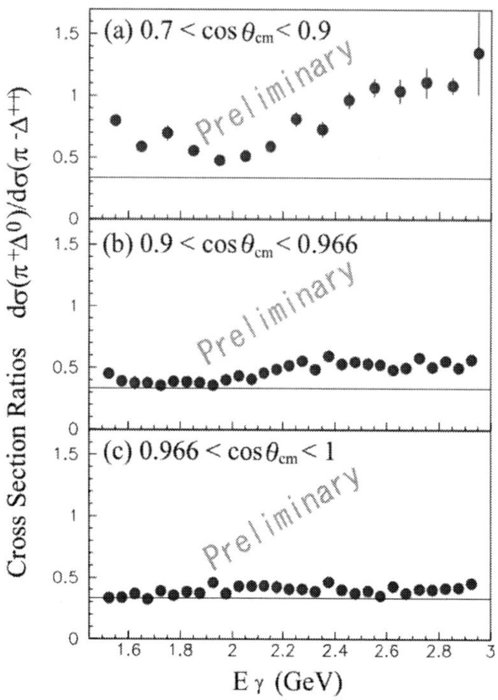

Figure 5. Preliminary cross section ratios $d\sigma(\pi^+\Delta^0)/d\sigma(\pi^-\Delta^{++})$ for (a) 0.7< $\cos\theta_{cm}$ <0.9, (b) 0.9< $\cos\theta_{cm}$ <0.966, and (c) 0.966< $\cos\theta_{cm}$ <1.

expected to advance the study of the nucleon and Δ resonances.

5 Acknowledgement

The authors gratefully acknowledge the staff of the SPring-8 facility for the support with excellent experimental conditions. The experiments were performed at the BL33LEP of SPring-8 with the approval of the Japan Synchrotron Radiation Research Institute (JASRI) as a contract beam line (Proposal No. BL33LEP/6001). This research was supported in part by the Ministry of Education, Science, Sports and Culture of Japan, the Ministry of Science and Technology of Taiwan, the National Research Foundation of Korea, and the U.S. National Science Foundation.

References

[1] Simon Capstick and Winston Roberts, Phys. Rev. D **49**, 4570 (1994)
[2] Simon Capstick and W. Roberts, Phys. Rev. D **58**, 074011 (1998)
[3] N. Muramatsu et al., Nucl. Instrum. Methods. Phys. Res. Sect. A **737**, 184 (2014)
[4] T. Nakano et al., Nucl. Phys. A **684**, 71 (2001)
[5] T. Mibe et al., Phys. Rev. Lett. **95**, 182001 (2005)
[6] T. Ishikawa et al., Phys. Lett. B **608**, 215 (2005)
[7] W. C. Chang et al., Phys. Lett. B **658**, 209 (2008)
[8] W. C. Chang et al., Phys. Rev. C **82**, 015205 (2010)
[9] W. C. Chang et al., Phys. Lett. B **684**, 6 (2010)
[10] S. Y. Ryu et al., Phys. Rev. Lett. **116**, 232001 (2016)
[11] W. C. Mizutani et al., Phys. Rev. C **96**, 062201(R) (2017)
[12] R. G. T. Zegers et al., Phys. Rev. Lett. **91**, 092001 (2003)
[13] M. Sumihama et al., Phys. Rev. C **73**, 035214 (2006)
[14] H. Kohri et al., Phys. Rev. Lett. **97**, 082003 (2006)
[15] M. Niiyama et al., Phys. Rev. C **78**, 035202 (2008)
[16] K. Hicks et al., Phys. Rev. Lett. **102**, 012501 (2009)
[17] N. Muramatsu et al., Phys. Rev. Lett. **103**, 012001 (2009)
[18] H. Kohri et al., Phys. Rev. Lett. **104**, 172001 (2010)
[19] A. O. Tokiyasu et al., Phys. Lett. B **728**, 616 (2014)
[20] S. H. Hwang et al., Phys. Rev. Lett. **108**, 092001 (2012)
[21] H. Kohri et al., Phys. Rev. Lett. **120**, 202004 (2018)
[22] H. Kohri et al., Phys. Rev. C **97**, 015205 (2018)
[23] S. H. Shiu et al., Phys. Rev. C **97**, 015208 (2018)
[24] G. Boschhorn et al., Phys. Rev. Lett. **17**, 1027 (1966)
[25] G. Boschhorn et al., Phys. Rev. Lett. **18**, 571 (1967)
[26] M. Dugger et al., Phys. Rev. C **79**, 065206 (2009)
[27] M. Guidal, J.-M. Laget, and M. Vanderhaeghen, Nucl. Phys. A **627**, 645 (1997)
[28] M. Dugger et al., Phys. Rev. C **88**, 065203 (2013)
[29] N. Sparks et al., Phys. Rev. C **81**, 065210 (2010)
[30] D. Elsner et al., Eur. Phys. J. A **33**, 147 (2007)
[31] P. Collins et al., Phys. Lett. B **771**, 213 (2017)
[32] C. Wu et al., Eur. Phys. J. A **23**, 317 (2005)
[33] J. D. Jackson, Nuovo Cimento **34**, 1644 (1964)
[34] A. M. Boyarski et al., Phys. Rev. Lett. **22**, 148 (1969)
[35] H. Kohri for LEPS collaboration, Few-Body Syst **59**, 88 (2018)
[36] H. Kohri, Int. Jour. of Mod. Phys. E **19**, 903 (2010)
[37] T. Ohta et al., Nucl. Instrum. Methods. Phys. Res. Sect. A **633**, 46 (2011)
[38] T. Ohta et al., Nucl. Instrum. Methods. Phys. Res. Sect. A **640**, 241 (2011)
[39] T. Ohta et al., Nucl. Instrum. Methods. Phys. Res. Sect. A **664**, 347 (2012)

KY electroproduction at CLAS12

Lucilla Lanza (for the CLAS Collaboration)[1,*]

[1]INFN, Istituto Nazionale di Fisica Nucleare, Sezione di Roma Tor Vergata

Abstract. An experimental program has been approved at the Thomas Jefferson National Accelerator Facility to measure the (ep,e'K^+)Y reactions using the CLAS12 setup in Hall B. Data have been obtained using a 7.5 GeV electron beam, impinging upon a liquid hydrogen target in the CLAS12 center. Scattered electrons have been detected in an angle range of 2.5° to 4.5° by the Forward Tagger (FT) and at angles greater than 5° in the CLAS12 Forward Detector, allowing to measure the KY electroproduction differential cross section and to probe the Q^2 evolution of the N* resonances electrocouplings in the Q^2 range from 0.05 GeV^2 to 3 GeV^2. The study of the Q^2 dependence of the electrocouplings will provide a crucial tool to investigate the possible hybrid nature of the N* resonances. Preliminary results from CLAS12 data are compared with simulated data using a realistic Gent Regge plus resonance event generator.

1 Introduction

Besides regular mesons and baryons a number of "exotic" hadrons have been conjectured, states having more valence quarks than the three quark configuration for baryons or the quark-antiquark one for mesons, namely tetraquarks and pentaquarks, or states having valence gluons in their structure, *hybrid mesons* and *hybrid baryons*.
Hybrid baryons are hypothetical three quark states with gluonic degrees of freedom. Their existence is allowed by Quantum Chromo Dynamics (QCD), and Lattice QCD (LQCD) calculations predict the existence of several hybrid states 1.3 GeV above the nucleon ground state [1].
The search for hybrid baryons is more difficult than for hybrid mesons because of issues in distinguishing a hybrid from a regular baryon. In fact, differently from hybrid mesons, hybrid and regular baryons can have the same quantum numbers. A useful approach has been recently employed in the study of the Roper resonance N(1440)$1/2^+$, that has been considered as a candidate for the lowest mass hybrid baryon [2]. To dismiss this possibility [3] [4] [5], meson electroproduction data were exploited, studying the photon virtuality (Q^2) evolution of the electrocouplings, that should present a different low Q^2 behavior for hybrids with respect to regular states. The ongoing program at the Thomas Jefferson National Accelerator Facility (Jefferson Laboratory) in Newport News (VA) employs photoproduction and electroproduction reactions to study the excitation of nucleons in the nucleon resonance mass range up to 3 GeV and proved to be successful; moreover a program concerning the search for hybrid baryons is in progress.
The task is to perform an extensive study of $K^+\Lambda$, $K^+\Sigma$ and $p\pi^+\pi^-$ electroproduction reactions, in the kinematic region that covers the mass range up to 3 GeV and in the Q^2 range from 0.05 GeV^2 to 2.0 GeV^2 with CLAS12 experimental setup. Once the existence of a new baryonic resonance is assessed, it will be possible to establish whether it is a hybrid or not by studying the Q^2 evolution of its electrocouplings. Data taking has started during Fall 2018 and the first set of data is at present under analysis.[1]

2 Theoretical background

First Lattice QCD predictions regarding hybrids were formulated in 2003 [6]. These calculations used the quenched approximation, i.e. the quantum loops of fermions in Feynman diagrams were neglected. The lowest hybrid baryon mass was predicted around 2 GeV, 1 GeV above the nucleon mass. In 2012 the first unquenched LQCD calculations were proposed [1]. Figure 1 shows the extracted spectrum of Nucleon and Delta states in the lowest mass range. The N* and Δ* denote excited states of the Nucleon and Delta, respectively. Grey boxes are regular states and blue boxes are states identified as hybrids. The corresponding negative parity hybrid states are predicted to be heavier.
The first group of hybrid baryons is expected to lie in a restricted energy band of ~ 300 MeV not occupied by ordinary baryons, 1.3 GeV above the nucleon ground state, with $J^P = 1/2^+$. If their width is not too broad it should be possible to observe them experimentally.
Besides the mass range prediction that has just been described, the signature to distinguish a hybrid baryon from an ordinary one arises from the study of their electromagnetic couplings which is described in section 2.1.

*e-mail: lucilla.lanza@roma2.infn.it

[1]The Hybrid Baryons program has been approved on July 2016 by Jefferson Laboratory Program Advisory Committee (PAC) 44.

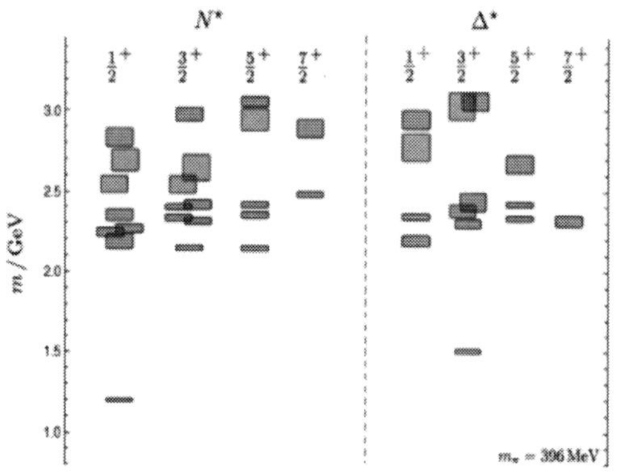

Figure 1. The light-quark baryon spectrum predicted in Lattice QCD at a pion mass of 396 MeV [1]. Grey boxes are conventional qqq states and the blue shaded boxes indicate hybrid baryons states. Note that both the mass of the nucleon ground state and of the $\Delta(1232)$ are shifted by nearly 300 MeV to higher masses.

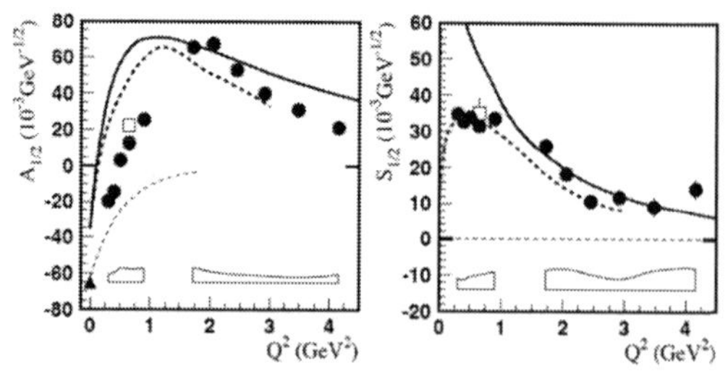

Figure 2. Helicity amplitudes for the $\gamma^* p \to N(1440)P_{11}$ transition. The full circles are the results obtained in [4] from CLAS data. The bands show the model uncertainties. The open boxes are the results of the combined analysis of CLAS single π and 2π electroproduction data [7]. The full triangle at $Q^2 = 0$ is the Review of Particle Physics (RPP) estimate [8]. The thick curves correspond to the results obtained in the LF relativistic quark models assuming that N(1440)P_{11} is a first radial excitation of the 3-quark ground state: [9] (dashed), [10] (solid). The thin dashed curves are obtained assuming that N(1440)P_{11} is a gluonic baryon excitation (q^3G hybrid state) [2].

2.1 Electromagnetic Couplings

Hybrid and regular baryons are expected to have the same quantum numbers, but the information from the electrocouplings can be exploited to identify their nature. In fact a distinctively different Q^2 evolution of the hybrid baryon electrocouplings is expected considering the different color-multiplet assignments for the quark-core in a regular versus a hybrid baryon, i.e. a color singlet versus octet. Theoretical models [2] predict a peculiar behavior of hybrid baryons electrocouplings at low Q^2: a fast drop of the transverse helicity amplitude $A_{1/2}(Q^2)$ and a suppression of the longitudinal helicity amplitude $S_{1/2}(Q^2)$.
Information from electrocouplings evolution with Q^2 has been exploited to solve the Roper puzzle, see [4] from which figure 2 is extracted. In this figure the full circles are the results obtained in [4] from CLAS data, the bands show the model uncertainties, the open boxes are the results of the combined analysis of CLAS single π and 2π electroproduction data [7], the full triangle at $Q^2 = 0$ is the Review of Particle Physics (RPP) estimate [8], the thick curves correspond to the results obtained in the light-front relativistic quark models (LF RQM) assuming that N(1440)P_{11} is a first radial excitation of the 3-quark ground state: [9] (dashed), [10] (solid). Finally the thin dashed curves are obtained assuming that N(1440)P_{11} is a hybrid baryon [2]. It is evident for both amplitudes that data points are inconsistent with the hybrid predictions, especially in the longitudinal helicity amplitude case, where in the hybrid case a suppression was expected and large values occur instead.

2.2 Hadronic Couplings

Unfortunately there is a lack of information regarding possible hadronic couplings of hybrid baryons. From the knowledge of the hadronic couplings one could select the best channel for an experimental study, but without theoretical information any experimental approach must rely on a set of assumptions. Final states with significant gluonic admixture, e.g. $B_G \to N\eta'$ [11] (the G subscript denotes a gluon component), should be preferential as well as final states containing $s\bar{s}$ contributions due to the $B_G \to s\bar{s}$ coupling, e.g. $B_G \to K^+\Lambda$, $B_G \to N^*(1535)\pi \to N\eta\pi$, $B_G \to \phi(1020)N$ and $B_G \to K^*\Lambda$, or final states to which the missing resonances are predicted to mainly decay, such as $B_G \to N\pi\pi$. Among all possible candidate final states, the K^+Y and the $N\pi\pi$ have been selected to study the sensitivity of the electroproduction reaction to the formation of a hybrid resonance, since they are the easiest channel among the listed ones for which a theoretical description is available. This work is focused on the $K^+\Lambda$ channel.

2.3 Signature for Hybrid Baryon States

Summarizing our understanding, based on available knowledge, we may state that the signature for the lower mass hybrid baryons may consist of:

- Resonances not predicted by conventional constituent quark models with I = $\frac{1}{2}$, $J^P = \frac{1}{2}^+$ and $J^P = \frac{3}{2}^+$, with masses from 1.8 GeV to 2.5 GeV decaying into $N\pi\pi$, $K^+\Lambda$ or $K^+\Sigma^0$ final states;
- Drop of the transverse helicity amplitudes $A_{1/2}(Q^2)$ and $A_{3/2}(Q^2)$ as a function of Q^2 faster than for ordinary three quark states, because of the extra glue-component in the valence structure;
- Suppressed longitudinal amplitude $S_{1/2}(Q^2)$ in comparison to transverse electro-excitation amplitudes.

3 CLAS12 and Forward Tagger at Jefferson Laboratory

The experiment is located at Jefferson Laboratory, in Hall B, with CLAS12 experimental setup [12].
The CLAS12 detector, shown in figure 3, has been used to take data with longitudinally polarized high energy electron beams scattering on a liquid-hydrogen target, with an operating luminosity of $L = 10^{35}cm^{-2}s^{-1}$ which corresponds to a total hadronic rate of 5×10^6 s^{-1}.
The electron beam has been provided by the CEBAF accelerator with three values of energy in different runs: 6.5 GeV, 7.5 GeV and 10.6 GeV.
The CLAS12 detector is highly hermetic, provides high

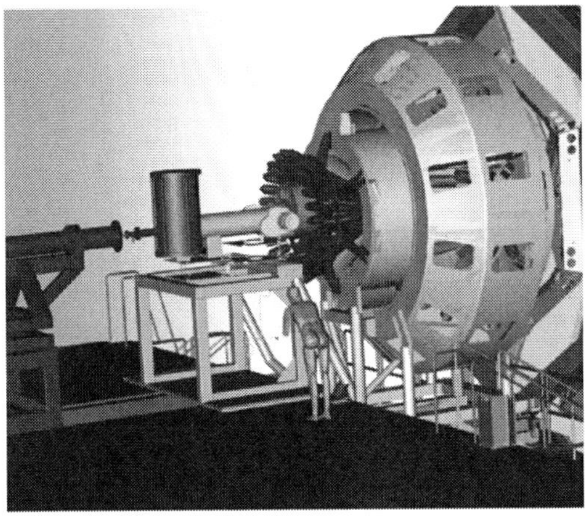

Figure 3. Sketch of the CLAS12 detector [12]: it consists of a central detector (CD) equipped with a superconducting solenoid and a forward detector (FD) equipped with a torus magnet, to cover polar angular ranges from 35° to 125° and from 5° to 35°, respectively.

multiplicity reconstruction, can run at high luminosity and allows to carry out the experiment under study. CLAS12 provides charged particle identification by correlating the particle speed β, measured by time of flight detectors, with momentum, the curvature of charged particle tracks in the high magnetic field. Cherenkov counters are used for electron-pion separation. CLAS12 consists of a Forward Detector, covering the forward polar angle range between 5° and 35°, and a Central Detector, covering larger angles up to 125°. The forward angular region is equipped with a torus magnet, while the central polar angle range of 35° to 125°. The forward angular region is equipped with a torus magnet, while the central polar angle range of 35° to 125° is provided with a 5T solenoid magnet.

3.1 Forward Detector

The Forward Detector is the part of the CLAS12 detector devoted to the measurement and identification of charged and neutral particles scattered between 5 and 35 degrees, in the full azimuthal angle. It consists of six identical magnetic spectrometers working in conjunction with a toroidal field, generated by a six coil superconducting magnet. Each spectrometer, corresponding to a sector of the Forward Detector, is composed of a high threshold Cherenkov counter, a set of drift chambers, a low threshold Cherenkov counter, scintillation counters and an electromagnetic calorimeter. The working principle of the Forward Detector is the following: charged particles are detected from their trajectory due to the torus magnet magnetic field and their momentum is reconstructed using the drift chambers. The Cherenkov counters and the time-of-flight detectors provide information useful for the discrimination among particles and the electromagnetic calorimeter gives the information about the energy. The design momentum resolution is $\frac{\sigma_E}{E} \simeq \frac{0.1}{\sqrt{E}}$ and the design parameters were chosen to allow for pions-kaons and for protons-kaons separation up to 3 GeV/c and up to 4 GeV/c, respectively.

3.2 Central Detector

The Central Detector is located around the target to detect particles emitted at low momentum, < 1.5 GeV/c, with large polar angles, between 35 and 125 degrees, in the full azimuthal range. It is composed of a 5T superconducting solenoid magnet, a Silicon Vertex Tracker, that surrounds the target, a time-of-flight detector and a Central Neutron Detector. Particle momentum is measured using information from the trackers within the strong solenoid magnetic field. Neutral particles may be detected by the Central Neutron Detector.

3.3 Forward Tagger

To extend the detection of scattered electrons to the smallest polar angles, CLAS12 includes a Forward Tagger (FT), shown in figure 4; it allows to detect electrons and photons emitted between 2.5 and 4.5 degrees. The design energy resolution is expected to be $2\%/\sqrt{E(\text{GeV})} \oplus 1\%$.
The FT is made of three components:

- Electromagnetic Calorimeter (FT-Cal), to measure the energy of the electromagnetic showers produced by electrons and photons and to provide a fast trigger signal;
- Hodoscope (FT-Hodo), to discriminate between electrons and photons that enter in the calorimeter;
- Tracker (FT-Trck), to measure the polar θ_e and azimuthal ϕ_e scattering angles of the electrons.

The electron beam interacts with the target via the exchange of a virtual photon, whose four momentum q = $(\nu, \mathbf{q})$ is the difference between the beam and the scattered electron momenta. It is possible to express the photon virtuality Q^2 as a function of the electron scattering angle θ_e and of the energy of the final electron $E_{e'}$:

$$Q^2 = 4E_eE_{e'}sin^2\frac{\theta_e}{2}. \quad (1)$$

http://doi.org/10.1051/epjconf/202024101011

Events for which the electron scattering occurs at very small angles, such as the ones detected by FT, also may have a low Q^2 (the electroproduction at this Q^2 can be considered for practical purposes photoproduction). By using electron beams of variable incoming energies E_e from 6.5 up to 10.6 GeV and detecting the scattered electron in the angular range from 2.5° to 135° it is possible to span Q^2 virtualities between 0.05 GeV^2 and 5 GeV^2. The lowest Q^2 range is reachable using the lowest beam energy and detecting scattered electrons in the FT.
Considering the importance of events at low Q^2, the FT is therefore a crucial component for this study.

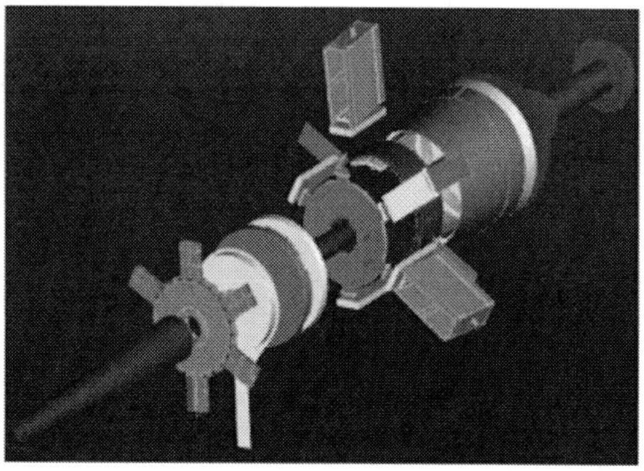

Figure 4. The Forward Tagger system. The FT provides electron and high energy photon detection in a range of polar angles θ = 2.5° − 4.5° and is fully integrated into the operation of CLAS12.

4 Simulation and Reconstruction of $K^+\Lambda$ Electroproduction Events in CLAS12

To determine the best running conditions for the experiment, simulations have been exploited using an Event Generator based on the Gent RPR-2011 Model [13] to produce electroproduction events and a Monte Carlo to simulate CLAS12 acceptance effects.

4.1 Regge-Plus-Resonance 2011 Model

The model employed to describe the KY electroproduction channel is the "Regge-Plus-Resonance" developed by the Gent group. It describes the reaction amplitude as the superposition of tree-level s-channel nucleon resonances on the non-resonant contribution. A set of 11 nucleon resonances has been considered to determine the optimum resonant contribution in the RPR p(γ, K^+)Λ framework. The model with the highest Bayesian evidence, called RPR-2011, includes the resonances S_{11} (1535), S_{11} (1650), F_{15} (1680), P_{13} (1720), P_{11} (1900), P_{13} (1900), D_{13} (1900) and F_{15} (2000). The non-resonant component is the sum of t-channel exchanges of Reggeized K^+ (494) and K^{*+} (892).

4.2 Selection of the Best Running Conditions

The best running conditions, in terms of beam energy and magnetic field strength, have been determined exploiting simulations.
In order to choose between a positive and a negative value for the torus current, events have been simulated with both the configurations and from the comparison it is possible to notice that the negative one, which corresponds to outbending electrons, provides a smaller acceptance gap between the regions covered by the FT and by CLAS12.
Additionally, by varying the intensity of the torus current in the range between -1500 A and -3375 A, it has been verified that the highest energy resolution obtained with the highest magnetic current configuration provides the best separation between the $K^+\Lambda$ and the $K^+\Sigma$ processes.
Finally, to select the best energy for the electron beam the efficiency curve (ratio between the number of detected events and the generated ones as a function of Q^2) has been studied for three different configurations, E_{beam} = 6.5 GeV, E_{beam} = 7.5 GeV and E_{beam} = 10.6 GeV. In all three cases the efficiency curve presents a drop between the regions covered by the Forward Tagger apparatus and CLAS12 detector, occurring in correspondence of different Q^2 ranges. Therefore the information from events taken with the three different beam energies allows to access complementary Q^2 ranges.

4.3 CLAS12 Acceptance Effects

Simulations performed employed a fast Monte Carlo program FASTMC that has been developed to estimate CLAS12 acceptance and event production rates. This code contains geometrical acceptances and parametrized resolution functions for the tracking systems, the electromagnetic calorimeters, the scintillation counters and the Forward Tagger. It also uses realistic field maps for the CLAS12 torus and solenoid, that can be scaled to study the effect of different magnet currents and polarities. The sensitive volumes of the tracking and detection regions have been designed to closely match the CLAS12 detector design. The high efficiency fiducial areas of the detector have been taken into account in the event reconstruction, which also provides a smearing of the final state charged particle momenta.
A full GEANT-based Monte Carlo and reconstruction software code has been developed in the meanwhile for CLAS12, called CLARA. Comparisons of the CLAS12 response using the FASTMC code suite and the full CLAS12 GEANT-4 code suite have shown reasonably good correspondence.

5 Preliminary Results from Physics Runs

Physics Runs for CLAS12 at Jefferson Laboratory started in February 2018. The first dedicated Run took data during Fall 2018 for the study of KY channels. Figure 5 shows the kaon missing mass spectrum, MM(e^-K^+), for 15 runs using a 7.5 GeV energy beam on a 5 cm long liquid-hydrogen target. This amounts to roughly 10% of

 http://doi.org/10.1051/epjconf/202024101011

the dedicated Run Fall 2018 7.5 GeV data set. The electron is detected in the Forward Electromagnetic Calorimeter and the K^+ is detected in the Forward Detector.
Peaks are visible for $\Lambda(1116)$, $\Sigma(1193)$, $\Sigma(1385)$, $\Lambda(1405)$ and $\Lambda(1520)$. In particular it is possible to distinguish the two peaks of $\Lambda(1116)$ and $\Sigma(1193)$. Events with the electron detected by Forward Tagger are at present under analysis, using the same mass range: while the missing mass resolution is expected to improve after momentum corrections and detector alignment optimization are applied, it is already sufficient to identify the reactions events.

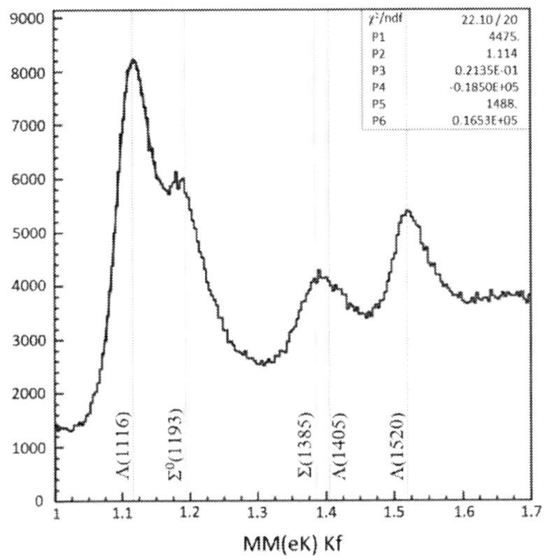

Figure 5. $MM(e^-K^+)$(GeV) plot using data from 15 Runs acquired during Fall 2018 requiring a K^+ detected in the Forward Detector. The vertical lines on the plot represent the nominal masses of the Λ and Σ hyperons. The blue line is the fit on the Λ^0 peak and the parameters are in the top right box. $\Lambda^0(1116)$ and $\Sigma^0(1193)$ baryons peaks can be distinguished. Courtesy of D. Carman.

6 Conclusions

This contribution reports preliminary results and analysis for KY electroproduction channels available from a subset of data collected by CLAS12 detector during Fall 2018 at Thomas Jefferson National Accelerator Facility. $MM(e^-K^+)$ plots obtained for events with the electron detected by CLAS12 and the K^+ detected in the Forward Detector present a clear separation between Λ and Σ^0 peaks. Next step will be to exploit new calibrated data that will be available soon. These data will be subjected to a full amplitude partial wave analysis, to study the resonance contribution to the KY electroproduction reactions.

References

[1] J. J. Dudek and R. G. Edwards, Phys Rev. D **85**, 054016 (2012)
[2] Z. P. Li and V. Burkert and Z. J. Li, Phys. Rev. D **46**, 70 (1992)
[3] I. G. Aznauryan et al. (CLAS Collaboration), Phys. Rev. C **78**, 045209 (2008)
[4] I. G. Aznauryan et al. (CLAS Collaboration), Phys. Rev. C **80**, 055203 (2009)
[5] I. G. Aznauryan and V. D. Burkert, Prog. Part. Nucl. Phys. **1**, 1-54 (2012)
[6] T. T. Takahashi and H. Suganuma, Phys. Rev. Lett. **90**, 182001 (2003)
[7] I. G. Aznauryan and V. D. Burkert et al., Phys. Rev. C **72**, 045201 (2005)
[8] Amsler et al. (Particle Data Group), Phys. Lett. B **667**, 1-6 (2008)
[9] S. Capstick and B. D. Keister, Phys. Rev. D **51**, 3598 (1995)
[10] I. G. Aznauryan, Phys. Rev. C **76**, 025212 (2007)
[11] E. Kou, Phys. Rev. D **63**, 054027 (2001)
[12] CLAS Collaboration, Tech. rep. Jefferson Laboratory (2008)
[13] L. De Cruz and J. Ryckebusch and T. Vrancx and P. Vancraeyveld, Phys. Rev. C **86**, 015212 (2012)

Studying time-like electromagnetic baryonic transitions with HADES in pion induced reactions

Béatrice Ramstein[1,*]
for the HADES collaboration

[1]Institut de Physique Nucléaire d'Orsay, CNRS-IN2P3, Univ. Paris Sud, Univ. Paris-Saclay, 91406 Orsay Cedex, France.

Abstract. A dedicated programme aiming at studying electromagnetic baryonic transitions in the time-like region has started at GSI using the pion beam and the High Acceptance Di-Electron Spectrometer (HADES) set-up. A pioneering experiment has been carried out in the second resonance region, at a center-of-mass energy of 1.49 GeV, using carbon and polyethylene targets, allowing for an analysis of the inclusive e^+e^- production and of the exclusive quasi-free $\pi^- p \to ne^+e^-$ reaction, in complement to hadronic channels. Predictions for the Dalitz decay of N(1520) and N(1535) based on real photon couplings strongly underestimate the e^+e^- yield at invariant masses larger than 300 MeV/c^2, which signals effects of time-like baryon transition form factors, in qualitative agreement with the Vector Dominance Models (VDM). A quantitative description of the observed e^+e^- yield is achieved by taking into account the contribution from off-shell ρs. The latter was provided by the preliminary results of a Bonn-Gatchina Partial Wave Analysis of the two-pion production channels, which were measured simultaneously in our experiment. A good agreement is also obtained using a covariant model for the time-like electromagnetic form factors for the N-N(1520) transition. The angular distributions for the e^+e^- production contain additional selective information on the spin structure of the different transitions. The measurements with the pion beam will be extended in future in the third resonance region. Hyperon Dalitz decay studies in proton induced reactions are also foreseen.

1 Introduction

Dilepton production is a favored probe for QCD matter studies. In particular, the emission of low mass dileptons gives insight on the properties of vector mesons in the hadronic matter. It allows to search for in-medium modifications of the ρ mesons [1], as predicted by hadronic many-body calculations [1] due to the dressing of the pions and the direct coupling of the ρ meson to baryons. These spectral functions are also expected to be affected by QCD chiral symmetry restoration in hot and dense matter. The main objective of experiments with the High Acceptance Di-Electron Spectrometer (HADES) [2] at GSI is the study of hadronic matter in the low temperature (T< 80 MeV) and large density (up to 3 times the normal nuclear matter density) region [3]. An experimental program using *pp*, "quasifree" *np* and more recently πN reactions is also developped, in order to provide reference spectra needed for the interpretation of medium effects and to isolate specific processes using exclusive channels, notably the baryon resonance Dalitz decays ($B \to Ne^+e^-$). An overview of HADES results is given in P. Salabura's report at this conference [4] and preliminary results concerning the hadronic channels measured in $\pi^- p$ collisions in the second resonance region can be found in the contribution of I. Ciepał to this conference [5]. This contribution is devoted to the study of electromagnetic channels measured with the pion beam in the same energy range.

2 Dalitz decay studies with HADES

Baryon resonance Dalitz decays ($B \to Ne^+e^-$) give insight into the electromagnetic structure of baryonic transitions for squared momentum transfer $q^2= M^2_{ee}$ in the time-like region ($q^2> 0$), in the $[4m_e^2, (M_B-M_N)^2]$ range, where M_{ee} is the e^+e^- invariant mass, m_e, M_B and M_N are the electron, baryon resonance and nucleon masses, respectively. As illustrated in figure. 1, this kinematical region is complementary to the space-like region (q^2 <0) probed in electron scattering, in particular with the CLAS detector at JLab [6]. The two regions are connected at the "photon point" (q^2=0), where information on the N-γ couplings is provided by meson photoproduction experiments. For larger positive q^2, the influence of the ρ and ω vector mesons is expected to play a significant role. Dalitz decay processes are therefore particularly relevant to investigate the applicability of Vector Dominance Model (VDM) in electromagnetic baryon transitions. HADES measurements in proton-proton collisions at an incident energy of 1.25 GeV provided the first determination of the Δ(1232) Dalitz decay branching ratio [7, 8] in the channel $pp \to ppe^+e^-$. At higher incident proton energies, the e^+e^- invariant mass spectra clearly demonstrate the effect of electromagnetic form factors of VDM type [9–11], which means that the

*e-mail: ramstein@ipno.in2p3.fr

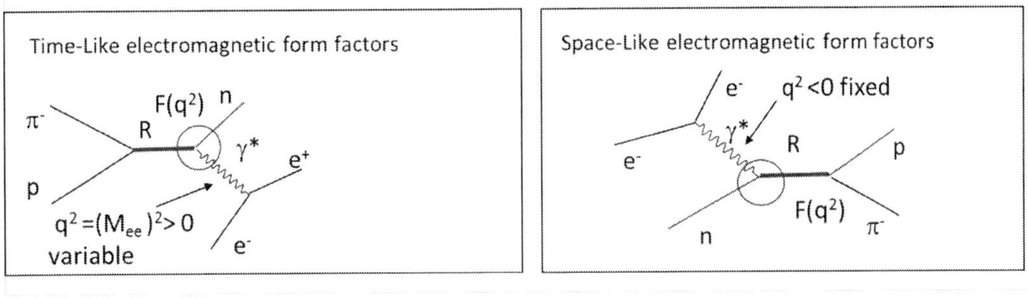

Figure 1. Left: Sketch of the Dalitz decay process of a resonance produced in the s-channel in a $\pi^- p$ reaction. The four-momentum transfer q^2 is positive and takes values between $4m_e^2$ and $(M_B - M_N)^2$, where M_{ee} is the e^+e^- invariant mass, m_e, M_B and M_N are the electron, baryon resonance and nucleon masses, respectively. Right: Excitation of a nucleon by electron scattering. The four-momentum transfer q^2 is negative and fixed by the energy and angle of the scattered electron.

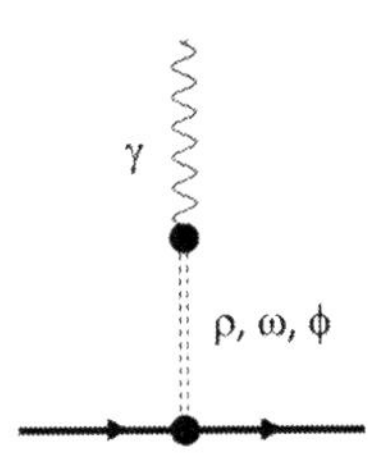

Figure 2. Sketch of the Vector Dominance Model. The electromagnetic interaction is mediated by vector mesons.

electromagnetic interaction is mediated by vector mesons as illustrated in figure 2. As discussed in [4], the results obtained in pp and pn reactions have been crucial to provide a reference for the study of off-shell ρ propagation in cold as well as hot and dense matter.

The extension of the program to experiments with the GSI pion beam offers a more direct way to investigate Dalitz decays. In πN reactions the baryon resonances are indeed produced with a fixed mass in the s channel ($M_B = \sqrt{s}$), so there is less overlap between several states and the πN interaction is better controlled. In addition, the exclusive channel π^-p→ ne^+e^- can be studied by selecting e^+e^- pairs in a missing mass range around the neutron mass, while in the case of the pp reaction, the request of a detected proton significantly reduces the detection acceptance in HADES. These reactions are in fact the reverse reaction from the pion electroproduction reaction, which is one of the experiments used to extract baryon transition form factors in the space-like region (see figure 1).

3 The pion beam line and the Hades detector

High Acceptance Di-Electron Spectrometer (HADES) is a versatile set-up [2] installed at SIS18 in GSI, Darmstadt. It provides excellent detection both for hadron and rare di-electron production, thanks to the high acceptance (30% for e^+e^- pairs in the vector meson region), the powerful particle ($p/K/\pi/e$) identification and the very good mass resolution (2 - 3% for $\omega \to e^+e^-$ decays). Pion beam experiments were since the very beginning part of the HADES scientific program and commissioning runs took place already in 1999 [12, 13]. To check the pion beam properties, a test experiment using the HADES detector was performed in 2014. The secondary pion beam is produced using an extracted primary ^{14}N beam provided by the SIS18 synchrotron with an intensity of (0.8 - 1.0)× 10^{11} ions/spill, close to the space-charge limit of the machine impinging on a 10 cm long beryllium target. The pions in a selected momentum range are transported through the beam line consisting of nine quadrupole and two dipole magnets from the production point up to the HADES target located 33 m downstream. The pion yield at the exit of the pion beam line reaches a maximum of about 10^6 pions/spill at p=1.0 GeV/c and is about twice smaller at p=0.7 GeV/c and 1.3 GeV/c.

Two silicon strip detectors along the beam line and a start detector right in front of HADES were used to measure the position of the beam particles. Beam transport calculations have been performed to optimize the transmission of the beam line and reconstruct event-by-event the pion momentum with a precision better than 0.3% (σ).

4 Data analysis

Data at four different pion beam momenta (0.654, 0.686, 0.738 and 0.787 GeV/c) in the second resonance region were collected on both polyethylene ($(C_2H_4)_n$) and carbon targets. The largest statistics was measured in the case of 0.69 GeV/c on the polyethylene target and this data set was used for dilepton production studies. As described in detail in [5], the precise normalization of the measurements is provided by the known π^-p elastic scattering differential cross sections [14], after subtraction of the carbon contribution, using a least-square tagging procedure, based on the comparison of measurements with the polyethylene and carbon targets. The data measured for the exclusive $\pi^- p \to \pi^+\pi^- n$ and $\pi^- p \to \pi^-\pi^0 p$ reactions have

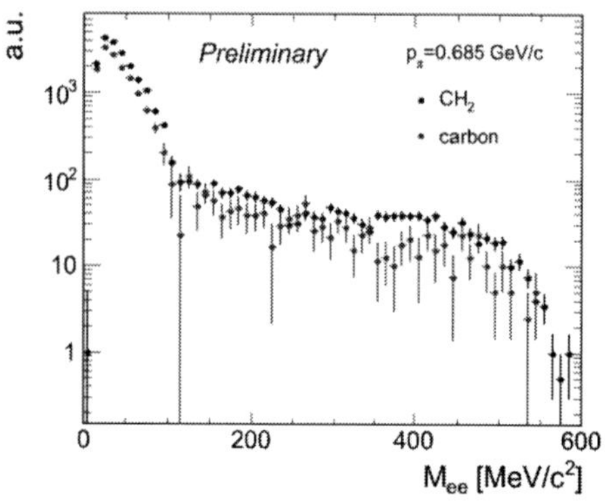

Figure 3. e^+e^- invariant mass distributions measured in the interaction of pions at p=0.685 GeV/c on polyethylene (black dots) and carbon (red dots) after normalization to the same incident flux.

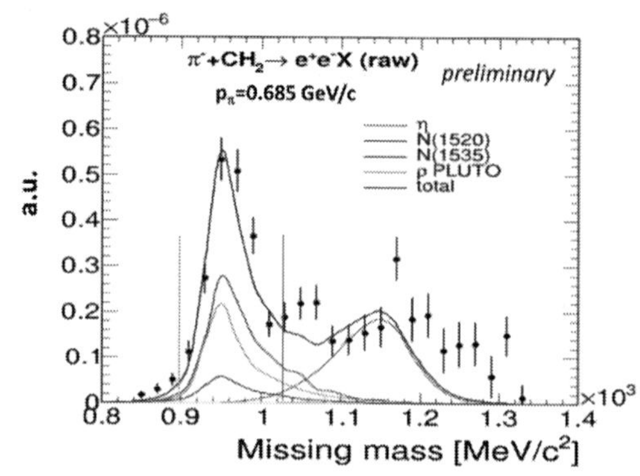

Figure 4. Distribution of the missing mass of e^+e^- events with invariant mass larger than 120 MeV/c^2. The results of the simulation (total shown as a blue curve) including the η (pink), N(1520) Dalitz decay (red), N(1535) Dalitz decay (dark green), ρ (light green) are compared to the simulation showing the different components.

then been included in the multichannel Partial Wave Analysis (PWA) developed by the Bonn-Gatchina group [15]. In this way, the different contributions to the total amplitude, such as $\Delta\pi$, ρN, σN can be extracted together with their couplings to the main contributing baryon resonances (N(1520), N(1535), N(1440),...). The ρN contribution deduced from this analysis is used as an input for the description of the dielectron channels, as described in sec. 5. For the e^+e^- channel, the subtraction of events produced by the interaction of the pion beam with carbon nuclei in the polyethylene target was not feasible, due to the too small statistics recorded on the carbon target. Specific analysis cuts were applied for the reconstruction of correlated e^+e^- pairs, in order to suppress tracks arising from photon conversion and to avoid their contribution to the combinatorial background [16]. The combinatorial background is then estimated using the like-sign pair yields measured in the same event. A threshold of 9 ° applied to the opening angle between the two leptons allowed for the complete rejection of correlated pairs from the photon conversion.

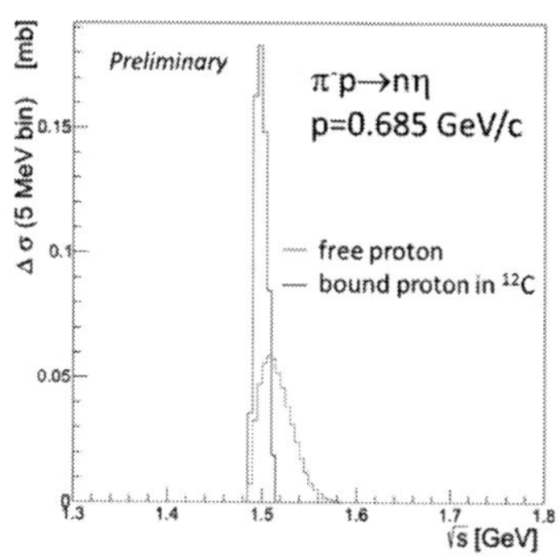

Figure 5. Distribution of cross sections for the $\pi^- p \to \eta n$ reaction in 5 MeV wide bins in $\sqrt{s}$ (center-of-mass energy) for the interaction with a free proton (blue) and a bound proton in the carbon target (red).

5 e^+e^- invariant mass and missing mass studies

The e^+e^- invariant mass distributions are displayed in figure 3 for both the carbon (red dots) and the polyethylene (black dots) targets after normalization to the same incident yields. The contribution due to interactions with the carbon nuclei is on average 70% for invariant masses below 300 MeV/c^2 and 50% above.

For e^+e^- invariant masses above the π^0 mass, the π^0 Dalitz decay does not contribute and the missing mass distribution clearly shows two structures (figure 4). The peak around the neutron mass can be attributed to the exclusive π^-p→ ne^+e^- reaction, either on a free proton or a bound proton in the carbon nuclei, in a quasi-free process. The contribution at larger missing masses is expected to be mainly due to the η Dalitz decay, where a photon is emitted in addition to the e^+e^- pair. The exclusive free or quasi-free π^-p→ ne^+e^- channel can then be selected by applying invariant mass (M_{ee} > 0.140 GeV/c^2) and missing mass (0.9 < MM < 1.03 GeV/c^2) cuts, as shown on figure 4. The number of selected events amounts to about 1500.

To describe the e^+e^- production in π^-+ $(C_2H_4)_n$ reactions, we developed simulations with the Pluto [17] event generator based on the available information for pion-nucleon interactions. For the π^0 production cross sections in the channels $\pi^-p \to \pi^0 n$, $\pi^-p \to \pi^0\pi^0 n$ and $\pi^-p \to \pi^0\pi^- p$, we use information from former data and from the SAID Partial Wave Analysis [14, 18–22]. The rapid variation of the η production cross section close to threshold has also been measured [23]. An essential ingredient of the simulation of the π^-p reactions is the pion beam momentum distribution, which is known from the pion beam

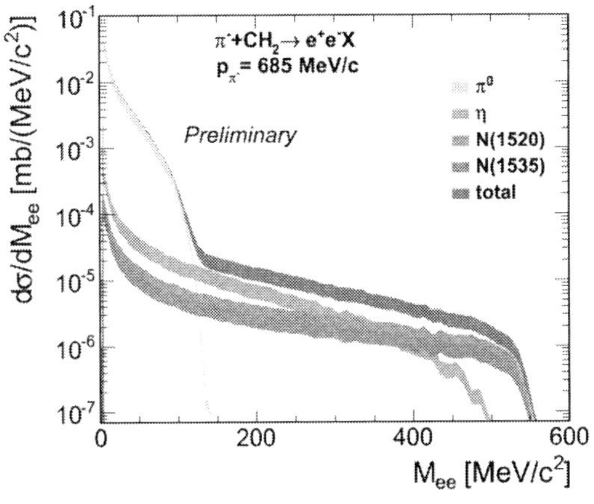

Figure 6. Differential cross sections as a function of the e^+e^- invariant mass for the different contributions used as inputs for the simulations of the π^-p→ ne^+e^-X reaction at an incident pion momentum of 0.685 GeV/c.

momentum reconstruction deduced from measured hits in the pion beam tracker using the optic parameters of the pion beam line. This distribution is centered at a value (p=0.686 GeV/c) very close to the η production threshold and can be approximated by a gaussian with a width of 1.7%. The η yield resulting from the folding of the η production excitation function and the beam momentum distribution is displayed in figure 5. The e^+e^- invariant mass distributions for the meson Dalitz decays are then deduced from a QED calculation using VDM form factors [24]. In addition to the Dalitz decay of the π^0 and η mesons, we can expect production of e^+e^- via the Dalitz decays of baryons. It is known from the PWA analysis of the two pion channels that the dominant resonant contributions in the s channel are due to N(1520) and N(1535) [5]. Starting from the QED calculation of [25], we found that the widths for the decay of the N(1520) (J^P=3/2$^-$) and N(1535) (J^P=1/2$^-$) into a nucleon and a virtual photon of mass M_{ee} can be deduced from the corresponding radiative decay widths by the same relation:

$$\Gamma^{N^*\to N\gamma^*}(q^2) = \frac{\sigma_+^3\sigma_-}{m_+^3 m_-}\frac{|G_T(q^2)|^2}{|G_T(0)|^2}\times\Gamma^{N^*\to N\gamma} \quad (1)$$

$$\text{where } q^2 = M_{ee}^2, \quad (2)$$

$$m_\pm = m_\star \pm m_N, \quad (3)$$

$$\sigma_\pm^2 = m_\pm^2 - M_{ee}^2, \quad (4)$$

$m_\star$ is the resonance mass, m_N is the nucleon mass and $G_T(q^2)$ is an effective form factor, which can be expressed as a function of the electric, magnetic and Coulomb form factors. The Dalitz decay differential decay width can then be calculated as a function of the e^+e^- invariant mass as

$$\frac{d\Gamma^{N^*\to Ne^+e^-}}{dM_{ee}} = \frac{2\alpha}{3\pi M_{ee}}\Gamma_{M_{N^*}}^{N^*\to N\gamma^*}(M_{ee}), \quad (5)$$

Assuming no dependence of the form factor on the four momentum transfer (point-like assumption), the differential Dalitz decay width of N(1520) and N(1535) can then be derived directly from the radiative decay width using Eq.1. To deduce the coresponding e^+e^- differential cross sections, we used the cross sections for the s-channel N(1520) and N(1535) resonant contributions in the $\pi^-p \to n\gamma$ reaction, which were deduced, using detailed balance, from the Bonn-Gatchina PWA of the $\gamma n \to \pi^-p$ channel. We also implemented the Dalitz decay of the $\Delta(1232)$, using the $\Delta\pi$ cross sections extracted from the PWA of the two pion production channels in the π^-p reaction at our energy. The partial decay width was calculated using a constant form factor (G_M=3, G_E=0, G_C=0) consistent with the radiative decay width [25].

Interactions with carbon nuclei were treated in a quasi-free approach, using the participant-spectator model and taking into account the nucleon momentum distribution in carbon. The available energy in the π^-p system, and hence the distribution of η yields is quite sensitive to this assumption (figure 5). For the different contributions, the cross sections for interaction of π^- beam particles with carbon nuclei (σ_C) were renormalized from the free cross sections ($s\sigma_H$) taking into account the ratio between yields measured respectively in the $(C_2H_4)_n$ and C targets. The total experimental and simulated yields are normalized as $\sigma_H + 0.5\sigma_C$ to take into account the different number of carbon and hydrogen atoms in polyethylene. The contributions of the different dilepton channels are then summed up incoherently. We consider two types of uncertainties to this dilepton cocktail. The first one is related to the knowledge of the free cross sections and the second one is due to the uncertainty of the carbon contribution. The cross section inputs for the simulation and the branching ratios to the e^+e^- emission are summarized in table 1. Figure 6 shows the results of the simulation for the full solid angle, with uncertainties on each component displayed as colored bands. The spectrum is strongly dominated by the π^0 Dalitz decay for e^+e^- invariant masses below 120 MeV/c^2, then the dominant contribution is the η Dalitz, up to invariant masses of about 280 MeV/c^2. For larger invariant masses, the Dalitz decay of N(1520) and N(1535) produced in the s-channel is found to be dominant. The $\Delta(1232)$ Dalitz decay contribution, which is about one order of magnitude lower than the N(1520) contribution and has a much steeper decrease as a function of e^+e^- invariant mass is not represented.

This cocktail can be considered as a reference for meson and baryon Dalitz decays in a point-like approach. It provides a reasonable description of the mesured dielectron yields for invariant masses below 300 MeV/c^2, where the π^0 and η Dalitz decay contributions dominate. However, it significantly underestimates the yield at higher invariant masses, where the baryon contribution is expected to take over. This deviation from a point-like behavior is even more striking in the exclusive π^-p → ne^+e^- channel, which clearly shows that time-like electromagnetic form factors need to be implemented. This was done in two different ways.

channel	σ(mb)			Branching ratios
	σ_H	σ_C	$\sigma_H + 0.5 \times \sigma_C$	
π^0n	9.0 ± 0.8	32.4 ± 2.9 ± 8.1	25.2 ± 2.3 ± 4.1	$\pi^0 \rightarrow \gamma\, e^+e^-$ 1.2 %
$\pi^0\pi^0$n	1.9 ± 0.1	6.8 ± 0.4 ± 1.7	5.3 ± 0.3 ± 0.85	
$\pi^-\pi^0$p	4.0 ± 0.5	14.4 ± 1.8 ± 3.6	11.2 ± 1.4 ± 1.8	
η n	0.63 ± 0.2	2.3 ± 0.1 ± 0.8	1.8 ± 0.25 ± 0.4	$\eta \rightarrow \gamma\, e^+e^-$ 6.9 10^{-3}
Δ^+ ±	0.78 ± 0.08	2.9 ± 0.3 ± 1.0	2.2 ± 0.23 ± 0.5	$\Delta(1232) \rightarrow N\, e^+e^-$4.2 10^{-5}
$\Delta^0\, \pi^0$	1. ± 0.1	3.7 ± 0.4 ± 1.3	2.9 ± 0.3 ± 0.65	
N(1520)	20.4 ± 2	75.5 ± 7 ± 27	58 ± 5.5 ± 13.5	$N(1520)^0 \rightarrow n\, e^+e^-$ 2.3 10^{-5}
N(1535)	9.7 ± 1	35.9 ± 3.7 ± 5.0	28 ± 2.9 ± 2.5	$N(1535)^0 \rightarrow n\, e^+e^-$ 3.5 10^{-5}

Table 1. Cross sections for the different channels implemented in the simulation for interactions of π^- with incident momentum of 0.685 GeV/c with proton, carbon and polyethylene targets. For the carbon and polyethylene targets, the first uncertainty is related to the elementary cross sections used in the participant-spectator approach, as given in the first column and the second to the ratio of carbon to proton cross section. The last column gives the branching ratio to the e^+e^- channel.

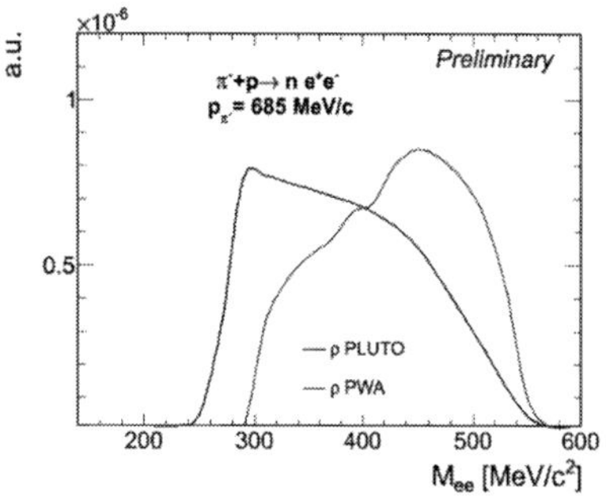

Figure 7. Simulations of the e^+e^- invariant mass distribution of the ρ decay contribution to the π^-p$\rightarrow$ ne^+e^- at 0.685 GeV/c with mass distributions from a Breit-Wigner (blue curve, labeled ρ PLUTO) or deduced from the PWA of hadronic channels (red curve, labeled ρ PWA).

The first approach was based on the VDM and took advantage of the simultaneous measurement of pionic channels. We used the off-shell ρ contribution from the PWA in the $\pi^- p$ reaction at an incident momentum of 0.69 GeV/c and converted it into a dilepton contribution, using VDM. It has to be noted that the spectral shape of this off-shell ρ contribution is quite different from the one expected for the decay of a ρ with a Breit-Wigner mass distribution (compare blue and red curves in figure 7). This is due to the coupling to the N(1520) which, according to the preliminary results of the PWA, plays a major role in the off-shell ρ production. Implementing the ρ contribution in the simulation results in a large dilepton yield above 300 MeV/c^2, which improves significantly the description of the data. As shown in figure 4, the simulation including this ρ contribution is able to reproduce the yield for the exclusive π^-p$\rightarrow$ ne^+e^- reaction. According to the PWA, the dominant resonant contribution to the off-shell ρ production is the N(1520) resonance. By this procedure, we demonstrate experimentally the validity of the VDM approach to take into account the hadronic structure of the electromagnetic transitions in the region of the N(1520). We can also notice that the yield for missing masses above 1.1 GeV/c^2, where the η Dalitz decay dominates is also fairly well reproduced, which confirms the consistency of our participant-spectator model.

In the second approach, we used the model of [26] which was developed for the time-like electromagnetic transition form factors of the N-N(1520) transition. In this model, baryons are described with a quark core and a meson cloud, with parameters fitted to the existing data for the N-N(1520) transition in the space-like region. The predictions of the model for the effective form factor $G_T(q^2)$ have been injected using Eq. 1 in our Dalitz decay simulation. The meson cloud contribution is strongly dominant in the time-like region and leads to an enhancement of the dilepton yield for invariant masses larger than 300 MeV/c^2. A detailed comparison with the data is in progress.

6 Angular distributions

Further information, in particular on the transverse and longitudinal virtual photon polarization can be obtained from the angular distributions. A convenient parametrization of the amplitudes $|A|$ at a given value of M_{ee} and emission angle of the virtual photon derives from the density matrix formalism [28]

$$|A|^2 \propto 4k^2[2\rho_{00}(1-\cos^2\theta) + 2\rho_{11}(1+\cos^2\theta) \\ +2\sqrt{2}\sin(2\theta)\cos\phi Re\rho_{10} \quad (6) \\ +2\sin^2\theta Re\rho_{1-1}\cos(2\phi)]. \quad (7)$$

Here, k^2, θ and ϕ denote the momentum, polar angle and azimuthal angles of the lepton in the virtual photon reference frame, respectively and ρ_{00}, ρ_{11}, ρ_{1-1} are the three independent density matrix coefficients which can be expressed as a function of the electromagnetic form factors or helicity amplitudes. This relation depends on the spin and parity of the transition. Nevertheless, the connection between ρ_{11} and the transverse $A_\perp$ and longitudinal $A_\parallel$ amplitudes can be simply written as:

$$\rho_{11} = \frac{A_\perp}{2A_\perp + A_\parallel} \quad (8)$$

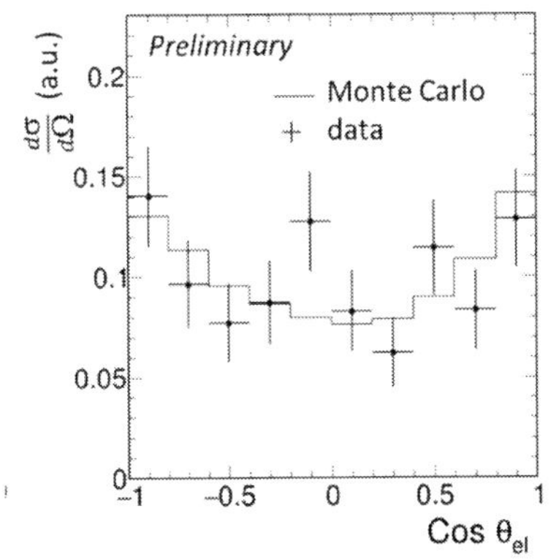

Figure 8. The exclusive $\pi^- p \to ne^+e^-$ reaction is selected using missing mass. The distribution of the lepton angle θ_{el} in the γ^* reference frame is shown for γ^* angles in the center-of-mass in the bin $0 < cos(\theta_{\gamma^*}^{CM}) < 0.5$. The red curve represents the result of the fit weighted by acceptance and efficiency [27].

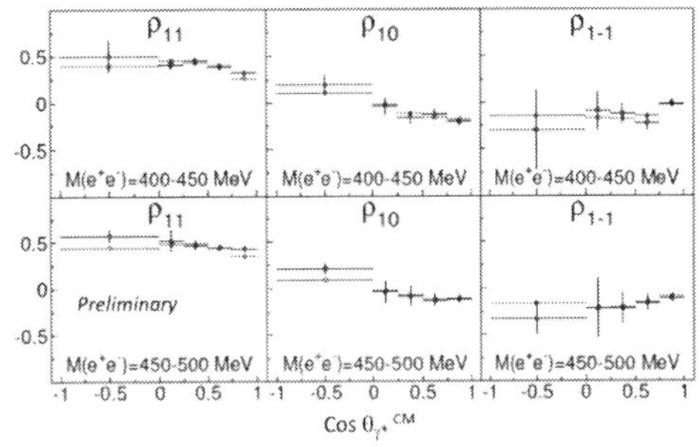

Figure 9. Expected precision on the spin density matrix elements ρ_{11}, ρ_{10}, ρ_{1-1} for 10^4 events as a function of $z = \cos\theta_{CM}$ and for two bins in dilepton invariant mass M_{ee}. The red points show results obtained in the full solid angle and the black points correspond to events measured within the HADES acceptance.

A method to extract spin density matrix elements from a fit to experimental data has been developed based on the maximum likelihood [29]. The coefficients can be extracted as weights of a Monte-Carlo simulation which takes into account the acceptance and efficiency of the detector, as illustrated in figure 8. Despite the low statistics which affects the precision of the data, the coefficients were studied in three bins of angle of the virtual photon, providing significant information on the spin structure of the transitions. For example, preliminary results indicate that, in some bins ρ_{11} is different from 0.5 and ρ_{10} is different from zero, which clearly demonstrates the contributions of virtual photons with longitudinal polarization, in contrast to real photons.

Moreover, the density matrix formalism can also be applied to analyze the angular distributions of pions from the off-shell ρ component of the PWA solution. The coefficients obtained from the two pion analysis should be consistent with the ones obtained from the dilepton distributions for $M_{ee} > 300$ MeV if the dominant channel is indeed the ρ meson decay. The studies are on-going and the results are expected soon. The density matrix formalism also allows for an easy comparison with model predictions. As explained in details in D. Nitt's contribution to this conference [30], we used the Lagrangian model of [28], which takes into account VDM form factors for the various baryonic transitions.

7 Project for future pion beam experiments

These results from a commissioning experiment provide very first and promising information about the baryon electromagnetic transitions in the time-like region. In the near future, an experiment is planned to explore the third resonance region ($\sqrt{s}$ = 1.7 GeV), where several baryonic resonances contribute (Δ(1620), Δ(1700), N(1650), N(1675), N(1630), N(1710), N(1720),...)[31]. Dielectron, strange and non strange meson production in $\pi^- p$ and $\pi^- C$ reactions will be measured. In this higher energy range, the sensitivity of electromagnetic form factors to VDM will be enhanced, since q^2 values closer to the Vector Meson poles can be reached, which will be very valuable e.g. to test predictions of $\rho - \omega$ interferences [32, 33]. Our goal is to reach a statistics for the signal 20 times larger than in the previous commissioning experiment, to allow for an extraction of the global longitudinal (related to the Coulomb form factor) and transverse (related to the electric and magnetic form factors) amplitudes for two bins in invariant mass and three bins in virtual photon angles (see figure 9). The measurements of various exclusive hadronic channels ($\pi^0\pi^-$n, $\pi^+\pi^-$p, ωn, ρn, ηn, KΛ, KΣ) in an energy scan are also foreseen. The objective is to increase by at least one of order of magnitude the existing data base for these hadronic decay channels. In this way, many poorly known hadronic couplings in the third resonance region could be determined with a precision at the percent level. Many open issues related to baryon structure also exist in this mass region (e.g. N(1710), Δ(1700)) and are in the focus of new meson beam facilities [34]. In relation to the general program of the HADES collaboration, the emphasis for future measurements will be put on improving the knowledge on ρ-N, ω-N and K-Λ couplings, which are needed as input for the modeling of the propagation of ρ mesons and strange particles in matter. The measurement of channels with neutral mesons (π^0, η) or real photons will be possible thanks to the new Electromagnetic Calorimeter, which has recently been put in operation. A higher electron efficiency is provided by the new RICH photon detector, developed in the context of the HADES-CBM FAIR Phase-0 program and which was also recently used successfully by the HADES collaboration in a measurement of Ag+Ag collisions. The acceptance will also be improved at the most forward angles (0.6-6.5°) thanks to a new detection system based in particular on straw tube trackers in the context of the HADES/PANDA FAIR Phase-0 program, which is currently under construction.

We also await improvements of the extraction efficiency of the primary ^{14}N or proton beams which could result in a significant increase of the pion beam flux.

Measurements on a carbon target, which are needed to extract the proton contribution in the polyethylene target, will also be used to study e^+e^- and meson production in π^-C reactions. The objective is to pin down possible cold matter effects on the ρ and ω spectral functions, as well as to quantify ω absorption.

Further in the future, the very poorly known fourth resonance region could be explored in experiments with a pion beam at $p_\pi \sim 1.3$ GeV/c. This kinematical region is well suited to provide accurate measurements of the $\Lambda(1405)$ which can be analyzed both in the ΣN neutral and charged decay modes.

8 Future studies of hyperon electromagnetic transitions

The HADES collaboration also plans to investigate the poorly known electromagnetic structure of hyperons. While elastic or transition form factors can be measured at large q^2 in e^+e^- annihilation reactions [35], the study of Dalitz decay of hyperons (Y'$\rightarrow$Ye^+e^-) is unique to investigate the low q^2 region. One open question which makes studies in this kinematical region very attractive is the relevance of VDM in the hyperon sector [36–38]. Due to the narrow width of most hyperons, they can be reconstructed with the invariant mass of their decay products and one can therefore take advantage of their abundant production in pp reactions (a few hundred μb at p$\geq$5 GeV/c). HADES can study hadronic decays, as well as radiative and Dalitz decays of hyperons. These measurements will also benefit from the on-going upgrade of the HADES set-up. In particular, the above-mentioned increase of the acceptance will be very beneficial for the detection of forward emitted protons. The measurement of radiative decays will also be possible with the new Electromagnetic Calorimeter. The foreseen increase of the acquisition rate (up to 200 kHz) is also needed to record enough statistics for the rare Dalitz decays. According to the dedicated simulations which have been performed for the $\Lambda(1520) \rightarrow \Lambda$ e^+e^- Dalitz decay, a signal rate of 550 Λe^+e^- events per day after all analysis cuts is expected. The comparison of the N(1520) and Λ(1520) Dalitz decays both measured with HADES will bring completely new information about the sensitivity of baryon electromagnetic structure to the strange quark content at finite q^2 in the time-like region. This program will be started at the present SIS18 facility and continued at higher energies at SIS100 (FAIR), in a complementary way to the studies foreseen at PANDA.

9 Conclusion

In complement to dense matter studies, the HADES collaboration developed an important program in elementary reactions to shed light on the role of off-shell ρ mesons as a mediator of the electromagnetic interaction. Recently, a new step was taken with an exploratory experiment using the GSI pion beam in the N(1520) region. It demonstrated the interest of pion beam experiments to perform direct studies of baryon Dalitz decays and to get insight into the completely unknown time-like electromagnetic structure of baryonic transitions at q^2 below the vector meson poles. The measurement of both e^+e^- and $\pi\pi$ channels allows for detailed checks of Vector Meson Dominance. Such studies will be pursued in the near future with the pion beam at GSI for heavier resonances. In addition, a program focussing on electromagnetic transitions between hyperons will be started at the present SIS18 facility. In further future, the HADES detector will exploit the proton and heavy ion beams of the SIS100 at the FAIR facility and pursue its complementary hadronic physics and dense matter studies at higher energies.

10 Acknowledgements

SIP JUC Cracow, Cracow (Poland), National Science Center, 2016/23/P/ST2/040 POLONEZ, 2017/25/N/ST2/00580, 2017/26/M/ST2/00600; TU Darmstadt, Darmstadt (Germany), VH-NG-823, DFG GRK 2128, DFG CRC-TR 211, BMBF:05P18RDFC1; Goethe-University, Frankfurt (Germany) and TU Darmstadt, Darmstadt (Germany), ExtreMe Matter Institute EMMI at GSI Darmstadt; TU München, Garching (Germany), MLL München, DFG EClust 153, GSI TMLRG1316F, BmBF 05P15WOFCA, SFB 1258, DFG FAB898/2-2; NRNU MEPhI Moscow, Moscow (Russia), in framework of Russian Academic Excellence Project 02.a03.21.0005, Ministry of Science and Education of the Russian Federation 3.3380.2017/4.6; JLU Giessen, Giessen (Germany), BMBF:05P12RGGHM; IPN Orsay, Orsay Cedex (France), CNRS/IN2P3; NPI CAS, Rez, Rez (Czech Republic), MSMT LM2015049, OP VVV CZ.02.1.01/0.0/0.0/16 013/0001677, LTT17003.

References

[1] R. Rapp, J. Wambach, Eur. Phys. J. **A6**, 415 (1999), hep-ph/9907502

[2] G. Agakichiev et al. (HADES), Eur. Phys. J. **A41**, 243 (2009), 0902.3478

[3] J. Adamczewski-Musch et al. (HADES), Nature Phys. **15**, 1040 (2019)

[4] P. Salabura *et al.* (HADES), *Exploring time like transitions in pp, π^-p and heavy ion reactions with HADES*, NSTAR2019

[5] I. Ciepal *et al.* (HADES), *Partial wave analysis of hades data for two-pion production in pion- nucleon reactions*, NSTAR2019

[6] V. Burkert, *N* Experiments and their Impact on Strong QCD*, NSTAR2019

[7] J. Adamczewski-Musch et al. (HADES), Phys. Rev. **C95**, 065205 (2017), 1703.07840

[8] M. Tanabashi et al. (Particle Data Group), Phys. Rev. D **98**, 030001 (2018)

[9] G. Agakishiev et al. (HADES), Phys. Rev. **C85**, 054005 (2012), 1203.2549

[10] G. Agakishiev et al. (HADES), Eur. Phys. J. **A48**, 64 (2012), 1112.3607
[11] G.. Agakishiev et al. (HADES), Eur. Phys. J. **A50**, 82 (2014), 1403.3054
[12] R. Simon et al., Prog. Part. Nucl. Phys. **247**, 42 (1999)
[13] J. Diaz et al., Nucl. Instrum. Meth. **A478**, 511 (2002)
[14] *CNS Data Analysis Center, SAID Home Page*, http://gwdac.phys.gwu.edu/
[15] A. Anisovich et al., Eur. Phys. J. **A24**, 111 (2005), hep-ph/0407211
[16] F. Scozzi (HADES), EPJ Web Conf. **137**, 05023 (2017)
[17] I. Fröhlich, F. Dohrmann, T. Galatyuk, R. Holzmann, P. Kahlitz et al., Eur. Phys. J. **A45**, 401 (2010), 0909.5373
[18] F. Bulos et al., Phys. Rev. Lett. **13**, 558 (1964)
[19] A. Starostin et al., Phys. Rev. **C72**, 015205 (2005)
[20] R.A. Burnstein, G.R. Charlton, T.B. Day, G. Quareni, A. Quareni-Vignudelli, G.B. Yodh, I. Nadelhaft, Phys. Rev. **137**, B1044 (1965)
[21] A.D. Brody et al., Phys. Rev. **D4**, 2693 (1971), [Erratum: Phys. Rev.D5,2899(1972)]
[22] S. Prakhov et al. (Crystal Ball Collaboration), Phys.Rev. **C69**, 045202 (2004)
[23] S. Prakhov et al., Phys. Rev. **C72**, 015203 (2005)
[24] I. Fröhlich et al., PoS **ACAT2007**, 076 (2007), 0708.2382
[25] M.I. Krivoruchenko, A. Faessler, Phys. Rev. **D65**, 017502 (2002)
[26] G. Ramalho et al., Phys. Rev. **D93**, 033004 (2016), 1512.03764
[27] B. Ramstein et al. (HADES), EPJ Web Conf. **199**, 01008 (2019)
[28] E. Speranza, M. Zetenyi, B. Friman, Phys. Lett. **B764**, 282 (2017), 1605.04954
[29] V.V. Sarantsev, private communication
[30] D. Nitt, *Dilepton production and anisotropy in pion-nucleon collisions*, NSTAR2019
[31] *Proposal for experiments at SIS18 during FAIR Phase0, Properties of hadron resonances and baryon rich matter*, (HADES Collab.), GSI 2017, http://repository.gsi.de/record/220071/files/HADES
[32] A.I. Titov, B. Kämpfer, Eur. Phys. J. **A12**, 217 (2001), nucl-th/0108019
[33] M. Lutz, B. Friman, M. Soyeur, Nuclear Physics A **713**, 97 (2003.)
[34] W.J. Briscoe et al., Eur. Phys. J. **A51**, 129 (2015), 1503.07763
[35] K. Schonning, *Hyperon physics with PANDA at FAIR*, NSTAR2019
[36] R.A. Williams, C.R. Ji, S.R. Cotanch, Phys. Rev. **C48**, 1318 (1993)
[37] G. Ramalho, K. Tsushima, Phys. Rev. **D88**, 053002 (2013), 1307.6840
[38] C. Granados, S. Leupold, E. Perotti, Eur. Phys. J. **A53**, 117 (2017), 1701.09130

Exploring time like tranistions in pp, πp and AA reactions with HADES

Piotr Salabura for the HADES collaboration[3,*]
and J. Adamczewski-Musch[4], B. Arnoldi-Meadows[8], A. Belounnas[16], A. Belyaev[7],A. Blanco[1], C. Blume[8], M. Böhmer[10], S Borisenko[13], L. Chlad[17], P. Chudoba[17], I. Ciepał[2], D. Dittert[5], J. Dreyer[6], W. Esmail[12], M. Dürr[11], L. Fabbietti[10,9], O. Fateev[7], P. Fonte[1,a], J. Friese[10], I. Fröhlich[8], J. Förtsch[19], T. Galatyuk[5,4], R. Gernhäuser[10], O. Golosov[15], M. Golubeva[13], R. Greifenhagen[6,b], F. Guber[13], M. Gumberidze[4], S. Harabasz[5,3], R. Hensch[8], C. Höhne[11,4], R. Holzmann[4], H. Huck[8], A. Ierusalimov[7], A. Ivashkin[13], B. Kämpfer[6,b], K-H. Kampert[19], B. Kardan[8], I. Koenig[4], W. Koenig[4], G. Kornakov[5], F. Kornas[5], R. Kotte[6], A. Kozela[2], I. Kres[19], J. Kuboś[2], A. Kugler[17], P. Kulesa[1]2, V. Ladygin[7], R. Lalik[3], A. Lebedev[14], S. Lebedev[11,7], S. Linev[4], L. Lopes[1], M. Lorenz[8], G. Lykasov[7], T. Mahmoud[11], A. Malige[3], J. Markert[4], T. Matulewicz[18], S. Maurus[10], V. Metag[11], J. Michel[8], S. Morozov[13,15], C. Müntz[8], L. Naumann[6], K. Nowakowski[3], J.-H. Otto[11], V. Patel[19], C. Pauly[19], V. Pechenov[4], O. Pechenova[4], O. Petukhov[13], D. Pfeifer[19], K. Piasecki[18], J. Pietraszko[4], A. Prozorov[17], W. Przygoda[3], K. Pysz[2], B. Ramstein[16], N. Rathod[3], A. Reshetin[13], L. Ritman[12], P. Rodriguez-Ramos[17], A. Rost[5], P. Salabura[3], F. Scozzi[5,16], F. Seck[5], I. Selyuzhenkov[4,15], I Shabanov[13], US Singh[3], J. Smyrski[3], S. Spies[8], H. Ströbele[8], J. Stroth[8,4], J. Stumm[8], O. Svoboda[17], M. Szala[8], P. Tlusty[17], M. Traxler[4], C. Ungethüm[5], O. Vazquez-Doce[10,9], V. Wagner[17], A.A. Weber[11], C. Wendisch[4], M.G. Wiebusch[8], P. Wintz[12], J. Wirth[10,9], A. Zhilin[14], P. Zumbruch[4]

[1]LIP-Laboratório de Instrumentação e Física Experimental de Partículas , 3004-516 Coimbra, Portugal
[2]Institute of Nuclear Physics, Polish Academy of Sciences, 31342 Kraków, Poland
[3]Smoluchowski Institute of Physics, Jagiellonian University of Cracow, 30-059 Kraków, Poland
[4]GSI Helmholtzzentrum für Schwerionenforschung GmbH, 64291 Darmstadt, Germany
[5]Technische Universität Darmstadt, 64289 Darmstadt, Germany
[6]Institut für Strahlenphysik, Helmholtz-Zentrum Dresden-Rossendorf, 01314 Dresden, Germany
[7]Joint Institute of Nuclear Research, 141980 Dubna, Russia
[8]Institut für Kernphysik, Goethe-Universität, 60438 Frankfurt, Germany
[9]Excellence Cluster 'Origin and Structure of the Universe' , 85748 Garching, Germany
[10]Physik Department E62, Technische Universität München, 85748 Garching, Germany
[11]II.Physikalisches Institut, Justus Liebig Universität Giessen, 35392 Giessen, Germany
[12]Forschungszentrum Juelich, 52428 Juelich, Germany
[13]Institute for Nuclear Research, Russian Academy of Science, 117312 Moscow, Russia
[14]Institute of Theoretical and Experimental Physics, 117218 Moscow, Russia
[15]National Research Nuclear University MEPhI (Moscow Engineering Physics Institute), 115409 Moscow, Russia
[16]Institut de Physique Nucléaire, CNRS-IN2P3, Univ. Paris-Sud, Université Paris-Saclay, F-91406 Orsay Cedex, France
[17]Nuclear Physics Institute, The Czech Academy of Sciences, 25068 Rez, Czech Republic
[18]Uniwersytet Warszawski - Instytut Fizyki Doświadczalnej, 02-093 Warszawa, Poland
[19]Bergische Universität Wuppertal, 42119 Wuppertal, Germany

[a] also at Coimbra Polytechnic - ISEC, Coimbra, Portugal
[b] also at Technische Universität Dresden, 01062 Dresden, Germany

Abstract. Radiative transition of an excited baryon to a nucleon with emission of a virtual massive photon converting to dielectron pair (Dalitz decays) provides important information about baryon-photon coupling at low q^2 in timelike region. A prominent enhancement in the respective electromagnetic transition Form-Factors (etFF) at q^2 near vector mesons ρ/ω poles has been predicted by various calculations reflecting strong baryon-vector meson couplings. The understanding of these couplings is also of primary importance for the interpretation of the emissivity of QCD matter studied in heavy ion collisions via dilepton emission. Dedicated measurements of baryon Dalitz decays in proton-proton and pion-proton scattering with HADES detector at GSI/FAIR are presented and discussed. The relevance of these studies for the interpretation of results obtained from heavy ion reactions is elucidated on the example of the HADES results.

*e-mail: piotr.salabura@uj.edu.pl

1 Introduction

QCD describes nucleons as colorless states of three nearly massless quarks confined to a small volume due to the strong force. The building blocks are light (u,d) quarks with current mass of only a few MeV emerging from the famed Higgs mechanism. Their much larger constituent mass is generated by the coupling of the quarks to the non-trivial QCD vacuum [1]. The underlying mechanism of the light quark mass generation is the spontaneous chiral symmetry breaking which leads to the appearance of Goldstone bosons, identified in the SU(2) flavour symmetry with pions, and chiral parity doublets (for example $\rho - a_1$ or nucleon- $N^*(1535)$) with a large mass difference. The order parameter of the chiral symmetry breaking is the non vanishing expectation value of the quark-anti quark condensate related to the pion decay constant by the Gell-Mann-Oaknes-Renner relation. Various model calculations predict a significant depletion of the expectation value of the quark condensate as a function of the temperature T and baryon-chemical potential μ_b. Lattice calculations performed in the limit of vanishing baryo-chemical potential predict a crossover phase transition from the hadron gas to the quark-gluon plasma at a pseudo-critical temperature $T_c \simeq 155MeV$ [2],[3] which is accompanied by the chiral symmetry restoration. The reduction of the chiral condensate expectation value predicted by the calculations is shown in Fig.1.

Experimentally, the structure of the QCD phase diagram is studied by means of heavy ion collisions over a broad energy range. Temperature and baryochemical potential are extracted from fits to hadronic yields measured by various detectors at LHC, RHIC, SPS and SIS18 at the lowest energy frontier. The extracted T, μ_b values obtained by the statistical hadronization model of [4] are presented in Fig.1. Since these temperatures characterize the late stage of the collision, where hadrons decouple from the strongly interaction system (so called freeze-out), it is expected that at highest energies initial temperatures are higher than the T_c and quark-gluon plasma is created. In the course of the collision evolution, QGP matter rapidly expands and when temperature drops to T_c, it hadronizes to a hadron gas. Dashed lines, shown in Fig. 1, depict hypothetical trajectories of collision at SPS energy of 158 AGeV and at 1.23 AGeV at SIS18.

At energies of around $1-2$ GeV per nucleon, where HADES experiment is operating, nuclear matter can be compressed to two-three times the ground state density and collective kinetic energy is dissipated into intrinsic degrees of freedom. As a result, some nucleons (about $20-30\%$) are excited to baryonic resonances and in the final state of the reaction mesonic states (mainly pions) are observed with increasing abundances as the center-of-mass energy of the collision rises. The system is characterized by a large baryo-chemical potential (baryon dominated matter in contrast to the meson dominated matter produced at higher energies) and moderate temperatures of about $50-70$ MeV. It resembles the matter created in neutron star mergers where similar temperatures can be reached [5]. Compressing nuclear matter can lead to a significant overlap of baryons which at some densities start to share their pion cloud content, hence creation of a new form of matter may be expected.

A scientific challenge, both to theory and experiment, is to understand the microscopic properties of such exotic matter over a broad range of the temperature and the baryo-chemical potential spanned by various experiments.

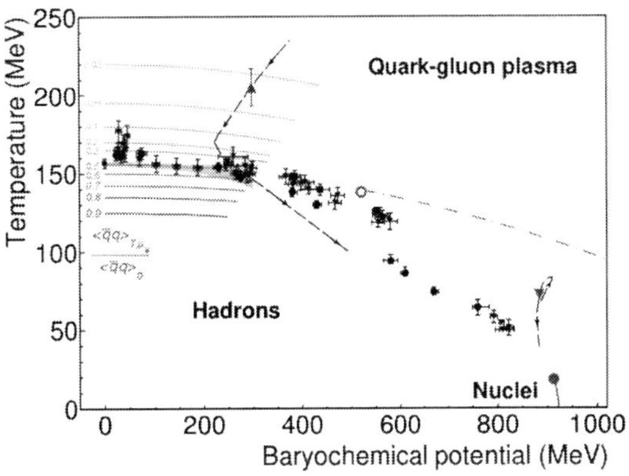

Figure 1. QCD phase diagramme. Experimental black points were obtained from a fit using statistical hadronization model to the hadron yields measured in various Heavy ion experiments [4]. The expectation values of the chiral condensate relative to the vacuum as constrained by the lattice calculations [2] are depicted as blue curves. The yellow band is the crossover region. The dotted–dashed purple curve shows the conjectured first-order phase transition, which terminates in a second-order QCD critical point (open purple circle), whose exact location is a matter of debate. The red upright triangle displays the temperature deduced from the dielepton measurements (see text for details). Lines with arrows show hypothetical trajectories for HI collisions at SPS and SIS18. Figure taken from [5]

One of the most promising probes used to examine properties of such exotic state of matter are dileptons. They are highly penetrating probes, not suffering strong final state interactions, hence providing insight into all phases of the collision. The dilepton radiation can be understood as emission of virtual photons in analogy to photon radiation from hot medium with the additional important parameter namely the mass. The spectral distribution of dileptons is strongly influenced by the temperature of collisions and the microscopic properties of medium. This connection can be understood by inspection of a formal expression for the differential thermal dielectron rate from a four volume of a medium characterized by T, μ_b : [6]

$$\frac{dR}{d^4qd^4x} = -\frac{\alpha^2}{\pi^3}\frac{L(M^2)}{M^2} Im\Pi_{em}(M, p, T, \mu_b) f^B(q_0, T) \quad (1)$$

In the formula $Im\Pi_{em}$ is the imaginary part the quark current-current correlator depending on the dilepton mass $M = q^2$ and the momentum p, $f^B(q_0, T)$ is the thermal Bose distribution function and $L(M^2) = (1 + \frac{2m_l^2}{M^2})\sqrt{1 - \frac{4m_l^2}{M^2}}$ is the lepton phase space element (it approaches one above dilepton threshold $M > 2m_{l^+l^-}$). $Im\Pi_{em}^{had}(M)$ is very well known in vacuum from e^+e^- annihilation into hadrons. It is flat in the Intermediate Mass Region (IMR): $3 > q^2 > 1.5$ GeV/c^2, as predicted for $e^+e^- \rightarrow q\bar{q}$ transition by perturbative QCD, and is dominated by $\rho/\omega/\phi$ resonances below (Low Mas region-LMR). The latter one demonstrates Vector Meson Dominance in the photon-hadron transitions, a phenomenon playing a prominent role in low energy hadron physics. In particular, the short lived ρ meson ($c\tau$ = 1.3 fm/c) with strong coupling to pions plays the most important role in the radiation from the fireball created in heavy ion collisions.

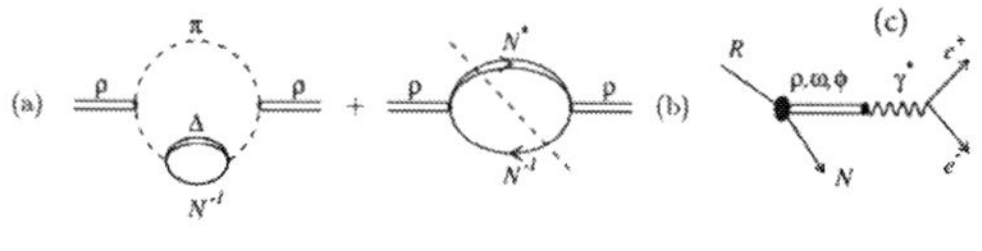

Figure 2. Schematic presentation of main graphs entering in-medium calculation of the ρ meson spectral function (a,b) and the relation to the elementary process of Dalitz decay of baryon resonance (c).

Various model calculations of the $Im\Pi_{em}$ for the non-vanishing T, μ_b predict its significant modifications in LMR due strongly modified ρ meson spectral function (for overview see [7]). The respective interactions can be grouped in two types of graphs (see Fig. 2) including pion-loop modifications (a) and baryon-ρ meson interactions (b). Systematic comparisons of the model calculations to the experimental data collected at SPS and RHIC reveals the dominant role of the baryon induced interactions [8],[9]. The modifications of ρ meson spectral function can be linked, though not directly, with chiral symmetry restoration by means of QCD and Weinberg sum rules. Indeed, the hadronic models do not include explicitly quantities characterizing QCD vacuum properties like quark or gluon condensates but the connection between the splitting of spectral functions of the chiral partners $\rho - a_1$ and the quark condensate and the pion decay constant is provided, as for example by the Weinberg sum rules [10].

$$\int ds \frac{1}{s}(D_\rho^V - D_{a1}^a) = f_\pi^2; \qquad (2)$$

$$\int ds(D_\rho^V - D_{a1}^a) = f_\pi{}^2 m_\pi^2 = -2m_q < q\bar{q} > \qquad (3)$$

Using the results of the hadronic models for the ρ meson in-medium spectral function and lattice results on the quark condensate evolution as a function of the temperature merging of the two spectral function has been demonstrated at $T = T_c$ and vanishing μ_b [11]. This signals the behaviour expected for the chiral symmetry restoration.

Another important benefit from dilepton measurements in heavy ion collisions is the ability to extract the temperature of the collision. Indeed, although the equation 1 holds for the radiation rate at the defined T, μ_b it has been demonstrated by comparison of various model calculations with the dilepton data that the time evolution of the fireball can be successfully implemented in the dilepton calculations (see [9],[12]). The two red triangles shown in Fig. 1 are the averaged temperatures of early collision phases extracted by analysis of the thermal radiation in IMR by the NA60 experiment at CERN and in the LMR by the HADES experiment at GSI, described in more details below. Note, that in both cases, the extracted temperatures are slightly higher than the one obtained from the analysis of the hadron spectra (black points).

Results of the hadronic models, however, strongly rely on the experimental input characterizing the vector meson-baryon couplings. The respective constraints on the calculations are provided by poorly known baryon resonance-vector meson branching ratios and constrains from photo-absorption data. The latter strongly rely on extrapolation from the photon point to the meson poles based on the Vector Meson Dominance. Both aspects, can be scrutinized more directly by measurements of baryon resonance radiative decays involving virtual timelike photons ($q^2 > 0$), described in more details in next section. This elementary process, shown in Fig.2 by the graph (c), is directly connected to the graph (b) by the respective cut. These decays are being investigated, for the first time, by HADES by studies of exclusive dielectron channels in proton and pion induced reactions, as described in this contribution.

2 Baryon transition form-factors

The matrix element for a radiative transition of an excited baryon to its ground state $\mathcal{M}(R \rightarrow N\gamma^*)$ by means of massive timelike photon can be expressed as a coherent sum of helicity amplitudes $A_{3/2}(q^2)$, $A_{1/2}(q^2)$ and $S_{1/2}(q^2)$, defined in the resonance decay frame [13]. The first two amplitudes are related to the transverse photon polarization while the last one is related to the longitudinal photon polarization. In the limit of $q^2 = 0$ (real photon) $S_{1/2}$ vanishes. The helicity amplitudes have been determined in space like region ($q^2 < 0$) for several resonances in electro- and photo- production experiment (for review see [14]). They are, however, unknown in the timelike region which is accessible by the Dalitz decays $R \rightarrow N\gamma^*$.

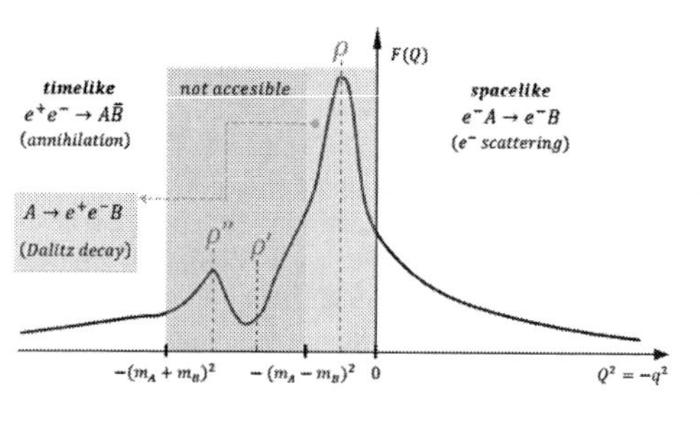

Figure 3. A schematic view of etFF evolution as a function of q^2 and kinematic conditions relevant for experiments in timelike and spacelike regions.

The respective differential decay width for the Dalitz decay $d\Gamma(R \rightarrow Ne^+e^-)/dM_{l^+l^-}$ depends on the resonance mass, spin and parity and is expressed as a function of the magnetic ($G_M(q^2)$), the electric ($G_E(q^2)$) and the Coulomb ($G_C(q^2)$) transition Form Factors (etFF). For the $J \geq 3/2$ (for example $\Delta(1232)$) there are three etFF and for $J < 3/2$ only two (for details see[13]). The relations between the helicity amplitudes and the etFF are given by linear transformations [13]. The calculations predict a strong dependence of the form-factors on the dielectron mass reflecting the important role of the intermediary vector meson fields in the transition, as predicted by VMD (see graph (c) in Fig. 2. Particularly the role of the broad ρ meson is visible in the strong enhancement of the transition rates for the dilepton masses even below the meson pole. Various implementations of VMD provide, however, different results, which calls for experimental constraints.

Fig. 3 displays a schematic view of the evolution of etFF as a function of q^2 and kinematic limits for measurements in spacelike and timelike regions

3 HADES experiment

HADES (High Acceptance DiElectron Spectrometer) is an apparatus installed at synchrotron SIS18 at GSI/FAIR in Darmstadt [15]. It has a large acceptance for charged particles and is optimized to identify very rare electrons and positrons in a hadron rich environment, which exceeds the electron signal by many orders of magnitude in multiplicity.

The HADES spectrometer consists of six identical sectors covering the full azimuth and polar angles from 18° to 85° measured relative to the beam direction. Each sector of the spectrometer contains a Ring-Imaging Cherenkov Detector (RICH) operating in a magnetic field-free region, two drift chambers (MDCs) in front of a toroidal magnetic field, and two outer MDCs behind the magnetic field, time-of-flight detectors, based for $\theta > 45°$ on scintillator rods (ToF) and for $\theta < 45°$ on resistive plates (RPC), combined with an electromagnetic calorimeter (ECAL) that replaces the Pre-shower detector used in previous campaigns. At forward angles ($\theta < 8°$) the detection of charged particles will be extended in 2019 by two straw tube tracking stations, developed for the PANDA detector, and a RPC ToF wall. This ensemble constitutes the Forward Detector.

Momentum measurement of charged particles in HADES is achieved by tracking the particles in front of and behind the toroidal field generated by six superconducting coils arranged symmetrically around the beam axis. A powerful and flexible trigger system selects events with defined multiplicity or topological pattern of charged hits in the TOF detectors. A major improvement of the spectrometer in terms of granularity and particle identification capability was achieved with the RPC time-of-flight detectors. The Forward Detector will significantly increase the acceptance for hyperon reconstruction in proton induced reactions.

The HADES detector operates with primary proton and Heavy Ion (HI) beams obtained from SIS18 and secondary pion beams. Runs were performed with proton (E_{kin} = 1.25, 2.2, 3.5 GeV), deuteron (E_{kin} = 1.25 GeV) beams on LH−2 target and with proton beam on Nb (E_{kin} = 3.5 GeV) target. The primary goal of these experiments was to provide a reference for HI reactions, particularly for Dalitz decays of baryon resonances and non-resonant nucleon-nucleon bremsstrahlung which had not been previously measured. Some highlights from these measurements are given in the next subsections.

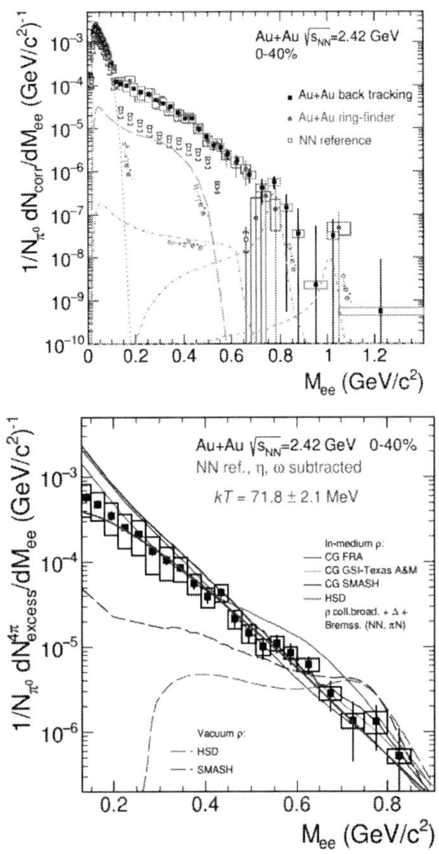

Figure 4. Top: Dielectron invariant mass distribution measured in Au+Au collisions compared to the freeze-out contribution from mesonic sources and the radiation from first chance baryonic contributions ($N-N$ -reference)(see text for details). Bottom: Dielectron excess yield extracted by subtracting the meson contributions as well as the $N-N$ reference normalized to the number of neutral pions. Red curve:thermal fit, dashed curves: ρ-meson vacuum line shape from HSD [18] and from SMASH [19] transport model calculations normalized to the respective number of neutral pions. Solid green curve: incoherent sum of all contributions: $\Delta \to Ne^+e^-$, $N-N$ and $\pi-N$ bremsstrahlung and in-medium ρ (collisional broadening scenario) from HSD . The dark-blue [19], blue[20] and pink [12] are curves show the results of three versions of coarse-grained calculations using in ρ medium-spectral functions from microscopic calculations [7].

3.1 Radiation from A+A collisions

Heavy ion experiments were conducted with $C+C$ (E_{kin} = 1.0, 2.0 AGeV), $Ar+KCl$ (E_{kin} = 1.76 AGeV) and $Au+Au$

E_{kin} = 1.25 AGeV collision systems. The results from the $C + C$ experiments demonstrated that the dilepton production in the light system can be successfully described by a superposition of nucleon-nucleon collisions, measured in separate runs, without any need to include in-medium effects [16].

However, in the medium heavy $Ar + KCl$ [17] and truly heavy $Au + Au$ system [5] a systematic excess of the radiation over the contributions expected from decays of neutral mesons (π, η) produced at freeze-out and from baryonic sources (Δ Dalitz decays and $N-N$ bremsstrahlung) determined from $N - N$ reactions (see next section). The latter is obtained as an average of the radiation measured in $p+p$ and $p + n$ collisions scaled by the respective number of pions. The scaling accounts for differences in the number of participating nucleons in the collision systems. The excess is clearly visible in comparison of the measured mass distribution to the aforementioned sources shown in top of Fig. 4

The excess radiation is shown in the lower part of Fig.4 as a function of the invariant mass in comparison to various model calculations (for references see [5]). The predictions of various models using thermal ansatz and applying Eq. 1 with the ρ meson spectral function obtained from the hadronic models are indicated as CG in Fig.4. The results explain the measured yield well and confirm a strong coupling of baryon resonances to the ρ meson and consequently its almost complete melting in dense baryonic medium (compare to the ρ vacuum spectral function shown by dashed lines). This conclusion justifies also a thermal fit (with function $dN/dM \simeq M_{ee}^{3/2} exp(-M_{ee}/kT)$ which provides the averaged temperature ($kT = 71.8 \pm 2.1$ MeV) of the radiating matter. The results of transport calculations (HSD) which use empirical scaling of the ρ meson width with the density also describe data reasonable well (except some overestimation at the meson pole). This calculation do include, however, explicit contributions from multistep processes in the dense phase of the collisions involving radiation from Δ and $N - N$ bremsstrahlung. These contributions are included in the hadronic model calculations in the spectral function. In particular these sources contribute below the two pion mass cut-off visible in the vacuum spectral function assumed in the HSD calculations (green-dashed line).

These HADES results constitutes an important reference for the planned beam energy scan within Compressed Baryonic Matter programme at FAIR and the lowest energy point in the measurement of dilepton excitation function.

3.2 Radiation from N-N collisions

The measurement with proton beams with E_{kin} = 1.25 GeV provides clean basis for studies of baryonic sources of dielectron pairs relevant for heavy ion reactions studied by HADES. The most important are the Dalitz decays of baryonic resonances $N^*(\Delta^*) \rightarrow Ne^+e^-$ and nucleon-nucleon bremsstrahlung. The meson contribution is limited to the π^0 Dalitz decay contributing to the dielectron spectrum with invariant masses smaller than the pion mass. The exclusive final states ppe^+e^- and pne^+e^-, separated in the $p+p$ and quasi-free $p+n$ collisions (from $p+d$) at the same energy allowed also for the direct comparison of the two collisions systems.

In $p + p$ collisions the dilepton pairs with invariant masses $M_{e^+e^-} > 0.15$ GeV/c^2 are consistently explained by the $\Delta^* \rightarrow Ne^+e^-$ decay, with only small $p - p$ bremsstrahlung contribution (see Fig. 5). This has been concluded from the combined analysis of the one pion ($pp\pi^0$, $pn\pi^+$) and the dielectron (ppe^+e^-, $ppe^+e^-\gamma$) exclusive final states [21],[22]. The latter allowed to access both $\Delta \rightarrow p\pi^0(e^+e^-\gamma)$ and $\Delta^+ \rightarrow pe^+e^-$ Dalitz decay channels. The Δ Dalitz decay was identified for the first time with the branching ratio $BR = (4.19 \pm 0.62 syst. \pm 0.34 stat.) \times 10^{-5}$. The Δ isobar production cross section was determined from the energy dependent Bn-Ga PWA analysis performed on the HADES one pion channels and 13 data sets measured at PNPI at lower energies [22]. The PWA analysis shows that at the energy of $\sqrt{s} = 2.42$ GeV the dominant contribution is given by the production of $\Delta(1232)$ in the intermediary state ($\simeq 70\%$ of the total cross section in the $pp\pi^0$ final state). The remaining part originates from the Roper excitation with a small non-resonant contribution.

Fig. 5 shows comparisons of the $p\pi^0$ invariant mass distribution reconstructed from the hadronic final state (open squares) and the dielectron $\Delta \rightarrow p\pi^0(e^+e^-\gamma)$ (black dots) to the PWA solutions (blue histogram). One can see a very good agreement between the data from the two reaction channels and also to the PWA solutions both in shape and in absolute scale (note that dielectron channel was not included in the PWA analysis). Similar conclusion was drawn from all studied differential distributions (invariant mass, angular distributions in the helicity and Gottfried-Jackson frames), for details see [21].

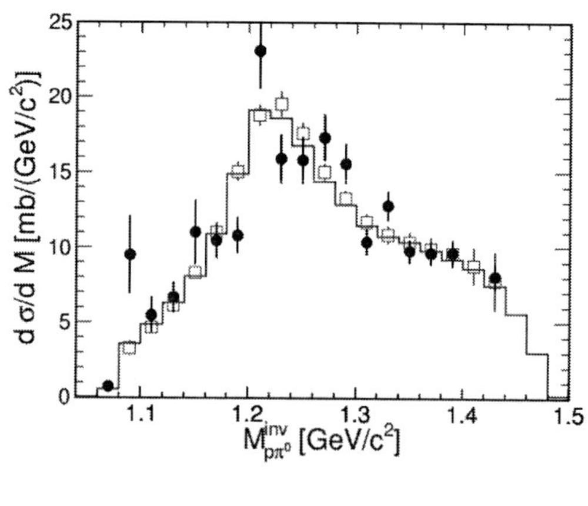

Figure 5. Dielectron data points after acceptance and branching ratio for $\pi^0 \rightarrow e^+e^-\gamma$ decay corrections (black dots) compared with the data from the exclusive $pp\pi^0$ channel (red open squares). The histogram shows total PWA solution (solid blue) obtained for the hadronic channel

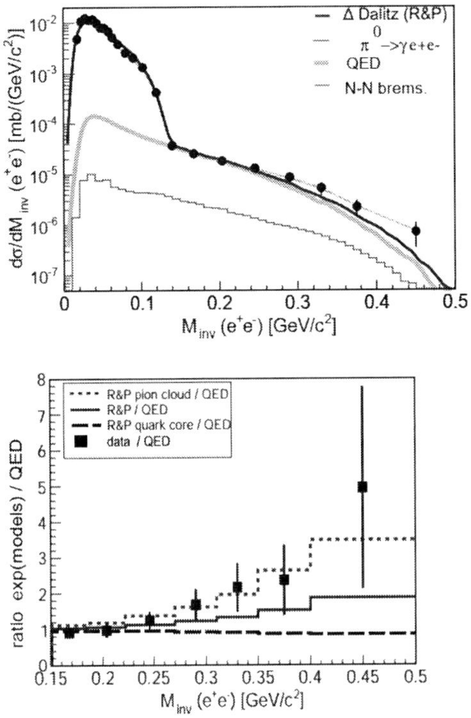

Figure 6. Upper: Dielectron invariant mass distribution from $p+p$ collisions at $E_{kin} = 1.25$ GeV compared to calculations using the Δ (model of Ramahlo and Pena (R&P)-blue [23] and for the point-like Δ (QED-cyan), π^0 Dalitz and $p+p$ bremsstrahlung. Bottom: Ratio of the measured dielectron invariant mass from the $\Delta(1232) \to pe^+e^-$ decay to the model of point-like particle (QED) compared to the Ramahlo and Pena (R&P) full model (blue) and assuming dominance of the photon-quark core (black dashed) and photon-pion-cloud contributions (red dashed) (see text and [21] for details).

From the measured differential mass distribution a ratio of the data to the calculations of Δ(1232) Dalitz decay assuming constant etFF $|G_M| = 3$ (value at the photon point) $G_E = G_C = 0$, labelled in Fig. 5 as QED, have been calculated. A slight increase of the effective etFF as a function of the photon virtuality ($q^2 = M_{e+e-}$) can be clearly seen in Fig.6. This is in agreement with the covariant quark model of [23] assuming significant contribution from the virtual photon-pion cloud coupling. In the model this coupling to the ρ meson is much stronger than to the quark core (compare dashed black and red dashed) and produces the aforementioned enhancement. Furthermore, since the $N-\Delta$ transition in the spacelike region is dominated by the spin-flip magnetic transition it can be expected that the dominant role in the measured low q^2 region is played by $G_M(q^2)$ and consequently that the virtual photons are preferentially transversally polarized.

This conclusion is corroborated by the analysis of the angular distributions of either of both leptons in the rest frame of the virtual photon (helicity frame). According to calculations [24], it should obey $\simeq 1 + Bcos^2\theta$ distribution, with B=1 for the transversely polarized photons. The anisotropy coefficient deduced from the data analysis $B = 1.17 \pm 0.34$ confirms the expectation.

However, the $n+p$ data exhibits a very different shape above π^0 mass as compared to the $p+p$ measured at the same energy (see Fig. 7) [25]. This can be clearly recognized as an apparent excess above the Δ contribution (red dashed line), which has a similar shape for the two reactions but explains the $p+p$ data very well. This finding was not completely unexpected since various calculations of N-N bremsstrahlung, defined here as the non-resonance contribution that has to be added coherently to the Δ excitation, performed already in the 90's in the framework of One Boson Exchange (OBE) models [26],[27], [28] predicted a stronger contribution of this non-resonance dilectron production in the neutron-proton interactions. However, the calculations showed a strong sensitivity of the dielectron yield on the off-shell electromagnetic form factors of nucleons, and on the etFF of Δ(1232). Also modelling of nucleon-nucleon interaction requires form-factors to dress the respective vertices and various nucleon-meson coupling constants which needs to be fixed from experimental data. While the latter could be fixed to some extent from the data on meson production close to the threshold, the first ones can only be inferred from the dilepton data. It is therefore not surprising that the model predictions could not explained both dielectron exclusive channels $p+p$ and $n+p$ because the relevant precision data were missing. This has changed with the new HADES data and the most recent calculations performed by [29] in OBE which account almost completely for the missing yield (dashed-dotted blue line in Fig. 7) and explain also $p+p$ data. In this calculations it was found that an important role is played by graphs accounting for the radiation from charged pion exchange line. It has been also emphasized that the electromagnetic pion form factor, dominated by the ρ meson, plays an essential role in the successful description of data. This is another example of sensitivity of dielectron radiation to the etFF's of hadrons in timelike region.

The alternative model of [30] assuming $\Delta\Delta \to \rho np \to npe^+e^-$ has been put forward for the explanation of $n+p$ data. It is shown by the solid green line in Fig.7. The salient feature of both calculations is the transition from the Δ Dalitz dominated region $0.14 < M_{e^+e^-} < 0.28$ GeV/c^2 to the higher mass region which is dominated by the off-shell ρ contribution.

Although both models do not provide more differential information, this behaviour seems to be corroborated by the electron angular distributions in the helicity frame. Whereas, see Fig. 8, in the lower mass region the distribution reveals the characteristic $1 + Bcos^2(\theta_l)$ shape with $B \simeq 1$, measured also for the Δ in the $p+p$, at larger invariant masses the distribution changes character and the anisotropy parameter takes the value $B = 0.25 \pm 0.35$. This can be interpreted as dominance of the emission from the meson line through the off-shell ρ meson. Indeed, the anisotropy parameters B extracted in the reference frame

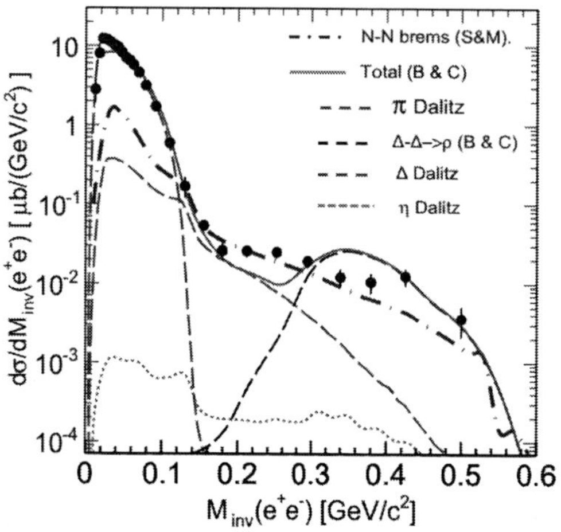

Figure 7. Invariant mass distribution of dielectrons measured in a quasi-free $n + p$ collisions at $E_{kin} = 1.25$ GeV compared to the $N - N$ bremsstrahlung model of [29] (blue-dashed) and the model of [30] (green) decomposed in the Δ Dalitz (red dashed), ΔΔ fusion and π^0 Dalitz (dashed-violet) contributions.

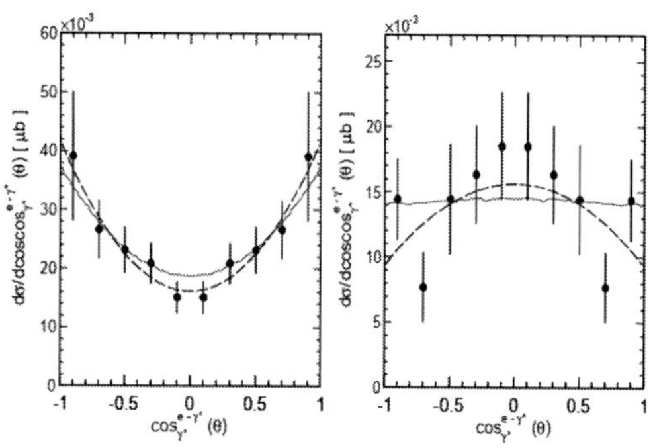

Figure 8. Angular distributions (black dots) of the leptons in the rest frame of the virtual photon for the dielectron mass of $0.14 < M_{e^+e^-} < 0.28$ GeV/c^2 (left panel), and with respect to the direction of the charged pion exchange for dielectrons with $M_{e^+e^-} > 0.28$ GeV/c^2 (right panel) corrected for the detector acceptance. The solid green curves display predictions from the simulations where the dominant source is the Δ(1232) in the low mass bin and the ρ in the high-mass bin (assumed to be isotropic). The dashed blue curve shows a fit with a function $A(1 + Bcos^2(\theta_e))$

with z axis fixed along the direction of the exchanged charged pion becomes slightly negative $B = -0.4 \pm 0.25$. This can can be compared to the value $B = -1$ expected for the pion annihilation [24].

Measurements of dielectron production in proton-proton collisions with $E_{kin} = 3.5$ GeV reveal importance of higher mass resonances [31, 32]. From the one pion exclusive channels $pp\pi^0$ and $pn\pi^+$, measured in the same experiment, cross sections for Δ and N^* resonances have been estimated by means of a resonance model assuming two step reaction mechanism with the intermediate resonance formation. The resonance model applied in the HADES analysis was inspired by the calculations presented in [33] applied in transport models for particle production at SIS18 energies. In the HADES model empirical angular distributions for the production of resonances showing a strong forward-backward peaking which is characteristic for peripheral reactions have been added in accordance to the measured distributions. A good description of the experimental data in the detector acceptance has been achieved allowing for an extrapolation to the full solid angle and an extraction of the pion production cross sections which are compatible with the word data (for details see [32]). Recently, the HADES resonance model has been successfully extended to two pion production (see contribution of A. Belounnas to this conference).

In the exclusive ppe^+e^- channel dielectron production from the baryon resonance Dalitz-decays and two-body $\rho/\omega \to e^+e^-$ meson decays have been concluded as dominant sources of the pair production. In particular, a significant yield below the vector meson pole has been identified and attributed to the Dalitz decays of Δ, N^* resonances proceeding through the intermediate ρ meson state, as explained in more details below.

Indeed, using estimates of production cross sections following form the pion analysis contributions to the dielectron spectrum was calculated. The dielectron yields, presented with dashed blue line in the upper panel of Fig.9, were obtained using constant eTFF, as predicted in [13] [34]. The uncertainty in the production cross sections of baryon resonances is presented by the grey band spanned around the dashed line. As one can see, the total contribution from the higher mass resonances (green line) is larger than from the Δ(1232), shown by the red line, and confirms importance of higher mass resonances. The model, however, cannot reproduce the measured yield and suggests strong off-shell vector meson couplings. These couplings modify the decay vertex of the resonance by means of the respective etFF (see graph (c) in Fig. 2). Since the mass dependence of the respective eTFF are not known, an alternative approach for the Dalitz decay of resonances was applied. It assumes a factorization of the $R \to pe^+e^-$ process into two steps $R \to \rho N$ and $\rho \to e^+e^-$ which could be calculated using the resonance branching ratios to the ρN from PDG (from 2014). Similar approach is also implemented in transport codes (GiBUU, UrQMD or recently SMASH) to account for the off-shell ρ meson contributions. Note that on-shell ρ meson contribution known from the two-pion production is included in the cocktail (violet line) but can not explain the apparent excess in the data, too. The GiBUU calculations utilizing the factorization scheme, shown in the lower panel by black line, come closer to the data and explain also the invariant mass distributions of the combined system pe^+e^-. A slightly better description is achieved using the same prescription but the cross section for resonance production extracted from the HADES resonance model (dotted line). The main dif-

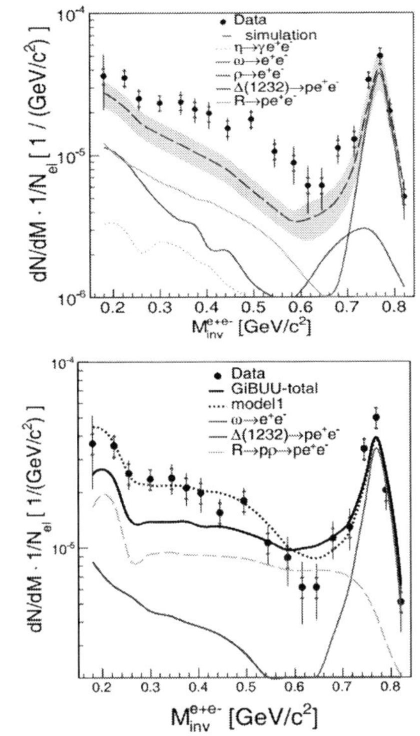

Figure 9. Top: Invariant mass distribution of dielectron pairs measured in the exclusive channel $p + p \to ppe^+e^-$ at $E_{kin} = 3.5$ GeV compared to cocktail of baryonic and mesonic sources. The contribution from baryon resonances is separated into contributions from Δ(1232) (red line) and higher mass states (green) assuming constant etFF. Bottom: same data compared to calculations using for the baryon resonance two step decay involving coupling to intermediate ρ meson state (for details see text).

ference in the HADES resonance model is a higher cross section for the $N^*(1520)$ (factor 8), slightly lower (factor 2) cross sections of the other resonances (for details see [32]) and lower branching ratio for the resonance decays (for details see [32]). The latter one was taken from the Bonn-Gatchina upper estimates of the $R \to N\rho$ instead of the respective branching ratios given in PDG (2014). One should note, that in the recent PDG those values have been removed and await new experimental input. The contribution of higher mass resonances (dashed green curve) is given by the decays of $N^*(1520)(38\%)$, $N^*(1720)(22\%)$, $\Delta(1620)(15\%)$ and $\Delta(1905)(6.5\%)$.

3.3 Pion beams in HADES

Recently, successful measurements were completed with HADES using secondary pion beams [35]. The pion beam was obtained from a primary ^{14}N beam with an intensity of 0.8-1.0×10^{11} ions/spill impinging on a beryllium (^9Be) target (see Fig.11). Beam intensity for negative pions as a function of the momentum scaled to the space charge limit of SIS18 are displayed in Fig. 10 [36].

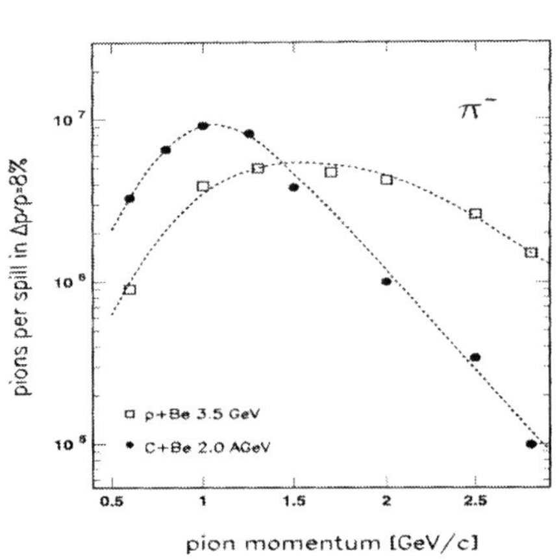

Figure 10. Negative pion yields at the HADES target point as a function of the momentum scaled to the space charge limit of SIS18.

Produced pions were transported by the beam line consisting of 9 quadrupole and 2 dipole magnets to the HADES target located 33 meters downstream from the production point (see Fig. 11). The maximum pion intensity of 10^6 pions/spill was achieved at a momentum $p = 1.0$ GeV/c. The intensity decreased by a factor 2 for the pion momentum range $p = 0.65 - 0.8$ GeV/c investigated in the measurement. Four reference beam momenta p = 656, 690, 748, 800 MeV/c were selected using different settings of currents in the magnets of the beam line. The differential transmission distribution of the pions as a function of the momentum can be approximated by Gaussian with a typical width of ≃ 1.7%. Further improvement to reduce the pion beam momentum resolution to $\delta p = 0.3\%$ was provided by a dedicated in-beam tracking system CERBEROS positioned in the beam line as shown in Fig.11. The detectors provide (x, y) coordinates of pion hits, with precision given by the 780μm pitch of strips. The hit coordinates were used to determine the optical properties of the beam line, which are needed for an event-by event momentum reconstruction, as described in details in [35]. The in-beam detector system included also a segmented START detector, made from mono-crystalline diamond material, placed in front of the HADES target.

Resonance formation in s-channel with a fixed mass presents an important advantage of pion induced reactions w.r.t reactions with proton beams where contributions from many resonances with different masses must be considered. The measurement was focused on two pion and dielectron production in the second resonance region with the main goal to characterize the role of off-shell ρ meson couplings to the $N^*(1520)$ and $N^*(1535)$ resonances [37]. A combination of two-pion with dielectron decay channels, that can be measured only by HADES, offers an unique possibility to explore baryon resonance decay channels into ρ and dielectron channels. To isolate the

http://doi.org/10.1051/epjconf/202024101013

ρ meson decay channels a combined Partial Wave Analysis of the hadronic channels in pion and photon induced reactions has been performed within Bonn- Gatchina approach. The excitation function of the different partial waves and $\Delta\pi$, $N\sigma$ and $N\rho$ isobar configurations has been determined. The N(1520) resonance is found to dominate the $N\rho$ final state with the branching ratio BR = $12.2 \pm 1.9\%$ [37].

Preliminary results are presented in reports of B. Ramstein and I.Ciepal to this conference.

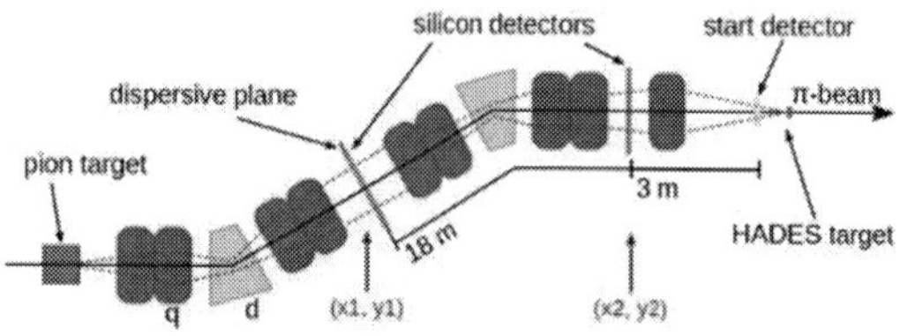

Figure 11. Pion beam line from the production point (pion target) to the HADES target position. Quadrupole and dipole magnets guide the pions (the dashed line) through the in-beam tracking detectors (indicated by the arrows).

4 Summary and outlook

Our recent studies of the baryon resonances in the exclusive channel $NN \rightarrow NNe^+e^-$ revealed patterns of far off-shell ρ production followed by dielectron decay. This raises the question to what extent the baryon meson cloud is instrumental in transferring excitation energy into the vector meson and to virtual photon with the invariant mass $m_{e^+e^-}$. This process is characterized by the interaction vertex described by the electromagnetic transition form-factors of the resonances in the (soft) timelike region where $0 < q^2 = (m_{e^+e^-})^2 < (M_R - M_N)^2$ with M_R and M_N the resonance and the nucleon mass, respectively. With the secondary pion beams [35] at GSI/FAIR and the HADES detector, one has an unique opportunity to explore radiative and Dalitz decays of baryon resonances into much more details. The first experiment, studying two pion and dielectron production in the second resonance region, confirms strong coupling for $N^*(1520)$ to the off-shell ρ meson in both two-pion and dielectron channel. The research programme formulated by HADES wihin FAIR Phase0 anticipates an energy scan in the third and fourth resonance region with high statistic measurements of neutral and charged mesons and dielectrons. This program is also extended to measurements of radiative decays of baryons containing strangeness [38]. In a similar way, as for the non-strange baryons Dalitz decays of excited hyperon states, Λ^*, Σ^* in pion and proton induced reactions at maximum SIS18 energies has been proposed. Especially, transitions of narrow $\Lambda(1520)$ and $\Sigma(1385)$ states to Λ and dielectron pair seems to be ideal candidates for such measurements. Also high statistics measurements of hadronic decays of $\Lambda(1405)$ and double strange cascades are considered. In particular, measurements of the controversial $\Lambda(1405)$ in pion-induced reactions, will provide important complementary information to the one already obtained by HADES in $p + p$ collisions to constrain the model calculations and to unravel the nature of $\Lambda(1405)$.

HADES results from heavy ion reactions demonstrate also the particular role played by vector meson-baryon interactions in the understanding of the dielectron radiation from dense baryonic matter. A better understanding of the coupling of baryonic resonances to intermediary ρ mesons is mandatory to validate emissivity calculations for hadron-resonance matter. Such studies can therefore pave the way for searching new states of matter with the upcoming Compressed Baryonic Matter program at FAIR's SIS100.

Acknowledgments

SIP JUC Cracow, Cracow (Poland), National Science Center, grants: 2016/23/P/ST2/04066 POLONEZ funded from the European Union's Horizon 2020 research and innovation programme under the Marie Sklodowska-Curie grant agreement No 665778, 2017/25/N/ST2/00580, 2017/26/M/ST2/00600; TU Darmstadt, Darmstadt (Germany), VH-NG-823, DFG GRK 2128, DFG CRC-TR 211, BMBF:05P18RDFC1; Goethe-University, Frankfurt (Germany) and TU Darmstadt, Darmstadt (Germany), ExtreMe Matter Institute EMMI at GSI Darmstadt; TU München, Garching (Germany), MLL München, DFG EClust 153, GSI TMLRG1316F, BmBF 05P15WOFCA, SFB 1258, DFG FAB898/2-2; NRNU MEPhI Moscow, Moscow (Russia), in framework of Russian Academic Excellence Project 02.a03.21.0005, Ministry of Science and Education of the Russian Federation 3.3380.2017/4.6; JLU Giessen, Giessen (Germany), BMBF:05P12RGGHM; IPN Orsay, Orsay Cedex (France), CNRS/IN2P3; NPI CAS, Rez, Rez (Czech Republic), MSMT LM2015049, OP VVV CZ.02.1.01/0.0/0.0/16 013/0001677, LTT17003.

References

[1] C.D Roberts, Few Body, Syst. **58**,5(2017) , see also contribution to this conference

[2] Borsanyi, S. et al., J.High Energy Phys. **73**, 2010:73 (2010)

[3] Bazavov, A. et al.,Phys.Lett.B **795**,15(2019)

[4] Andronic, A. et al., Nature **561**, 321 (2018).

[5] Adamczewski-Musch, J. at al. , Nature Physics (2019) doi.org/10.1038/s41567-019-0583-8

[6] L. McLerran and R. D. Pisarski, Nucl. Phys. A **796**, 83 (2007)

[7] R. Rapp and J. Wambach, Adv. Nucl. Phys. **25**, 1 (2000)

[8] H. van Hees and R. Rapp, Nucl. Phys. A **806** (2008) 339

[9] R. Rapp and H. van Hees, Phys. Lett. B **753** (2016) 586

[10] S. Weinberg Phys. Rev. Lett **18**, 507 (1967)

[11] P. M. Hohler and R. Rapp, Phys. Lett. B **731** (2014) 103

[12] S. Endres, H. van Hees and M. Bleicher, Phys. Rev. C **92**, 014911 (2015), Phys. Rev. C **93** (2016) no.5, 054901,

[13] M. I. Krivoruchenko, B. V. Martemyanov, A. Faessler and C. Fuchs, Annals Phys. **296**, 299 (2002)

[14] I. G. Aznauryan and V. D. Burkert, Prog. Part. Nucl. Phys. **67**, 1 (2012)

[15] G. Agakishiev *et al.* [HADES Collaboration], Eur. Phys. J. A **41** (2009) 243 doi:10.1140/epja/i2009-10807-5 [arXiv:0902.3478 [nucl-ex]].

[16] G. Agakishiev *et al.* [HADES Collaboration], Phys. Lett. B **690**, 118 (2010)

[17] G. Agakishiev *et al.* [HADES Collaboration], Phys. Rev. C **84** (2011) 014902 doi:10.1103/PhysRevC.84.014902 [arXiv:1103.0876 [nucl-ex]].

[18] Bratkovskaya, E. L. et al., Phys. Rev. C **87**, 064907 (2013)

[19] Staudenmaier, J., et al. Phys. Rev. C **98**, 054908 (2018)

[20] Galatyuk, T., Eur. Phys. J. **52**A, 131 (2016)

[21] J. Adamczewski-Musch *et al.* [HADES Collaboration], Phys. Rev. C **95** (2017) no.6, 065205

[22] G. Agakishiev et al. (HADES Collaboration), Eur. Phys. J. A **51**, 137 (2015).

[23] G. Ramalho and M. T. Pena, Phys. Rev. D **85**, 113014 (2012)

[24] E. L. Bratkovskaya, O. V. Teryaev, and V. D. Toneev, Phys. Lett. B **348**, 283 (1995)

[25] J. Adamczewski-Musch *et al.* [HADES Collaboration], Eur. Phys. J. A **53**, no. 7, 149 (2017) doi:10.1140/epja/i2017-12341-3 [arXiv:1703.08575 [nucl-ex]].

[26] M. Schafer, H. C. Donges, A. Engel and U. Mosel, Nucl. Phys. A **575**, 429 (1994)

[27] M. Schafer, H. C. Donges and U. Mosel, Phys. Lett. B **342**, 13 (1995)

[28] F. de Jong and U. Mosel, Phys. Lett. B **392**, 273 (1997)

[29] R. Shyam and U. Mosel, Phys. Rev. C **79**, 035203 (2009) doi:10.1103/PhysRevC.79.035203 [arXiv:0811.0739 [hep-ph]].

[30] M. Bashkanov and H. Clement, Eur. Phys. J. A **50**, 107 (2014)

[31] G. Agakishiev *et al.* [HADES Collaboration], Eur. Phys. J. A **48**, 64 (2012)

[32] G. Agakishiev *et al.*, Eur. Phys. J. A **50**, 82 (2014)

[33] S. Teis et al., Z. Phys. A bf 356, 421 (1997).

[34] M. Zetenyi and G. Wolf, Heavy Ion Phys. **17**, 27 (2003) doi:10.1556/APH.17.2003.1.5 [nucl-th/0202047].

[35] J. Adamczewski-Musch *et al.* [HADES Collaboration], Eur. Phys. J. A **53**, no. 9, 188 (2017).

[36] J.Diaz *et al.*, Nucl. Instrum. Meth. A **478**, 511 (2002). doi:10.1016/S0168-9002(01)00895-6

[37] J. Adamczewski-Musch *et al.* [HADES], Phys. Rev. C **102** (2020) no.2, 024001 doi:10.1103/PhysRevC.102.024001 [arXiv:2004.08265 [nucl-ex]].

[38] R. Lalik [HADES Collaboration], J. Phys. Conf. Ser. **1137** (2019) no.1, 012057. doi:10.1088/1742-6596/1137/1/012057

EPJ Web of Conferences **241**, 01014 (2020)
NSTAR 2019
http://doi.org/10.1051/epjconf/202024101014

$K^+\Lambda(1405)$ photoproduction at the BGO-OD experiment

Georg Scheluchin[1,*], *Stefan* Alef[1], *Patrick* Bauer[1], *Reinhard* Beck[2], *Alessandro* Braghieri[13], *Philip* Cole[3], *Rachele* Di Salvo[4], *Daniel* Elsner[1], *Alessia* Fantini[4,5], *Oliver* Freyermuth[1], *Francesco* Ghio[6,7], *Anatoly* Gridnev[8], *Daniel* Hammann[1], *Jürgen* Hannappel[1], *Thomas* Jude[1], *Katrin* Kohl[1], *Nikolay* Kozlenko[8], *Alexander* Lapik[9], *Paolo* Levi Sandri[10], *Valery* Lisin[9], *Giuseppe* Mandaglio[11,12], *Roberto* Messi[4,5], *Dario* Moricciani[4], *Vladimir* Nedorezov[9], *Dmitry* Novinsky[8], *Paolo* Pedroni[13], *Andrei* Polonski[9], *Björn-Eric* Reitz[1], *Mariia* Romaniuk[4], *Hartmut* Schmieden[1], *Victorin* Sumachev[8], *Viacheslav* Tarakanov[8], and *Christian* Tillmanns[1]

[1]Rheinische Friedrich-Willhelms-Universität Bonn, Physikalisches Institut, Nußallee 12, 53115 Bonn, Germany
[2]Helmholtz-Institut fuer Strahlen- und Kernphysik, Universitaet Bonn, Nussallee 1-16, D-53115 Bonn Germany
[3]Lamar University, Department of Physics, Beaumont, Texas, 77710, USA
[4]INFN Roma "Tor Vergata", Rome, Italy
[5]Università di Roma "Tor Vergata", Via della Ricerca Scientifica 1, 00133 Rome, Italy
[6]INFN sezione di Roma La Sapienza, P.le Aldo Moro 2, 00185 Rome, Italy
[7]Istituto Superiore di Sanita, Viale Regina Elena 299, 00161 Rome, Italy
[8]Petersburg Nuclear Physics Institute, Gatchina, Leningrad District, 188300, Russia
[9]Russian Academy of Sciences Institute for Nuclear Research, prospekt 60-letiya Oktyabrya 7a, Moscow 117312, Russia
[10]INFN - Laboratori Nazionali di Frascati, Via E. Fermi 40, 00044 Frascati, Italy
[11]INFN sezione Catania, 95129 Catania, Italy
[12]Universita degli Studi di Messina, Via Consolato del Mare 41, 98121 Messina, Italy
[13] INFN sezione di Pavia, Via Agostino Bassi, 6 - 27100 Pavia, Italy

Abstract. Since the discovery of the $\Lambda(1405)$, it remains poorly described by conventional constituent quark models, and it is a candidate for having an "exotic" meson-baryon or "penta-quark" structure, similar to states recently reported in the hidden charm sector.
The $\Lambda(1405)$ can be produced in the reaction $\gamma p \rightarrow K^+\Lambda(1405)$. The pure I=0 decay mode into $\Sigma^0\pi^0$ is prohibited for the mass-overlapping $\Sigma(1385)$. Combining a large aperture forward magnetic spectrometer and a central BGO crystal calorimeter, the BGO-OD experiment is ideally suited to measure this decay with the K^+ in the forward direction. Preliminary results are presented.
*Supported by DFG (PN 388979758, 405882627).

1 Introduction

For decades hadron spectroscopy has been used to investigate the strong interaction. Experiments on protons showed that baryons consist of three valance quarks and sea-quarks. While three quark models show a good agreement between calculated and measured for most of the states, the $\Lambda(1405)$ deviates from the predictions. It is lighter in mass than its non-strange counterpart, $N(1535)$, even thought it has a strange quark in its composition. Furthermore the mass distribution, also called line shape, does not follow a Breit-Wigner distribution. The initial inconsistencies elevated $\Lambda(1405)$ as an unconventional state candidate since the discovery more than 50 years ago.

Nowadays there are more theoretical models for $\Lambda(1405)$ having a molecule like structure of $N\bar{K}$. Lattice QCD calculations give more support for the molecule like structure compared to a genuine three quark state [1][2]. A study using a chiral unitarity model, where the resonance is generated dynamically from $N\bar{K}$ interactions with other channels constructed from the octets of baryons and mesons, shows that the line shape would depend on the

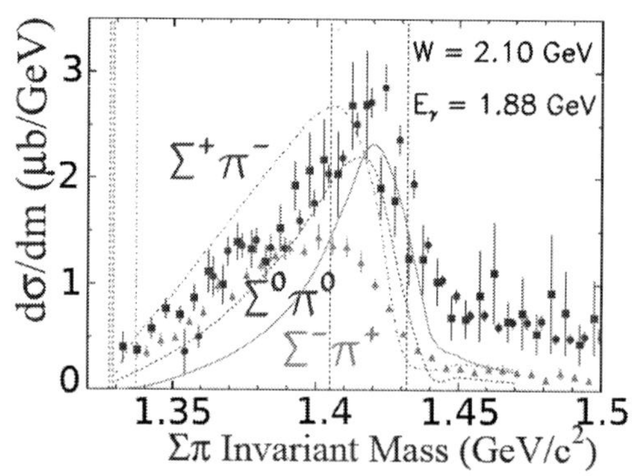

Figure 1. Line shape results from the CLAS experiment [4][5] for different decay modes. Colored lines show model predictions of Nacher et al. [3]. Figure taken from [4].

*e-mail: scheluchin@physik.uni-bonn.de

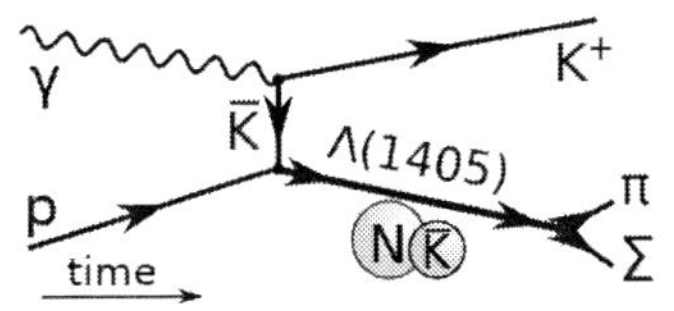

Figure 2. Possible photoproduction scheme of the Λ(1405).

decay mode [3]. The results from the CLAS experiment are seen in figure 1 [4]. The measurements show that the line shape does differ for the decay modes. On close inspection it is visible that the measurement and predictions for the charged decays differ from the predictions and a second experiment could help to resolve this discrepancy.

Λ(1405) can be produced via $\gamma p \rightarrow K^{+}\Lambda(1405)$ as seen in figure 2. With the assumption of a molecular like state, one can assume that the cross section for Λ(1405) is increased if the transferred momentum to the baryon is low. This means that the K^{+} needs to take most of the photon momentum, which corresponds to extreme forward angles in a fixed target experiment, while the Λ(1405) decays almost at rest isotropically. The BGO-OD[1] experiment with its central calorimeter and forward spectrometer is ideally suited to measure such kinematics, which were not yet explored by other experiments for the $\Sigma^{0}\pi^{0}$ decay. In the following chapters the experimental setup and particle identification are described in more detail. The experiment is explained in section 2 and the reaction identification in section 3. Preliminary results are shown in section 4.

[1] **B**ismuth **G**ermanum **O**xide calorimeter with a **O**pen **D**ipole magnet spectrometer

2 The BGO-OD experiment

The BGO-OD experiment is located at the ELSA[2] facility in Bonn. ELSA is an electron accelerator with a quasi continuous beam up to 3.2 GeV [6]. In figure 3 the general overview of the accelerator facility is shown. It is a three stage electron accelerator. Using a linear accelerator LINAC the electrons are accelerated to relativistic velocities before the injection into the booster synchrotron. Here the electrons are accelerated to an energy of 0.5 GeV to 1.6 GeV before they are guided to the stretcher ring. The stretcher ring needs to be filled multiple times by the synchrotron to be completely full. After this the electrons are accelerated up to 3.2 GeV. The electrons are only extracted partially to either the CBELSA/TAPS[8] or the BGO-OD experiment each revolution. This results in a quasi continuous electron beam for the duration of 5 s to 12 s until the ring is empty and the process starts anew.

In figure 4 the BGO-OD experiment is shown in more detail. The electron beam hits a radiator inside the goniometer tank, which creates a real photon beam via bremsstrahlung. The energy of the photons is determined by measurement of the bremsstrahlung electron with the Tagger detector. The photon beam interacts with the target cell inside the BGO ball, which is filled with liquid hydrogen or deuterium. The final state particles of the reaction are detected with the BGO calorimeter using bismuth germinate oxide crystals between polar angles 25° to 155°. Charged particles traveling in θ <12° are detected in the forward detector. The track trajectory before the open-dipole magnet is measured with the MOMO and SciFi2 scintillating fibre detectors, and the drift chambers measure the trajectory after the magnet. The measured track curvature is used to determine the momentum, while the ToF walls at the end of the experiment measure the veloc-

[2] The **E**lectron **S**tretcher **A**ccelerator in Bonn(Germany)

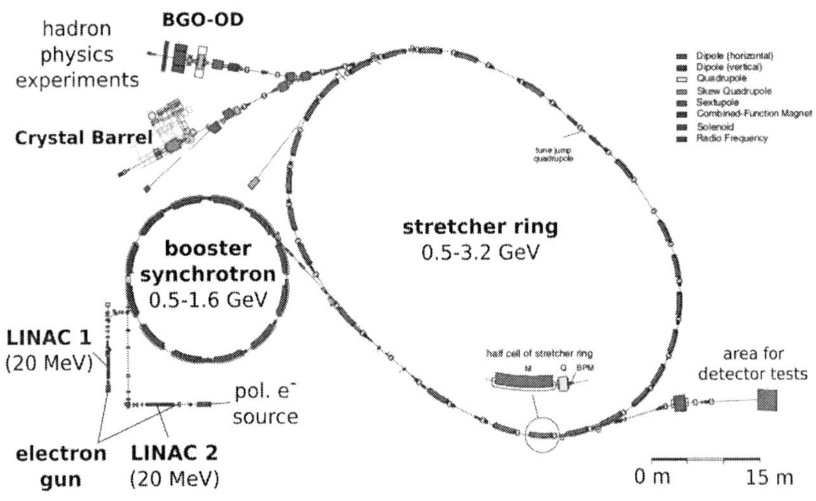

Figure 3. Overview of the Electron Stretcher Accelerator [7] in Bonn, showing the main components.

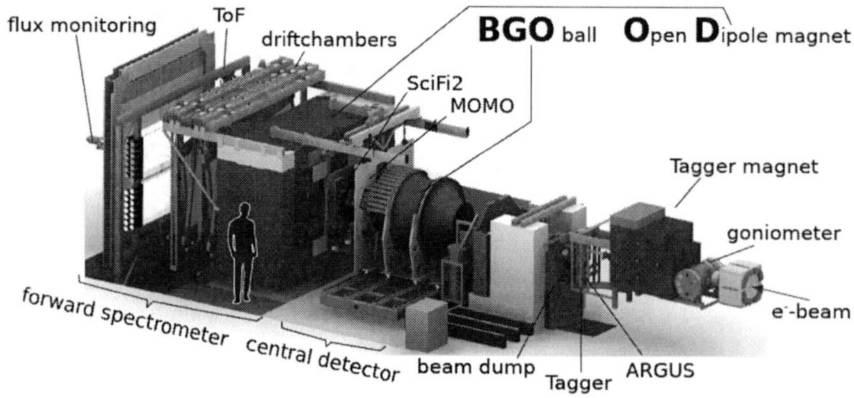

Figure 4. Overview of the BGO-OD experiment.

ity via time of flight. Both measured quantities can be used for particle identification as seen in figure 5.

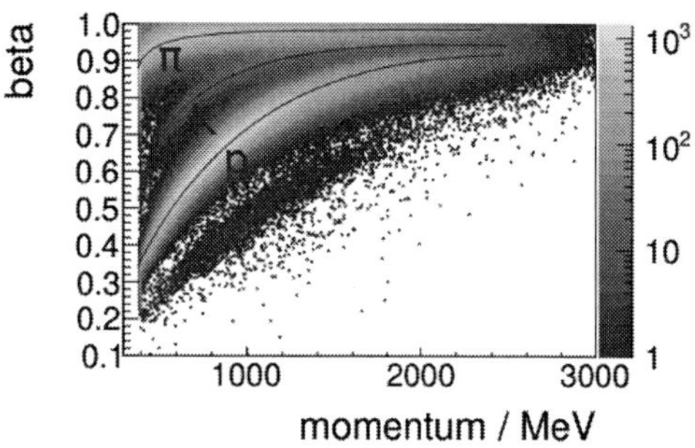

Figure 5. Particle velocity β against momentum p detected in the forward spectrometer. Due to the mass difference the charged particles π, K and p can be distinguished indicated by lines.

3 $\Lambda(1405)$ identification

$\Lambda(1405)$ was identified via the $\Sigma^0\pi^0$ decay, which is prohibited for $\Sigma(1385)$. The complete reactions is $\gamma p \rightarrow K^+\Lambda(1405) \rightarrow K^+\Sigma^0\pi^0$. The reaction can be identified by detecting K^+ and π^0, while the Σ^0 is identified via missing mass techniques. This can be achieved with the K^+ detected in the forward spectrometer and is described in section 3.1. An additional technique is used by detecting the full final state increasing the K^+ polar angle acceptance in exchange for lower statistics, which is described in section 3.2.

3.1 K^+ at extreme forward angles

The K^+ can be identified in the forward spectrometer using the measured velocity and momentum of the charged particles, as seen in figure 5. A π^0 can be identified by combining two measured photons in the central calorimeter and select events with invariant mass close to the π^0 mass. This leaves the Σ^0 particle to be identified through missing mass techniques. Figure 6 shows the missing mass to K^+ and $K^+\pi^0$ systems. The red lines indicate the possible missing hyperons in the reaction. The events with a Λ mass originate from the prominent $\Sigma(1385)^0 \rightarrow \Lambda\pi^0$ decay. While it looks like there are some events with missing Σ^0, studies showed that most of these events are combinatorial background from $\Sigma(1385)$ as seen in figure 7. First preliminary results for the differential cross section can be seen in section 4.2. Further analysis is needed to improve the signal to background ratio.

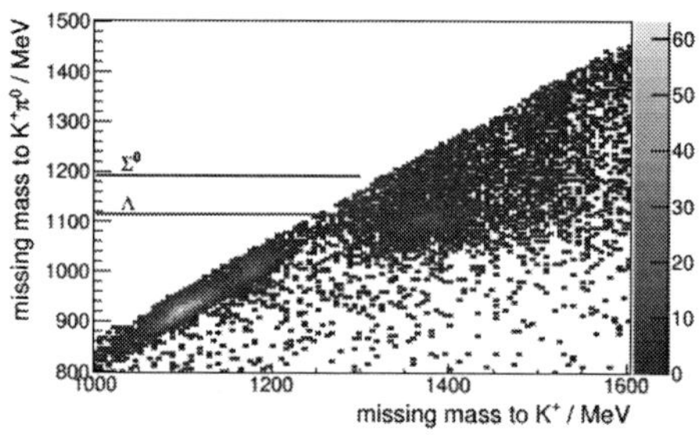

Figure 6. Missing mass to $K^+\pi^0$ against missing mass to K^+, while the K^+ was detected in the forward spectrometer.

3.2 Full topology measurement

Measuring all final state particles in the reaction:

$$\gamma p \rightarrow \Lambda(1405)K^+ \quad (1)$$
$$\rightarrow \Sigma^0\pi^0 K^+ \rightarrow \Lambda\gamma\pi^0 K^+ \quad (2)$$
$$\rightarrow p\pi^-\gamma\gamma\gamma K^+, \quad (3)$$

allows identification with K^+ polar angles outside the forward spectrometer acceptance. However the lack of particle identification increases the number of combinatorial background. Therefore after creating all possible combinatorics, a kinematics fit is used to improve the mass

http://doi.org/10.1051/epjconf/202024101014

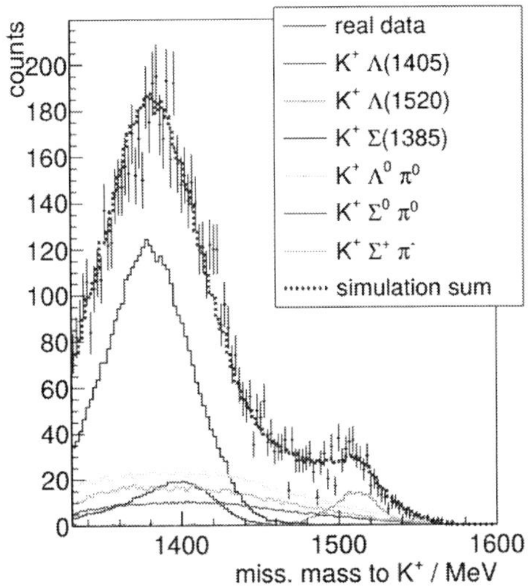

Figure 7. Missing mass to K^+. This is a projection of the two dimensional histogram in figure 6. The colored lines show the contributions of different simulated channels fitted to the data.

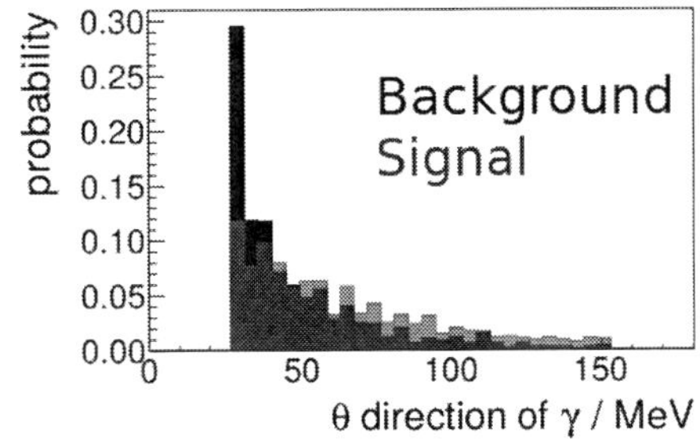

Figure 8. Angular distribution of photon compared between signal in red and background in black.

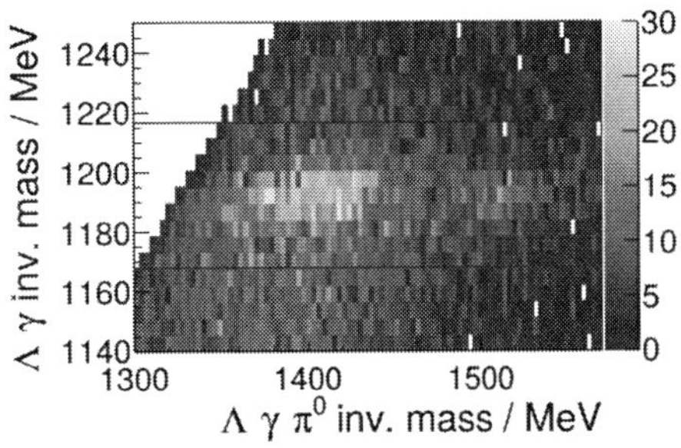

Figure 9. Invariant mass of $\gamma\Lambda$ against $\gamma\Lambda\pi^0$ in the reaction $\gamma p \to \Sigma^0\pi^0K^+$.

resolution and reduce background via confidence level selection cuts. In figure 8 the angular distribution of the photons is shown for signal and background simulation after the kinematic fit. The combinatorial background passes the selection cuts predominantly with low photon angles. Events with low photon angles are removed from the analysis to improve the signal to noise ratio at the cost of statistics. The $\gamma\Lambda$ against $\gamma\Lambda\pi^0$ invariant mass distribution is plotted in figure 9. The signal and background can be distinguished, as only the signal shows a correlation to the Σ^0 mass in the $\Lambda\gamma$ invariant mass. A peak for $\Lambda(1405)$ and $\Lambda(1520)$ is visible as both decay to $\Sigma^0\pi^0$. To subtract the background, the mass outside the signal region, marked by the red lines is used to modulate the amplitude of different simulated reactions, for example $K^+\Sigma(1385)$. In figure 10 the projection of the $\Lambda\gamma\pi^0$ mass is shown, where the signal with $\Lambda\gamma$ mass close to Σ^0 is subtracted. With this fit the background distribution is modeled and can be extrapolated to the region of the signal. In figure 11, the $\Lambda\gamma$ projection of the two dimensional fit is shown. The signal region was not included in the fit. Subtracting the fitted background distribution from the data yields the invariant mass distribution for pure $\Sigma^0\pi^0$ reactions, which is plotted in figure 12. The contribution of non $\Lambda(1405)$ events is estimated by the green line. At this point the line shape and differential cross section can be extracted by subtraction of the simulated contributions. The preliminary results are shown in section 4.

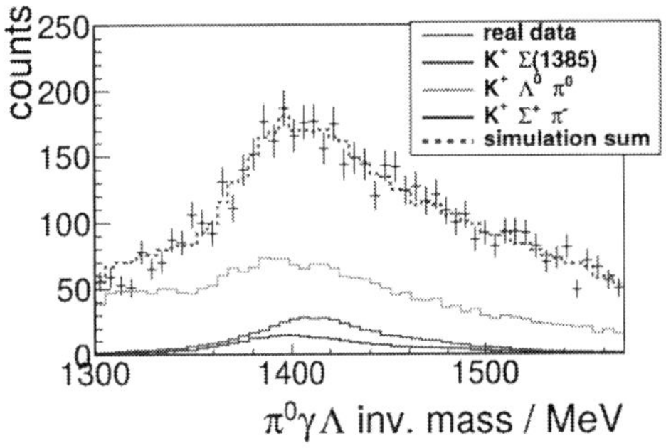

Figure 10. $\Lambda\gamma\pi^0$ mass projection of figure 9, without events where $\Lambda\gamma$ mass is close to Σ^0 mass. The colored lines show example reactions contributing to the background shape, while the red dotted line shows the total fit including also reactions not shown in figure.

4 Preliminary results for $\Lambda(1405) \to \Sigma^0\pi^0$ line shape and differential cross section

Using the analysis steps from the previous section the line shape and differential cross section was determined. For now only the $\Sigma^0\pi^0$ decay was analyzed, as the BGO-OD setup is well suited to measure the decay particles. In sec-

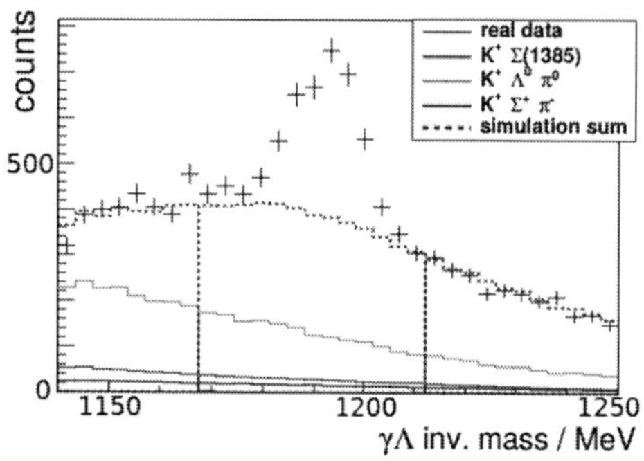

Figure 11. $\Lambda\gamma$ mass projection of figure 9. The colored lines show example reactions contributing to the background shape, while the red dotted line shows the total fit including also reactions not shown in figure.

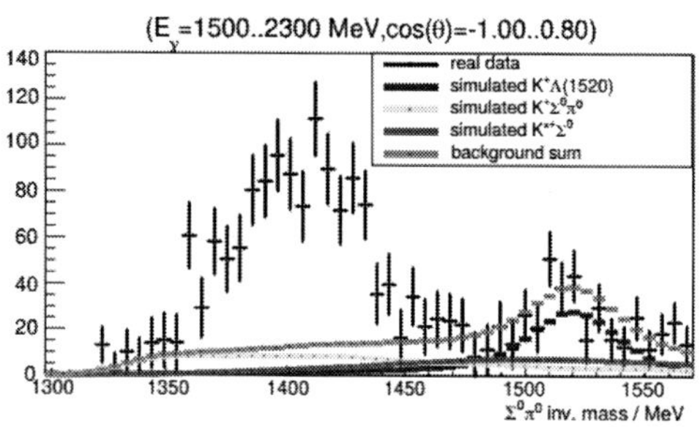

Figure 12. Invariant mass of $\Sigma^0\pi^0$ after all analysis steps. The colored lines show the estimation for not $\Lambda(1405)$ reactions.

tion 4.1 the line shape results are shown, while section 4.2 shows the differential cross section.

4.1 Line shape

The preliminary results on the line shape are compared to other experimental results in figure 13. The $\Lambda(1520)$ signal was not subtracted for consistency checks. The results are statistically comparable to the CLAS experiment, which derives from the better acceptance for this decay mode compared to the CLAS setup. Comparing the ANKE and BGO-OD results indicates a double peak structure in the line shape, which agree to the two pole structure of $\Lambda(1405)$[10]. This seems not to be present in the CLAS data and the statistical error is too big to make a definite statement. All experimental results agree statistically to the predicted line shape in the figure. As mentioned in the introduction the deviations to the predictions show up in the charged decay modes, which will be analyzed in the future.

4.2 Differential cross section

The results on the cross section are plotted in figure 14. In general the cross section agrees to the CLAS

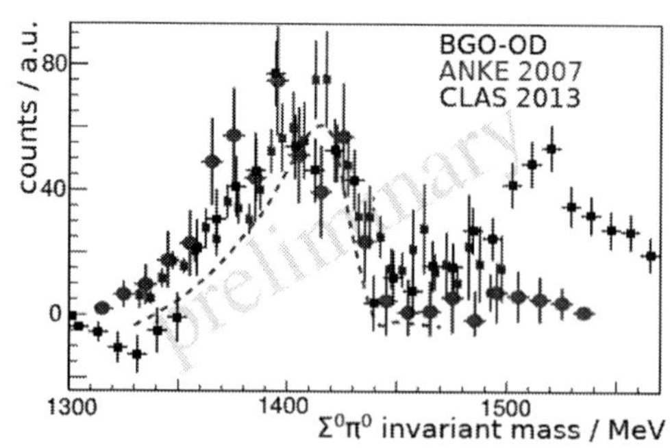

Figure 13. Preliminary line shape of $\Lambda(1405)$ for E_γ = 1500 to 2300 MeV compared to the CLAS experiments in blue[5] and ANKE in red[9]. The dashed blue line marks the predictions by J.C.Nacher[3]. The BGO-OD and ANKE results are in yield with no normalization. To compare the line shape, the results of the experiments were scaled such that the maximum values agree.

results. The forward spectrometer analysis extends the CLAS results to extreme forward angles, which correlates to minimum transfer momentum. While the results agree for higher energies, it suggest the photon energy bin E_γ = 1550..1750 MeV shows a more flat distribution compared to the CLAS results. The charged decay modes results of CLAS also show this behavior compared to the their measured neutral decay mode. Further investigation is needed to clarify this.

5 Summary and Outlook

The BGO-OD experiment is ideally suited to investigate the formation of unconventional states. Since more than 50 years $\Lambda(1405)$ is a potential candidate for such a state. Models describe $\Lambda(1405)$ as a $N\bar{K}$ molecule-like structure, which is formed below the free $N\bar{K}$ production threshold. This is also used as an explanation of the sudden cut-off in the invariant mass distribution (line shape) above 1426 MeV as this marks the free production threshold. Thus the line shape is of great interest for the study of $\Lambda(1405)$. Preliminary results for the decay mode $\Lambda(1405) \rightarrow \Sigma^0\pi^0$ line shape and differential cross section agree within statistics to the CLAS experiment, while the differential cross section could be extended to extreme forward angles thanks to the forwards spectrometer of the BGO-OD experiment. The charged decay modes of $\Lambda(1405)$ will be investigated in the future.

Acknowledgements

I thank the ELSA group for operating and maintaining of the electron accelerator. For the strong support on the maintenance and improvement of our experimental setup I thank the technical staff of the contributing institutions.

This work was supported by the Deutsche Forschungsgemeinschaft project numbers 388979758 and 405882627. Our Russian collaborators thank the

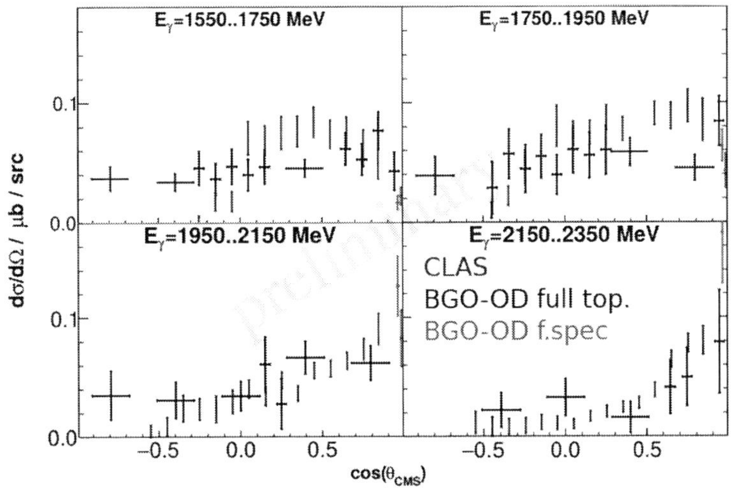

Figure 14. Preliminary differential cross section against angle compared to the CLAS experiment[5] in blue. The black points mark the full topology reconstruction while magenta shows the K^+ in the forward spectrometer analysis.

Russian Scientific Foundation (grant RSF number 19-42-04132) for financial support.

References

[1] J.M.M. Hall et al., Physical Review Letters **114**, doi:10.1103/PhysRevLett.114.132002 (2015)

[2] R. Molina and M. Döring, Physical Review Letters D **94**,056010 (2016)

[3] J.C.Nacher et al., Physical Review Letters B **455**,55 (1999)

[4] K.Moriya et al.(CLAS Collaboration), Physical Review Letters C **87**,035206 (2013)

[5] K.Moriya et al.(CLAS Collaboration), Physical Review Letters C **88**,045201 (2013)

[6] W. Hillert, Europ. Phys. Jour. A **28**, 139 (2006)

[7] On basis of website. url: http://www-elsa.physik.uni-bonn.de/index_en.html

[8] M.Nanova et al. (CB-ELSA Collaboration), Phys. Rev. C, **94** (2016)

[9] I.Zychor et al., Physical Review Letters B **660**, 167-171 (2008)

[10] J.A. Oller and U.-G. Meißner, Phys. Lett. B **500**, 263-272 (2001)

Hyperon Physics with PANDA at FAIR

Karin Schönning[1,*] for the PANDA collaboration

[1]Department of Physics and Astronomy, Uppsala University, Sweden

Abstract.
Hyperons provide new angles on two of the most challenging problems in contemporary physics: a coherent and quantitative description of the strong interaction, and the matter-antimatter asymmetry of the Universe. The production dynamics and the electromagnetic structure of strange hyperons, as well as hyperon spectroscopy, give insights into the strong interaction in the confinement domain. Furthermore, two-body decays of strange hyperons provide clean tests of CP symmetry, an essential piece to the matter-antimatter puzzle. The future experiment PANDA at FAIR offers unique possibilities to study different aspects of hyperons using antiproton beams. In particular, the hitherto almost unexplored multi-strange sector will be addressed. The expected large production cross sections of hyperons and the versatile, near 4 π detector design makes PANDA a veritable hyperon factory already from the first phase of operation. In these proceedings, the opportunities for hyperon physics with PANDA will be outlined. I will also address how we can benefit from the weak hyperon decays, that provide straight-forward access to the full spin density matrix.

1 Introduction

In this proceedings, I will outline how hyperons can shed light on some of the most challenging, unresolved problems of contemporary physics. These problems are all related to one of the most abundant building blocks of the Universe: the nucleon.

Despite being known for a century, many of the nucleon features remain a mystery. Until this day, we are struggling to understand from first principles (a) its abundance, (b) its spin, (c) its mass, (d) its intrinsic structure and (e) its size: (a) There is much more matter (nucleons) than anti-matter (anti-nucleons) in the Universe. If this was not fine-tuned at the Big Bang, then this asymmetry must be of dynamical origin [1]. (b) The sum of the measured valence quarks' contribution to the nucleon spin is only about half of the total spin [2]. (c) In the chiral limit, where the (current) quark masses are zero, the mass of the resulting nucleon would still amount to more than 90% of the physical nucleon mass [3]. Thus, in contrast to the electron, the nucleon mass is not governed by the Higgs mechanism, but is dynamically generated of the strong interaction. To phrase it differently: More than 90% of the visible mass of the Universe is caused by the strong interaction. (d) The neutron has an intriguing charge distribution: negative in the center and at the rim, positive in between [4]. For a ground-state wave function this amount of wiggles is surprising. (e) The electric radius of the proton has been determined from electron-proton scattering, from the spectrum of electronic hydrogen and from the spectrum of muonic hydrogen. Most measurements using electrons resulted in a larger proton radius than the results obtained with muons [5]. This is known as the proton radius puzzle which puts lepton universality into question. In a recent measurement of electronic hydrogen, a smaller radius was obtained that was in agreement with measurements with muons [6]. However, the internal disagreements of the electron data remain to be understood.

A successful approach in various areas of physics when you do not understand a system, is to make a small change and see how it reacts. For instance[7]

1. Scatter on it
2. Excite it
3. Replace one of the building blocks

In these proceedings, we will focus on the third item. Replacing a building block – in this case a light quark – in a nucleon with a different one – a strange, charm or bottom quark – results in a hyperon. The question *"What happens if you replace one of the light quarks in a nucleon, with a heavier one?"* constitutes the common ground of hyperon physics. By combining two or three of the items, research fields such as hyperon structure (item 1 and 3) and hyperon spectroscopy (item 2 and 3) emerge.

2 Hyperons

The scale probed in a given reaction depends on the mass of the interacting entities. In the production of a system with strangeness, the scale is governed by the mass of the strange quark, $m_s \approx 95$ MeV/c^2 which is close to the cut-off $\Lambda_{QCD} \approx 200$ MeV/c^2, defining the scale where quarks are confined into composite systems. As a consequence,

*e-mail: karin.schonning@physics.uu.se

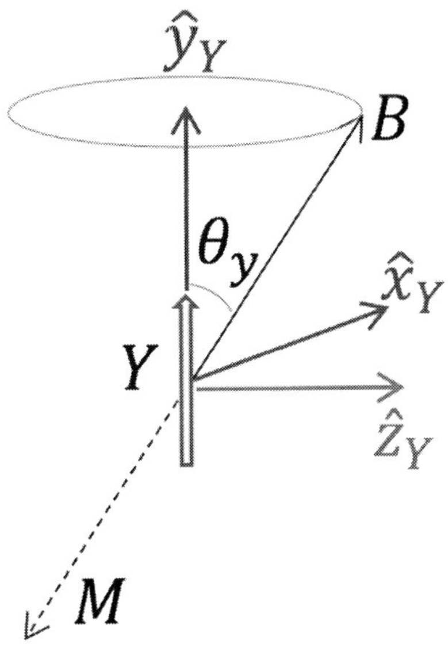

Figure 1. The $Y \to BM$ decay, with the spin direction of Y along the y-axis.

the relevant degrees of freedom of such processes are unclear: quarks and gluons, or hadrons? Strange hadrons, *e.g.* hyperons, therefore probe the confinement domain of the strong interaction. This is in contrast to the charm quark $/m_c \approx 1275$ MeV/c^2) which is more than ten times heavier than the strange quark. This corresponds to the scale where perturbative QCD breaks down and these processes therefore probe the transition region between perturbative and non-perturbative QCD.

Hyperons have an advantage compared to nucleons: their weak, parity violating and thereby self-analysing decays. This means that the daughter particles from a hyperon decay are emitted according to the direction of the spin of the mother hyperon. Figure 1 illustrates the two-body decay $Y \to BM$, where a spin 1/2 hyperon Y decays into a spin 1/2 baryon B and a pseudo-scalar meson M. The angular distribution $W(\cos\theta_B)$ of B in the rest system of Y can be expressed in terms of the polarisation $P_y(\cos\theta_Y)$ with respect to some reference axis $\vec{y}$ as a function of the Y scattering angle [8, 9]

$$W(\cos\theta_B) = \frac{1}{4\pi}(1 + \alpha P_y(\cos\theta_Y)\cos\theta_B). \quad (1)$$

P_y carries information about the production process and depends on the energy. The decay asymmetry α is related to the interference between the parity violating and the parity conserving decay amplitudes, T_s and T_p [10]. Equation 1 shows how parameters with physical meaning can be extracted from measurable quantities and demonstrates the unique potential of the hyperon as a diagnostic tool.

3 PANDA at FAIR

The future experiment Proton antiproton ANnihilations at DArmstadt (PANDA) at the Facility for Antiproton and Ion Research (FAIR) in Darmstadt, Germany, offers unique opportunities for hyperon physics. PANDA will be an integrated part of the High Energy Storage Ring (HESR). It is a multipurpose detector designed for a broad physics program, with focus on the strong interaction [11].

3.1 The HESR

The antiprotons will be delivered from the HESR within a momentum range from 1.5 GeV/c up to 15 GeV/c. In the start-up phase, or *Phase One*, the HESR will be able to accumulate up to 10^{10} antiprotons within a time span of 1000 s. In a final stage, *Phase Three*, a dedicated ring *i.e.* the Recuperated Experimental Storage Ring (RESR) will allow up to 10^{11} antiprotons to be injected and stored in the HESR. HESR offers stochastic cooling that results in a beam momentum spread of better than $5 \cdot 10^{-5}$. The antiproton beam will impinge on a hydrogen cluster jet or pellet target. During the Phase One, the HESR will provide a luminosity of $\approx 10^{31}$ cm^2s^{-1} whilst the design luminosity of $\approx 2 * 10^{32}$ cm^2s^{-1} will be achieved during Phase Three.

3.2 The PANDA detector

The PANDA detector [11] is divided into a target spectrometer part (TS) and a forward spectrometer part (FS). The TS provide precise vertex tracking by the micro vertex detector (MVD), straw tube trackers (STT) and gas electron multiplier detectors (GEM). Time-of-flight detectors (TOF) and detection of internally reflected Cherenkov light (DIRC) offer particle identification and an electromagnetic calorimeter (EMC) will measure energies. The trajectories in the TS are bent by the field of a solenoid magnet providing a field of 2.0 T. The FS consists of straw tube stations for tracking, a dipole magnet, a ring imaging Cherenkov counter (RICH) detector and a TOF for particle identification and a Shashlyk electromagnetic calorimeter. The PANDA detector is shown in Fig. 3.

PANDA will be one of the first experiments featuring a time-based data acquisition system (DAQ) where data are read out as a continuous stream using an entirely software-based data selection scheme. This is in contrast to the present paradigm where data are acquired from the detectors by means of a hardware trigger. This change of paradigm is driven by the increasing reaction rates and large amounts of data to be stored. However, the software filter will provide a challenge since it needs to reconstruct events, tracks and clusters in real time to make decisions. At the design luminosity, the event rate will be up to 20 MHz which requires solutions beyond the state of the art in data processing and transmission.

The feasibility studies as well as future data analysis are performed within the common simulation and analysis framework PandaROOT [12]. It comprises the complete simulation chain, including Monte Carlo event generation, particle propagation and detector response, hardware digitization, reconstruction and calibration, and data analysis. PandaROOT is derived from the FairROOT framework [13] which in turn is based on ROOT [14].

Figure 2. Schematic overview of the full PANDA setup. The antiproton beam will run from left to right, and the target jets/pellets from top to bottom. The left part of the detector surrounds the interaction point and is the Target Spectrometer, whereas the right part is the Forward Spectrometer.

3.3 Advantages of PANDA

PANDA has several advantages compared to currently operating or planned facilities:

- The production cross sections of hyperon-antihyperon final states are generally large in antiproton-proton annihilations [15]. This results in large production rates.
- Multi-strange hyperons and antihyperons can be produced in two-body reactions, which makes the kinematics simple and the dynamics straight-forward to parameterise.
- The momentum range covers the full multi-strange hyperon spectrum.
- The symmetric hyperon-antihyperon final state enables better control of systematics.
- The near 4π acceptance allows for exclusive reconstruction which minimises background.

4 Hyperon topics in PANDA

The combination of the PANDA features mentioned in Section 3.3 opens up for a plenty of hyperon physics topics that are difficult to study elsewhere. Two main fundamental questions, *i.e.* QCD in the confinement domain and the matter-antimatter asymmetry of the Universe, manifest themselves in different aspects of the nucleon. By utilising hyperons as a diagnostic tool, we address these questions through four research topics: hyperon production, hyperon spectroscopy, hyperon structure and hyperon decays. This is illustrated in Fig. 3.

4.1 Hyperon Production

The production of strange hyperons probe the strong interaction the confinement domain, as pointed out in Section 2. For a coherent understanding of the strong interaction, studies of exclusive hyperon-antihyperon production are crucial. The large amount of high-quality data on single-strange hyperons provided by the PS185 collaboration in the 1990 [15, 16] resulted in an extensive development of theoretical models. These were either based on quark-gluon degrees of freedom [17], kaon exchange [18] or a combination of the two [19]. In particular, spin observables are proven to be sensitive to the production mechanism. While most models can explain basic features such as the total cross section and the scattering angle distribution, no model was able to completely describe the spin structure of the studied reactions [15, 20]. It would be illuminating to see how spin observables such as polarisation and spin correlations behave in the multi-strange sector. However, so far only a few bubble-chamber events have been found for Ξ^- and Ξ^0 production in $\bar{p}p$ annihilations [21]. Triple-strange hyperon production has not been observed at all in $\bar{p}p$ and the same is true for single-charmed hyperons. Measurements of the spin structure of $\bar{p}p \to \overline{Y}Y$ for ground-state multi-strange and single-charmed hyperons can serve as important guidance for the development of models for strangeness production, which in turn can shed light on the strong interaction in this energy domain.

The remarkably large production cross sections for various hyperons in $\bar{p}p \to \overline{Y}Y$ [15] imply that we can collect large hyperon samples also with the initial modest luminosity. Previous simulation studies, using a simplified Monte Carlo framework, have shown promising results regarding efficiency and background reduction [22–24]. Recently, more realistic simulation studies have been performed for single- and double strange hyperons. These simulations were performed using the PandaROOT framework, with ideal pattern recognition and particle identification. Additional selection criteria were applied that mimic the effects from a full, realistic implementation.

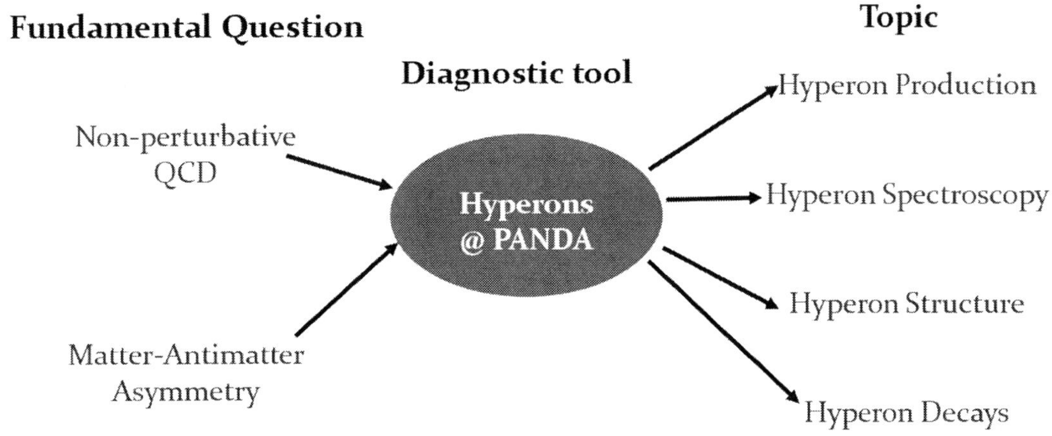

Figure 3. Conceptual map of hyperon physics in PANDA. It shows how the two over-arching fundamental questions (left) can be approached by a common diagnostic tool, *i.e.* hyperons (middle), from which the different subtopics emerge (right).

- $\bar{p}p \to \bar{\Lambda}\Lambda, \bar{\Lambda} \to \bar{p}\pi^+, \Lambda \to p\pi^-$ at p_{beam} = 1.64 GeV/c [25]
- $\bar{p}p \to \bar{\Sigma}^0\Lambda, \bar{\Sigma}^0 \to \bar{\Lambda}\gamma, \bar{\Lambda} \to \bar{p}\pi^+, \Lambda \to p\pi^-$ at p_{beam} = 1.77 GeV/c and 6.0 GeV/c [26].
- $\bar{p}p \to \bar{\Xi}^+\Xi^-, \bar{\Xi}^+ \to \bar{\Lambda}\pi^+, \bar{\Lambda} \to \bar{p}\pi^+, \Xi^- \to \Lambda\pi^-, \Lambda \to p\pi^-$ at p_{beam} = 4.6 GeV/c and 7.0 GeV/c [25].

All channels include intermediate Λ and $\bar{\Lambda}$, that were identified by combining the reconstructed final state pions and protons/antiprotons. Vertex fits and mass window criteria were applied to select good Λ and $\bar{\Lambda}$ candidates. The background was further reduced by requiring the decay vertex to be displaced by a certain distance from the interaction point. To identify $\bar{\Sigma}^0$, the $\bar{\Lambda}$ candidates were combined with final state photons. The $\bar{\Lambda}\Lambda$ and $\bar{\Sigma}^0\Lambda$ final states were finally identified by applying a four-constraint fit. The $\Xi^-/\bar{\Xi}^+$ candidates were obtained by combining $\Lambda/\bar{\Lambda}$ candidates by pions that do not originate from a $\bar{\Lambda}\Lambda$ decay. A decay tree fitter was then applied to the full sequential decay of the Ξ^- and $\bar{\Xi}^+$ hyperons. The resulting signal efficiencies are given in Table 1 and show that also at the reduced Phase One luminosity of 10^{31}cm^{-2}s^{-1}, the exclusive hyperon reconstruction rates will be high. This, in combination with the estimated low background level, demonstrates the potential of PANDA as a hyperon factory.

4.2 Hyperon Spectroscopy

Baryon excitation spectra provide a different angle to the strong interaction. In particular, they address the question about how the forces between the quarks bind them together into baryons. Are quark degrees of freedom relevant or should baryons rather be described in terms of *e.g.* di-quark/quark configurations? Does baryon-meson dynamics play a role? Are there exotic states, such as pentaquarks and dibaryons? Hyperons address the additional question of SU(3) flavour symmetry. It is interesting to compare the findings from world-wide efforts in N^* and Δ spectroscopy [29, 30], such as missing resonances and level ordering, with those of the single-strange sector. In particular, the lightness of the $\Lambda(1405)$ is difficult to explain in terms of the simple Constituent Quark Model but has instead been suggested to be a molecular state, see *e.g.* Ref.[31]. It would be illuminating to see how these observations carry over to the double- and triple strange hyperon spectra. However, the existing data bank is very scarce. Only one excited Ξ state and no excited Ω states are considered well established, *i.e.* labeled with $***$ or $****$ within the PDG classification scheme. In fact, even for the ground state Ξ and Ω, the parity has not been determined experimentally. Furthermore, the spin determination of the Ω relies on assumptions on the Ξ_c and Ω_c spin [32].

Whereas current baryon spectroscopy addresses either light-, single-strange, charmed or bottom baryons, there is a gap to fill in the multi-strange sector. PANDA is an excellent candidate to fill this gap thanks to the large foreseen hyperon production rates, the possibility of producing multi-strange hyperons in two-body reactions and the symmetric hyperon-antihyperon conditions. Recent, dedicated simulation studies on $\Xi^*(1690)$ and $\Xi^*(1820)$ production, presented by J. Pütz at this conference, demonstrates the excellent prospects of the multi-strange hyperon spectroscopy programme with PANDA [33].

4.3 Hyperon Structure

For nucleons, there are several ways to study structure in both soft and hard processes. In the soft region, Electro-

http://doi.org/10.1051/epjconf/202024101015

Table 1. Results from simulation studies of the various production reactions of ground state hyperons [25, 26]. The efficiencies are for exclusive reconstruction, *i.e.* considering all final state particles. The S/B denotes signal-to-background ratio and the asterisk * the upper limit of a 90% confidence interval.

$p_{\bar{p}}$ (GeV/c)	Reaction	σ (μb)	Eff (%)	Decay	S/B	Rate (s^{-1}) at 10^{31}cm^{-2}s^{-1}
1.64	$\bar{p}p \to \bar{\Lambda}\Lambda$	64.0 [16]	15.7 [25]	$\Lambda \to p\pi^-$	114 [25]	44
1.77	$\bar{p}p \to \bar{\Sigma}^0\Lambda$	10.9 [16]	5.3 [26]	$\Sigma^0 \to \Lambda\gamma$	> 11* [26]	2.4
6.0	$\bar{p}p \to \bar{\Sigma}^0\Lambda$	20.0 [27]	6.1 [26]	$\Sigma^0 \to \Lambda\gamma$	21 [26]	5.0
4.6	$\bar{p}p \to \bar{\Xi}^+\Xi^-$	1.0 [28]	8.2 [25]	$\Xi^- \to \Lambda\pi^-$	274 [25]	0.3
7.0	$\bar{p}p \to \bar{\Xi}^+\Xi^-$	0.3 [28]	7.9 [25]	$\Xi^- \to \Lambda\pi^-$	165 [25]	0.1

Magnetic Form Factors (EMFFs) quantify the inner, dynamical structure. Elastic lepton-nucleon scattering gives access to the space-like EMFFs, as shown to the left in Fig. 4. Here, the interpretation is fairly intuitive: the form factors are related to the charge- and magnetization densities [34]. Unfortunately, since hyperons are unstable, they are unfeasible as beams or targets and thus difficult to study in elastic scattering. As a consequence, the space-like region is hard to access experimentally. Instead, the time-like form factors constitute the most viable structure observables for hyperons [35]. They can be studied in various processes shown to the right of the y-axis in Figure 4. From the measured time-like form factors, the more intuitive space-like quantities (*e.g.* charge- and magnetisation densities) can be calculated using dispersion relations [36]. Time-like EMFFs can be complex with a relative phase, reflecting the existence of intermediate states. This phase polarises the hyperons, even if the initial state is unpolarised [37].

The time-like region covers two distinct parts with respect to the momentum transfer squared, q^2, of the virtual photon:

- The high-q^2 part, explored when an e^+e^- pair annihilates to form a baryon-antibaryon pair ($e^+e^- \to B_1\bar{B}_2$, see rightmost part of Fig. 4). This region covers $q^2 > (m_{B_1} + m_{B_2})^2$.
- The low-q^2 part, probed in baryon Dalitz decays ($B_1 \to B_2 e^+e^-$, see middle-left part of Fig. 4) and covering $q^2 < (m_{B_1} - m_{B_2})^2$.

If $B_1 = B_2$, the form factors are *direct*, and if $B_1 \neq B_2$ they are *transition* form factors.

So far, only a few experimental studies of hyperon structure exist. The vast majority are carried out in the high-q^2 part. Cross section measurements were performed by BaBar [38], CLEO-c [39] and BESIII [40] and the complete time-like structure was pinned down for the first time in a recent measurement by BESIII [41]. Some structure observables of the charmed Λ_c^+ has been extracted by Belle [42] and BESIII [43].

The low-energy region ($\approx$ 200 MeV) is rather unexplored territory. Low-energy quantities, such as magnetic and charge radii, are unknown for most ground-state hyperons [9] and the same holds for the transition form factors, that can be accessed in Dalitz decays. A pioneering measurement in the non-strange baryon sector was recently reported by the HADES collaboration for $\Delta^+(1232) \to \gamma^* p \to e^+e^-p$ [44]. In the strange sector, the only existing measurement is the weak branching fraction of the $\Xi^0 \to \Lambda e^+e^-$ decay that was found to be of the order 10^{-6} [45].

The major challenge in studying hyperon Dalitz decays is the small predicted branching fractions, which are typically within 10^{-3} - 10^{-6}. However, the expected large hyperon production rates in PANDA can compensate for the small branching fractions. For example, the simulations in Table 1 show that at 6 GeV/c, $\bar{\Sigma}^0$ hyperons can be reconstructed at a rate of 5 events per second. This corresponds to $\approx$ 400 000 events per day. Provided the reconstruction efficiency of $\bar{\Sigma}^0 \to \bar{\Lambda}e^+e^-$ is of the same order of magnitude as $\bar{\Sigma}^0 \to \bar{\Lambda}\gamma$, even the branching fraction predicted by QED of 10^{-3} would mean that a fairly large data sample could be collected within a few weeks. This enables a test of one of the most long-standing QED predictions from Ref. [46], as well as a study of the q^2-dependence of the magnetic form factor $|G_M|$. The latter is related to one of the low-energy constants of the χPT Lagrangian [7].

Thanks to an upgrade of the HADES detector [47], including forward tracker planes from PANDA, hyperon Dalitz decay studies can be undertaken already during *FAIR Phase 0* [48].

4.4 Hyperon Decays

According to today's paradigm, equal amounts of matter and antimatter were created in the Big Bang. However, according to all observations, our visible Universe consists mostly of matter. Where did the antimatter go? The dynamical enrichment of matter with respect to antimatter is called *Baryogenesis* [1]. It is only possible if the following criteria are fulfilled: i) processes exist which violate baryon number conservation ii) there are reactions in which charge conjugation (C) and charge conjugation and parity (CP) symmetry are violated and iii) the processes in i) and ii) occurred outside thermal equilibrium.

CP violation is incorporated in the SM by the Cabibbo-Kobayashi-Maskawa mechanism [49]. In the meson sec-

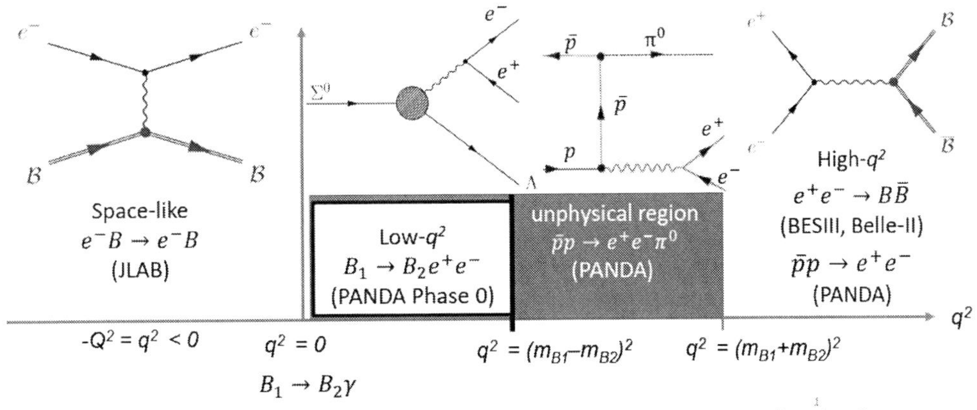

Figure 4. Processes for extracting EMFF in the space-like (left) and time-like (right) region. The low-q^2 ($4m_e^2 < q^2 < (m_{B1} - m_{B2})^2$) part of the time-like region is studied by Dalitz decays, the unphysical region ($4m_{\pi 0}^2 < q^2 < (m_{B1} + m_{B2})^2$) by $\bar{p}p \to e^+e^-\pi^0$ and the high-q^2 region ($q^2 > (m_{B1} + m_{B2})^2$) by $B\bar{B} \leftrightarrow e^+e^-$. Note that the unphysical region is only accessible for protons.

tor, it is established since long by experimental observations of K^0 and B^0 decays. Very recently, the LHCb collaboration reported the first evidence of CP violation in decays of the charmed D meson [50]. However, our visible Universe consists largely of spin carrying baryons, for which the situation is very unclear. The only indication of CP violation observed so far for baryons was observed in the four-body decay of the bottom Λ_b hyperon in LHCb [51]. The deviation is consistent with the SM. In total, no CP violating effects beyond the SM have ever been seen neither in the meson nor in the baryon sector. Furthermore, all deviations are too small to explain the observed matter-antimatter asymmetry of the Universe. Indeed, baryogenesis requires physics beyond the Standard Model [52].

The strange baryon sector is interesting since Super-Symmetry (SuSy) [53] predicts CP violating effects that are one to two orders of magnitudes larger than in the SM [54]. Precise measurements of hyperon decays can therefore distinguish these. Furthermore, strange hyperons decay predominatly into two-body states for which CP-odd observables are straight-forward to define. CP conservation means that particles and antiparticles decay in the same way, but with inverted spatial coordinates. Hence for a weak, two-body hyperon decay, the asymmetry parameter α in Eq. 1 should be the same as the $-\bar{\alpha}$ of the corresponding antihyperon. Then one can define

$$A_{CP} = \frac{\alpha + \bar{\alpha}}{\alpha - \bar{\alpha}}, \quad (2)$$

for which a non-zero value indicates CP violation [8].

In the case of multi-strange hyperons, *e.g.* the Ξ^-, the sequential weak decays provide access to the additional decay asymmetry parameters β and ϕ. From β and $\bar{\beta}$, the asymmetries B_{CP}, B'_{CP} and $\Delta\phi$ can be defined in a similar manner as for A_{CP} in Eq. 2. These are more sensitive to CP violation by a factor of 100 and 10, respectively [55].

The most precise experimental CP test for strange baryons so far is provided by the HyperCP experiment at Fermilab and concerns the sequential decay of the double-strange Ξ^- hyperon [56]. The most precise measurement for the Λ hyperon was recently achieved by the BESIII collaboration [57]. However, to test the SM and beyond SM theories, the precision needs to be improved by several orders of magnitude. This can be achieved in the third phase of PANDA, when the HESR has reached its design luminosity.

Hyperon-antihyperon ($Y\bar{Y}$) pair production from $p\bar{p}$ annihilations provides a clean environment for CP tests compared to the meson sector, since no mixing occurs between the Y and the $\bar{Y}$. The large production rates in $p\bar{p}$ annihilations ensure excellent statistical precision. With the design luminosity of $2 \cdot 10^{32} cm^{-2}s^{-1}$, and the efficiencies of Table 1, $10^7 - 10^8$ exclusive $\bar{\Lambda}\Lambda$ events can be collected within one day. For $\bar{\Xi}^+\Xi^-$ pairs, the corresponding number is $10^5 - 10^6$. This means that PANDA has potential to reach the sensitivities required for putting SM, as well as theories beyond SM, at test.

5 Spin formalism in binary hyperon reactions

The weak decay of ground-state hyperons give access to spin variables, as demonstrated in Eq. 1. We have briefly discussed how this can be utilised in studies of the production mechanism, the structure and symmetry tests of the decay. Taking the full, joint decay chain into account also provides additional information in spectroscopy studies.

Hyperons produced in strong (*e.g.* $\bar{p}p$ annihilations) and electromagnetic (*e.g.* e^+e^-) with unpolarised beam(s) and targets can be polarised along the normal (y_Λ in Fig. 5) of the decay plane. In addition, the spin of the hyperon is correlated with that of the antihyperon. The polarisation and spin correlations, *i.e.* the *spin variables*, are related to the production and are therefore reaction specific functions of the CMS energy and the hyperon scattering angle. The number of spin variables depend on the spin of the produced hyperons.

For the $e^+e^- \to \bar{Y}Y$ reaction, vector exchange (either virtual photon or a vector resonance, *e.g.* J/Ψ) dominates the production process. For spin 1/2 hyperons, the joint angular distribution of the full process, including the subsequent hyperon decays, can be expressed in terms of two production parameters. In Ref.[58], the angular distribution parameter η and the phase $\Delta\Phi$ are chosen. These are *global* in the sense that they do not depend on the hyperon scattering angle. Equivalent global parameterisations are s- and d partial waves, spin flip or non-flip amplitudes, or electric and magnetic form factors. Due to this simple feature of the $e^+e^- \to \bar{Y}Y$ production mechanism, the spin variables have a well-defined dependence on the hyperon scattering angle. The cross section depends on five measurable quantities. In Ref. [58] these are chosen to be the hyperon scattering angle in the CMS system and the helicity angles of the proton and antiproton from the hyperon and antihyperon decays. This multi-dimensional approach is nearly model-independent, since the only assumption is that the process is dominated by vector exchange, and maximises the use of measured information. As a consequence, the statistical precision for a given sample size is increased. The method was utilised in a recent measurement by the BESIII collaboration in a precision measurement of the Λ decay asymmetry parameters [57] and in the first complete measurement of the Λ time-like structure [41]. The formalism has been extended to include also sequentially decaying spin 1/2 hyperons [60, 61] and spin 3/2 baryons [62].

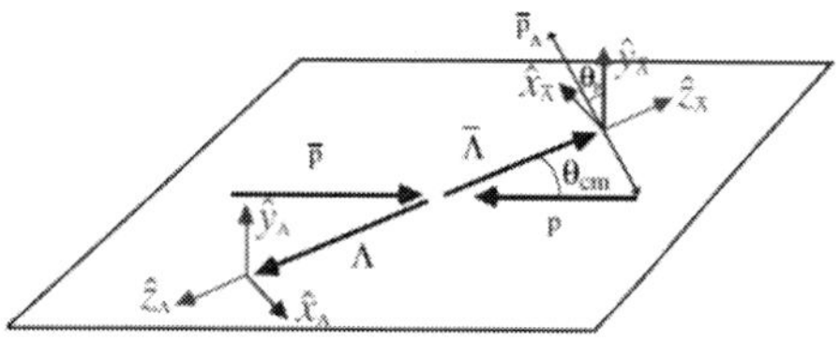

Figure 5. The reference system of the $\bar{p}p \to \bar{Y}Y$ reaction.

The $\bar{p}p \to \bar{Y}Y$ reaction occurs through contributions from many different amplitudes. The angular dependence of the spin variables can therefore not be parameterised in terms of global parameters as in the e^+e^- case but need to be measured and compared to theory predictions. In the case of a spin 1/2 hyperon Y (antihyperon $\bar{Y}$) produced in $\bar{p}p \to \bar{Y}Y$ and decaying into a spin 1/2 baryon B (antibaryon B) and a meson M (antimeson $\bar{M}$), the angular distribution of the decay baryon and antibaryon can be parameterised at each hyperon scattering angle $\cos\theta_Y$:

$$I(\theta_B,\theta_{\bar{B}})_{\theta_Y} = N[1 + \alpha \sum_i P^Y_{i,\theta_Y} \cos\theta_{Bi} + \bar{\alpha} \sum_j P^{\bar{Y}}_{j,\theta_Y} \cos\theta_{\bar{B}j} + \alpha\bar{\alpha} \sum_{ij} C^{Y\bar{Y}}_{ij,\theta_Y} \cos\theta_{Bi}\theta_{\bar{B}j}] \quad (3)$$

where $i, j = x, y, z$. With the unpolarised beam and target foreseen with PANDA and the reference system defined in Fig. 5, most spin variables must be zero due to parity conservation. The only non-zero spin variables are P^Y_y, $P^{\bar{Y}}_y$, $C^{Y\bar{Y}}_{xz}$, $C^{Y\bar{Y}}_{zx}$, $C^{Y\bar{Y}}_{xx}$, $C^{Y\bar{Y}}_{yy}$ and $C^{Y\bar{Y}}_{zz}$ [22, 63]. Of these, only five are independent since $P^Y_y = P^{\bar{Y}}_y$ and $C^{Y\bar{Y}}_{xz} = C^{Y\bar{Y}}_{zx}$.

The $\bar{p}p \to \bar{\Omega}^-\Omega$ reaction is more complicated; assuming it is correct that the Ω has spin 3/2, it has fifteen polarisation parameters $r^L_M(\cos\theta_\Omega)$, where $L = 0, 1, 2, 3$ and $M = 0, \pm1, \pm2, \pm3$. Three of these can be accessed in the Ω decay [22], the other four can be accessed through the sequential decay chain $\Omega \to \Lambda K^-, \Lambda \to p\pi^-$, and similarly for $\bar{\Omega}^+$ [64, 65]. From the full decay chain, structure observables such as form factors can be determined, as well as fundamental properties such as spin.

Most existing Partial Wave Analysis (PWA) frameworks for baryons consider decay processes. However, by including the weak sequential decay chain of ground state hyperons, it should be possible to extract more information from the available data, provided the samples are large enough. Previous chapters have demonstrated the potential of the various PANDA phases in this regard and the development of combined, model-independent framework will further enhance the physics scope of PANDA.

6 Acknowledgement

The author is grateful for the support by the local organisers of NSTAR 2019, the Swedish Research Council (No. 2013-04278) and the Knut and Alice Wallenberg foundation, Sweden (No. 2016.0157).

References

[1] A. D. Sakharov, Pisma Zh. Eksp. Teor. Phys. Fiz. **5**, 32 (1967).
[2] C. A. Aidala *et al.*, Rev. Mod. Phys. **85**, 655 (2013).
[3] S. Scherer and M. R. Schindler, Lect. Notes Phys. **830**, 1 (2012).
[4] G. A. Miller, Phys. Rev. Lett. **99**, 112001 (2007).
[5] R. Pohl *et al.*, Nature **466** (2010) 213; C. E. Carlson, Prog. Part. Nucl. Phys. **82**, 59 (2015).
[6] N. Bezginov *et al.*, Science **365**, Issue 6457, 1007-1012 (2019).
[7] C. Granados, S. Leupold and E. Perotti, Eur. Phys. J. **A 53**, 17 (2017).
[8] I. I. Bigi, A. I. Sanda, *CP violation* Cambridge Monographs on Particle Physics, Nuclear Physics and Cosmology, **vol. 9**, Cambridge University Press (2000).
[9] M. Tanabashi *et al.*, Phys. Rev. D **98**, 030001 (2018).
[10] T. D. Lee and C. N. Yang, Phys. Rev. **108**, 1645 (1957).
[11] The PANDA Collaboration, PANDA Technical Progress Report, FAIR-ESAC (2005).
[12] S. Spataro *et al.*, *The PandaRoot framework for simulation, reconstruction and analysis*, J. Phys.: Conf. Ser. **331**, 032031 (2011).
[13] M. Al-Turany *et al.*, *The FairRoot Framework*, J. Phys.: Conf. Ser. **396**, 022001 (2012).
[14] R. Brun and F. Rademakers, *ROOT - An Object Oriented Data Analysis Framework*, Nucl. Inst. Meth. in Phys. Res. A **389**, 81-86 (1997).

[15] T. Johansson, *Proceedings of LEAP 2002* 95 (2003).
[16] P. D. Barnes *et al.* Nucl. Phys. A **526**, 575 (1991);P. D. Barnes *et al.* Phys. Rev. C **62**, 055203 (2000);K. D. Paschke *et al.*, Phys. Rev. C **74**, 015206 (2006).
[17] M. Kohno and W. Weise, Phys. Lett. B **179** 15 (1986); H. R. Rubinstein and H. Snellman Phys. Lett. B **165** 187 (1985); S. Furui and A. Faessler, Nucl. Phys. A **486** 669 (1987); M. Burkardt and M. Dillig Phys. Rev. C **37** 1362 (1988); M. A. Alberg *et al.* Z. Phys. A **331** 207 (1988).
[18] F. Tabakin and R. A. Eisenstein Phys. Rev. C **31** 1857 (1985); M. Kohno and W. Weise, Phys. Lett. B. **179** 15 (1985); P. La France *et al.* Phys. Lett. B **214** 317 (1988); R. G. E. Timmermans *et al.* 1992 Phys. Rev. D **45** 2288 (1992); J. Haidenbauer *et al.* Phys. Rev. C **46** 2516 (1992).
[19] P. G. Ortega *et al.* Phys. Lett. B. **696** 352 (2011).
[20] E. Klempt, F. Bradamante, A. Martin, J.-M. Richard, Phys. Rep. **368**, p 119 (2002).
[21] B. Musgrave and G. Petmezas, Nuovo Cim. **35**, 735 (1965);C. Baltay *et al.*, Phys. Rev. B **140**, 1027(1965).
[22] E. Thomé, *Multi-Strange and Charmed Antihyperon-Hyperon Physics for PANDA*
Ph. D. Thesis, Uppsala University (2012).
[23] The PANDA collaboration, Physics Performance Report (2009).
[24] S. Grape, *Studies of PWO Crystals and Simulations of the* $\overline{p}p \to \overline{\Lambda}\Lambda, \overline{\Lambda}\Sigma^0$ *Reactions for the PANDA experiment*
Ph.D. Thesis, Uppsala University (2009).
[25] W. Ikegami-Andersson, *Exploring the Merits and Challenges of Hyperon Physics with PANDA at FAIR*, PhD Thesis, Uppsala University (2020).
[26] G. Perez Andrade, *Production of the* Σ^0 *hyperons in the PANDA experiment at FAIR*, Master thesis, Uppsala University (2019).
[27] H. Becker *et al.*, Nucl. Phys. B **141**, 48 (1978).
[28] A. B. Kaidalov and P. E. Z. Volkovitsky, Phys. C **63** 51 (1994).
[29] E. Klempt and J. M. Richard, Rep. Prog. Phys. **82** 1095 (2010).
[30] V. Crede and W. Roberts, Rep. Prog. Phys. **76** 076301 (2013).
[31] J. M. M. Hall *et al.*, Phys. Rev. Lett. **114**, 132002 (2015).
[32] B. Aubert *et al.* (BaBar Collaboration), Phys. Rev. Lett. **97**, 112001 (2006).
[33] J. Puetz, these proceedings (2019).
[34] V. Punjabi *et al.*, Eur. Phys. J. A **51**, 79 (2015).
[35] S. Pacetti *et al.*, Phys. Rep. **550-551**, 1 (2015).
[36] M. A. Belushkin et al., Phys. Rev. C **75** 035202 (2007).
[37] A. Z. Dubnickova *et al.*, Nuovo Cim. A **109** 241 (1996).
[38] B. Aubert *et al.* (BaBar Collaboration), Phys. Rev. D **76**, 092006 (2007).
[39] S. Dobbs *et al.*, Phys. Lett. B **739**, 90 (2014); S. Dobbs *et al.*, Phys. Rev. **D 96**, 092004 (2017).
[40] M. Ablikim *et al.*(BESIII Collaboration), Phys. Rev. D **97**, 032013 (2018).
[41] M. Ablikim *et al.* (BESIII Collaboration), Phys. Rev. Lett. **123**, 122003 (2019).
[42] G. Pakhlova *et al.*, Phys. Rev. Lett. **101**, 172001 (2008).
[43] M. Ablikim *et al.* (BESIII Collaboration), Phys. Rev. Lett. **120**, 132001 (2018).
[44] J. Adamczewski-Musch *et al.*, Phys. Rev. C **95**, 065205 (2017).
[45] J. R. Batley *et al.*, Phys. Lett. B **650**, 1 (2007).
[46] J. Bernstein *et al.*, Phys. Rev. **139**, 1650 (1965);G. Feldman and T. Fulton, Nucl. Phys. **8**, 106 (1958).
[47] G. Agakishiev *et al.*, Eur. Phys. J. A **41**, 243 (2009).
[48] J. Adamczewski-Musch *et al.*, Proposal for experiments at SIS18 during FAIR Phase 0 (2017).
[49] M. Kobayashi and T. Maskawa, Prog. Theor. Phys. **49**, 652 (1973); N. Cabibbo, Phys. Rev. Lett. **10** (1963) 531.
[50] R. Aaij (LHCb collaboration), Phys. Rev. Lett. **122**, 211803 (2019)
[51] R. Aaij (LHCb collaboration), Nature Phys. **13**, 391 (2017).
[52] L. Canetti *et al.*, New J. Phys. **14**, 095012 (2012).
[53] X.-G. He *et al.*, Phys. Rev. D **61**, 071701 (2000).
[54] J. Tandean *et al.*, Phys. Rev. D **67**, 056001 (2003).
[55] J. F. Donogue *et al.*, Phys. Rev. D **74** 3927 (1995).
[56] HyperCP Collaboration, Phys. Rev. Lett. **93** 262001 (2004).
[57] M. Ablikim *et al.* (BESIII Collaboration), Nature Phys. **15**, p. 631–634 (2019).
[58] G. Fäldt and A. Kupsc, Phys. Lett. B **772**, 16 (2017).
[59] G. Fäldt, Phys. Rev. D **97**, 053002 (2018).
[60] G. Fäldt and K. Schönning, *Sequential hyperon decays in the reaction* $e^+e^- \to \Sigma^0\bar{\Sigma}^0$, arXiv:1908.04157 (2019).
[61] P. Adlarson and A. Kupsc, *CP symmetry tests in the cascade-anticascade decay of charmonium*, arXiv:1908.03102 (2019).
[62] E. Perotti *et al.*, Phys. Rev. D **99**, 056008 (2019).
[63] K. D. Paschke *et al.*, Phys. Rev. C **74**, 015206 (2006).
[64] K. Schönning and E. Thomé, J. Phys. Conf. Ser. 556, 012007 (2014).
[65] E. Perotti, J. Phys. Conf. Ser. **1024** no.1, 012019 (2018).

Measurement of polarization observables in neutral double pion photoproduction off the proton with the CBELSA/TAPS experiment

Tobias Seifen[1,*] for the CBELSA/TAPS collaboration

[1]Helmholtz-Institut für Strahlen- und Kernphysik, Universität Bonn, Nussallee 14-16, 53115 Bonn, Germany

Abstract. Data on the reaction $\gamma p \to p\pi^0\pi^0$ were taken with the CBELSA/TAPS experiment at ELSA where linearly polarized photons impinged on a transversely polarized butanol target. Target asymmetries and double polarization observables were extracted and entered the BnGa-PWA. Systematic differences in the branching ratios of N^* and Δ^* resonances hint at the internal structure of these states.

1 Introduction

The excited states in hadron spectroscopy are broad and overlapping. Hence, a partial wave analysis (PWA) needs to be performed where polarization observables are essential to resolve ambiguities in the PWA solutions. Resonance contributions cannot be extracted from the (unpolarized) cross section alone. At higher center-of-mass energy the importance of multi-meson final states increases and cascading decays via intermediate resonances occur.

The neutral double pion channel is especially advantageous compared to the charged pion channel, since it suffers much less from non-resonant contributions: There is no diffractive ρ production, no Kroll-Ruderman term (direct $\Delta\pi$ production) and t-channel processes are supressed, thus enhancing the sensitivity to resonance contributions.

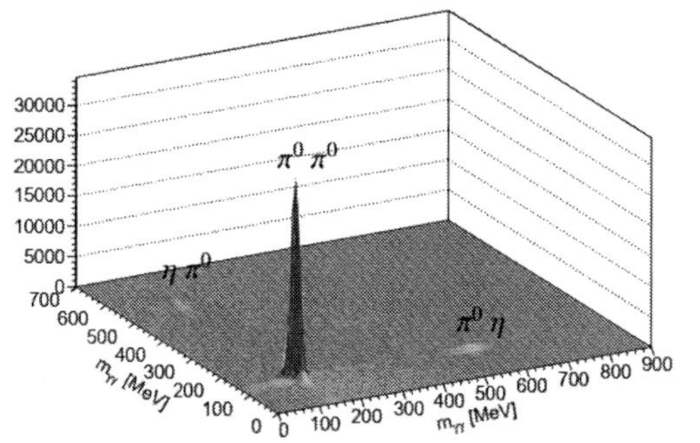

Figure 1. Invariant mass of one photon pair versus the other after the angular cuts and a cut on the missing proton mass.

2 Experimental Setup

The CBELSA/TAPS experiment allows the measurement of double polarization observables. Electrons with an energy of 3.2 GeV from the Electron Stretcher Accelerator ELSA [1] produced linearly polarized photons (maximal polarization degree of 66 % at E_γ = 850 MeV) by means of coherent bremsstrahlung from a diamond crystal [2]. The energy-tagged photons hit a transversely polarized butanol (C_4H_9OH) target [3] (mean polarization degree of 74 %). The Crystal Barrel calorimeter [4] and the Forward Plug, consisting out of 1320 CsI(Tl) crystals, surround the target. Together with the TAPS calorimeter [5] (216 BaF_2 crystals), for polar angles below 12°, nearly the full solid angle was covered.

For the identification of charged particles, a scintillating fiber detector [6] (513 fibers in three layers) surrounding the target and plastic scintillator counters in front of the Forward Plug and TAPS crystals were used.

Furthermore, a gas-cherenkov detector located between the Crystal Barrel and TAPS calorimeters was used to supress electro-magnetic background on the trigger level.

*e-mail: seifen@hiskp.uni-bonn.de

3 Data Selection

To select events stemming from the reaction $\gamma p \to p\pi^0\pi^0 \to p4\gamma$ five hits were required. Four of these hits had to be neutral and be registered in the calorimeters. For the proton candidate a hit in one of the charge-sensitive detectors, giving only a direction, was sufficient. Cuts on the proton direction relative to the vector sum of the photon momenta were applied, requiring the proton and the photons to be back to back in the center of mass system. The invariant mass of one $\gamma\gamma$-pair plotted versus the other is depicted in Fig. 1. The $p\pi^0\pi^0$ final state, as well as the $p\pi^0\eta$ finale state are clearly visible. The final selection employs a kinematic fit including an anti-cut on the $p\pi^0\eta$ final state, with which the combinatorial background still present in Fig. 1 can be eliminated. A cut on the confi-

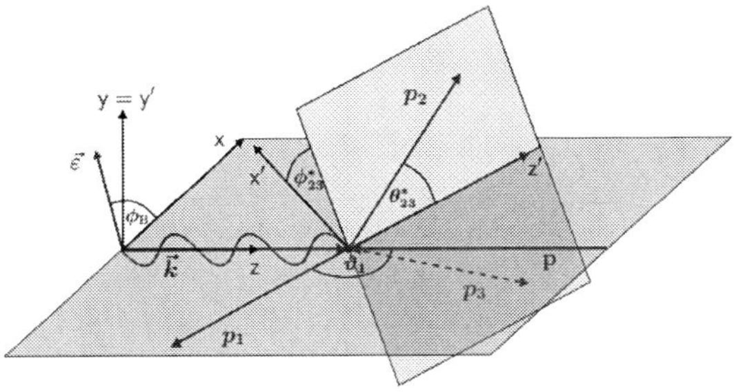

Figure 2. Kinematics of the 3-body final state, showing the reaction plane spanned by the incoming photon and one of the final state particles (p_1) and the decay plane spanned by the remaining final state particles (p_2 and p_3). For the $p\pi^0\pi^0$ final state there are two choices for particles 1(23): $p(\pi^0\pi^0)$ and $\pi^0(p\pi^0)$.

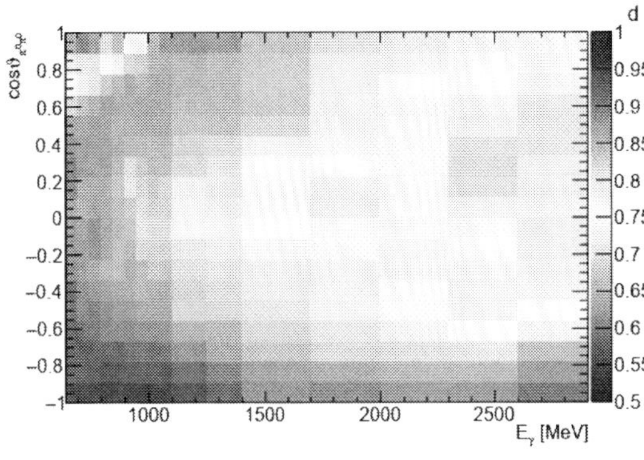

Figure 3. Dilution factor d depending on E_γ and $\cos\vartheta_{\pi^0\pi^0}$.

dence level (> 0.1) acted as a joint mass cut on the pions and the missing proton.

After applying all cuts, the amount of background varies kinematically dependent from well below 1 % to about 5 %. Globally, a background contamination of about 1.5 % was reached.

4 Analysis

When considering multi-meson final states, like in the reaction $\gamma p \rightarrow p\pi^0\pi^0$, additional kinematic variables are needed to describe the kinematics compared to the case of single meson photoproduction. In total five kinematic variables are used (cf. Fig. 2): Apart from the beam energy E_γ and the scattering angle $\cos\vartheta_1$ those are the invariant mass m_{23} and two angles, ϕ^*_{23} and θ^*_{23}.

The polarized cross section of double pseudoscalar meson photoproduction (only taking linear beam polarization and transverse target polarization into account) is given by [7]:

$$\begin{aligned}\frac{d\sigma}{d\Omega} = \frac{d\sigma_0}{d\Omega} \cdot \Big\{ & 1 + \Lambda_x \cdot P_x + \Lambda_y \cdot P_y \\ & + \delta_\ell \sin(2\phi_B) \cdot I^s_{\text{eff}} + \delta_\ell \cos(2\phi_B) \cdot I^c_{\text{eff}} \\ & + \Lambda_x \delta_\ell \sin(2\phi_B) \cdot P^s_x + \Lambda_y \delta_\ell \sin(2\phi_B) \cdot P^s_y \\ & + \Lambda_x \delta_\ell \cos(2\phi_B) \cdot P^c_x + \Lambda_y \delta_\ell \cos(2\phi_B) \cdot P^c_y \Big\},\end{aligned} \quad (1)$$

where Λ_x, Λ_y are the target polarizations in / perpendicular to the reaction plane, and δ_ℓ the beam polarization (with angle ϕ_B to the reaction plane). The beam asymmetries I^s_{eff} and I^c_{eff} contain the beam asymmetries from the bound protons in the carbon and oxygen nuclei and are therefore not considered here further.

In a quasi 2-body approach, where p_2 and p_3 are added up, half of the polarization observables (namely P_x, P^c_x, and P^s_y) vanish exactly, while the others are often identified with their single meson counter parts ($P_y = T$, $P^s_x = -H$, and $P^c_y = -\tilde{P}$).

4.1 Dilution factor

Since the butanol target also contains bound nucleons in the carbon and oxygen nuclei which can not be polarized, the target polarization is effectively reduced by the fraction of polarizable (free) protons. This so-called dilution factor depends on the kinematic variables as well as the applied cuts, since the bound nucleons are subject to Fermi motion. Therefore, the effectively remaining carbon and oxygen dilution was determined experimentally with the use of a carbon foam target of approximately the same area density as the carbon and oxygen nuclei in the butanol target.

The resulting dilution factor is depicted in Fig. 3 as a function of beam energy E_γ and $\cos\vartheta_{\pi^0\pi^0}$. Especially for lower energies ($E_\gamma \lesssim 1100$ MeV) it reaches values of around 0.9.

4.2 Results

The polarization observables were determined in each kinematic bin by means of an event-based maximum-likelihood fit to the ϕ-distribution of the events. Possible detector asymmetries were expanded in a Fourier series and coefficients which interfere with the polarization observables were fitted.

As an example, results for the target asymmetry T in the quasi 2-body approach are shown in Fig. 4 as a function of $\cos\vartheta_{\pi^0\pi^0}$ in the beam energy range of 650 MeV to 2600 MeV. The data points are plotted together with predictions from the 2π-MAID [8] and the BnGa-PWA [9], as well as a new BnGa solution which contains this data. Clear deviations are seen to the MAID predictions over the whole energy range while the 2014 BnGa solution describes the data fairly well at lower energies. At higher energies (above 1700 MeV) larger deviations become apparent. The new BnGa solution describes the target asymmetry quite well over the whole energy range.

Fig. 5 shows the target asymmetry P_x as a function of $\phi^*_{p\pi^0}$. Due to parity conservation P_x has to be odd under the transformation $\phi^* \mapsto 2\pi - \phi^*$. Thus P_x vanishes when integrating over ϕ^*. The open symbols in Fig. 5 make use

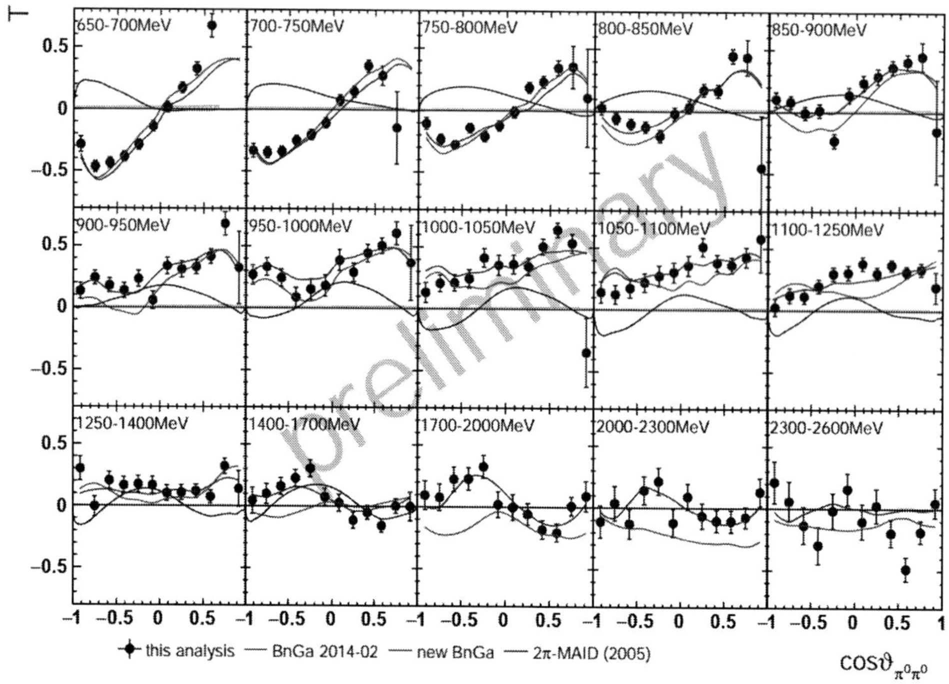

Figure 4. Targetasymmetry T as a function of beam energy E_γ and $\cos\vartheta_{\pi^0\pi^0}$.

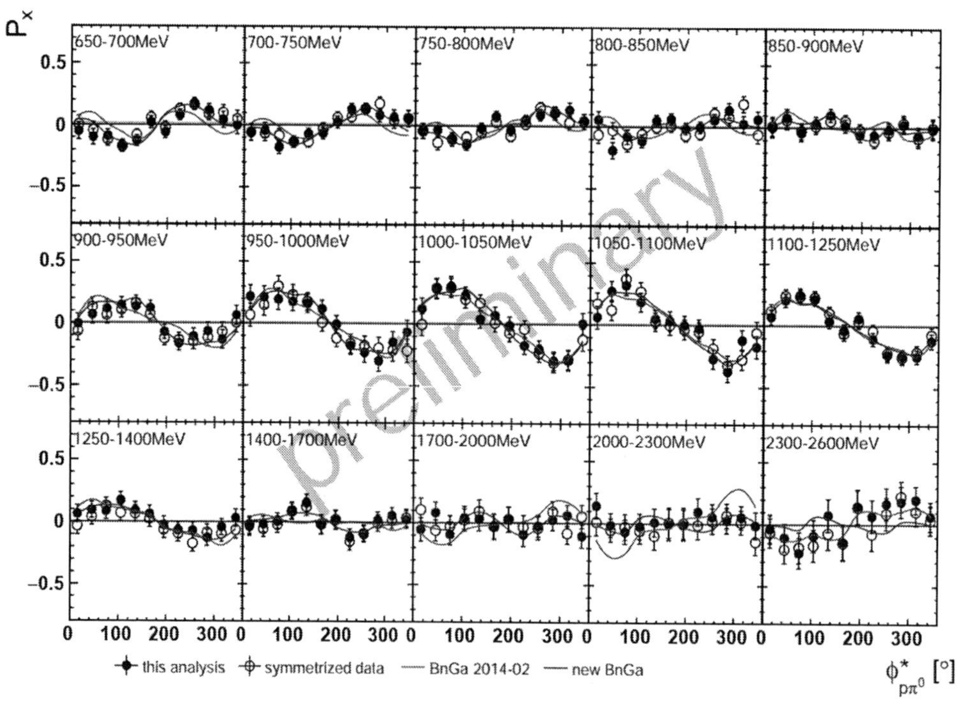

Figure 5. Targetasymmetry P_x as a function of beam energy E_γ and $\phi^*_{p\pi^0}$.

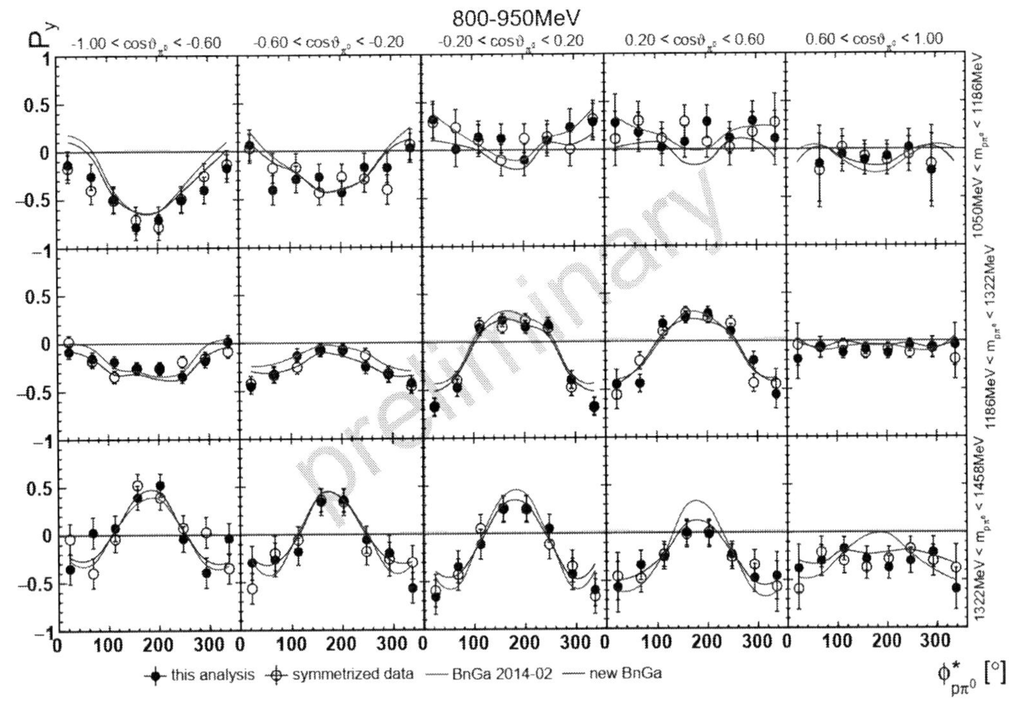

Figure 6. Targetasymmetry P_y as a function of $\phi^*_{p\pi^0}$ in the energy bin (800 – 950) MeV. Within a row the $\cos\vartheta_{\pi^0}$ is varied, within a column the invariant mass $m_{p\pi^0}$. Only a single energy bin is shown here.

of the symmetry property, which is clearly fulfilled in the data.

For beam energies below 1250 MeV the statistics were high enough to determine the target asymmetries in a 4-dimensional analysis. Fig. 6 shows an example of the target asymmetry P_y in one energy bin (800 MeV < E_γ < 950 MeV), where $\phi^*_{p\pi^0}$ is plotted on the x-axis, $\cos\vartheta_{\pi^0}$ is varied within the rows and $m_{p\pi^0}$ within the columns. It becomes immediately apparent that the observables can change significantly from bin to bin. This information is not fully contained in the partially integrated observables.

4.3 PWA

The extracted polarization observables were considered in a new solution of the BnGa coupled-channel partial wave analysis which includes all major data of pion and photo-induced reactions off protons and neutrons [10]. Investigating the decay modes of the excited baryon states, it becomes apparent that some resonances have a significant branching ratio into excited hadrons while others decay dominantly into ground state nucleons or deltas. The structure of the baryon wave function might explain this behavior.

Expanding the spatial wave function of an excited baryon in a harmonic oscillator basis (with oscillators λ and ϱ) allows to group the wave functions into three classes. First, there are wave functions where only one

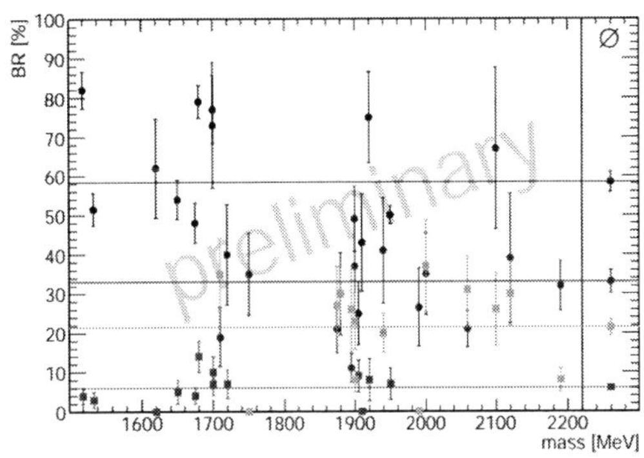

Figure 7. Branching ratios of resonances included in the BnGa-PWA are illustrated. One-oscillator resonances decaying into the ground states (Nπ or $\Delta\pi$) are shown as black dots, decays into excited resonances (N(1520)π, N(1535)π, N(1680)π, or Nσ) as red squares. Mixed-oscillator states decaying into the ground states are shown as blue dots, those decaying into the excited states as green squares. Additionally, the average branching ratios are shown on the right and as colored lines.

oscillator (either λ or ϱ) is excited, e.g. $\ell_\lambda = 2, \ell_\varrho = 0$. Then, there are those where both oscillators are excited simultaneously, e.g. $\ell_\lambda = 1 = \ell_\varrho$. And lastly, there is the coherent sum of the previous two classes, i.e. part of the wave function has only one oscillator excited and part of the wave function has both oscillators excited simultaneously.

Resonances of the second class, which only have a spatial wave function where both oscillators are in an excited state (belonging to the antisymmetric SU(6) 20-plet), have not been found experimentally yet.

Those resonances with only one excited oscillator were found to decay dominantly (nearly 60 % on average) into the ground states Nπ or $\Delta\pi$ (black dots in Fig. 7) while the branching ratios for decays into excited states (N(1520)π, N(1535)π, N(1680)π, or Nσ) are small (red squares), 6 % on average. On the other hand, resonances which additionally have a part in their wave function with both oscillators excited, have a similar branching ratio for decays into the ground states (blue dots in Fig. 7) and excited states (green squares), about 33 % to about 22 % on average.

5 Summary

New results on the measurement of (double) polarization observables in the reaction $\gamma p \rightarrow p\pi^0\pi^0$ with the CBELSA/TAPS experiment were reported, utilizing linearly polarized photons and a transversely polarized butanol target. The data were included in the data base of the BnGa coupled channel partial wave analysis. The new PWA solution hints at the structure of the wave function: Compared to resonances whose wave function have only one excited oscillator, resonances whose wave function have parts with both oscillators in an excited state, have a significant branching ratio into an excited baryon or meson followed by a subsequent decay into the three-particle final state.

References

[1] W. Hillert, *Eur. Phys. J. A* **28**, 139-148 (2006).
[2] D. Elsner *et al.*, *Eur. Phys. J. A* **39**, 373 (2009).
[3] Ch. Bradtke, H. Dutz, H. Peschel *et al.*, *Nucl. Instr. Meth. A* **436**, 430-442 (1999).
[4] E. Aker *et al.*, *Nucl. Instr. Meth. A* **321**, 69-108 (1992).
[5] R. Novotny, *IEEE Trans. Nucl. Sci. NS* **38**, 379-385 (1991).
[6] G. Suft *et al.*, *Nucl. Instr. Meth. A* **531**, 416-424 (2005)
[7] W. Roberts, T. Oed, *Phys. Rev. C* **71**, 055201 (2005).
[8] H. Arenhövel, A. Fix, *Eur. Phys. J. A* **25**, 115-135 (2005).
[9] V. Sokhoyan *et al.*, *Eur. Phys. J. A* **51**, no.8, 95 (2015) Erratum: [Eur. Phys. J. A **51** no.12, 187 (2015)]
[10] Webpage on baryon spectroscopy of the BnGa group: pwa.hiskp.uni-bonn.de/baryon_x.htm

Photoproduction of mesons and Compton scattering on the proton: Selected results from the A2 Collaboration at MAMI

Vahe Sokhoyan for the A2 Collaboration at MAMI

[1]University of Mainz, Institute for Nuclear Physics

Abstract. The A2 Collaboration performs a manifold research program using real photons in the Crystal Ball/TAPS experiment at the MAMI accelerator facility in Mainz. The experiments take advantage of high-intensity unpolarized, linearly or circularly polarized photon beams, and unpolarized or polarized targets. The detector setup provides almost complete coverage in solid angle and is well suited for the detection of multi-particle final states. In order to probe the internal structure of the nucleon, the spectrum of baryon resonances is studied via measurements of unpolarized cross sections and various polarization observables in single and double meson photoproduction. The program aiming to determine the scalar and spin polarizabilities of the nucleons with high precision is performed with the Compton scattering experiments. In 2017, the focal plane detector used in the tagging system of the Crystal Ball/TAPS experiment was completely renewed, allowing new measurements with unprecedentedly high precision. This paper presents recent selected results from the A2 Collaboration at MAMI.

1 Introduction

The program carried out by the A2 Collaboration covers a broad range of topics in hadron physics. The experiments are performed with various combinations of unpolarized, linearly or circularly polarized tagged photons with unpolarized or polarized targets. The polarization of recoil nucleons can be measured with an available recoil polarimeter. In addition to the proton targets, various experiments are carried out with light and heavy nuclei for the studies of mesons photoproduction on neutrons and search for in-medium modifications of hadrons in the nuclear medium. The experimental setup consisting of the Crystal Ball and TAPS calorimeters provides a nearly 4π acceptance coverage.

At energies below pion threshold (80–140 MeV), new data sets on Compton scattering on the proton were recently acquired. Based on these data sets, the scalar polarizabilities of the proton will be extracted with an unprecedented precision. Moreover, the beam asymmetry Σ_3 was measured for Compton scattering below pion threshold for the first time, in order to obtain complementary information for the extraction of the proton scalar polarizabilities, in addition to the unpolarized cross section.

At higher energies, various measurements are performed for single and double meson photoproduction in order to study the dynamics and properties of baryon resonances. The study of double meson final states is mainly focused on photoproduction of $2\pi^0$ and $\pi^0\eta$ pairs. Recently, we obtained a new data set for differential cross sections and beam helicity asymmetry for the $\gamma p \rightarrow p\pi^0\eta$ reaction at photon energies from the threshold up to 1.45 GeV. Moreover, the beam helicity asymmetry was measured for the first time for the photoproduction of $\pi^0\eta$ pairs on heavy nuclei.

This paper includes recent results from the proton scalar polarizability measurement and the new results on the production of $\pi^0\eta$ pairs on a free proton and nuclear targets.

*e-mail: sokhoyan@uni-mainz.de

2 Experimental Setup

The experiments were performed with a tagged photon beam at the MAMI accelerator in Mainz [1, 2]. The electron beam impinged on an amorphous or a diamond radiator, the electrons were deflected in the magnetic field and detected by the focal plane detector [3–5]. The recent upgrade of the focal plane detector with new plastic scintillators, equipped with Si-PMTs significantly extended the capability of the tagging system in handling high rates. The Crystal Ball/TAPS calorimeter system, used to detect particles in the final state is shown schematically in Fig. 1. The Crystal Ball calorimeter covers polar angles from 20° to 160° and consists of 672 NaI crystals [6], whereas TAPS covers the forward angles (1°–20°) and is built of 366 BaF_2 and 72 $PbWO_4$ crystals [7, 8]. Both calorimeters have a full coverage in azimuthal angle. The Crystal Ball calorimeter is additionally equipped with a Particle Identification Detector (PID), made of 24 scintillator bars [9] and with two Multiwire Proportional Chambers [10], which are used for identification and tracking of charged particles. In the forward region, covered by the TAPS calorimeter, plastic veto scintillators are used to distinguish between charged and neutral particles.

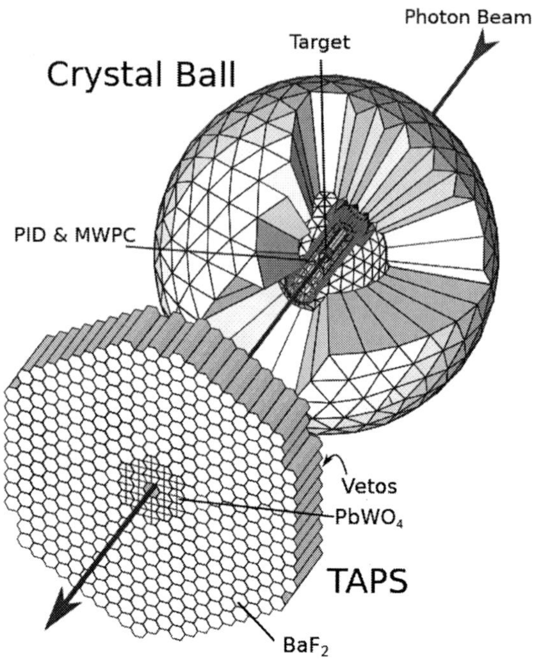

Figure 1. Schematic view of the Crystal Ball/TAPS detection system.

3 Measurement of the proton scalar polarizabilities

The scalar polarizabilities are fundamental properties of the proton. They are of great importance for nuclear and atomic physics, and other related fields. Moreover, these polarizabilities are currently a significant source of uncertainty in the determination of the proton charge radius from muonic hydrogen Lamb shift [11]. Additionally, the precise determination of the scalar polarizabilities is crucial for the extraction of the spin polarizabilities. Previously, the proton scalar polarizabilites were extracted from the unpolarized cross section of Compton scattering on the proton at low energies (see, e.g. Refs. [12–15]). The largest of the data sets for Compton scattering was previously obtained with the TAPS setup at MAMI [12].

Recently, we acquired new data sets on Compton scattering for the photon energy range 80–140 MeV, measured the unpolarized cross section for Compton scattering with unprecedented precision and performed the first ever measurement of the beam asymmetry Σ_3 below pion production threshold. As discussed in Ref. [16], the measurement of the beam asymmetry provides a potentially complementary sensitivity to the scalar polarizabilities of the proton in addition to the measurement of the unpolarized cross section.

Experimentally, the beam asymmetry Σ_3 is measured with linearly polarized photon beam and unpolarized target and is defined as:

$$\Sigma_3 \equiv \frac{\sigma_{\parallel} - \sigma_{\perp}}{\sigma_{\parallel} + \sigma_{\perp}}, \qquad (1)$$

where the cross sections $\sigma_{\parallel}$ and $\sigma_{\perp}$ are extracted with the photon polarization being either parallel or perpendicular to the scattering plane.

The results of the first pilot experiment with the Crystal Ball/TAPS setup [17] proved the feasibility of a high-precision measurement of the proton scalar polarizabilities α_{E1} and β_{M1} with the A2 setup. Figure 2 shows an example missing mass distribution from the pilot experiment (the scattered photon was detected in the Crystal Ball). The signal is clearly identified and the shape of the experimental spectra is in good agreement with the distribution obtained from Monte Carlo simulation, indicating a low background contamination.

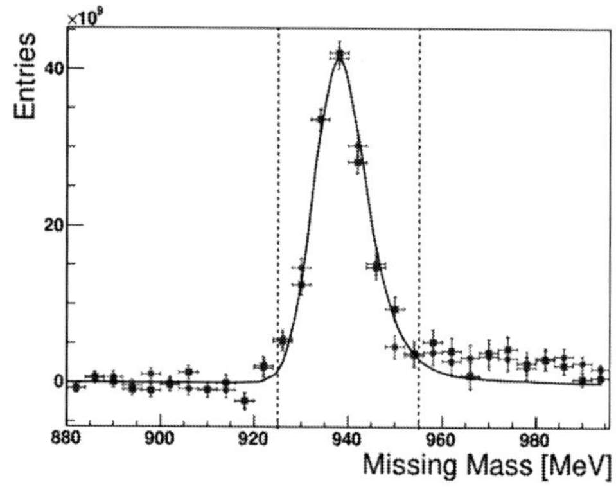

Figure 2. Compton scattering on the proton: Missing mass for the incoming photon energy range 79–98 MeV and polar angles for the scattered photon covering the range $30° < \theta < 155°$. The data were taken with linearly polarized photons with polarization planes either parallel (red circles) or perpendicular to the horizontal plane (blue squares). The black curve shows the corresponding distribution obtained from our Monte Carlo simulation. The dashed lines indicate the cut on the missing mass applied in the analysis. Figure taken from Ref. [17].

After the upgrade of the tagging system, new data were taken by the A2 Collaboration in 2018 with a linearly polarized photon beam and an unpolarized liquid hydrogen target and the analysis is underway (PhD of E. Mornacchi, Mainz) [18, 19]. The new data set shows an improvement of about six times in statistics compared to the pilot experiment. In total, more than 1.2×10^6 Compton scattering events were acquired below pion photoproduction threshold in the range $E_\gamma = 80 - 140$ MeV, constituting by far the highest statistics data set compared to all previous measurements.

Figures 3 and 4 show preliminary results for the unpolarized cross section and beam asymmetry Σ_3, respectively. Comparison of the new cross section data [18, 19] with the highest statistics data set available till now — the

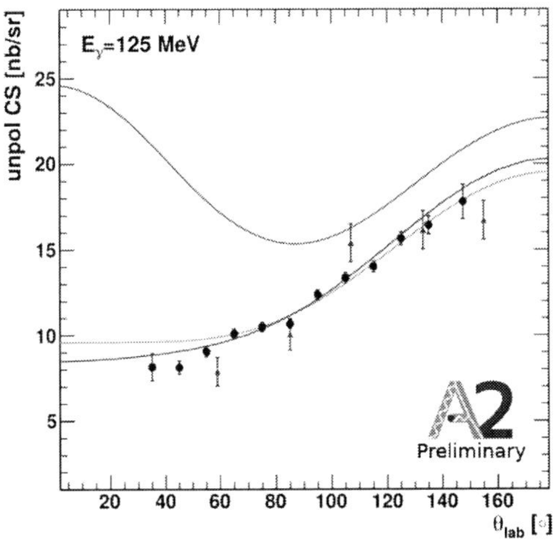

Figure 3. Preliminary unpolarized cross section (black circles) [18], compared to the previous results from Ref. [12] (red triangles). The curves correspond to the Born contribution (brown), Dispersion Relation calculation (magenta) [20, 21], and heavy baryon ChPT calculation (green) [22]. The Dispersion Relation and heavy baryon ChPT calculations were performed for $\alpha_{E1} = 10.65 \times 10^{-4}$ fm^3 and $\beta_{M1} = 3.15 \times 10^{-4}$ fm^3.

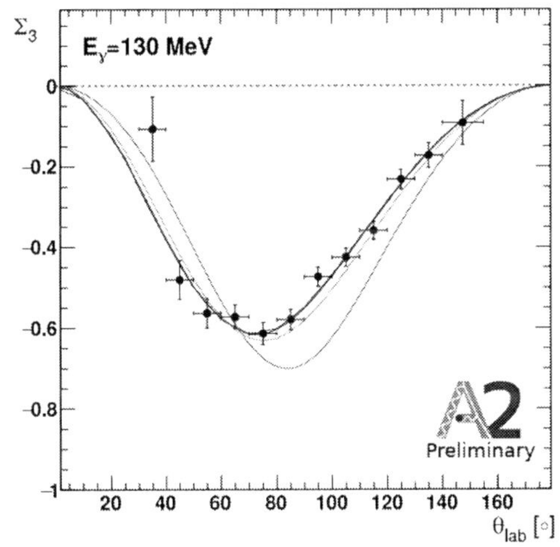

Figure 4. Preliminary beam asymmetry Σ_3 (black circles) [18]. The curves correspond to the Born contribution (brown), Dispersion Relation calculation (magenta) [20, 21], heavy baryon ChPT calculation (green) [22], and ChPT (blue) [23]. All theoretical calculations were performed for $\alpha_{E1} = 10.65 \times 10^{-4}$ fm^3 and $\beta_{M1} = 3.15 \times 10^{-4}$ fm^3.

MAMI 2001 data [12] — indicates the very significant improvement in statistical accuracy. Moreover, the new data allow us to extend the angular coverage to smaller scattered photon angles. The beam asymmetry Σ_3 from the new measurement is in good agreement with the theoretical calculations. The systematic accuracy is improved compared to the pilot experiment [17] due to the upgrades made to the experimental apparatus, including a more stable linear polarization, improved efficiency of the new tagging system, continuous monitoring of the photon flux, and enhanced performance of the refurbished data acquisition system.

4 Photoproduction of $\pi^0\eta$ pairs on the proton

Within the last years, a significant progress has been made at different experiments studying the production of $\pi^0\eta$ pairs on the proton and [24–31] and light nuclei (deuteron and helium) [32, 33]. At energies below 1.5 GeV, the D_{33} wave plays an important role. In the region of the threshold the contribution of the $\Delta(1700)3/2^-$ resonance dominates, whereas the importance of the $\Delta(1940)3/2^-$ resonance increases at higher energies [24–29, 34–36]. Thus, the production of $\pi^0\eta$ pairs is well-suited for studying the $\Delta(1700)3/2^-$ resonance at low energies without being strongly affected by other contributions. Furthermore, an additional selectivity in transitions between decays N^* or two Δ^* resonances is introduced due to the fact that the η meson acts as an isospin filter.

In 2018, we published a new data set for the reaction $\gamma p \to p\pi^0\eta$ covering photon energies from the threshold up to E$_\gamma$ = 1.45 GeV [37]. Figures 5 and 6 show the new differential cross section for the reaction $\gamma p \to p\pi^0\eta$ compared with previously existing data and model calculations. Our new data agree well with results from the previous measurements with the A2 [24] and CBELSA/TAPS experiments [29]. The new data set is described reasonably by the new version of the Mainz model [37] and Bonn-Gatchina Partial Wave Analysis (BnGa PWA) [29]. Some discrepancies present at low and high energies (see Ref. [37] for further details) indicate potential impact of the data for model calculations. After fitting the new data within the framework of the Mainz model, the parameters of the dominant $\Delta(1700)3/2^-$ resonance remained practically the same as in the previous version of the Mainz model. However, some changes were obtained for other resonances (e.g. for $\Delta(1940)3/2^-$). Further details on the influence of the new data on the resonance parameters can be found in Ref. [37]. In the future, the new data can be added to partial wave analyses on an event-by-event basis using available 4-vectors.

5 Production of $\pi^0\eta$ pairs on carbon, aluminum and lead

Due to the selectivity in terms of contributions of baryon resonances and their decays, photoproduction of $\pi^0\eta$ pairs on nuclei is well suited for studying possible modifications

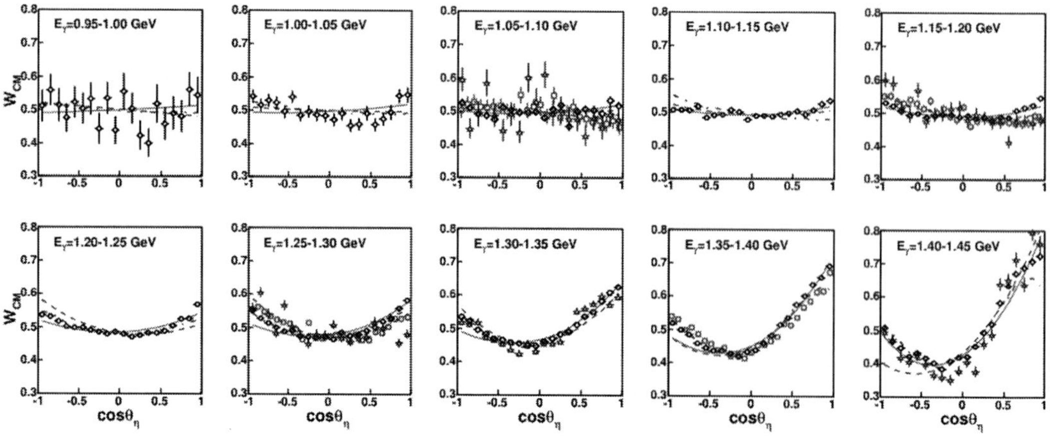

Figure 5. Angular distributions $W_{CM}(\theta_\eta)$ (normalized to unity) for the reaction $\gamma p \to p\pi^0\eta$: New A2 data (open circles), data from CBELSA/TAPS [27, 29] (blue stars, combination of results from both papers), and previous A2 data [24, 35] (magenta open squares) at energies close to the ones from the new data set. Theoretical curves: Predictions of the BnGa PWA [29] (blue dash-dotted curve), previous version of the Mainz model [26] (red dashed curve), and the new version of the Mainz model where the new A2 data were included (solid green curve). Figure taken from Ref. [37].

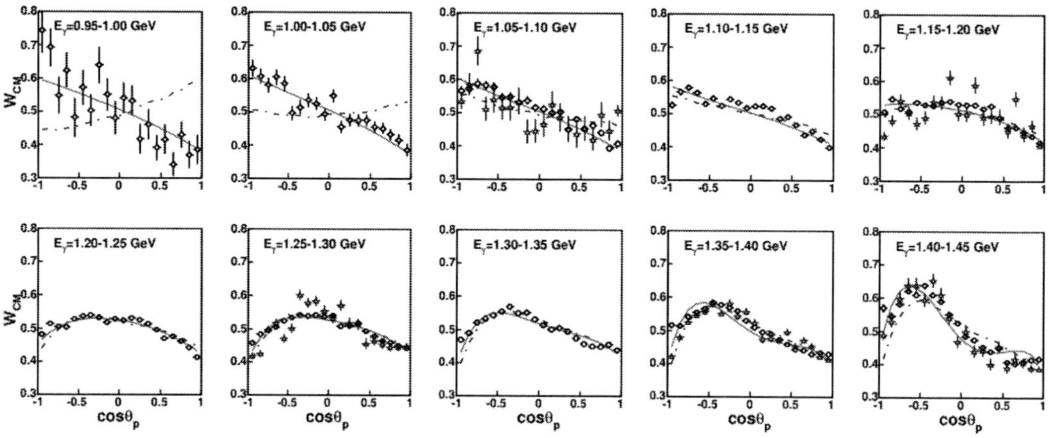

Figure 6. The same notation as for Fig. 5, but the angular distributions for the proton are shown. Figure taken from Ref. [37].

of the contributing resonances in the nuclear medium. The previous results obtained for the proton target [24, 25, 27–31, 37] and recent results for the deuteron from the A2 collaboration [32, 33] confirmed the dominance of the D_{33} partial wave for the production $\pi^0\eta$ pairs on the proton and neutron. Recently, we obtained new data on the photoproduction of $\pi^0\eta$ pairs for carbon, aluminum, and lead targets and measured the beam helicity asymmetry $I^{\odot}$ for these nuclei for the first time [38]. These measurements allowed testing whether the contribution of the D_{33} partial wave still dominates the $\pi^0\eta$ photoproduction mechanism on heavy nuclei. As shown previously for the deuteron target [32], the beam helicity asymmetry is very similar for the free proton and deuteron target, although the unpolarized cross section is significantly reduced. Therefore, one could assume that the beam helicity asymmetry used for studying the production of the D_{33} partial wave on heavy nuclei would not be affected significantly by the Final State Interaction (FSI) effects.

The asymmetry $I^{\odot}$, originating due to the contribution of the D_{33} wave can be expressed [25]:

$$I^{\odot}(\Phi_\pi) = A_1 \sin\Phi_\pi + A_2 \sin 2\Phi_\pi, \qquad (2)$$

where the angle Φ_π, is defined as the azimuthal pion angle in the πN rest frame with respect to the plane determined by the momenta of the incident photon and πN system in the center-of-mass (c.m.) frame.

Due to this feature, the observable $I^{\odot}$ can be used to study the modifications of the amplitudes in the nuclear environment and compared to the production on a free nucleon. Beyond conventional (FSI) effects, the most likely

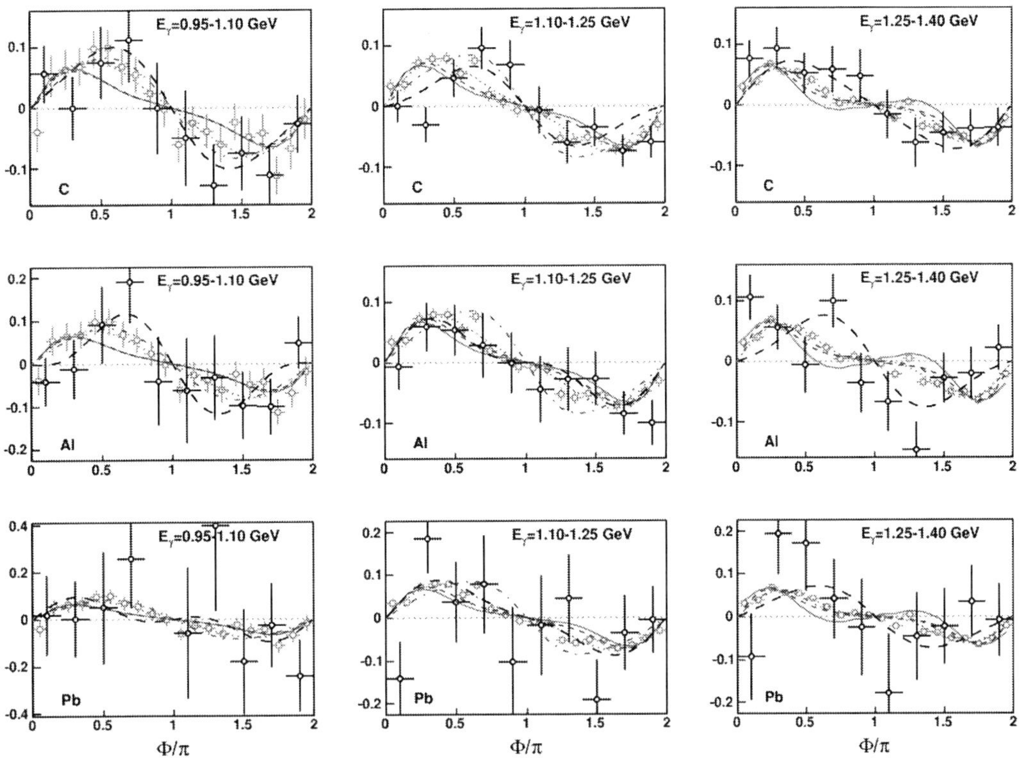

Figure 7. Beam-helicity asymmetry $I^{\odot}$ for the carbon, aluminum, and lead targets (open circles) and the free proton data (cyan open squares) [37]. Curves: Previous version of the Mainz model [25] (magenta dash-dotted line), latest version of the Mainz model from Ref. [37] for a free proton (red dashed line) and for the carbon, aluminum, or lead targets (solid green line). Fits to the experimental data points are performed according to function from Eq. (2) for the nuclear targets (long-dashed black line) and for the hydrogen target (cyan dotted line). The results for carbon, aluminum, and lead are shown in top, middle, and bottom rows, respectively. The data point in the bin $0.8 < \Phi/\pi < 1.0$ and $0.95 < E_\gamma < 1.10$ GeV for Pb is at $I^{\odot} = -0.65 \pm 0.30$. Figure taken from Ref. [38].

modifications in the nuclear medium could be the suppression or an increase of the width of the D_{33} states. Due to the inelastic mechanisms, the relative contributions of contributing terms can be modified, leading to modifications of the shape of the observable $I^{\odot}(\Phi_\pi)$ described by Eq. (2).

In the new experiment, we measured the observable $I^{\odot}(\Phi_\pi)$ for carbon, aluminum, and lead targets. The beam helicity asymmetry is measured using circularly polarized photons and unpolarized target and can be expressed as:

$$I^{\odot}(\Phi) = \frac{d\sigma^+ - d\sigma^-}{d\sigma^+ + d\sigma^-} = \frac{1}{P_\gamma}\frac{N^+ - N^-}{N^+ + N^-}, \quad (3)$$

where $d\sigma^\pm$ corresponds to the differential cross sections as a function of angle Φ (denoted as Φ_π in our case) for the two helicity states of the incoming photon, P_γ is the degree of circular polarization of the photon, and $N^\pm$ represents intensities for each of the two helicity states at the angle Φ.

The beam helicity asymmetry was extracted from events with four or five photon candidates in the final state (for further analysis details see Ref. [38]). The $\pi^0\eta$ events corresponding to the production on protons and neutrons were not separated in the analysis because it was not possible to distinguish between these channels for four-cluster events. Also, it was assumed in the analysis that the target particle had nucleon mass and zero momentum.

Figure 7 shows the beam helicity asymmetry $I^{\odot}(\Phi)$ for the obtained $\gamma N \to 4\gamma X$ events for nuclear targets in comparison with the previous A2 measurement on a free proton [37]. The experimental data for the nuclear targets are generally in good agreement with the free proton data. The data are also compared with the calculations performed within the previous [25] and the latest [37] versions of the Mainz model. The calculations shown for the Mainz model for the nuclear targets are described in Ref. [38] and include FSI effects, showing very similar behavior for the nuclear targets and free proton. Moreover, the model calculations are in agreement with the data obtained with nuclear and free proton targets.

These comparisons are demonstrated in a more quantitative way in Fig. 8, where shown are the coefficients A_1 and A_2 from Eq. (2), extracted from fitting the $I^{\odot}(\Phi_\pi)$ distributions with a function from Eq. (2). These coefficients obtained for the nuclear targets are compared to each other,

 http://doi.org/10.1051/epjconf/202024101017

to the values obtained for the free proton, and to the corresponding predictions of the Mainz model. The coefficients A_1, corresponding to the contribution of the D_{33} wave only, are in good agreement for the nuclear targets and free proton (both for the data and model calculations). The coefficients A_2, describing the interference of the D_{33} wave with other waves, are significantly smaller than A_1, again showing a reasonable agreement between nuclear targets and for free proton data within the statistical errors.

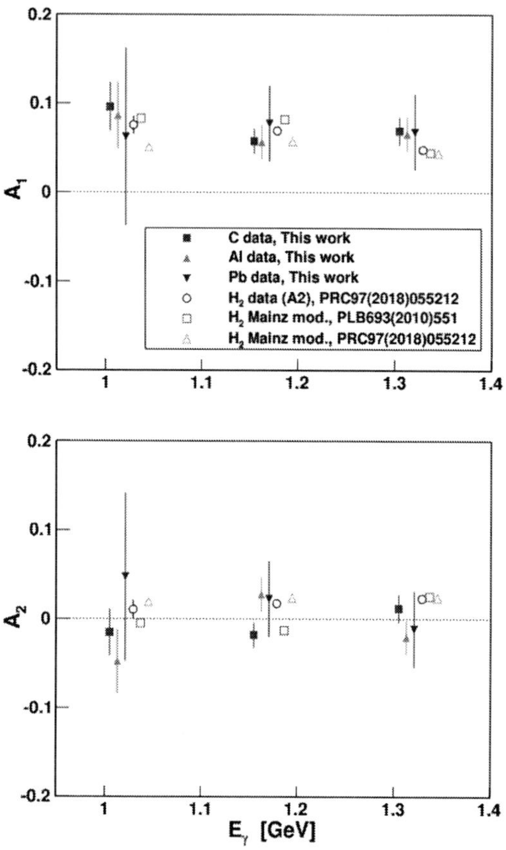

Figure 8. Coefficients A_1 and A_2 obtained from fitting $I^{\odot}(\Phi)$ distributions with a function from Eq. (2). Shown are the coefficients obtained for the nuclear targets, free proton data [37], and results from the previous [25] and the latest Mainz model [37]. Figure taken from Ref. [38].

The good agreement for $I^{\odot}(\Phi)$ measured for the nuclear targets and with the free proton indicates similar mechanisms for the photoproduction of $\pi^0\eta$ pairs (dominated by contribution of the D_{33} wave in our energy range). Interestingly, this observation is consistent with the measurements performed earlier by the A2 Collaboration with a deuterium target [32], where the results for $I^{\odot}(\Phi)$ on a quasi-free proton and neutron were found to be in agreement with the free-proton data. In analogy with the deuteron results, our new data indicate that the asymmetry $I^{\odot}(\Phi)$ is not significantly affected by FSI effects for heavier nuclear targets.

The obtained results stimulate further searches for in-medium modifications of baryon resonances by measuring polarization observables in general and observable $I^{\odot}(\Phi)$ in particular. Moreover, this study represents the first measurement of the production of $\pi^0\eta$ pairs on heavy nuclei, opening a route for further studies of this kind.

References

[1] H. Herminghaus *et al.*, IEEE Trans. Nucl. Sci. **30**, 3274 (1983).

[2] K.-H. Kaiser *et al.*, Nucl. Instrum. Methods Phys. Res. A **593**, 159 (2008).

[3] I. Anthony *et al.*, Nucl. Instrum. Methods Phys. Res. A **301**, 230 (1991).

[4] S. J. Hall *et al.*, Nucl. Instrum. Methods Phys. Res. A **368**, 698 (1996).

[5] J. C. McGeorge *et al.*, Eur. Phys. J. A **37**, 129 (2008).

[6] A. Starostin *et al.*, Phys. Rev. C **64**, 055205 (2001).

[7] R. Novotny, IEEE Trans. Nucl. Sci. **38**, 379 (1991).

[8] A. R. Gabler *et al.*, Nucl. Instrum. Methods Phys. Res. A **346**, 168 (1994).

[9] D. Watts *et al.*, Chin. Phys. C **33**, 1183-1188 (2009).

[10] D. Watts, The Crystal Ball and TAPS detectors at the MAMI electron beam facility, in Proceedings of the 11th nternational Conference on Calorimetry in Particle Physics (Color2004), Perugia, Italy 2004 (World Scienti
c, 2005) (World Scientific, 2004), Vol. 560.

[11] A. Antognini et al., Science **339**, 417 (2013).

[12] V. Olmos de Leon *et al.*, Eur. Phys. J. A **10**, 207-2013215 (2001).

[13] F. J. Federspiel *et al.*, Phys. Rev. Lett. **67**, 1511 (1991).

[14] *B. E. MacGibbon et al.*, Phys. Rev. C **52**, 2097 (1995).

[15] A. Zieger *et al.*, Phys. Lett. B **278**, 34 (1992).

[16] N. Krupina, V. Pascalutsa, Phys.Rev.Lett. **110**, 262001 (2013).

[17] V. Sokhoyan *et al.*, Eur. Phys. J. A **53**, 14 (2017).

[18] E. Mornacchi (University of Mainz), PhD thesis in preparation.

[19] E. Mornacchi [A2 Collaboration], EPJ Web Conf. **199**, 05020 (2019).

[20] D. Drechsel, M. Gorchtein, B. Pasquini, M. Vanderhaeghen, Phys. Rev. C **61**, 015204 (1999).

[21] B. Pasquini, D. Drechsel, M. Vanderhaeghen, Phys. Rev. C **76**, 015203 (2007).

[22] J.A. McGovern, D.R. Philips, H.W. Grießhammer, Eur. Phys. J. A **490**, 12 (2013).

[23] V. Lensky and V. Pascalutsa, Eur. Phys. J. C **65**, 195 (2010).

[24] V. L. Kashevarov *et al.*, Eur. Phys. J. A **42**, 141 (2009).

[25] V. L. Kashevarov *et al.*, Phys. Lett. B **693**, 551 (2010).

[26] A. Fix, V. L. Kashevarov, A. Lee, and M. Ostrick, Phys. Rev. C **82**, 035207 (2010).
[27] I. Horn *et al.*, Eur. Phys. J. A **38**, 173 (2008).
[28] E. Gutz *et al.*, Phys. Lett. B **687**, 11 (2010).
[29] E. Gutz *et al.*, Eur. Phys. J. A **50**, 74 (2014).
[30] J. Ajaka *et al.*, Phys. Rev. Lett. **100**, 052003 (2008).
[31] T. Nakabayashi *et al.*, Phys. Rev. C **74**, 035202 (2006).
[32] A. Käser *et al.*, Eur. Phys. J. A **52**, 272 (2016).
[33] A. Käser *et al.*, Phys. Lett. B **786**, 305 (2018).
[34] M. Döring, E. Oset, and D. Strottman, Phys. Rev. C **73**, 045209 (2006).
[35] A. Fix, M. Ostrick, and L. Tiator, Eur. Phys. J. A **36**, 61 (2008).
[36] A. Fix, V.L. Kashevarov, M. Ostrick, NP A **909**, 1 (2013).
[37] V. Sokhoyan *et al.*, Phys. Rev. C **97**, 055212 (2018).
[38] V. Sokhoyan *et al.*, Phys. Lett. B **802**, 135243 (2020).

Non-strange baryons (CLAS)

Steffen Strauch[1,*] for the CLAS Collaboration

[1]University of South Carolina, Columbia, SC 29208

Abstract. Baryon spectroscopy is an essential tool in the study of nucleon resonances. The use of polarization observables can greatly clarify the spectrum of broad and overlapping nucleon excitations. The N^* program with the CEBAF Large Acceptance Spectrometer (CLAS) at the Thomas Jefferson National Accelerator Facility includes experimental studies with linearly- and circularly-polarized tagged-photon beams, longitudinally- and transversely-polarized nucleon targets, and recoil polarizations. An overview of these experimental studies and recent results are presented.

1 Introduction

The CLAS Collaboration is pursuing an experimental study of baryon resonances with two components. The first is the spectroscopy of excited baryon states with photoproduction experiments. These experiments aim to search for new states and improve the understanding of known states. The experimental data are used in phenomenological analyses or theoretical models to extract resonance information. The extraction is a problem encompassing many short-lived, wide, and overlapping resonances with multiple decay-channels. It is particularly important to include in the analysis data from a variety of excitation and decay channels, as some of the missing states may couple only weakly to, *e.g.*, the πN final state. The CLAS experiments that are reported here include the $\pi^0 p$, $\pi^+ n$, ηp, $\eta' p$, ωp, $K^+\Lambda$, $K^*\Lambda$, and $\pi^+\pi^- p$ reaction channels for reactions off the proton, and $\pi^- p$, $K^0\Lambda$, and $K^0\Sigma^0$ for quasi-free reactions off the neutron. The determination of a set of polarization observables, along with cross-section measurements, is essential in placing stringent constraints on the amplitude of the reaction. The second component of the CLAS N^* program is the measurement of transition form factors of known resonances in electroproduction experiments to study their internal structure and confining potential. R. Gothe presented the CLAS and CLAS12 transition-form-factor experiments at this NSTAR 2019 conference.

The following gives a brief overview of recent examples of photoproduction experiments that were performed with the CEBAF Large Acceptance Spectrometer (CLAS) detector [1] at the Jefferson National Accelerator Facility (JLab). In these experiments, unpolarized, circularly-, or linearly-polarized tagged [2] Bremsstrahlung photon beams were incident on unpolarized, longitudinally, or transversally polarized hydrogen or deuterium targets. The newly developed polarized targets were key to the success of the campaign. The frozen-spin target (FROST) [3] provides dynamically polarized protons in TEMPO-doped butanol. In the g9a run, the average proton polarization was approximately 82% [4]. The dilution factor, the fraction of the number of free to the total number of protons that contribute to the reaction, depends on the reaction channel and kinematics; for the $\gamma p \to \pi^+ n$ reaction, it was typically well above 0.5. Spin-lattice relaxation times up to 4000 h were observed with FROST [3]. The polarized solid deuterium-hydride (HDice) [5, 6] target provided polarized neutrons. In the g14 run, the D polarization of the HDice target averaged 25% [7]. The spin relaxation time of the HDice target can reach years. Final-state particles were detected in the CLAS or reconstructed through invariant and missing masses. The measured observables include total and differential cross-sections, as well as single- and double-polarization observables with combinations of a polarized beam, target, and (in the case of hyperon production) the polarization of the recoiling baryon.

2 Photoproduction off the proton

Measurements of the beam-target observable, E, in the $\gamma p \to \pi^+ n$ reaction with circularly-polarized photons incident on longitudinally-polarized protons were the first results from the FROST experiment [4]. Over 900 data points were obtained in a center-of-mass energy range from W = 1.25 GeV to 2.23 GeV. For most of the covered energy range, previous partial-wave analyses (PWA) showed significant deviations; only at low energies, the agreement with the data was fair. The inclusion of the data in new analyses led to a better description of the data and an improved amplitude determination. These more tightly constrained amplitudes also help fix the πN components of broader multi-channel analyses. One particularly interesting result is the strengthened the evidence for the, at the time, poorly known $\Delta(2200)7/2^-$ resonance by improving the Bonn-Gatchina (BnGa) fit at the highest energies [8]. The mass of the $\Delta(2200)7/2^-$ resonance is significantly

*e-mail: strauch@sc.edu

higher than the mass of its parity partner $\Delta(1950)7/2^+$, which is the lowest-mass Δ^* resonance with spin-parity $J^P = 7/2^+$. The observation conflicts with the prediction of models assuming a phase transition in high-mass resonances. The $\Delta(1950)7/2^+$ is now listed as a three-star resonance by the Particle Data Group [9].

Two CLAS analyses of single-pion photoproduction data from FROST have recently been completed and are being prepared for publication. The first is the extraction of the beam, T, and beam-target, F, polarization observables in $\gamma p \to \pi^0 p$ from data with circularly-polarized photons and transversally-polarized protons [10]. The results are in good agreement with previous experimental data but extend to much higher energies, up to about $W = 2.5$ GeV. Although these results have not yet been included in fits of recent partial-wave analyses, the PWA solutions are already in fair agreement with the data, even at the highest covered energies. The second analysis determined the beam-target observable, G, from data with linearly polarized photons on a longitudinally polarized proton target in the $\pi^0 p$ and $\pi^+ n$ channels for energies from $W = 1.425$ GeV to 2.290 GeV [11]. While present PWA predictions are in fair agreement with the data, they do not describe the G data well at higher energies. Figure. 1 shows example angular distributions for G in the $\pi^+ n$ channel for two out of the 23 energy bins in the experiment.

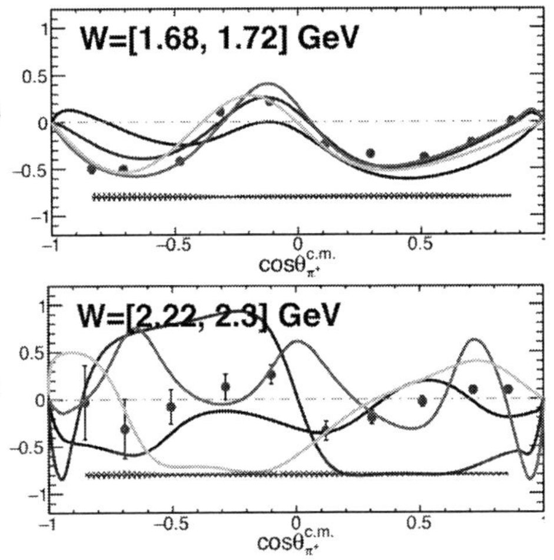

Figure 1. Beam-target polarization observable $-G$ in the $\gamma p \to \pi^+ n$ reaction from FROST for two selected energy bins out of 23 (red data points) [11]. The curves are predictions from SAID MA19 (blue) [12], MAID (magenta), GW-Jülich (cyan) [13], and BnGa (black) [14, 15].

The η meson is an isoscalar particle ($I = 0$), and its production in the s-channel process can proceed via N^* states with $I = \frac{1}{2}$ but not via Δ^* resonances with $I = \frac{3}{2}$; the channel ηp serves as an isospin filter. The polarization observable E has been measured with FROST in the $\gamma p \to \eta p$ reaction from threshold to $W = 2.15$ GeV [16]. Initial investigation of these results with the Jülich-Bonn dynamical coupled-channel model show pronounced changes in the description of this observable when these new CLAS data are included. This analysis gave a good description of the data without the need for an additional and previously suggested narrow resonance near 1.68 GeV. The CLAS Collaboration has also measured the linearly-polarized photon-beam asymmetry, Σ, in the $\gamma p \to \eta p$ and $\gamma p \to \eta' p$ reactions. The data significantly extend the range of previous measurements to energies up to about $W = 2.08$ GeV [17]. Based on the Jülich-Bonn approach and the BnGa partial-wave formalism, initial analyses of the data corroborate the evidence for the nucleon resonances $N(1895)1/2^-$, $N(1900)3/2^+$, $N(2100)1/2^+$, and $N(2120)3/2^-$ [17]. The analysis by Anisovich *et al.* [18] indicates that those four resonances are the most significant contributors to the η' photoproduction process. This analysis of the CLAS data also allowed for the determination of the decay branching ratios of these resonances into $N\eta'$ final states. The Particle Data Group now recognizes these states as three- and four-star resonances [9].

Like the η and η', ω mesons have isospin zero. The ω photoproduction selects isospin-$\frac{1}{2}$ resonances in the nucleon resonance spectrum. The CLAS Collaboration has studied cross-sections and numerous polarization observables in the $\gamma p \to \omega p$ reaction using the $\omega \to \pi^+\pi^-\pi^0$ decay mode. In a measurement with circularly polarized photons and the longitudinally polarized FROST target, the helicity asymmetry E was determined from threshold to about $W = 2.3$ GeV [19]. An analysis of the data within the BnGa framework found as leading partial waves at the reaction threshold the $3/2^+$, identified with the sub-threshold $N(1720)3/2^+$ resonance, and $5/2^+$ partial waves. Pomeron t-channel exchange was seen to grow with energy and to dominate all other contributions above $W \approx 2$ GeV; t-channel pion exchange was found to remain small in the studied energy range [19]. The magnitudes of the interference terms between the leading Pomeron-exchange and the resonant portion of the $J^P = 3/2^+$ partial waves, as well as the smaller waves with $J^P = 1/2^-, 3/2^-$, and $5/2^+$, were fixed with the help of recent CLAS measurements of the photon-beam asymmetry Σ [20]. The data significantly increase the size of the database for this observable and extend its coverage to energies up to $W = 2.1$ GeV. First-time measurements of the target asymmetry T in the $\gamma p \to \omega p$ reaction were obtained with the FROST target from the reaction threshold up to 2.8 GeV [21]. The observed asymmetries are rich in structures and significant; with $|T|$ reaching 0.3–0.4 in the third-resonance region. The FROST data also allowed to determine a new set of results for the observable Σ [21] that were found to be in overall fair agreement with previous data, including the earlier CLAS data [20]. Finally, the double-polarization observables F, H, and P in ω photoproduction have been obtained from the FROST experiment using circularly polarized and linearly polarized photons [22]. The results of partial-wave analyses, including these new data, provide evidence for several poorly known states; the recently added $N(1875)3/2^-$ and $N(2120)3/2^-$ states, the new $N(1880)1/2^+$, and the poorly established $N(2000)5/2^+$. The latter state was also observed in a single-channel partial-wave analysis by the CLAS Collaboration [23].

Hyperon photoproduction is an important tool in the study of nucleon resonances. The parity-violating weak decay of hyperons allows access to recoil polarization observables through the measurement of the decay-proton angular distribution. Earlier CLAS measurements of the hyperon-photoproduction reactions $\gamma p \to K^+\Lambda$ and $\gamma p \to K^+\Sigma^0$ included the determination of the Λ recoil polarization P from measurements with unpolarized photon beams [24, 25], as well as the polarization-transfer observables C_x and C_z with circularly-polarized beams [26]. Among other data, these CLAS findings were particularly important in an analysis by the BnGa group [27, 28] that established further evidence for the, at the time, poorly known $N(1900)3/2^+$ resonance. This resonance is predicted by symmetric three-quark models but is not expected to exist in some quark-diquark models [28]. By now, this resonance is recorded as a firmly established four-star resonance by the Particle Data Group [9]. New hyperon-photoproduction data for the reactions $\gamma p \to K^+\Lambda$ and $\gamma p \to K^+\Sigma^0$ from CLAS were extracted from measurements with linearly-polarized photons off an unpolarized proton target with energies up to $W \approx 2.2$ GeV [29]. The linearly-polarized beam, together with the recoil polarization of the hyperon, give access to five polarization observables: Σ, P, T, O_x, and O_z. These data greatly increase the world database in accuracy and kinematic coverage; the beam-recoil observables, O_x and O_z, in the $K^+\Sigma^0$ channel have not been reported before. An earlier 2014 solution of the BnGa partial-wave analysis did not describe most data from this experiment. An updated solution that included the new CLAS data and additional $N^*(3/2^+)$ and $N^*(5/2^+)$ resonances describes the $K\Lambda$ data well, whereas the agreement with the $K\Sigma$ data is fair [29]. Hyperon photoproduction data from the FROST experiment have been analyzed [30]; final results are not yet available.

Furthermore, the CLAS Collaboration reported first measurements of the spin-density matrix elements of $K^{*+}(892)$ mesons in the reaction $\gamma p \to K^{*+}\Lambda$ where the K^{*+} was identified in the observation of $K^{*+} \to K_S\pi^+$ with a subsequent $K_S \to \pi^+\pi^-$ [31]. The measurement covered a photon-energy range from threshold to 3.9 GeV. The results from this experiment — differential cross-sections, spin density matrix elements, and the Λ recoil polarization — were analyzed in the BnGa framework. This analysis found very significant contributions of the nucleon resonances $N(1895)1/2^-$ and $N(2100)1/2^+$ to the reaction, and some indications for three new resonances that decay into $K^*\Lambda$ [31].

Highly excited nucleons, in the mass region above 1.6 GeV, have substantial branching ratios into the $\pi\pi N$ final state; the decay proceeds typically through either the $\pi\Delta$ or ρN intermediate states. The study of $\pi\pi N$ photoproduction is, therefore, a sensitive tool in the search for new baryon states that couple weakly to other decay channels, especially to πN. The CLAS Collaboration recently reported on the first measurement of nine single-differential cross sections for the $\gamma p \to \pi^+\pi^- p$ reaction in the energy range from $W = 1.6$ GeV to 2.0 GeV [32]. The various cross-sections are differential in the invariant masses, as well as polar and azimuthal angles for different pairs of final-state particles. The extensive data set was analyzed with the updated Jefferson Lab-Moscow State University meson-baryon reaction model, and a good description of the data was achieved [32]. The analysis provided the first extraction of the photocouplings of all prominent nucleon resonances in this mass region from $\pi^+\pi^- p$ photoproduction data. Compared to previous πN analyses, the accuracy of the photocouplings of the resonances with dominant decay into the $\pi\pi N$ final state, $\Delta(1620)1/2^-$, $\Delta(1700)3/2^-$, $N(1720)3/2^+$, and $\Delta(1905)5/2^+$, were greatly improved [32].

Beyond differential cross sections, polarization observables provide further constraints on the extraction of the amplitudes of the photoproduction process. A review of the polarization observables for the two-pion production off the nucleon is given in Ref. [33]. The CLAS Collaboration was first to study the $\pi^+\pi^- p$ photoproduction with a circularly polarized beam and to measure the beam-helicity asymmetry, $I^\odot$, in the energy range between $W = 1.35$ and 2.30 GeV [34]. The use of linearly polarized photons and the FROST target allows for the extraction of a large number of polarization observables. The FROST group is working on the determination of twelve polarization observables determined from $\gamma p \to p\pi^+\pi^-$ data; preliminary results are reported in Refs. [35, 36]. Figure 2 illustrates the quality of data that is expected from those analyses. Here, the azimuthal angular distributions of the double-polarization observable $P_y^\odot$ are shown for various polar angular ranges for one center-of-mass energy bin raging from $W = 1550$ MeV to 1650 MeV [36]. The data are compared with model calculations by Fix [37] and are fitted with a low-order polynomial in $\sin\Phi^*$. This particular observable is accessible in measurements with circularly polarized photons off transversally polarized protons. Other observables accessible with these experimental conditions include $I^\odot$, $P_x^\odot$, P_x and P_y. [33].

3 Photoproduction off the neutron

Compared to the scale of photoproduction data with proton targets, the world database for neutron targets is sparse. However, data from both proton and neutron targets are needed, for example, to completely specify the amplitude of the $\gamma N \to \pi N$ reaction. The amplitude of the reaction contains two parts with isospin $I = \frac{1}{2}$ and $I = \frac{3}{2}$, respectively. While the excitation of $I = \frac{3}{2}$ Δ^* states can be entirely determined from proton target data, measurements with both neutron and proton targets are required to deduce the isospin $I = \frac{1}{2}$ amplitudes, and separate γpN^* and γnN^* couplings [38]. The couplings provide essential guides to nucleon-structure models. A complication associated with the extraction of neutron data from experiments on nuclear targets, like the deuteron, is the model-dependent interpretation of the results.

The CLAS Collaboration measured the differential cross sections for the quasi-free $\gamma d \to \pi^- p(p)$ reaction (with proton spectator) in the energy range from $W = 1.31$ GeV to 2.37 GeV. After corrections of the γd data

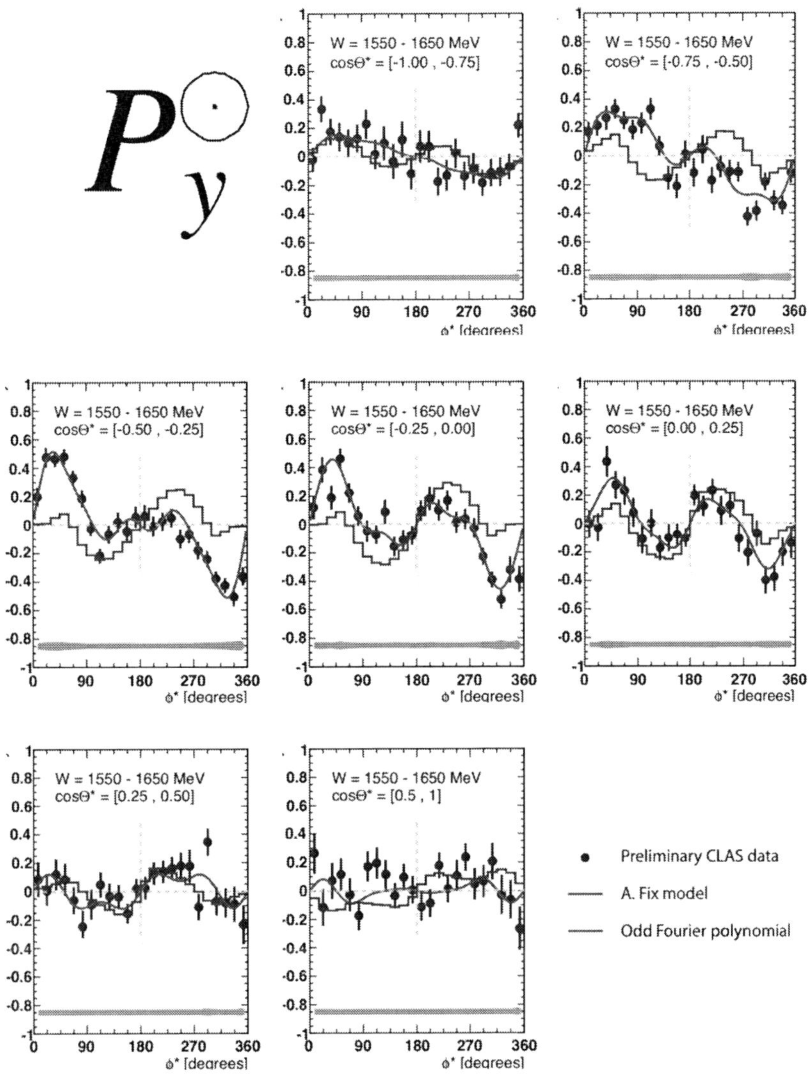

Figure 2. Preliminary Φ^* azimuthal angular distributions of the beam-target polarization observable $P_y^{\odot}$ for the $\gamma p \to \pi^+\pi^- p$ reaction in the energy range W = 1550 MeV to 1650 MeV [36]. The seven panels correspond to bins in the polar angle Θ^*. The solid green histograms are CLAS-acceptance averaged results of model calculations from A. Fix [37]. To guide the eye, the data are also fitted with low-order polynomials in $\sin\Phi^*$ (red curves). The angles Φ^* and Θ^* are, respectively, the azimuthal and polar angles of the final-state π^+; see Ref. [36] for details.

for final-state interactions, 8428 differential cross-section data points were extracted for the $\gamma n \to \pi^- p$ reaction, nearly tripling the corresponding world statistics [12]. A SAID multipole analysis was performed on the data set and resulted in the extraction of several photo-decay amplitudes $N^* \to \gamma n$ with small uncertainties, and, in particular, it allowed the first-ever determination of the excited neutron multipoles for the $N(1440)1/2^+$, $N(1535)1/2^-$, $N1650)1/2^-$, and $N(1720)3/2^+$ resonances [12]. CLAS data with a linearly polarized photon beam have been used to extract about 1200 data points of the polarization observable Σ in $\gamma d \to \pi^- p(p)$ covering an energy range from W = 1.62 GeV to 2.36 MeV [39]. The final results are not yet available.

The CLAS Collaboration has taken production data with circularly- and linearly-polarized photons off a polarized solid deuterium-hydride target (HDice) [5, 6] up to center-of-mass energies of $W \approx 2.3$ GeV. The run conditions were optimized for polarized neutron reactions. A first result is the determination of the beam-target double-polarization asymmetry E in the $\gamma d \to \pi^- p(p)$ reaction [7]. New partial-wave analyses that included the data resulted in updated $\gamma n N^*$ photocouplings [7]. Additional work on the experimental data is ongoing to extract the polarization observables Σ and G. Preliminary results for Σ are consistent with existing partial-wave analyses, whereas the results for G deviate strongly [40]. The data hint at the prospect of providing valuable constraints for new partial-wave analyses.

Hyperon photoproduction data off the proton were key to elucidate the nucleon resonance spectrum. Now, the CLAS Collaboration has reported on the first measurement of the differential and total cross sections for the $\gamma d \to K^0\Lambda(p)$ reaction with energies up to $W = 2.5$ GeV [41]. Consistent cross-sections were extracted from two different CLAS experiments with a partial overlap in the photon-energy range. First, partial-wave analyses were performed that describe the data without the introduction of new resonances. Two independent solutions from the BnGa-fit describe the cross-section data equally well [41]. The induced polarization P and the polarization-transfer coefficients C_x and C_z were also extracted for the $K^0\Lambda(p)$ channel from data with circularly polarized photons on a deuterium target in the energy range up to $W = 2.4$ GeV [42]. The preliminary results in Fig. 3 indicate that $C_z \approx 1$ at forward kaon-angles for photon energies up to 1.4 GeV. The preliminary data have been compared to results from the ANL Osaka and BnGa groups, and the significant deviations hint at the potential of these data to further constrain the N^* amplitude analysis. The beam-target helicity asymmetry E in $K^0\Lambda$ and $K^0\Sigma^0$ photoproduction on the neutron were extracted for the first time [43]. The experiment made use of the HDice target and the identification of $\pi^+\pi^-\pi^-p(X)$ with intermediate hyperons in the final state in the CLAS. The data cover an energy range from $W = 1.70$ GeV to 2.34 GeV. Regrettably, this analysis is limited by small statistics.

4 Summary

Much of the progress in the understanding of the baryon spectrum is owed to a large amount of new photoproduction data. The data include reaction cross sections and polarization observables from experiments with polarized beams or targets, as well as recoil polarization observables. Many final states have been studied in the experiments. The progress in the field is evident in the increased number of known N and Δ resonances, as reported by the Particle Data Group [9]. The N^* program of CLAS has been a critical contributor to these studies, and ongoing, and future analyzes of CLAS data will continue to elucidate the nucleon resonance spectrum.

References

[1] B.A. Mecking et al., Nucl. Instrum. Meth. **A503**, 513 (2003)
[2] D. Sober et al., Nucl. Instrum. Meth. A **440**, 263 (2000)
[3] C. Keith, J. Brock, C. Carlin, S. Comer, D. Kashy, J. McAndrew, D. Meekins, E. Pasyuk, J. Pierce, M. Seely, Nucl. Instrum. Meth. A **684**, 27 (2012)
[4] S. Strauch et al. (CLAS), Phys. Lett. **B750**, 53 (2015), 1503.05163
[5] C.D. Bass et al., Nucl. Instrum. Meth. **A737**, 107 (2014)
[6] M.M. Lowry et al., Nucl. Instrum. Meth. **A815**, 31 (2016), 1901.04548
[7] D. Ho et al. (CLAS), Phys. Rev. Lett. **118**, 242002 (2017), 1705.04713
[8] A.V. Anisovich, V. Burkert, J. Hartmann, E. Klempt, V.A. Nikonov, E. Pasyuk, A.V. Sarantsev, S. Strauch, U. Thoma, Phys. Lett. **B766**, 357 (2017), 1503.05774
[9] M. Tanabashi et al. (Particle Data Group), Phys. Rev. **D98**, 030001 (2018)
[10] H. Jiang, S. Strauch (CLAS), Few Body Syst. **59**, 125 (2018)
[11] N. Zachariou, D. Watts, L. Zana, J. McAndrew, CLAS Analysis
[12] P.T. Mattione et al. (CLAS), Phys. Rev. **C96**, 035204 (2017), 1706.01963
[13] D. Rönchen, M. Döring, U.G. Meißner, Eur. Phys. J. **A54**, 110 (2018), 1801.10458
[14] V. Sokhoyan et al. (CBELSA/TAPS), Eur. Phys. J. **A51**, 95 (2015), [Erratum: Eur. Phys. J.A51,no.12,187(2015)], 1507.02488
[15] A.V. Anisovich et al., Eur. Phys. J. **A52**, 284 (2016), 1604.05704
[16] I. Senderovich et al. (CLAS), Phys. Lett. **B755**, 64 (2016), 1507.00325
[17] P. Collins et al. (CLAS), Phys. Lett. **B771**, 213 (2017), 1703.00433
[18] A.V. Anisovich, V. Burkert, P.M. Collins, M. Dugger, E. Klempt, V.A. Nikonov, B.G. Ritchie, A.V. Sarantsev, U. Thoma, Phys. Lett. **B772**, 247 (2017), 1706.05144
[19] Z. Akbar et al. (CLAS), Phys. Rev. **C96**, 065209 (2017), 1708.02608
[20] P. Collins et al. (CLAS), Phys. Lett. **B773**, 112 (2017), 1706.04280
[21] P. Roy et al. (CLAS), Phys. Rev. **C97**, 055202 (2018), 1711.05176
[22] P. Roy et al. (CLAS), Phys. Rev. Lett. **122**, 162301 (2019), 1812.02106
[23] M. Williams et al. (CLAS), Phys. Rev. **C80**, 065209 (2009), 0908.2911
[24] J.W.C. McNabb et al. (CLAS), Phys. Rev. **C69**, 042201 (2004), nucl-ex/0305028
[25] M.E. McCracken et al. (CLAS), Phys. Rev. **C81**, 025201 (2010), 0912.4274
[26] R.K. Bradford et al. (CLAS), Phys. Rev. **C75**, 035205 (2007), nucl-ex/0611034
[27] A.V. Sarantsev, V.A. Nikonov, A.V. Anisovich, E. Klempt, U. Thoma, Eur. Phys. J. **A25**, 441 (2005), hep-ex/0506011
[28] V.A. Nikonov, A.V. Anisovich, E. Klempt, A.V. Sarantsev, U. Thoma, Phys. Lett. **B662**, 245 (2008), 0707.3600

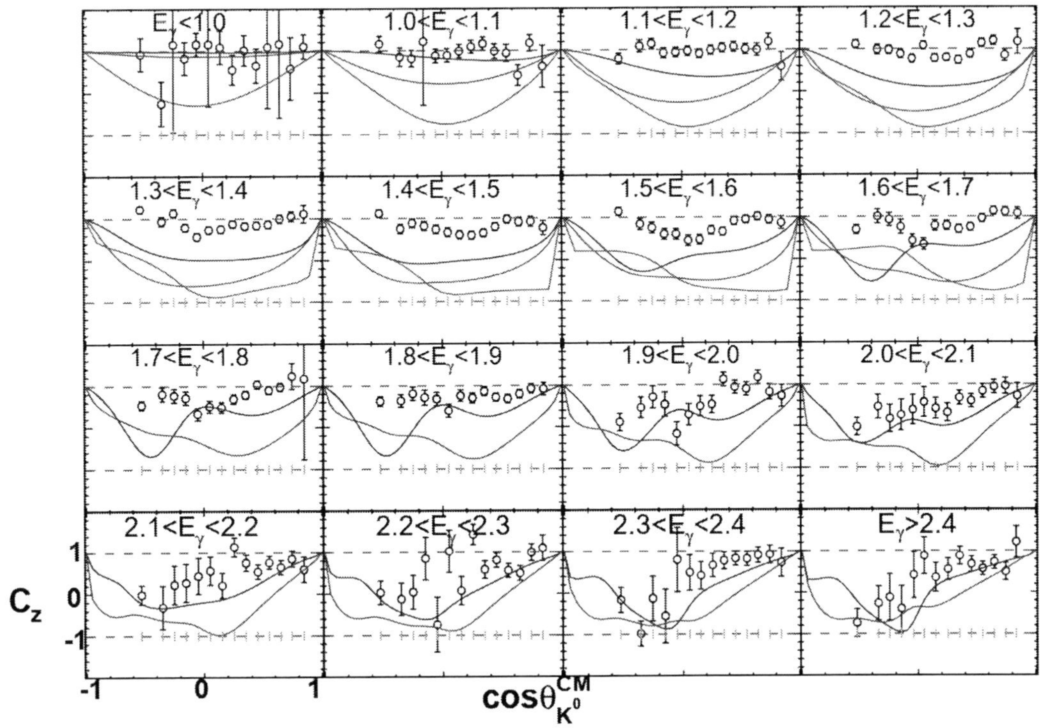

Figure 3. Preliminary K^0-angular distributions of the polarization-transfer observable C_z in the $\gamma d \to K^0\Lambda(p)$ reaction (open circles) [42, 44]. The panels show results from various photon-energy bins where the maximum energy is about 2.6 GeV. The preliminary data are compared with analysis solutions. The blue and the red curves show results from the BnGa group; the pink curve is from ANL Osaka [45].

[29] C.A. Paterson et al. (CLAS), Phys. Rev. **C93**, 065201 (2016), 1603.06492

[30] N.K. Walford, JPS Conf. Proc. **17**, 062005 (2017)

[31] A.V. Anisovich et al. (CLAS), Phys. Lett. **B771**, 142 (2017)

[32] E. Golovatch et al. (CLAS), Phys. Lett. **B788**, 371 (2019), 1806.01767

[33] W. Roberts, T. Oed, Phys.Rev. **C71**, 055201 (2005), nucl-th/0410012

[34] S. Strauch et al. (CLAS Collaboration), Phys. Rev. Lett. **95**, 162003 (2005), hep-ex/0508002

[35] S. Park (CLAS), AIP Conf. Proc. **1432**, 401 (2012)

[36] L.A. Net, S. Strauch (CLAS), Few Body Syst. **59**, 123 (2018)

[37] A. Fix, H. Arenhovel, Eur. Phys. J. **A25**, 115 (2005), nucl-th/0503042

[38] A.M. Sandorfi, Few Body Syst. **60**, 10 (2019)

[39] D. Sokhan, CLAS Analysis

[40] H.Y. Lu (CLAS, g14 Analysis Group), Few Body Syst. **60**, 18 (2019)

[41] N. Compton et al. (CLAS), Phys. Rev. **C96**, 065201 (2017), 1706.04748

[42] C. Gleason, Y. Ilieva (CLAS), Few Body Syst. **59**, 129 (2018)

[43] D.H. Ho et al. (CLAS), Phys. Rev. **C98**, 045205 (2018), 1805.04561

[44] C. Gleason, private communication (2019)

[45] H. Kamano, S.X. Nakamura, T.S.H. Lee, T. Sato, Phys. Rev. **C88**, 035209 (2013), 1305.4351

Search for the $N(1685)$ in $\eta\pi$-Photoproduction

Dominik Werthmüller[1,*] on behalf of the A2 Collaboration at MAMI

[1]Department of Physics, University of York, Heslington, York, YO10 5DD, UK

Abstract. The nucleon-like member $N(1685)$ of the speculative baryon antidecuplet denotes one possible explanation for the narrow peak-structure around $W = 1.68$ GeV observed in the total cross section of η-photoproduction off the neutron. If this baryon existed, it would likely to be seen in other reactions as well. While the aforementioned peak, whatever its nature is, was confirmed by several experiments, claims for signatures of the $N(1685)$ in other reactions and observables are mainly made by V. Kuznetsov *et al.* using GRAAL data. Their latest work suggests signals of both $N(1685)$ charge states in all isospin channels of $\eta\pi$- photoproduction off the proton and neutron. This contribution reports on challenging these claims with data from the A2 at MAMI experiment employing photon beam energies from E_γ=1.43–1.58 GeV. The $\eta\pi^0 p$ and $\eta\pi^+ n$ final states produced from a hydrogen target were studied and new analysis cuts were tested in order to enhance a possible signal.

1 Introduction

The search for exotic states denotes a major focus of contemporary hadron physics. Such bound states of the strong interaction are not built from ordinary $q\bar{q}$ (for mesons) and qqq (for baryons) configurations in terms of the quark model but involve $q\bar{q}q\bar{q}$ (tetraquark), $qqqq\bar{q}$ (pentaquark), etc. as well as gluonic degrees of freedom to constitute a color singlet.

1.1 Exotic Pentaquarks

The most recent experimental evidence for pentaquark states was claimed in 2015 by the LHCb collaboration in the $J/\psi p$ final state of Λ_b^0 decays hinting at an $uudc\bar{c}$ quark content [1]. In the meantime, a total of three states are thought to be identified [2]. GlueX, a first independent experiment using J/ψ-photoproduction as an alternative production mechanism could not see any evidence of these pentaquark states [3]. More data from GlueX and other experiments are being taken and analyzed, and theoretical efforts to understand the production mechanisms are ongoing [4].

Before the evidence for pentaquarks containing heavy charm quarks emerged, in 2003 the LEPS collaboration claimed evidence for a $uudd\bar{s}$ pentaquark state named Θ^+ [5]. In the peculiar episode that followed the announcement, several experiments initially confirmed the existence of a structure around $W = 1540$ MeV in the KN system. In the end, however, large statistics measurements, mainly from the CLAS experiment, could not confirm the initial findings (see [6] for an overview). Nevertheless, as the detector acceptances of CLAS and LEPS are quite different, new LEPS results using an improved setup and analysis are still awaited eagerly [7]. In addition, there are newer positive claims from the DIANA experiment [8] and a subgroup of the CLAS collaboration [9].

Pentaquark states were studied using the quark model long before the Θ^+ claim [10, 11]. The search for such a state at LEPS, although, was motivated by the striking prediction of the Chiral Quark-Soliton Model [12], a Skyrme model in which baryons are described as solitons of the chiral field. This model not only describes the known octet and decuplet but also predicts the antidecuplet depicted in figure 1. The exotic Z^+ (later renamed to Θ^+) with isospin $I = 0$ and strangeness $S = 1$ was predicted to be as narrow as 1 MeV and to decay into K^+n or K^0p, which matched the LEPS results.

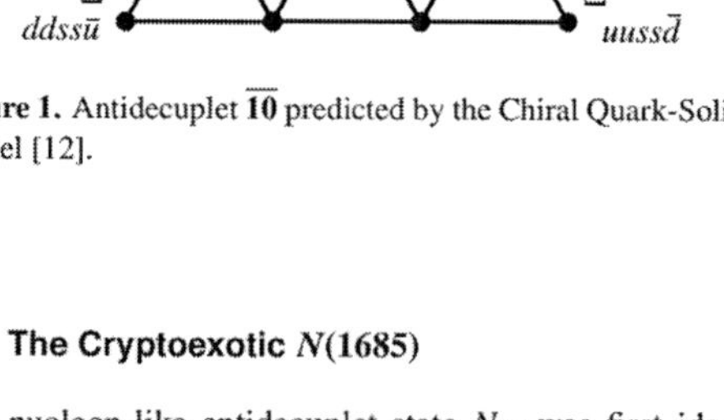

Figure 1. Antidecuplet $\overline{\mathbf{10}}$ predicted by the Chiral Quark-Soliton Model [12].

1.2 The Cryptoexotic $N(1685)$

The nucleon-like antidecuplet state $N_{\overline{10}}$ was first identified with the nucleon resonance $N(1710)1/2^+$ (although

*e-mail: dominik.werthmueller@york.ac.uk

the width of the $N_{\overline{10}}$ was predicted to be much more narrow) in order to calculate the masses of the other members of the multiplet [12]. The $N_{\overline{10}}$ (as well as the $\Sigma_{\overline{10}}$) is cryptoexotic, i.e., its quantum numbers can also be realized in an ordinary qqq configuration. It was quite a surprise when in 2007, Kuznetsov *et al.* (members of the GRAAL collaboration) presented evidence for a $N_{\overline{10}}$ candidate with matching properties [13]: The bump at $W = 1685$ MeV was observed in the cross section of η-photoproduction on the neutron in agreement with the leading $N_{\overline{10}} \to \eta N$ decay [12]. Secondly, the width of the structure was very narrow (around 30 MeV) [14] compared to usual nucleon resonances. Thirdly, the absence of any bump in the corresponding cross section on the proton matched the predicted small photocoupling compared to the neutron ('neutron anomaly') [15].

While first there were doubts about the presence of the structure in the η-photoproduction cross section, the bump was observed by several independent experiments, such as LNS [16], CBELSA/TAPS [17, 18] and A2 at MAMI [19–21]. Similar structures are also claimed to be present in different reaction channels and observables, namely the beam asymmetry Σ in $\gamma p \to \eta p$ [22], and in Compton scattering off the neutron (cross section [23]) and off the proton (beam asymmetry Σ [24]). These results all originate from the GRAAL experiment and remain unconfirmed by any other experiment so far.

In view of some early analyses of the available data supporting the scenario of a new narrow resonance [14, 25–28], a $N(1685)$ with unknown quantum numbers was added as a one-star resonance to the 2012 Review of Particle Physics [29] by the Particle Data Group (PDG). The entry was removed in later editions when theoretical studies and partial-wave analyses showed that the bump around $W = 1685$ MeV in the η-photoproduction cross section on the neutron could also be due to coupled-channel effects of known nucleon resonances [30, 31], effects from strangeness threshold openings [32] and cusps [33], and S_{11} interferences [27, 34, 35]. More insight will be gained from the ongoing measurements of polarization observables, with first results for the observable E already available [36–38]. This work revealed that a calculation of the Bonn-Gatchina (BnGa) model [35] including a narrow P_{11} resonance gave a slightly better description of the experimental data than the model without a narrow state. Due to the limited statistics, the significance of this finding is still under debate [39].

1.3 Motivation for this work

The latest claim of a $N_{\overline{10}}$ signature as a narrow $N(1685)$ resonance is again made by Kuznetsov *et al.* using GRAAL data in the $\gamma N \to \eta\pi N$ channels [40]. Here, the $N_{\overline{10}}$ could be produced in the decay of a heavier N or Δ resonance R via the emission of a pion $R \to \pi N_{\overline{10}}$ followed by the decay $N_{\overline{10}} \to \eta N$. Therefore, in contrast to the direct production potentially observed in $\gamma n \to \eta n$, the signal could also be observed using a proton target despite the suppressed photon coupling of the $N_{\overline{10}}$. Indeed, signals in all four isospin channels of $\gamma N \to \eta\pi N$ are claimed in [40], whereas only the peak in $\gamma p \to \eta\pi^0 p$ (measured on the free proton) is statistically significant.

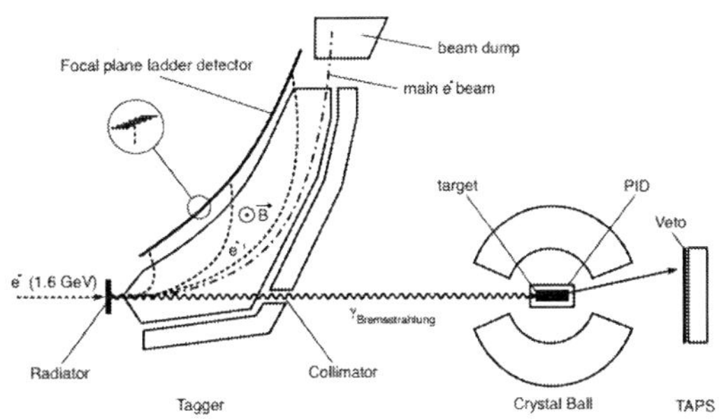

Figure 2. Schematic view of the setup at the tagged-photon beam experiment A2 at MAMI.

The $\gamma N \to \eta\pi N$ reactions have been previously measured at several experiments, such as LNS [41], GRAAL [42], CBELSA/TAPS [43–45] and A2 at MAMI [46–50]. The main findings were that for $E_\gamma < 1.6$ GeV, the reaction is dominated by the excitation of the $\Delta(1700)3/2^-$ and the $\Delta(1940)3/2^-$ resonances with subsequent decays to $\eta\Delta(1232)3/2^+$. Smaller contributions come from the $\pi N(1535)1/2^-$ isobar and photoproduction of the $a_0(980)$ meson. Evidence of a $N(1685)$ was not found in any of these studies as statistics was either limited or the event selection was not optimized to reveal a potential signal. The goal of this work is therefore to measure $\gamma p \to \eta\pi^0 p$ and $\gamma p \to \eta\pi^+ n$ in order to look for any signs of a narrow structure around $m(\eta N) \sim 1685$ MeV. The available A2 data for the energy range E_γ=1.43–1.58 GeV have not been used so far to study these reactions.

2 Experimental Setup

The experimental data were obtained at the tagged-photon beam facility A2 at the electron-accelerator facility MAMI in Mainz, Germany. Out of three data sets of similar size, one data set of 310 hours was used for this work. Figure 2 shows an overview of the experimental setup. In contrast to 'standard' experiments, the Endpoint-Tagger (EPT) device allowing to access photon-beam energies E_γ=1.43–1.58 GeV close to the energy of the incoming 1.6 GeV electron beam was installed. The 10-cm liquid hydrogen target was installed in the center of the Crystal Ball detector (CB) consisting of 672 NaI(Tl) crystals, which cover polar angles from 20–160 degrees with almost full azimuthal acceptance. Particles with polar angles from 5–20 degrees were detected in the TAPS calorimeter wall installed 1.5 m downstream from the target comprising 366 BaF_2 crystals. Thin plastic scintillator detectors arranged as a barrel around the target in CB and in front of individual TAPS crystals allowed charged-particle vetoing and particle identification via a dE/E-analysis in both calorimeters. Particles detected in TAPS could also be discriminated by time-of-flight and pulse-shape analyses, the latter utilizing the two scintillation-light components in the

BaF_2 crystals. The experimental trigger required an energy deposition higher than ~550 MeV in CB as the primary goal of the EPT-experiments were studies involving the η'-meson. More details about the setup can be found in [51].

3 Data Analysis

The following reactions were analyzed in this work:

1) $\gamma p \rightarrow \eta\pi^0 p$ with $\eta \rightarrow 2\gamma$
2) $\gamma p \rightarrow \eta\pi^+ n$ with $\eta \rightarrow 2\gamma$
3) $\gamma p \rightarrow \eta\pi^+ n$ with $\eta \rightarrow 3\pi^0 \rightarrow 6\gamma$

3.1 Event Selection

As for now, exclusive analyses were performed, i.e., all particles in the final state were required to be detected. The thresholds for clusters in the CB and TAPS calorimeters were set to 20 MeV. Using the detected particles, all combinations to form the corresponding final states were checked with a series of cuts. In addition, a kinematic fit was performed to optimize the measured angles and energies of the particles using 4-momentum and meson invariant-mass constraints. Particle identification was performed with dE/E (CB) and time-of-flight (TAPS) analyses for pion and neutron candidates in reactions 2) and 3). BaF_2 pulse-shape cuts were used in all reactions to separate photons from massive particles. Next, cuts on the reaction kinematics were implemented. Neutral pions and η-mesons were identified in the corresponding $m(\gamma\gamma)$ invariant masses by applying 3σ-cuts around their nominal mass. All particles were required to lie in the same reaction plane — this was ensured by cutting on the three azimuth-angle differences between the $p_i p_j$ and p_k systems ($p \in \{\eta, \pi, N\}$). A further cut was employed on the angular difference between the detected and fitted direction of the nucleon. The confidence level (CL) provided by the kinematic fit allowed an additional cut (CL > 2.7×10^{-3}) to reject background events. Having applied all cuts, the number of events with multiple valid particle combinations were small. In order to account for them in the signal extraction, large data sets of Monte-Carlo (MC) simulations are needed. Therefore, for the current work only the combination with the highest CL was kept for further analysis.

3.2 Signal Extraction

The data were analyzed in bins of photon-beam energy E_γ and η-nucleon invariant mass $m(\eta N)$. In order to remove any background remaining after the previously discussed cuts, the $\eta\pi$-missing mass (in case of reaction 1) and 2), nucleon mass subtracted) or the $m(3\pi^0)$-invariant mass (in case of reaction 3)) were fitted with signal and background contributions obtained from MC simulations using a Geant4-based [52] model of the A2 experiment. The reason for the choice of different fit variables is the fact that very little background is present in analysis 3) and the $m(3\pi^0)$-invariant mass provided more stable and

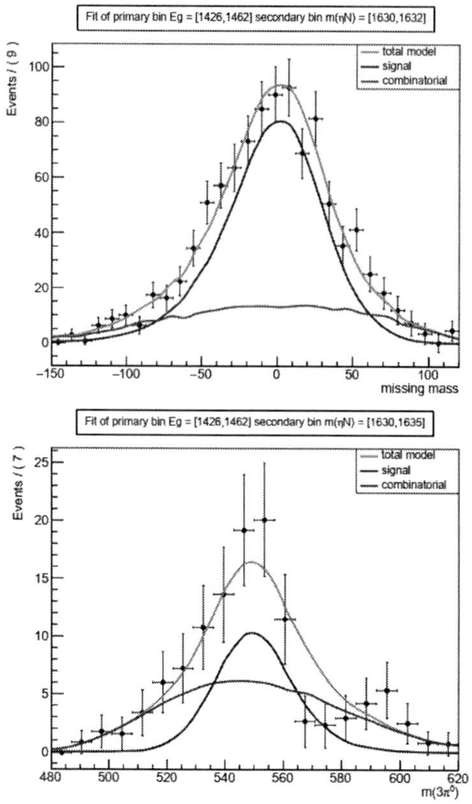

Figure 3. Example unbinned likelihood fits for the same E_γ, $m(\eta N)$-bin. Top: Fit of the $\eta\pi^0$-missing mass in the analysis of $\gamma p \rightarrow \eta\pi^0 p$ with $\eta \rightarrow 2\gamma$. Bottom: Fit of the $m(3\pi^0)$-invariant mass in the analysis of $\gamma p \rightarrow \eta\pi^+ n$ with $\eta \rightarrow 3\pi^0$. Signal and background contributions were obtained from MC simulations.

unambiguous fits compared to the $\eta\pi$-missing mass. Example fits of both types are shown in figure 3. Unbinned likelihood fits were performed which allowed to make use of the sPlot-technique [53] to obtain weights for the signal and background contribution on an event-by-event basis. This enables to unfold signal and background in any variable that is not correlated with the fit variable.

3.3 Van Hove Plots and Longitudinal Phase-Space

The use of longitudinal phase-space allows to reduce the phase-space dimensionality of multi-particle final states by assuming that the transverse momenta can be neglected in high-energy particle collisions [54]. In case of a three-particle final state, the longitudinal phase-space has two dimensions and Van Hove plots containing six sectors can be constructed. The sectors represent the possible combinations of the longitudinal momenta of all particles, i.e., whether they are going forward (direction of beam) or backward (direction of target) in the center-of-mass frame. As different reaction mechanisms will populate different

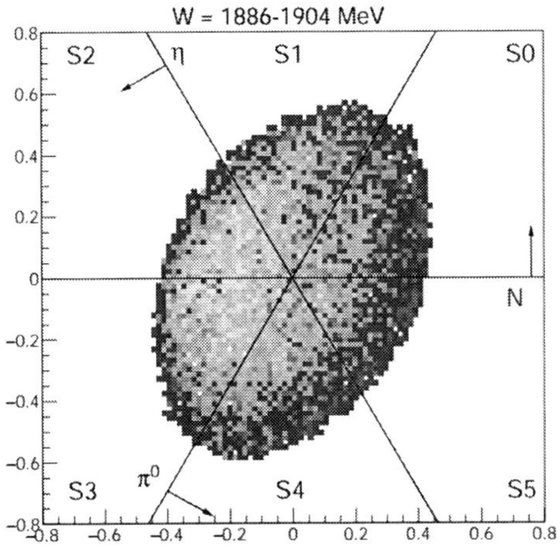

Figure 4. Van Hove plot for $\gamma p \rightarrow \eta\pi^0 p$ in the energy range E_γ=1426–1462 MeV. The six sectors S0 to S5 represent different combinations of forward or backward going particles (see text).

sectors, Van Hove plots can be used to disentangle different processes.

The aim of the work performed here was to investigate the possibility of using Van Hove plots to separate the potential production of the $\pi N(1685)$ isobar from the dominating $\eta\Delta(1232)$ isobar. The main caveat is that the in the studied energy range the transverse momenta cannot be neglected as required by the longitudinal phase-space approximation. Nevertheless, a partial separation could already lead to a sufficient improvement of the signal-to-background ratio.

Figure 4 shows the Van Hove plot for the first energy bin in the $\gamma p \rightarrow \eta\pi^0 p$ analysis. The sector with the largest number of events is S3, consistent with the dominating production of the $\eta\Delta(1232)$ isobar, with a forward going η-meson and $\pi^0 p$ going backward. Similarly, a possible $\pi N(1685)$ isobar is expected to be predominantly located in sector S5 assuming a production mechanism which yields a backward-going $N(1685)$ and a forward-going π-meson. As mentioned before, due to the relatively low beam energy, the locii for the different production mechanisms are not confined to one sector and strongly overlap.

4 Preliminary Results

Preliminary results of the observable of interest, the $m(\eta N)$ invariant mass, are shown in figures 5–8 and will be discussed in the following.

4.1 Uncut Distributions

First, the invariant mass distributions were studied using the standard analysis cuts discussed in section 3.1. In case of the $\gamma p \rightarrow \eta\pi^0 p$ reaction, the results shown in figure 5 are compared to the distributions obtained from phase-space MC simulations using event weights from the BnGa partial-wave analysis of [45], which were scaled to the data using the ratio of the distribution integrals. No acceptance correction has been applied here. There is an overall good agreement between the two distributions with a small discrepancy at the maximum, where the data shows a slightly sharper shape. No excess around 1685 MeV is observed. Only the $\eta \rightarrow 2\gamma$ decay has been analyzed so far, as the $\eta \rightarrow 3\pi^0$ decay leads to a total number of nine particles in the final state, which will degrade the detection efficiency resulting in a smaller data set.

Figure 6 shows the ηn-invariant mass for $\gamma p \rightarrow \eta\pi^+ n$. Both decays of the η-meson were analyzed and are compared to each other by scaling the $\eta \rightarrow 3\pi^0$ results to the results of the $\eta \rightarrow 2\gamma$ data sample. Despite the lack of an individual efficiency correction, the overall features of the distributions are rather similar. Statistically significant finer structures are hard to identify due to the limited statistics. The main reasons for the worse statistical quality are the lower ($\sim 1/3$) detection efficiency for neutrons compared to protons and the fact that the experimental trigger required a high total energy (~550 MeV) deposited in the CB detector. Since neutrons deposit on average less energies than protons, and the π^+ does not contribute to the detected energy with its whole rest mass in contrast to the π^0, the trigger is less efficient resulting in a smaller data set.

4.2 Enhancing a Potential $N(1685)$-Signal

In order to enhance a possible signal of a narrow $N(1685)$ state, additional analysis cuts have to be applied. At the moment, only the $\gamma p \rightarrow \eta\pi^0 p$ reaction was studied regarding further cuts as statistics of the current $\gamma p \rightarrow \eta\pi^+ n$ data sample is too poor. This is expected to improve when all available data will have been analyzed.

In order to suppress contributions from the $\eta\Delta(1232)$ isobar, a cut requiring $m(\pi^0 p) < 1180$ MeV was applied. The resulting $m(\eta p)$ distributions are shown in figure 7. The data are again compared to a MC simulation weighted with the BnGa-model that was scaled to the data by the ratio of integrals. In bins two to four, the overall agreement is quite good considering the statistical uncertainties of both data and model. The statistical errors of the model are not shown directly but fluctuations are clearly visible. These fluctuations will be removed by generating a larger MC data sample in the final analysis. In the first bin, there are some noticeable deviations of the data from the model: For $m(\eta p)$=1640–1680 MeV, the model seems to overestimate the data, which form a peak-like structure around 1690 MeV. The statistical significance of these features is low, though. Also, the scaling of the model to the data will need to be implemented using absolute normalizations in the final analysis rather than using a factor that averages over the whole distributions.

Figure 8 shows the $m(\eta p)$ distributions including the $m(\pi^0 p) < 1180$ MeV cut and requiring in addition that the event belongs to sector S5 in the Van Hove plot. As discussed in section 3.3, this is expected to be the predominant sector for the potential $\pi N(1685)$ isobar. Statistics

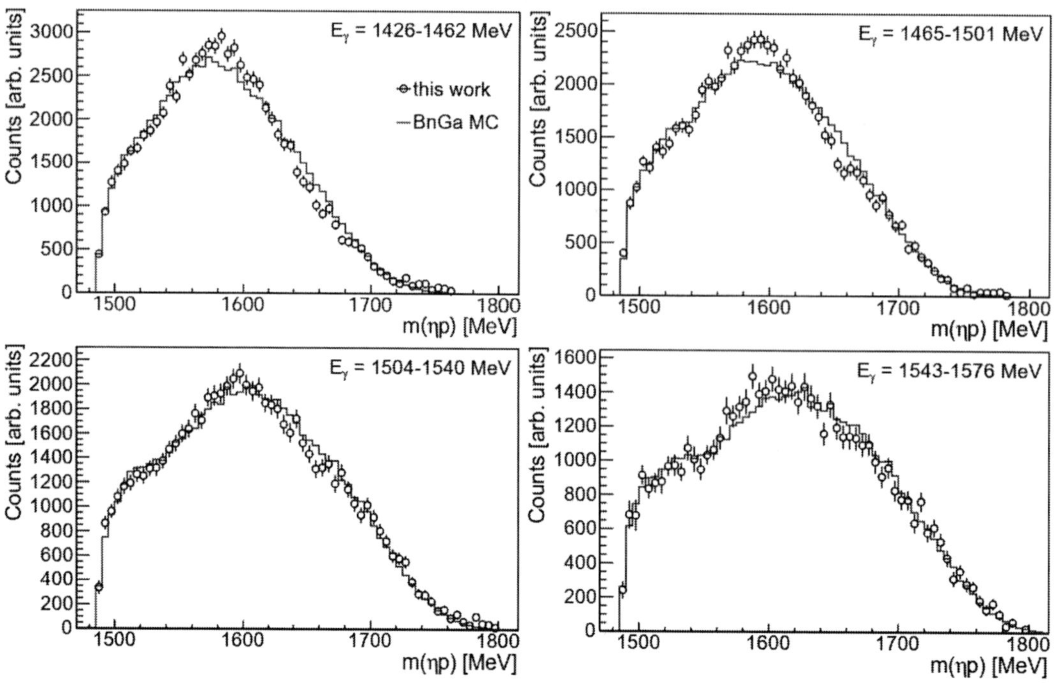

Figure 5. Invariant mass $m(\eta p)$ of $\gamma p \to \eta\pi^0 p$ for four bins of photon-beam energy (no acceptance correction). The data (black circles) are compared to a phase-space MC simulation (blue curves, normalized by integral ratio) using event weights from the BnGa partial-wave analysis of [45].

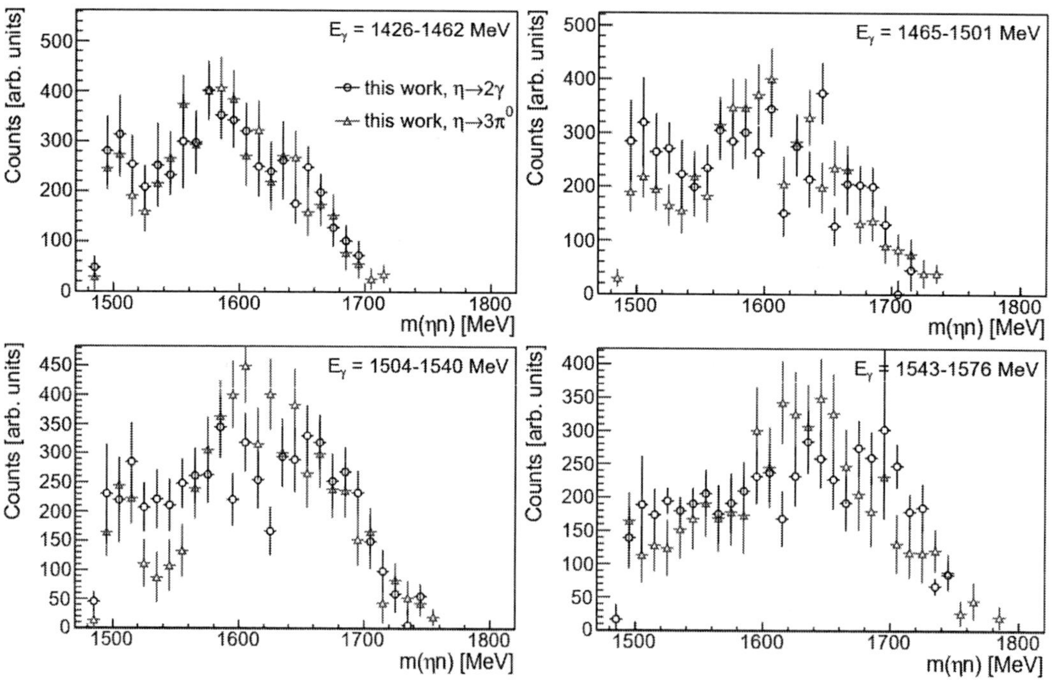

Figure 6. Invariant mass $m(\eta n)$ of $\gamma p \to \eta\pi^+ n$ for four bins of photon-beam energy (no acceptance correction). Two data samples using the $\eta \to 2\gamma$ decay (black circles) and the $\eta \to 3\pi^0$ decay (red triangles, normalized by integral ratio) have been analyzed.

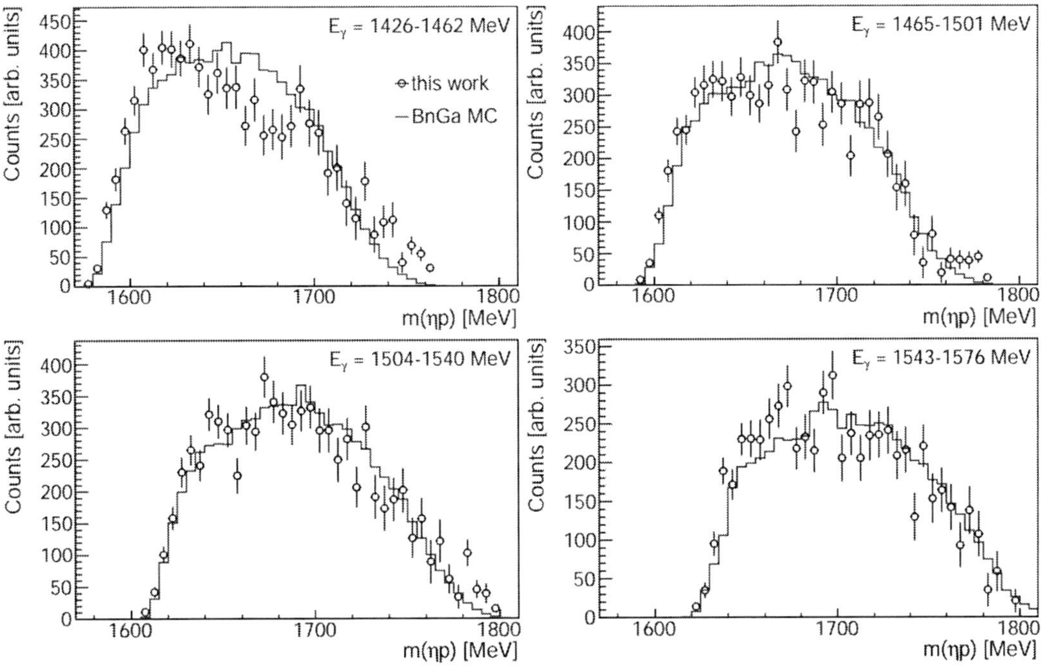

Figure 7. Invariant mass $m(\eta p)$ of $\gamma p \rightarrow \eta\pi^0 p$ for four bins of photon-beam energy with $m(\pi^0 p) < 1180$ MeV (no acceptance correction). The data (black circles) are compared to a phase-space MC simulation (blue curves, normalized by integral ratio) using event weights from the BnGa partial-wave analysis of [45].

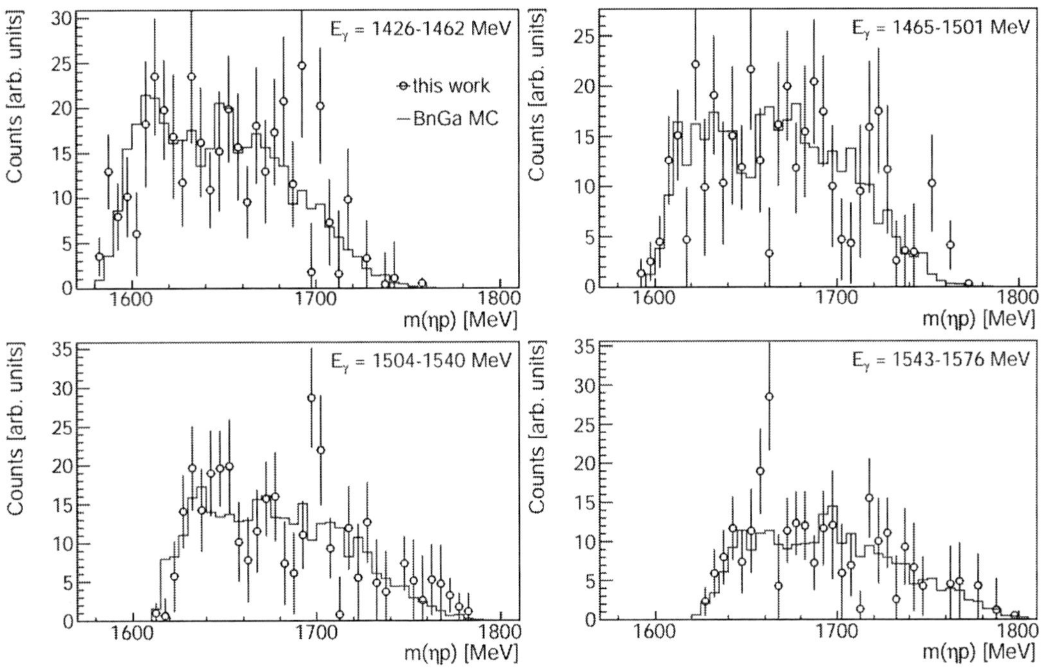

Figure 8. Invariant mass $m(\eta p)$ of $\gamma p \rightarrow \eta\pi^0 p$ for four bins of photon-beam energy with $m(\pi^0 p) < 1180$ MeV for events in the Van Hove plot-sector S5 (no acceptance correction). The data (black circles) are compared to a phase-space MC simulation (blue curves, normalized by integral ratio) using event weights from the BnGa partial-wave analysis of [45].

is obviously further reduced and fluctuations are also significant in the BnGa-model. Nevertheless, overall the data and model distribution agree rather well. Finer structures cannot be identified due to the limited statistics, although there a few bins with more entries around 1700 MeV, again not statistically significant.

5 Summary and Outlook

The reactions $\gamma p \to \eta\pi^0 p$ and $\gamma p \to \eta\pi^+ n$ in the energy range E_γ=1.43–1.58 GeV have been analyzed using data obtained at the A2 at MAMI experiment. The goal of this work was to check recent claims [40] of a signature of the $N(1685)$-antidecuplet baryon in the $m(\eta p)$ and $m(\eta n)$ invariant masses of the reactions $\gamma N \to \eta\pi N$.

For $\gamma p \to \eta\pi^0 p$, our data show a good agreement with previous measurements represented by the BnGa partial-wave analysis [45]. When a cut requesting $m(\pi^0 p) < 1180$ MeV to suppress the dominating $\eta\Delta(1232)$ background is applied, there are some deviations between the data and the model around the region of interest for photon-beam energies E_γ=1426–1462 MeV. The statistical significance of these deviations is small, nevertheless this finding needs to be investigated further by increasing the statistical quality of both data and model. Having more data will also allow to get better statistics when using an additional cut on the preferred signal sector in the Van Hove plot, which could help isolating a small $\pi N(1685)$ contribution to the reaction.

The $\gamma p \to \eta\pi^+ n$ analysis currently lacks the statistical quality for a refined analysis regarding a narrow $N(1685)$. All of the available data need to be analyzed in combination with an optimized event selection to come to a final conclusion here.

Next steps will include optimizations in the analysis ($\eta \to 3\pi^0$ decay for $\eta\pi^0 p$ final state, cuts, test of inclusive analyses), the use of all available data, the generation of a larger MC data sample using the BnGa-model, and acceptance correction with absolute normalization and comparison with the Mainz model [50].

References

[1] R. Aaij et al., Phys. Rev. Lett. **115**, 072001 (2015)
[2] R. Aaij et al., Phys. Rev. Lett. **122**, 222001 (2019)
[3] A. Ali et al., Phys. Rev. Lett. **123**, 072001 (2019)
[4] D. Winney et al., Phys. Rev. D **100**, 034019 (2019)
[5] T. Nakano et al., Phys. Rev. Lett. **91**, 012002 (2003)
[6] K.H. Hicks, Eur. Phys. J. H **37**, 1 (2012)
[7] M. Yosoi, EPJ Web Conf. **199**, 01020 (2019)
[8] V.V. Barmin et al., Phys. Rev. C **89**, 045204 (2014)
[9] M.J. Amaryan et al., Phys. Rev. C **85**, 035209 (2012)
[10] R.L. Jaffe, Topical Conference on Baryon Resonances, Oxford, SLAC-PUB-1774 (1976)
[11] H. Högaasen, P. Sorba, Nucl. Phys. B **145**, 119 (1978)
[12] D. Diakonov, V. Petrov, M. Polyakov, Z. Phys. A **359**, 305 (1997)
[13] V. Kuznetsov et al., Phys. Lett. B **647**, 23 (2007)
[14] R.A. Arndt et al., Phys. Rev. C **69**, 035208 (2004)
[15] M.V. Polyakov, A. Rathke, Eur. Phys. J. A **18**, 691 (2003)
[16] F. Miyahara et al., Prog. Theor. Phys. Suppl. **168**, 90 (2007)
[17] I. Jaegle et al., Phys. Rev. Lett. **100**, 252002 (2008)
[18] I. Jaegle et al., Eur. Phys. J. A **47**, 89 (2011)
[19] D. Werthmüller et al., Phys. Rev. Lett. **111**, 232001 (2013)
[20] D. Werthmüller et al., Phys. Rev. C **90**, 015205 (2014)
[21] L. Witthauer et al., Eur. Phys. J. A **49**, 154 (2013)
[22] V. Kuznetsov et al., Acta Phys. Polon. B **39**, 1949 (2008)
[23] V. Kuznetsov et al., Phys. Rev. C **83**, 022201 (2011)
[24] V. Kuznetsov et al., Phys. Rev. C **91**, 042201 (2015)
[25] K.S. Choi et al., Phys. Lett. B **636**, 253 (2006)
[26] A. Fix, L. Tiator, M.V. Polyakov, Eur. Phys. J. A **32**, 311 (2007)
[27] A.V. Anisovich et al., Eur. Phys. J. A **41**, 13 (2009)
[28] M. Shrestha, D.M. Manley, Phys. Rev. C **86**, 055203 (2012)
[29] J. Beringer et al. (Particle Data Group), Phys. Rev. D **86**, 010001 (2012)
[30] V. Shklyar, H. Lenske, U. Mosel, Phys. Lett. B **650**, 172 (2007)
[31] R. Shyam, O. Scholten, Phys. Rev. C **78**, 065201 (2008)
[32] M. Döring, K. Nakayama, Phys. Lett. B **683**, 145 (2010)
[33] L. Tiator et al., EPJ Web Conf. **199**, 01019 (2019)
[34] A.V. Anisovich et al., Eur. Phys. J. A **49**, 67 (2013)
[35] A.V. Anisovich et al., Eur. Phys. J. A **51**, 72 (2015)
[36] L. Witthauer et al., Phys. Rev. Lett. **117**, 132502 (2016)
[37] L. Witthauer et al., Eur. Phys. J. A **53**, 58 (2017)
[38] L. Witthauer et al., Phys. Rev. C **95**, 055201 (2017)
[39] A.V. Anisovich et al., Phys. Rev. C **95**, 035211 (2017)
[40] V. Kuznetsov et al., JETP Letters **106**, 693 (2017)
[41] T. Nakabayashi et al., Phys. Rev. C **74**, 035202 (2006)
[42] J. Ajaka et al., Phys. Rev. Lett. **100**, 052003 (2008)
[43] I. Horn et al., Eur. Phys. J. A **38**, 173 (2008)
[44] E. Gutz et al., Phys. Lett. B **687**, 11 (2010)
[45] E. Gutz et al., Eur. Phys. J. A **50**, 74 (2014)
[46] V.L. Kashevarov et al., Eur. Phys. J. A **42**, 141 (2009)
[47] V.L. Kashevarov et al., Phys. Lett. B **693**, 551 (2010)
[48] J.R.M. Annand et al., Phys. Rev. C **91**, 055208 (2015)
[49] A. Käser et al., Eur. Phys. J. A **52**, 272 (2016)
[50] V. Sokhoyan et al., Phys. Rev. C **97**, 055212 (2018)
[51] P. Adlarson et al., Phys. Rev. C **92**, 024617 (2015)
[52] S. Agostinelli et al., Nucl. Instrum. Methods Phys. Res., Sect. A **506**, 250 (2003)

[53] M. Pivk, F.R. Le Diberder, Nucl. Instrum. Methods Phys. Res., Sect. A **555**, 356 (2005)

[54] L. Van Hove, Nucl. Phys. B **9**, 331 (1969)

Towards a combined analysis of inclusive/exclusive electroproduction

Astrid N. Hiller Blin[1,*], *Vitaly V.* Chesnokov[3], and *Victor I.* Mokeev[2]

[1]Institut für Kernphysik & PRISMA+ Cluster of Excellence, Johannes Gutenberg Universität, D-55099 Mainz, Germany
[2]Thomas Jefferson National Accelerator Facility, Newport News, VA 23606, USA
[3]Skobeltsyn Nuclear Physics Institute and Physics Department at Moscow State University, 119899 Moscow, Russia

Abstract. In view of the major advances achieved by the CLAS experiments in studying the N^* electroexcitation amplitudes, as well as further extension of these studies in the experiments with CLAS12, we present an approach for the evaluation of the resonant contributions to inclusive electron scattering off protons. For the first time, the resonant contributions are determined from the experimental results on N^* electrocouplings available from the data analyses of exclusive meson electroproduction off protons. This is a useful benchmark for future endeavours on understanding the transition region between low and high-energy regions, strongly related to tests on quark-hadron duality.

1 Introduction

Due to their connection to the nucleon parton distribution functions (PDFs), which were recently related to the QCD Lagrangian (see Refs. [1–3] and references therein for an overview), and in view of the exploration of the quark-hadron duality [4], the interest in inclusive electron scattering reactions off nucleons has been strong. The Jefferson Lab (JLab) data played an important role in these studies [5–9]. The experiments in Halls A/C [6, 7] provided precise information on inclusive electron scattering at the invariant masses of the final hadrons of $W < 4.0$ GeV and photon virtualities $Q^2 < 6.0$ GeV^2 allowing us to disentangle the transverse and longitudinal parts of the inclusive cross sections. The CLAS detector offers a unique possibility of obtaining the inclusive structure functions F_2 in the broad range of $W < 2.5$ GeV at given photon virtuality $Q^2 < 4.5$ GeV^2 [10]. With the CLAS12 experiments, the largest kinematic coverage ever achieved in the resonance region, with photon virtualities up to $Q^2 = 12$ GeV^2 will be reached [11]. The broad coverage in W at given Q^2 is particularly important in the resonance region: due to the presence of resonant structures in the observable kinematics, with an intricate behavior as functions of W and Q^2, the interpolation over W and Q^2 to extract the structure functions becomes challenging in the resonance region. In addition, the analyses of CLAS experiments on exclusive meson electroproduction enabled to determine the nucleon resonance N^* electro-excitation amplitudes in terms of the transverse, $A_{1/2}(Q^2)$, $A_{3/2}(Q^2)$ and longitudinal, $S_{1/2}(Q^2)$ electrocouplings [12, 13]. This information has become available for most excited nucleon states in the mass range $M_{N^*} < 1.8$ GeV and at $Q^2 < 5.0$ GeV^2, and substantial evidence of a new baryon state $N'(1720)\ 3/2^+$ has been found recently [14, 15]. In the near future, the data on $\pi^+\pi^-p$ electroproduction with CLAS will provide electrocouplings of most N^* in the mass range up to 2.0 GeV and at $Q^2 < 5.0$ GeV^2 [16].

The availability of electrocoupling data on individual nucleon resonances makes it possible to, in this work, evaluate the resonant contributions to the inclusive electron scattering observables, in a fashion independent of the inclusive data, with the use of a relativistic Breit-Wigner ansatz [17]. In a next step, together with a non-resonant background, this will allow to determine the transition from low to high energies and photon virtualities Q^2, and consequently between resonance bound valence quarks and asymptotically free partons.

2 Formalism

Following the approach in Ref. [17] and definitions therein, we describe the contributions of the N^* and Δ resonances to the transverse (longitudinal) inclusive virtual photon-proton cross section within the relativistic Breit-Wigner ansatz

$$\sigma^R_{T(L)}(W,Q^2) = \frac{\pi}{q_\gamma^2} \times \sum_{N^*}(2J_r+1)\frac{M_r^2\Gamma_{\rm tot}(W)\Gamma_\gamma{}^{T(L)}(M_r,Q^2)}{(M_r^2-W^2)^2+M_r^2\Gamma^2_{\rm tot}(W)}, \qquad (1)$$

where the resonance masses, spins and total hadronic decay widths are given by M_r, J_r and $\Gamma_{\rm tot}(W)$, respectively. The resonance electromagnetic decay widths to the final states, $\Gamma_\gamma^{T(L)}$ are related to the resonance electrocouplings

*e-mail: hillerbl@uni-mainz.de

$A_{1/2}(Q^2)$, $A_{3/2}(Q^2)$, and $S_{1/2}(Q^2)$ as

$$\Gamma_\gamma^T(W=M_r,Q^2) = \frac{q_{\gamma,r}^2(Q^2)}{\pi}\frac{2M_N}{(2J_r+1)M_r} \times \left(|A_{1/2}(Q^2)|^2 + |A_{3/2}(Q^2)|^2\right),$$
$$\Gamma_\gamma^L(W=M_r,Q^2) = 2\frac{q_{\gamma,r}^2(Q^2)}{\pi}\frac{2M_N}{(2J_r+1)M_r}|S_{1/2}(Q^2)|^2, \quad (2)$$

with $q_{\gamma,r} = q_\gamma|_{W=M_r}$. We use the information on the resonance electrocouplings, including the new $N'(1720)\ 3/2^+$ state [14, 15], from interpolation and extrapolation of the CLAS data [12, 13] over Q^2 as described in Ref. [17]. The interpolation/extrapolation tools developed by CLAS [18] are used for the central values of the electrocouplings from $Q^2 = 0.5$ GeV^2 to $Q^2 = 5.0$ GeV^2, and their uncertainties are estimated from the experimental data error bars. These uncertainties are propagated to the inclusive electron scattering observables via a bootstrap based approach [19]. At the resonant point $W{=}M_r$ the resonance total decay widths $\Gamma_{tot}(W)$ were taken from the PDG [20]. Their W-dependencies were evaluated as described in Ref. [17], assuming that the centrifugal barrier penetration factor shapes the W-evolution of the total resonance decay widths. The resonant contributions to the unpolarized inclusive cross section is then given by [7]

$$\sigma_U^R(W,Q^2) = \sigma_T^R(W,Q^2) + \epsilon_T\sigma_L^R(W,Q^2), \quad (3)$$

$$\epsilon_T = \left(1 + 2\frac{\nu^2+Q^2}{Q^2}\tan^2\frac{\theta_e}{2}\right)^{-1}, \quad (4)$$

with θ_e the electron scattering angle and ν the energy transfer to the target. The resonant pieces of the inclusive structure functions are computed from the cross sections in Eq. 3 as [21]

$$F_1^R(W,Q^2) = \frac{KW}{4\pi^2\alpha}\sigma_T^R,$$
$$F_2^R(W,Q^2) = \frac{KW}{4\pi^2\alpha}\frac{2x}{1+\frac{Q^2}{\nu^2}}\left(\sigma_T^R + \sigma_L^R\right) = \frac{KW}{4\pi^2\alpha}\frac{2x}{1+\frac{Q^2}{\nu^2}}\frac{1+R_{LT}}{1+\epsilon_T R_{LT}}\sigma_U, \quad (5)$$

where $x = \frac{Q^2}{2M_N\nu}$, $K = \frac{W^2-M_N^2}{2M_N}$ and $R_{LT} = \sigma_L/\sigma_T$. We refer to Ref. [17] for further details.

A special procedure has been developed for the interpolation of the experimental results on inclusive structure functions in the region of 1.07 GeV< W <4.0 GeV and $Q^2 < 7.0$ GeV^2 [10]. This procedure is also used for the interpolation of the resonant contributions. In Ref. [22], the inclusive structure functions, virtual photon and electron scattering inclusive cross sections can be computed online for the beam energies defined by the user. A user-friendly interface is available to provide the computed results both as numerical files and as sets of plots.

3 Results and discussion

In Figs. 1 and 2 we show the transverse (unpolarized) cross sections, $\sigma_{T(U)}^R$ for an electron beam energy of 10.6 GeV, and at representative values of photon virtualities in the range of $Q^2 \geq 1$ GeV^2, where partons become active degrees of freedom in inclusive electron scattering. In Fig. 1, the decomposition into the individual contributions of each resonance is shown, while in Fig. 2 the resonant contribution from all resonances combined are compared with the full unpolarized cross section data [10, 22]. Although the resonances clearly cluster into three regions, their tails give important contributions to the neighboring ones. Even the first region, clearly dominated by the single $\Delta(1232)\ 3/2^+$ resonance at $Q^2 \approx 1.0$ GeV^2, is also affected by the tail of the $N(1440)\ 1/2^+$. This becomes increasingly relevant at higher Q^2, since the transverse $\Delta(1232)\ 3/2^+$ electrocouplings decrease with Q^2 much faster than those of the $N(1440)\ 1/2^+$ [22]. In the second resonance region, the $N(1535)\ 1/2^-$ gives the largest contribution at $Q^2 > 2.0$ GeV^2 due to the slower decrease of its $A_{1/2}$ electrocoupling with Q^2 in comparison with that of $N(1520)3/2^-$ [12, 13]. For the same reason, its tail gives increasingly noticeable contributions with Q^2 also to the third resonance region. Because of the large total decay width of the $N(1440)1/2^+$ resonance, which amounts to 300 MeV, its complementary contribution is observed in all three resonance regions at $W < 1.7$ GeV. The contribution from several overlapping resonances combined generates the peak seen in the third resonance region of the cross sections. The largest contributions in the third region are assigned to the $N(1680)\ 5/2^+$ and the $N'(1720)\ 3/2^+$ candidate. The contribution of the $N'(1720)\ 3/2^+$ is needed in order to reproduce the third resonance maximum seen in the virtual photon inclusive cross section.

Overall, the transverse resonant part gives the largest contribution to the resonant cross sections. The size of the longitudinal piece increases with W, but does not exceed 30% of the total cross section in the kinematical region shown here. The evolution of the unpolarized resonant cross sections with Q^2 is rather complex: both the first and the third regions show a stronger fall-off with Q^2 than the second peak, mainly due to the $N(1535)\ 1/2^-$ contribution in the second region. The different resonance states thus seem to display rather considerable differences in the Q^2 evolution of their electrocouplings, thus underlying the importance of dedicating separate analyses to each of them.

The predicted resonant contributions to the longitudinal virtual photon cross sections are shown in Figs. 3. The longitudinal part of the cross section is sensitive to high-lying N^*. The largest contributions to the longitudinal exclusive cross sections come from the resonances with masses above 1.6 GeV. The contributions from each resonance demonstrate a pronounced dependence on Q^2.

Future experimental results from CLAS on resonance electrocouplings in the mass range from 1.6 GeV to 2.0 GeV and 0.4 GeV^2 < Q^2 < 5.0 GeV^2 expected from the data [23–26] will improve our estimates on the contributions from excited nucleon states in the mass range above 1.6 GeV. Furthermore, they will provide the first results on electrocouplings of the resonances in the mass range above 1.9 GeV, where the transition from the resonance to deep inelastic scattering regions takes place.

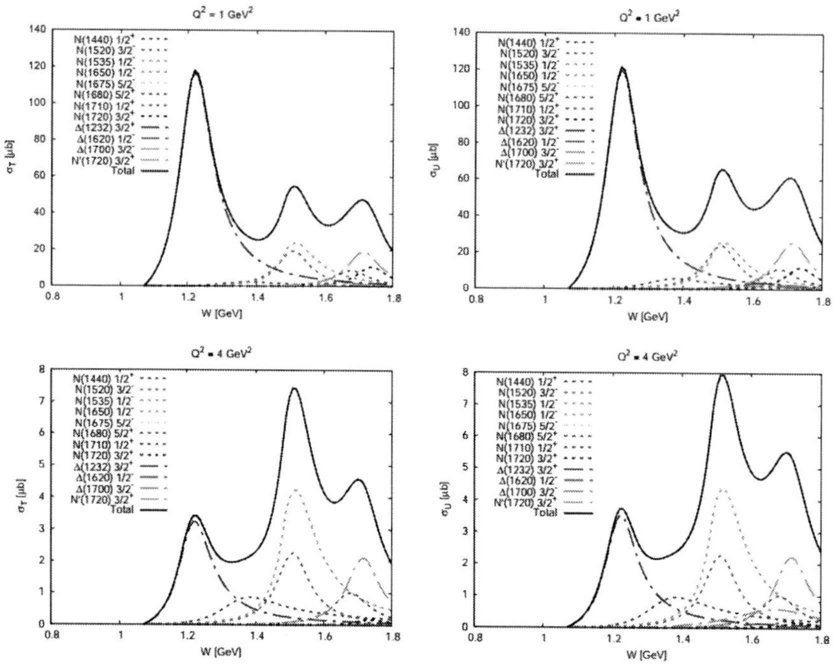

Figure 1. Decomposition of the transverse (unpolarized) resonant cross sections, shown as thick black curves in the left (right) column, into the separate contributions of each resonance included in the model, at $Q^2 = 1$ and 4 GeV^2, for an electron beam energy of 10.6 GeV.

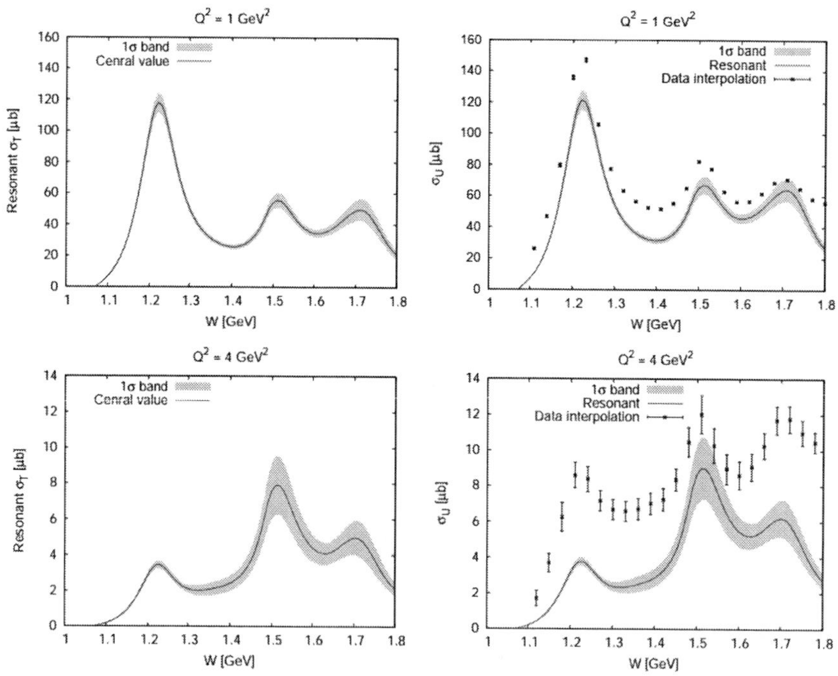

Figure 2. Resonant contributions (green curves with uncertainty bands) to the transverse (unpolarized) virtual photon-proton cross sections $\sigma_{T(U)}$ are shown on the left (right) column, at $Q^2 = 1$ and 4 GeV^2, for an electron beam energy of 10.6 GeV. The data points are from CLAS [10, 22].

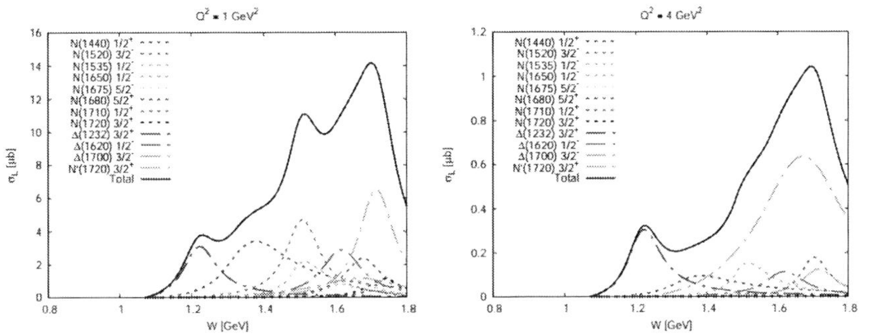

Figure 3. Decomposition of the longitudinal resonant cross sections, shown as thick black curves, into the separate contributions of each resonance included in the model, at $Q^2 = 1$ and 4 GeV^2.

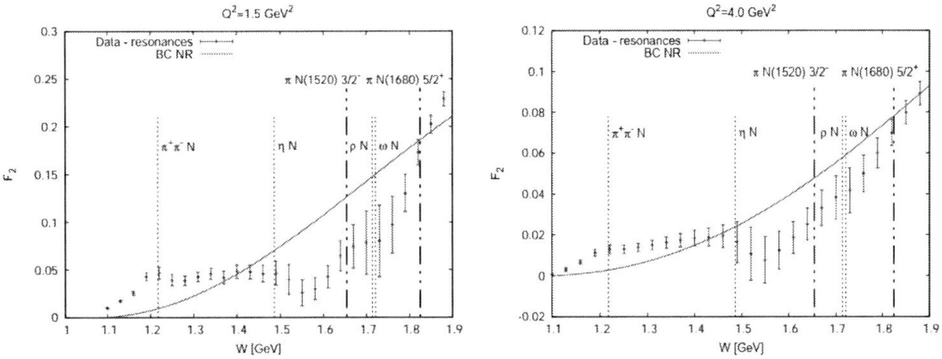

Figure 4. Pseudo-data (purple points) for the differences between the F_2 structure function data [10, 22] and the estimated resonant contributions, compared to the background used in Ref. [7] (green curve). The dotted vertical lines show the opening of meson-nucleon electroproduction channels. The dash-dotted vertical lines show the opening of π N^* channels.

Since our evaluation of the resonant contributions depends only on exclusive electroproduction data, it enables to give a first estimate of the non-resonant contributions to the inclusive observables. This estimate consists in determining the difference between the inclusive data and the resonant contributions estimated independently from exclusive meson electroproduction data. This is shown in Fig. 4, and can be interpreted as pseudo-data for the background of the structure function F_2 measured by CLAS [5]. Both the data and the model uncertainties are propagated into the final error bars shown. We also show the comparison of the contributions inferred from the data with the background model from Ref. [7]. The non-resonant contributions determined by realistically accounting for the nucleon resonances demonstrate several structures and a sharp increase with W from 1.6 GeV to 1.7 GeV seen in all Q^2-bins. One can observe several kinks in the W dependence of the background for F_2. It appears, however, that each of them is associated with the opening of a meson-baryon channel, namely $\pi\pi N$ at 1.21 GeV, ηN at 1.49 GeV, and ωN at 1.72 GeV. We also show the W values for the opening of the $\pi^+ N(1520)\ 3/2^-$, $\pi^+ N(1680)\ 5/2^+$ and ρN channels, calculated at the resonance central masses. Because of the appreciable decay widths ($\gtrsim$ 100 MeV) of the unstable final states, instead of kinks these channels produce sharp growths in the W dependence of the F_2 structure function at 1.6 GeV $\lesssim W \lesssim$ 1.7 GeV, seen in all Q^2 bins. These features are not presented in the extrapolation of the non-resonant contributions from the deep inelastic scattering region into the resonance region in Ref. [7]. Therefore, the realistic estimate of the resonance contributions from the experimental electrocoupling data may have impact on the insight into the parton distributions in the ground-state nucleon at Bjorken variable x values close to unity, within the resonance region. This is the subject of our future work.

Preliminary results on inclusive electron scattering with a 10.6 GeV electron beam have recently become available from the CLAS12 detector. The achieved kinematic coverage is shown in Fig. 5 (top). The Q^2 dependence of the normalized inclusive event yield shown in Fig. 5 (bottom) is in good agreement with the expectation from the interpolation of the CLAS and world data on inclusive electron scattering [10]. The approach presented

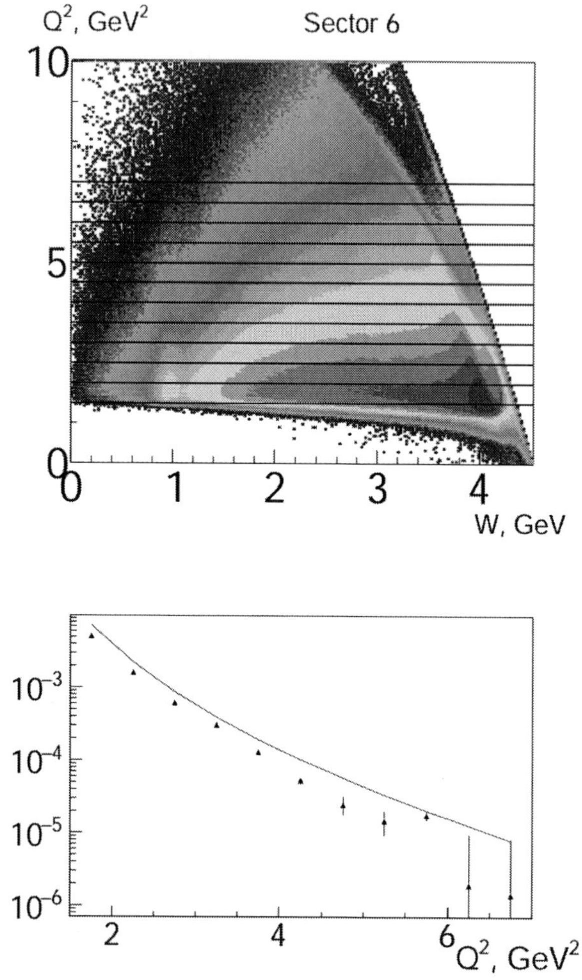

Figure 5. Preliminary Q^2 vs W distribution of inclusive electron scattering events measured with the CLAS12 detector (top). Normalized yield (arb. units) as a function of Q^2 (bottom) at W=1.22 GeV in comparison with parameterization developed in [10]

in these proceedings provides a phenomenological tool for the analyses of the CLAS12 data on inclusive electron scattering at high photon virtualities of $Q^2 > 5.0$ GeV^2 in the resonance region.

4 Summary

In this work, for the first time the resonant contributions to unpolarized electron scattering observables have been evaluated in the first three resonance regions, using as input the CLAS data on resonance electrocouplings [12, 13]. For all three resonance peaks, we observed substantial contributions from the resonance tails in the neighboring regions. The behaviour in the Q^2 evolution of the resonance structures is non-trivial: the first and third peaks decrease strongly with Q^2, while the second region has a softer decrease: in fact, its ratio with respect to the background stays relatively constant. These distinctively different structural features underline the importance of the separate studies of the nucleon resonances for the exploration of strong QCD.

The comparison between the resonant contributions and data from CLAS [5] allowed to give an estimate for the separate contribution of the background at different values of Q^2. The electrocouplings of nucleon resonances in an even broader mass and Q^2 range will soon become available [24, 27], and with the CLAS12 endeavour on both exclusive and inclusive electron-induced reactions, the data base will be extended to the largest Q^2 interval ever achieved, from 0.05 GeV^2 to 12 GeV^2. The present study can thus also be used as a phenomenological tool for the analysis of CLAS12 data. Ultimately, our goal is to model resonant and background contributions in a combined fashion in wide Q^2 regions, in order to describe the data at low and high energies. By using the experimental results on the nucleon resonance electrocouplings we are also planning on gaining insight into the parton distributions in the ground-state nucleons at Bjorken variable x values close to unity, within the resonance region.

This work was supported by the Deutsche Forschungsgemeinschaft (DFG, German Research Foundation), in part through the Collaborative Research Center [The Low-Energy Frontier of the Standard Model, Projektnummer 204404729 - SFB 1044], and in part through the Cluster of Excellence [Precision Physics, Fundamental Interactions, and Structure of Matter] (PRISMA+ EXC 2118/1) within the German Excellence Strategy (Project ID 39083149). It was also supported by the U.S. Department of Energy, Office of Science, Office of Nuclear Physics under contract DE-AC05-06OR23177 and Skobeltsyn Nuclear Physics Institute and the Physics Department at Moscow State University.

References

[1] H.W. Lin et al., Prog. Part. Nucl. Phys. **100**, 107 (2018), 1711.07916

[2] P. Jimenez-Delgado, W. Melnitchouk, J.F. Owens, J. Phys. **G40**, 093102 (2013), 1306.6515

[3] J. Gao, L. Harland-Lang, J. Rojo, Phys. Rept. **742**, 1 (2018), 1709.04922

[4] E.D. Bloom, F.J. Gilman, Phys. Rev. Lett. **25**, 1140 (1970)

[5] M. Osipenko et al. (CLAS), Phys. Rev. **D67**, 092001 (2003), hep-ph/0301204

[6] S.P. Malace et al. (Jefferson Lab E00-115), Phys. Rev. **C80**, 035207 (2009), 0905.2374

[7] M.E. Christy, P.E. Bosted, Phys. Rev. **C81**, 055213 (2010), 0712.3731

[8] C.E. Keppel, M.I. Niculescu et al. (2000), JLab experiment E00-002

[9] Y. Liang et al. (Jefferson Lab Hall C E94-110) (2004), nucl-ex/0410027

[10] A.A. Golubenko, V.V. Chesnokov, B.S. Ishkhanov, V.I. Mokeev (2019), 1902.02900

[11] V.D. Burkert, Ann. Rev. Nucl. Part. Sci. **68**, 405 (2018)

[12] I.G. Aznauryan, V.D. Burkert, Prog. Part. Nucl. Phys. **67**, 1 (2012), 1109.1720

[13] V.I. Mokeev (CLAS), Few Body Syst. **59**, 46 (2018), 1801.09750

[14] V.I. Mokeev, I. Aznauryan, V. Burkert, R. Gothe, EPJ Web Conf. **113**, 01013 (2016), 1508.04088

[15] V.D. Burkert, V.I. Mokeev, B.S. Ishkhanov (2019), 1901.09709

[16] V.I. Mokeev, *The 12th International Workshop on the Physics of Excited Nucleons - N*2019, (2019). See these proceedings.*

[17] A.N. Hiller Blin et al., Phys. Rev. **C100**, 035201 (2019)

[18] *Fits of the resonances electrocouplings*, https://userweb.jlab.org/~isupov/couplings/, https://userweb.jlab.org/~isupov/couplings/

[19] W.H. Press, S.A. Teukolsky, W.T. Vetterling, B.P. Flannery (1992)

[20] M. Tanabashi et al. (Particle Data Group), Phys. Rev. **D98**, 030001 (2018)

[21] D. Drechsel, B. Pasquini, M. Vanderhaeghen, Phys. Rept. **378**, 99 (2003), hep-ph/0212124

[22] *Structure functions and cross-sections Database*, http://clas.sinp.msu.ru/strfun/, http://clas.sinp.msu.ru/strfun/

[23] A. Trivedi, Few Body Syst. **60** (2019)

[24] E.L. Isupov et al. (CLAS), Phys. Rev. **C96**, 025209 (2017), 1705.01901

[25] N. Markov, K. Joo, M. Ungaro, L.C. Smith, V. Mokeev, Few Body Syst. **59**, 134 (2018)

[26] N. Markov et al. (CLAS), *Exclusive $\pi^0 p$ electroproduction off protons in the resonance region at photon virtualities 0.4 GeV^2 $\leq Q^2 \leq$ 1 GeV^2* (2019), 1907.11974

[27] G.V. Fedotov et al. (CLAS), Phys. Rev. **C98**, 025203 (2018), 1804.05136

On $SU(3)_F$ positive-parity octet and decuplet baryons

Chen Chen[1,*]

[1]Institut für Theoretische Physik, Justus-Liebig-Universität Gießen, 35392 Gießen, Germany

Abstract. A continuum approach to the three valence-quark bound-state problem in quantum field theory, employing parametrisations of the necessary kernel elements, is used to compute the spectrum and Poincaré-covariant wave functions for all flavour-$SU(3)$ octet and decuplet baryons and their first positive-parity excitations. Such analyses predict the existence of nonpointlike, dynamical quark-quark (diquark) correlations within all baryons; and a uniformly sound description of the systems studied is obtained by retaining flavour-antitriplet–scalar and flavour-sextet–pseudovector diquarks. The analysis predicts the existence of positive-parity excitations of the Ξ, Ξ^*, Ω baryons, with masses, respectively (in GeV): 1.84(08), 1.89(04), 2.05(02). These states have not yet been empirically identified. This body of analysis suggests that the expression of emergent mass generation is the same in all *u, d, s* baryons and, notably, that dynamical quark-quark correlations play an essential role in the structure of each one. It also provides the basis for developing an array of predictions that can be tested in new generation experiments.

1 Introduction

The analysis of a baryon as a three–valence-body bound-state problem in continuum quantum field theory became possible following the formulation of a Poincaré-covariant Faddeev equation [1–5], which is depicted in Fig. 1. This approach assumes that dynamical, nonpointlike diquark correlations play an important role in baryon structure. Whilst the diquarks do not survive as asymptotic states, *viz.* they do not appear in the strong interaction spectrum [6, 7], the attraction between the quarks in the $\bar{3}$ channel draws a picture in which two quarks are always correlated as a colour-$\bar{3}$ diquark pseudoparticle, and binding is effected by the iterated exchange of roles between the bystander and diquark-participant quarks. This approach to the spectrum and interactions of baryons has been applied widely with phenomenological success, *e.g.* Refs. [8–24].

A spectrum of flavour-$SU(3)$ octet and decuplet baryons, their parity partners, and the radial excitations of these systems, was computed in Ref. [10, 17, 30] using a symmetry-preserving treatment of a vector × vector contact-interaction (SPCI) [25–29] as the foundation for the relevant few-body equations. With such an approach: the diquark bound-state amplitudes are momentum-independent but the diquarks are dynamical degrees-of-freedom with nonzero electromagnetic radii [30]; and one obtains a spectrum that is qualitatively equivalent to the quark-model and lQCD results.

In Ref. [19] we employed a QCD-kindred framework which used elsewhere [18] to perform a realistic study of the four lightest ($I = 1/2, J^P = 1/2^{\pm}$) baryon isospin-doublets; extended it to flavour-$SU(3)$ and delivered therewith a spectrum of octet and decuplet baryons and their

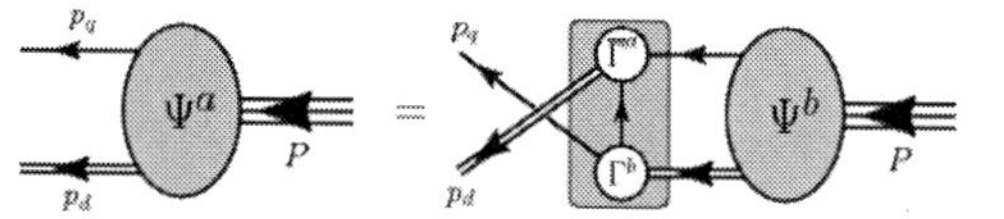

Figure 1. Poincaré covariant Faddeev equation: a linear integral equation for the matrix-valued function Ψ, being the Faddeev amplitude for a baryon of total momentum $P = p_q + p_d$, which expresses the relative momentum correlation between the dressed-quarks and -nonpointlike-diquarks within the baryon. The shaded rectangle demarcates the kernel of the Faddeev equation: *single line*, dressed-quark propagator (Sec. 2.2); Γ, diquark correlation amplitude (Sec. 2.3); and *double line*, diquark propagator (Sec. 2.4).

first positive-parity excitations along with structural insights drawn from analyses of their Poincaré-covariant wave functions.

We emphasise that the continuum analyses indicated above form part of the body of Dyson-Schwinger equation (DSE) studies of hadron structure [29, 31–49]. In this approach, the challenge is a need to employ a truncation so as to define a tractable problem. Much has been learnt; and one may now separate DSE studies into three classes. *Class-A*. model-independent statements about QCD; *Class-B*. illustrations of such statements using well-constrained model elements and possessing a traceable connection to QCD; *Class-C*. QCD-inspired analyses whose elements have not been computed using a truncation that preserves a systematically-improvable connection with QCD. The analysis described herein lies within Class-C.

We describe our approach to the baryon bound-state problem in Sec. 2; and detail and explain the character of

*e-mail: Chen.Chen@theo.physik.uni-giessen.de

the solutions for the octet and decuplet baryons and their first positive-parity excitations in Sec. 3. Section 4 provides a summary and indicates some new directions.

2 Baryon Bound State Problem

2.1 Faddeev equation

In its general form, the Faddeev equation sums all possible exchanges and interactions that can take place between the three dressed-quarks that express a baryon's valence-quark content. Used with a realistic quark-quark interaction [50–56], it predicts the appearance of soft (nonpointlike) fully-interacting diquark correlations within baryons, whose characteristics are greatly influenced by dynamical chiral symmetry breaking (DCSB) [13]. Consequently, the problem of determining a baryon's mass, internal structure, etc., is transformed into that of solving the linear, homogeneous matrix equation depicted in Fig. 1.

2.2 Dressed quarks

Regarding flavour-$SU(3)$ octet and decuplet baryons and their radial excitations, the Faddeev kernel in Fig. 1 involves three basic elements, *viz.* the dressed light-quark propagators, $S_f(p)$, $f = u, d, s$, and the correlation amplitudes and propagators for all participating diquarks. Much is known about $S_f(p)$, and in constructing the kernel we use the algebraic forms described in Ref. [19], which have proven efficacious in the explanation and unification of a wide range of hadron observables [11–14, 57, 58]. (*N.B.* We assume isospin symmetry throughout, *i.e.* u- and d-quarks are mass-degenerate and described by the same propagator. Consequently, all diquarks in an isospin multiplet are degenerate.)

2.3 Correlation amplitudes

In Fig. 1, all participating diquarks are colour-antitriplets because they must combine with the bystander quark to form a colour singlet.

Diquark isospin-spin structure is more complex. Accounting for Fermi-Dirac statistics, five types of correlation are possible in a $J = 1/2$ bound-state: flavour-$\bar{3}$–scalar, flavour-6–pseudovector, flavour-$\bar{3}$–pseudoscalar, flavour-$\bar{3}$–vector, and flavour-6–vector. However, only the first two are important in positive-parity systems [15, 17, 18]; and the associated leading correlation amplitudes are, respectively:

$$\Gamma^j_{0^+}(k;K) = g^j_{0^+}\, \vec{H}\, T^j_{\bar{3}_f}\, \gamma_5 C\, \mathcal{F}(k^2/[\omega^j_{0^+}]^2)\,, \tag{1a}$$

$$\Gamma^g_{1^+\mu}(k;K) = i g^g_{1^+}\, \vec{H}\, T^g_{6_f}\, \gamma_\mu C\, \mathcal{F}(k^2/[\omega^g_{1^+}]^2)\,, \tag{1b}$$

where: K is the total momentum of the correlation, k is a two-body relative momentum, $\mathcal{F}(x) = (1 - e^{-x})/x$, $\omega^{j,g}_{J^P}$ are size parameters, and $g^{j,g}_{J^P}$ are couplings into the channel, fixed by normalisation; $\vec{H} = \{i\lambda^7_c, -i\lambda^5_c, i\lambda^2_c\}$, with $\{\lambda^k_c, k = 1, \ldots, 8\}$ denoting Gell-Mann matrices in colour-space, expresses the diquarks' colour antitriplet character; $C = \gamma_2\gamma_4$ is the charge-conjugation matrix;

$$\{T^j_{\bar{3}_f}, j = 1,2,3\} = \{i\lambda^2, i\lambda^5, i\lambda^7\}\,, \tag{2a}$$

$$\begin{aligned}\{T^g_{6_f}, g = 1,\ldots,6\} = \{&s_0\lambda^0 + s_3\lambda^3 + s_8\lambda^8, \lambda^1, \lambda^4,\\ &s_0\lambda^0 - s_3\lambda^3 + s_8\lambda^8, \lambda^6, s_0\lambda^0 - 2s_8\lambda^8\}\,,\end{aligned} \tag{2b}$$

with $s_0 = \sqrt{2/3}$, $s_3 = 1/\sqrt{2}$, $s_8 = 1/\sqrt{6}$, $\{\lambda^k, k = 1, \ldots, 8\}$ denoting flavour-$SU(3)$ Gell-Mann matrices, $\lambda^0 = \text{diag}[1,1,1]$, and all flavour matrices left-active on column$[u,d,s]$.

Turning to decuplet baryons, since it is not possible to combine a $\bar{3}_f$ diquark with a 3_f-quark to obtain a member of the symmetric 10_f representation of $SU(3)_f$, decuplet baryons only contain 6_f–axial-vector diquarks, which are associated with the amplitudes in Eq. (1b).

2.4 Diquark propagators, masses, couplings

A propagator is associated with each quark-quark correlation in Fig. 1; and we use [11, 18]:

$$\Delta^{0^+j}(K) = \frac{1}{[m^j_{0^+}]^2}\,\mathcal{F}(K^2/[\omega^j_{0^+}]^2)\,, \tag{3a}$$

$$\Delta^{1^+g}_{\mu\nu}(K) = \left[\delta_{\mu\nu} + \frac{K_\mu K_\nu}{[m^g_{1^+}]^2}\right]\frac{1}{[m^g_{1^+}]^2}\,\mathcal{F}(K^2/[\omega^g_{1^+}]^2)\,. \tag{3b}$$

These algebraic forms ensure that the diquarks are confined within the baryons, as appropriate for coloured correlations: whilst the propagators are free-particle-like at spacelike momenta, they are pole-free on the timelike axis. This is sufficient to ensure confinement via the violation of reflection positivity (see, e.g. Ref. [43], Sec. 3).

The diquark masses and sizes are related via

$$m^{j,g}_{J^P} = \sqrt{2}\,\omega^{j,g}_{J^P}\,. \tag{4}$$

This identification accentuates the free-particle-like propagation characteristics of the diquarks within the baryon [11]. The mass-scales are constrained by numerous studies; and we use (in GeV):

$$m^{(1)}_{0^+} = 0.80\,,\; m^{(2,3)}_{0^+} = 0.95\,, \tag{5a}$$

$$m^{(4,5,7)}_{1^+} = 0.90\,,\; m^{(6,8)}_{1^+} = 1.05\,,\; m^{(9)}_{1^+} = 1.20\,, \tag{5b}$$

where the values of $m^{(j=1)}_{0^+}$ and $m^{(g=4,5,7)}_{1^+}$ are drawn from Refs. [11, 12], because they provide for a good description of numerous dynamical properties of the nucleon, Δ-baryon and Roper resonance; and the masses $m^{(j=2,3)}_{0^+}$, $m^{(g=6,8)}_{1^+}$ and $m^{(g=9)}_{1^+}$ are derived therefrom via an equal-spacing rule, *viz.* replacing one light-quark by a s-quark brings an extra 0.15 GeV $\approx M^E_s - M^E_u$ [19]. And for completeness, we will subsequently display results with all diquark masses varied by ±5%.

Using canonical normalization condition [19] and the masses in Eqs. (5):

$$g^{(1)}_{0^+} = 14.75\,,\; g^{(2,3)}_{0^+} = 9.45\,, \tag{6a}$$

$$g^{(4,5,7)}_{1^+} = 12.73\,,\; g^{(6,8)}_{1^+} = 7.62\,,\; g^{(9)}_{1^+} = 3.72\,. \tag{6b}$$

Given that it is the coupling-squared which appears in the Faddeev kernels, scalar diquarks will dominate the Faddeev amplitudes of $J = 1/2$ baryons; but pseudovector diquarks must also play a material role because $g_{1^+}^2/g_{0^+}^2 \approx 0.7$, which agrees with the previous studies on the nucleon, see, *e.g.* Ref. [67, 68].

2.5 Faddeev amplitudes and wave functions

In solving the Faddeev equation, Fig. 1, one obtains both the mass-squared and bound-state amplitude of all baryons with a given value of J^P. In fact, it is the form of the Faddeev amplitude which fixes the channel. A baryon is described by

$$\Psi^B = \psi_1^B + \psi_2^B + \psi_3^B\,, \tag{7}$$

where the subscript identifies the bystander quark, *i.e.* the quark that is not participating in a diquark correlation, $\psi_{1,2}^B$ are obtained from $\psi_3^B =: \psi^B$ by a cyclic permutation of all quark labels.

For an octet baryon ($B_{\underline{8}} = N, \Lambda, \Sigma, \Xi$; $J^P = 1/2^+$),

$$\begin{aligned}
&\psi^{\underline{8}}(p_i,\alpha_i,\sigma_i) \\
&= \sum_{j\in B_{\underline{8}}} [\Gamma^j_{0^+}(k;K)]^{\alpha_1\alpha_2}_{\sigma_1\sigma_2}\,\Delta^{0^+j}(K)\,[\varphi^{8j}_{0^+}(\ell;P)u(P)]^{\alpha_3}_{\sigma_3} \\
&+ \sum_{g\in B_{\underline{8}}} [\Gamma^g_{1^+\mu}]^{\alpha_1\alpha_2}_{\sigma_1\sigma_2}\,\Delta^{1^+g}_{\mu\nu}\,[\varphi^{8g}_{1^+\nu}(\ell;P)u(P)]^{\alpha_3}_{\sigma_3}\,,
\end{aligned} \tag{8}$$

where $(p_i, \sigma_i, \alpha_i)$ are the momentum, spin and isospin labels of the quarks constituting the bound state; $P = p_1+p_2+p_3 = p_d+p_q$ is the total momentum of the baryon; $k = (p_1 - p_2)/2$, $K = p_1 + p_2 = p_d$, $\ell = (-K + 2p_3)/3$; j and g are the labels in Eqs. (2) and the sums run over those flavour-combinations permitted in $B_{\underline{8}}$; and $u(P)$ is a Euclidean spinor. The remaining elements in Eq. (8) are the following matrix-valued functions:

$$\varphi^{8j}_{0^+}(\ell;P) = \sum_{k=1}^{2} \mathfrak{s}^j_{\underline{8}k}(\ell^2,\ell\cdot P)\,\mathcal{S}^k(\ell;P)\,, \tag{9a}$$

$$\varphi^{8g}_{1^+\nu}(\ell;P) = \sum_{k=1}^{6} \mathfrak{a}^{g+3}_{\underline{8}k}(\ell^2,\ell\cdot P)\,\gamma_5\mathcal{A}^k_\nu(\ell;P)\,, \tag{9b}$$

where

$$\begin{aligned}
&\mathcal{S}^1 = \mathbf{I}_{\rm D}\,,\ \mathcal{S}^2 = i\gamma\cdot\hat\ell - \hat\ell\cdot\hat P\mathbf{I}_{\rm D}\,, \\
&\mathcal{A}^1_\nu = \gamma\cdot\ell^\perp\hat P_\nu\,,\ \mathcal{A}^2_\nu = -i\hat P_\nu\mathbf{I}_{\rm D}\,,\ \mathcal{A}^3_\nu = \gamma\cdot\hat\ell^\perp\hat\ell^\perp_\nu\,, \\
&\mathcal{A}^4_\nu = i\hat\ell^\perp_\nu\mathbf{I}_{\rm D}\,,\ \mathcal{A}^5_\nu = \gamma^\perp_\nu - \mathcal{A}^3_\nu\,,\ \mathcal{A}^6_\nu = i\gamma^\perp_\nu\gamma\cdot\hat\ell^\perp - \mathcal{A}^4_\nu\,,
\end{aligned} \tag{10}$$

with $\hat\ell^2 = 1$, $\hat P^2 = -1$, $\ell^\perp = \hat\ell_\nu + \hat\ell\cdot\hat P\hat P_\nu$, $\gamma^\perp = \gamma_\nu + \gamma\cdot\hat P\hat P_\nu$.

Owing to the symmetry-prescribed absence of flavour-$\bar 3$ components, the structure of decuplet baryons is simpler ($B_{\underline{10}} = \Delta, \Sigma^*, \Xi^*, \Omega$; $J^P = 3/2^+$):

$$\begin{aligned}
&\psi^{\underline{10}}_\mu(p_i,\alpha_i,\sigma_i) \\
&= \sum_{g\in B_{\underline{10}}} [\Gamma^g_{1^+\mu}]^{\alpha_1\alpha_2}_{\sigma_1\sigma_2}\,\Delta^{1^+g}_{\mu\nu}\,[\varphi^{10g}_{\nu\rho}(\ell;P)u_\rho(P)]^{\alpha_3}_{\sigma_3}\,,
\end{aligned} \tag{11}$$

where $u_\rho(P)$ is a Rarita-Schwinger spinor; and, with $\mathcal{S}^k$ and $\mathcal{A}^k_\nu$ in Eq. (10),

$$\varphi^{10g}_{\nu\rho}(\ell;P) = \sum_{k=1}^{8} \mathfrak{a}^{g+3}_{\underline{10}k}(\ell^2,\ell\cdot P)\,\mathcal{D}^k_{\nu\rho}(\ell;P)\,, \tag{12a}$$

$$\mathcal{D}^k_{\nu\rho} = \mathcal{S}^k\,\delta_{\nu\rho}\,, \qquad k = 1,2\,, \tag{12b}$$

$$\mathcal{D}^k_{\nu\rho} = i\gamma_5\,\mathcal{A}^{k-2}_\nu\,\ell^\perp_\rho\,, \quad k = 3,\ldots,8\,. \tag{12c}$$

The (unamputated) Faddeev wave function can be computed from the amplitude specified by Eqs. (8) – (12) simply by attaching the appropriate dressed-quark and diquark propagators. It may also be decomposed in the form of Eqs. (9), (12). Naturally, the scalar functions are different, and we label them $\tilde{\mathfrak{s}}^j_{\underline{8}k}$, $\tilde{\mathfrak{a}}^g_{\underline{8}k}$, $\tilde{\mathfrak{a}}^g_{\underline{10}k}$.

Both the Faddeev amplitude and wave function are Poincaré covariant, *i.e.* they are qualitatively identical in all reference frames. Naturally, each of the scalar functions that appears is frame-independent, but the frame chosen determines just how the elements should be combined. Consequently, the manner by which the dressed-quarks' spin, S, and orbital angular momentum, L, add to form a particular J^P combination is frame-dependent: L, S are not independently Poincaré invariant.[1] The set of baryon rest-frame quark-diquark angular momentum identifications can be found in Refs. [19, 62, 63].

3 Solutions and their Properties

3.1 Masses of the dressed-quark cores

We can now report results obtained by solving the Faddeev equations for octet and decuplet baryons and their first positive-parity excitations. Our computed masses are listed in Table 1: the uncertainties indicate the response of a given mass to a coherent 5% increase/decrease in the mass-scales associated with the diquarks and dressed-quarks.

It is worth noting the emergence of a Σ-Λ mass-splitting despite the fact that we have assumed isospin symmetry, *i.e.* mass-degenerate u- and d-quarks, described by the same propagator, so that all diquarks in an isospin multiplet are degenerate. Whilst the Λ^0 and Σ^0 baryons are associated with the same combination of valence-quarks, their spin-flavour wave functions are different: the $\Lambda^0_{I=0}$ contains more of the lighter $J = 0$ diquark correlations than the $\Sigma^0_{I=1}$. It follows that the Λ^0 must be lighter than the Σ^0. Therefore, our result agrees with the prediction of the Gell-Mann–Okubo mass formula [69–71].

The computed masses and corresponding empirical results are depicted in Fig. 2. It is apparent that the computed masses are uniformly larger than the corresponding empirical values. This is because our results should be viewed as those of a given baryon's dressed-quark core, whereas the empirical values include all contributions, including

[1] The nature of the combination is also scale dependent because the definition of a dressed-quark and the character of the correlation amplitudes changes with resolving scale, ζ, in a well-defined manner [60, 61]. Our analysis is understood to be valid at $\zeta \simeq 1$ GeV.

Table 1. Computed dressed-quark-core masses of ground-state octet and decuplet baryons, and their radial excitations. Row 1: Baryon ground-states. Row 3: First positive-parity excitations of the ground-states. Masses in rows labelled "expt." are taken from Ref. [65]. A hyphen in any position indicates that no empirically known resonance can confidently be associated with the theoretically predicted state. (All dimensioned quantities are listed in GeV.)

	Row		N	Λ	Σ	Ξ	Δ	Σ^*	Ξ^*	Ω
n=0	1	DSE	1.19(13)	1.37(14)	1.41(14)	1.58(15)	1.35(12)	1.52(14)	1.71(15)	1.93(17)
	2	expt.	0.94	1.12	1.19	1.31	1.23	1.38	1.53	1.67
n=1	3	DSE	1.73(10)	1.85(09)	1.88(11)	1.99(11)	1.79(12)	1.93(11)	2.08(12)	2.23(13)
	4	expt.	1.44(03)	$1.60^{+0.10}_{-0.04}$	1.66(03)	-	1.57(07)	1.73(03)	-	-

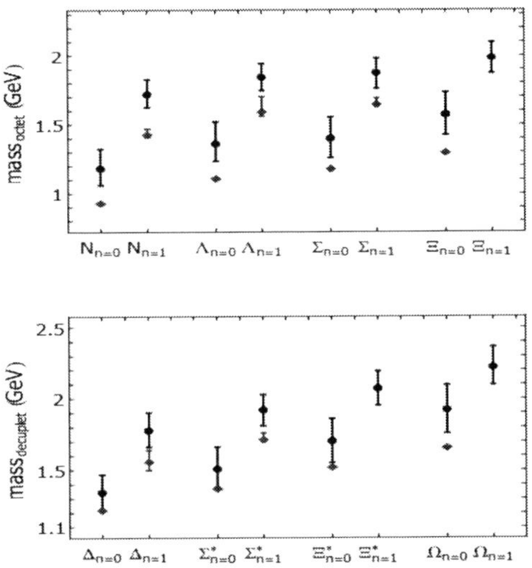

Figure 2. Upper panel: Pictorial representation of octet masses in Table 1. *Circles* (black) – computed masses. The vertical riser indicates the response of our predictions to a coherent ±5% change in the mass-scales associated with the diquarks and dressed-quarks. *Diamonds* (green) – empirical Breit-Wigner masses [65]. The horizontal axis lists a particle name with a subscript that indicates whether it is ground-state ($n = 0$) or first positive-parity excitation ($n = 1$). **Lower panel**: Analogous plot for the decuplet masses in Table 1. Where noticeable, the estimated uncertainty in the location of a resonance's Breit-Wigner mass is indicated by an error bar.

meson cloud effects, which typically generate a measurable reduction [64]. This was explained and illustrated in a study of the nucleon, its parity-partner and their radial excitations in Ref. [18]; and has also been demonstrated using the SPCI [10, 17]. Identifying the difference between our predictions and experiment, then one finds that such effects are fairly homogeneous across the spectrum:

$$\text{mean}[m^{8}_{\text{DSE}} - m^{8}_{\text{expt.}}] = 0.26(4)\,, \tag{13a}$$

$$\text{mean}[m^{10}_{\text{DSE}} - m^{10}_{\text{expt.}}] = 0.19(5)\,. \tag{13b}$$

Namely, they act to reduce the mass of ground-state octet and decuplet baryons and their first positive-parity excitations by 0.23(6) GeV.

3.2 Diquark content

The Faddeev amplitude of each baryon can be decomposed into a sum of N_{qq} terms, $\{\mathcal{F}_i, i = 1,\ldots,N_{qq}\}$, each one of which is directly identifiable with a particular diquark component. The value of N_{qq} depends on the baryon's spin-flavour structure:

	N	Λ	Σ	Ξ	Δ	Σ^*	Ξ^*	Ω
N_{qq}	14	10	14	14	8	16	16	8

. (14)

In connection with each term, we define

$$\mathbb{D}_i = \int \frac{d^4\ell}{(2\pi)^4} |\mathcal{F}_i(\ell^2, \ell\cdot P)|^2 \tag{15}$$

and subsequently compute

$$\mathbb{Q}_{\mathfrak{t}} = \mathbb{W}^{-1} \sum_{i\in\mathfrak{t}} \mathbb{D}_i\,,\; \mathbb{W} = \sum_{i=1}^{N_{qq}} \mathbb{D}_i\,, \tag{16}$$

where $\mathfrak{t}$ ranges over the $\mathfrak{s}(=0^+)$ and $\mathfrak{a}(=1^+)$ components of the baryon considered. Here, $\mathbb{W}$ defines a four-dimensional L^2-norm of the baryon's Faddeev amplitude and the ratios $\mathbb{Q}_{\mathfrak{t}=\mathfrak{s}^{\prime},\mathfrak{a}^{\prime\prime}}$ express the relative size of the contribution from each diquark correlation to this norm. The values of these fractions are one indication of the relative strengths of the various diquark components within a baryon. They are listed in Table III of Ref.[19] and depicted for the more complex octet states in Fig. 3–upper panel.

An alternative gauge is to consider the relative contribution to a given baryon's mass owing to each of the diquark components in its Faddeev amplitude. We evaluate this by computing the hadron's mass in the absence of all except the dominant diquark correlation (often the lightest possible contributor) and then introducing the remaining correlations in their order of importance, which is determined by trial-and-error. Typically, the mass obtained using only the dominant correlation is larger than the all-correlation result. (Otherwise, the Faddeev equation would suppress any new correlation.) We therefore define the relative mass-contribution of a given correlation as follows. Suppose two diquarks contribute: qq_1, qq_2, with qq_1

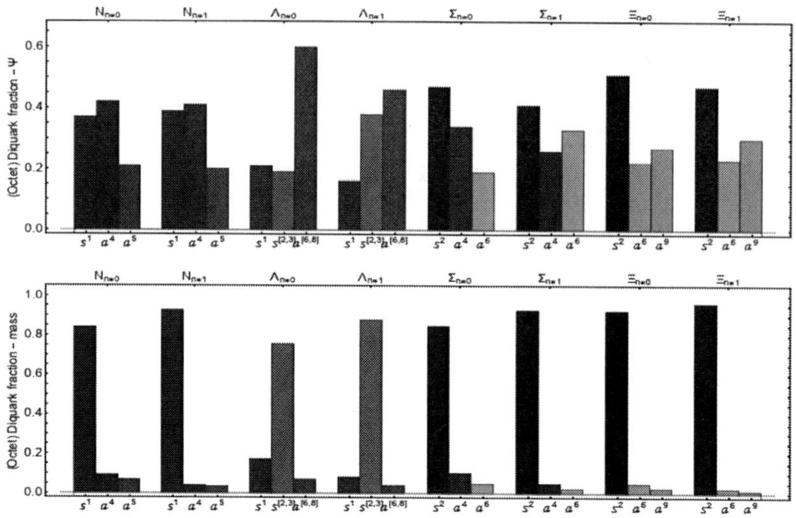

Figure 3. *Upper panel* – Relative strengths of various diquark components within the indicated baryon's Faddeev amplitude, defined via Eqs. (15), (16). *Lower panel* – Relative contribution to a baryon's mass from a given diquark correlation in that baryon's Faddeev amplitude, defined in association with Eqs. (17).

dominant. Further suppose that the qq_1-only baryon has mass m_1 and adding qq_2 gives $m_{\rm final} = m_2 < m_1$, then

$$P_1 = m_1/(m_1 + |m_2 - m_1|)\,, \tag{17a}$$

$$P_2 = |m_2 - m_1|/(m_1 + |m_2 - m_1|)\,. \tag{17b}$$

This procedure has a clear generalisation to systems with more than two diquark correlations; and the results obtained in this way are listed in Table III of Ref.[19] and depicted for the more complicated octet states in Fig. 3–lower panel.

As observed elsewhere [18], the difference between the upper and lower panels of Fig. 3 is marked. In each of the octet cases depicted in the lower panel, there is a single dominant diquark component; namely, a scalar diquark; and each new correlation adds binding, reducing the computed mass. On the other hand, measuring the relative strength of diquark correlations using the Faddeev amplitude decomposition, drawn in Fig. 3–upper panel, one arrives at a somewhat different picture; but such differences can largely be attributed to the lack of interference between diquark components in the measure defined by Eqs. (15), (16). Notwithstanding these facts, comparisons between baryons using any single measure are meaningful, *e.g.* using either scheme, the nucleon and its first positive-parity excitation possess very similar diquark content.

3.3 Rest-frame orbital angular momentum

Drawing upon Sec. 2.3, we now expose the rest-frame orbital angular momentum content of each baryon. Connected with each matrix, there is a scalar function[19, 62, 63], the collection of which we denote as $\{\mathcal{Y}_i, i = 1,\ldots,N_{qq}\}$, *e.g.* the five rest-frame ${}^2\mathbb{S}$-components in the proton are identified with $\mathcal{Y}_{1,\ldots,5}$ and these functions are

$$\tilde{\mathbb{s}}^1_{N1},\ \tilde{\mathbb{a}}^{4,5}_{N2},\ [\tilde{\mathbb{a}}^{4,5}_{N3} + 2\tilde{\mathbb{a}}^{4,5}_{N5}]/3. \tag{18}$$

Using this decomposition, we compiled Table 2. Plainly every one of the systems considered is primarily S-wave in nature, since they are not generated by the Faddeev equation unless S-wave components are contained in the wave function. This observation provides support in quantum field theory for the constituent-quark model classifications of these systems, so long as here angular momentum is understood at the hadronic scale to be that between the quark and diquark. Notwithstanding that, Table 2 reveals that P-wave components play a measurable role in octet ground-states and their first positive-parity excitations: they are attractive in ground-states and repulsive in the excitations. Regarding decuplet systems: the ground-state masses are almost completely insensitive to non-S-wave components; and in the first positive-parity excitations, P-wave components generate a little repulsion, some attraction is provided by D-waves, and F-waves have no measurable impact.

In order to further elucidate these remarks, we turn our attention to the Faddeev wave functions themselves and, for each baryon, compute

$$\mathbb{L}_i = \int \frac{d^4\ell}{(2\pi)^4}\, |\mathcal{Y}_i(\ell^2, \ell\cdot P)|^2, \tag{19}$$

and subsequently define the following rest-frame angular momentum strengths:

$$\mathbb{S} = \mathbb{T}^{-1}\sum_{i\in S}\mathbb{L}_i\,,\quad \mathbb{P} = \mathbb{T}^{-1}\sum_{i\in P}\mathbb{L}_i\,,$$
$$\mathbb{D} = \mathbb{T}^{-1}\sum_{i\in D}\mathbb{L}_i\,,\quad \mathbb{F} = \mathbb{T}^{-1}\sum_{i\in F}\mathbb{L}_i\,, \tag{20a}$$

$$\mathbb{T} = \sum_{i=1}^{N_{qq}}\mathbb{L}_i\,. \tag{20b}$$

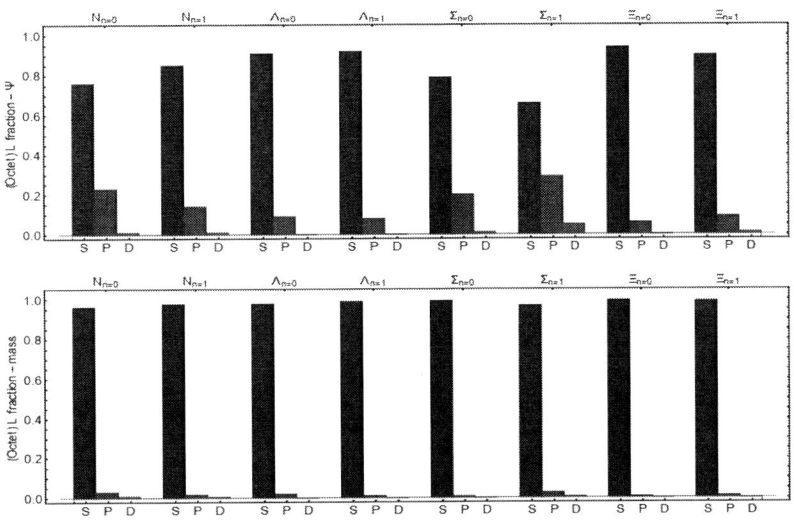

Figure 4. Octet baryons and their first positive-parity excitations. **Upper panel** – Baryon rest-frame quark-diquark orbital angular momentum fractions, as defined in Eqs. (20). **Lower panel** – Relative contribution of various quark-diquark orbital angular momentum components to the mass of a given baryon.

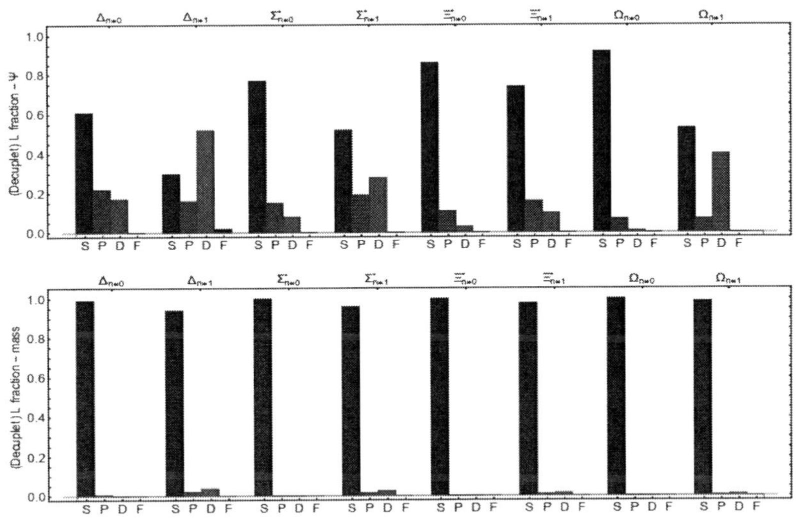

Figure 5. Decuplet baryons and their first positive-parity excitations. **Upper panel** – Baryon rest-frame quark-diquark orbital angular momentum fractions, as defined in Eqs. (20). **Lower panel** – Relative contribution of various quark-diquark orbital angular momentum components to the mass of a given baryon.

Constructed thus, $\mathbb{T}$ defines a four-dimensional L^2-norm of the baryon's rest-frame Faddeev wave function and the ratios $\mathbb{S}$, $\mathbb{P}$, $\mathbb{D}$, $\mathbb{F}$ express the relative size of the contribution from each angular momentum component to this norm. Our results are depicted in the upper panels of Figs. 4, 5.

As with our examination of the diquark content, another gauge of the relative importance of different partial waves within a baryon is to depict their contributions to a baryon's mass, which can be computed using the information in Table 2. The results are depicted in the lower panels of Figs. 4, 5. Once again, even though some contributions are repulsive and others attractive, we draw all bars as positive, following a procedure analogous to that described in connection with Eqs. (17).

Fig. 4 reveals that, concerning the rest-frame quark-diquark orbital angular momentum fractions in octet baryons, both measures deliver the same qualitative picture of each baryon's internal structure. It follows that there is little mixing between partial waves in the computation of a baryon's mass. Regarding Fig. 5, this is also

Table 2. Computed baryon quark-core masses: **upper panel**, flavour-octet; and **lower panel**, flavour-decuplet. Row 1, each panel: results obtained using the complete Faddeev wave function, *i.e.* with all angular momentum components included. Subsequent rows: masses obtained when the indicated rest-frame angular momentum component(s) is(are) excluded from the Faddeev wave function. Empty locations indicate that a solution is not obtained under the conditions indicated. (All dimensioned quantities are listed in GeV.)

L content	$N_{n=0}$	$N_{n=1}$	$\Lambda_{n=0}$	$\Lambda_{n=1}$	$\Sigma_{n=0}$	$\Sigma_{n=1}$	$\Xi_{n=0}$	$\Xi_{n=1}$
S,P,D	1.19	1.73	1.37	1.85	1.41	1.88	1.58	1.99
$-,P,D$	–	–	–	–	–	–	–	–
$S,-,D$	1.24	1.71	1.40	1.83	1.42	1.84	1.59	1.97
$S,P,-$	1.20	1.74	1.37	1.85	1.41	1.89	1.58	1.99
$S,-,-$	1.24	1.71	1.40	1.83	1.42	1.84	1.59	1.97

L content	$\Delta_{n=0}$	$\Delta_{n=1}$	$\Sigma^*_{n=0}$	$\Sigma^*_{n=1}$	$\Xi^*_{n=0}$	$\Xi^*_{n=1}$	$\Omega_{n=0}$	$\Omega_{n=1}$
S,P,D,F	1.35	1.79	1.52	1.93	1.71	2.08	1.93	2.23
$-,P,D,F$	–	–	–	–	–	–	–	–
$S,-,D,F$	1.36	1.75	1.52	1.90	1.71	2.06	1.93	2.22
$S,P,-,F$	1.35	1.82	1.52	1.95	1.71	2.09	1.93	2.24
$S,P,D,-$	1.35	1.79	1.52	1.93	1.71	2.08	1.93	2.23
$S,-,-,-$	1.35	1.80	1.52	1.93	1.71	2.08	1.93	2.23

true for the decuplet states, but there are greater quantitative dissimilarities. It is nevertheless evident in both panels that S-wave strength is shifted into D-wave contributions within decuplet positive-parity excitations, as has previously been observed [16, 59].

It is here worth contrasting these results for low-lying positive-parity baryons with those obtained elsewhere [18] for the two lightest $(I, J^P) = (1/2, 1/2^-)$ partners of the nucleon. For those systems: no solution is obtained unless P-waves are present; P-waves are the largest component of the rest-frame wave function and dominant in determining the mass, with S-waves bringing some attraction; and D-waves are negligible.

4 Summary

We computed the spectrum and Poincaré-covariant wave functions for all flavour-$SU(3)$ octet and decuplet baryons and their first positive-parity excitations (Sec. 3.1). A basic prediction of such Faddeev equation studies is the presence of strong nonpointlike, fully-interacting quark-quark (diquark) correlations within all baryons; and our analysis confirms that for a realistic description of these states, it is necessary and sufficient to retain only flavour-$\bar{3}$–scalar and flavour-6–pseudovector correlations (Sec. 3.2). Namely, negative-parity diquarks are negligible in these positive-parity baryons. Moreover, in its rest-frame, every system considered may be judged as primarily S-wave in character (Sec. 3.3).

In arriving at these conclusions, we draw a similar picture to quark model descriptions of these systems, so long as rest-frame orbital angular momentum is identified with that existing between dressed-quarks and -diquarks, which are the correct strong-interaction quasiparticle degrees-of-freedom at the hadronic scale and on a material domain extending beyond. In addition, we confirm the quark-model result that each ground-state octet and decuplet baryon possesses a radial excitation and consequently predict the existence of positive-parity excitations of the Ξ, Ξ^*, Ω baryons.

Acknowledgements

I am grateful to R. W. Gothe for convening the session in which this perspective was presented. The material contained herein is the product of collaborations with many people, to all of whom I am greatly indebted. This work was partly supported by the Helmholtz International Center for FAIR within the LOEWE program of the State of Hesse and by the DFG grant FI 970/11-1.

References

[1] R. T. Cahill, C. D. Roberts and J. Praschifka, Austral. J. Phys. **42**, 129 (1989).
[2] C. J. Burden, R. T. Cahill and J. Praschifka, Austral. J. Phys. **42**, 147 (1989).
[3] R. T. Cahill, Austral. J. Phys. **42**, 171 (1989).
[4] H. Reinhardt, Phys. Lett. B **244**, 316 (1990).
[5] G. V. Efimov, M. A. Ivanov and V. E. Lyubovitskij, Z. Phys. C **47**, 583 (1990).
[6] A. Bender, C. D. Roberts and L. von Smekal, Phys. Lett. B **380**, 7 (1996).
[7] M. S. Bhagwat, A. Höll, A. Krassnigg, C. D. Roberts and P. C. Tandy, Phys. Rev. C **70**, 035205 (2004).
[8] C. Chen *et al.*, Phys. Rev. D **99**, 034013 (2019).
[9] C. Mezrag, J. Segovia, L. Chang and C. D. Roberts, Phys. Lett. B **783**, 263 (2018).
[10] C. Chen, L. Chang, C. D. Roberts, S.-L. Wan and D. J. Wilson, Few Body Syst. **53**, 293 (2012).
[11] J. Segovia, I. C. Cloët, C. D. Roberts and S. M. Schmidt, Few Body Syst. **55**, 1185 (2014).
[12] J. Segovia *et al.*, Phys. Rev. Lett. **115**, 171801 (2015).
[13] J. Segovia, C. D. Roberts and S. M. Schmidt, Phys. Lett. B **750**, 100 (2015).
[14] J. Segovia and C. D. Roberts, Phys. Rev. C **94**, 042201(R) (2016).
[15] G. Eichmann, C. S. Fischer and H. Sanchis-Alepuz, Phys. Rev. D **94**, 094033 (2016).
[16] G. Eichmann, Few Body Syst. **58**, 81 (2017).
[17] Y. Lu *et al.*, Phys. Rev. C **96**, 015208 (2017).
[18] C. Chen *et al.*, Phys. Rev. D **97**, 034016 (2018).
[19] C. Chen, G. I. Krein, C. D. Roberts, S. M. Schmidt and J. Segovia, Phys. Rev. D **100**, no. 5, 054009 (2019).
[20] Z. F. Cui *et al.*, Phys. Rev. D **102**, no.1, 014043 (2020).
[21] J. Rodríguez-Quintero *et al.*, [arXiv:1909.13793 [nucl-th]].

[22] J. Segovia, C. Chen, Z. F. Cui, Y. Lu and C. D. Roberts, AIP Conf. Proc. **2249**, no.1, 020010 (2020).

[23] I. G. Aznauryan *et al.*, Int. J. Mod. Phys. E **22**, 1330015 (2013).

[24] M. Y. Barabanov *et al.*, [arXiv:2008.07630 [hep-ph]].

[25] C. Chen, L. Chang, C. D. Roberts, S. M. Schmidt, S. Wan and D. J. Wilson, Phys. Rev. C **87**, 045207 (2013).

[26] S. S. Xu, C. Chen, I. C. Cloet, C. D. Roberts, J. Segovia and H. S. Zong, Phys. Rev. D **92**, no. 11, 114034 (2015).

[27] J. Segovia, C. Chen, I. C. Cloët, C. D. Roberts, S. M. Schmidt and S. Wan, Few Body Syst. **55**, 1-33 (2014).

[28] J. Segovia, C. Chen, C. D. Roberts and S. Wan, Phys. Rev. C **88**, no.3, 032201 (2013).

[29] P. L. Yin, C. Chen, G. Krein, C. D. Roberts, J. Segovia and S. S. Xu, Phys. Rev. D **100**, no. 3, 034008 (2019).

[30] H. L. L. Roberts, A. Bashir, L. X. Gutiérrez-Guerrero, C. D. Roberts and D. J. Wilson, Phys. Rev. C **83**, 065206 (2011).

[31] C. D. Roberts and A. G. Williams, Prog. Part. Nucl. Phys. **33**, 477 (1994).

[32] C. D. Roberts and S. M. Schmidt, Prog. Part. Nucl. Phys. **45**, S1 (2000).

[33] P. Maris and C. D. Roberts, Int. J. Mod. Phys. E **12**, 297 (2003).

[34] C. D. Roberts, M. S. Bhagwat, A. Höll and S. V. Wright, Eur. Phys. J. ST **140**, 53 (2007).

[35] P. Boucaud *et al.*, Few Body Syst. **53**, 387 (2012).

[36] C. D. Roberts, J. Phys. Conf. Ser. **706**, 022003 (2016).

[37] A. C. Aguilar, D. Binosi and J. Papavassiliou, Front. Phys. China **11**, 111203 (2016).

[38] M. R. Pennington, J. Phys. G **43**, 054001 (2016).

[39] C. Mezrag, H. Moutarde and J. Rodríguez-Quintero, Few Body Syst. **57**, 729 (2016).

[40] G. Eichmann *et al.*, Prog. Part. Nucl. Phys. **91**, 1 (2016).

[41] C. S. Fischer, Prog. Part. Nucl. Phys. **105**, 1 (2019).

[42] C. Chen, L. Chang, C. D. Roberts, S. Wan and H. S. Zong, Phys. Rev. D **93**, no. 7, 074021 (2016).

[43] T. Horn and C. D. Roberts, J. Phys. G. **43**, 073001 (2016).

[44] F. E. Serna, C. Chen and B. El-Bennich, Phys. Rev. D **99**, no. 9, 094027 (2019).

[45] A. C. Aguilar *et al.*, Eur. Phys. J. A **55** (2019) no.10, 190.

[46] J. Segovia, C. Chen, Z. F. Cui, Y. Lu and C. D. Roberts, [arXiv:1911.04923 [nucl-th]].

[47] C. Shi *et al.*, Phys. Rev. D **92** (2015), 014035.

[48] C. Chen and C. Shi, EPJ Web Conf. **113** (2016), 05013.

[49] S. X. Qin, C. Chen, C. Mezrag and C. D. Roberts, Phys. Rev. C **97**, no. 1, 015203 (2018).

[50] M. Hopfer, A. Windisch and R. Alkofer, PoS **ConfinementX**, 073 (2012).

[51] D. Binosi, L. Chang, J. Papavassiliou and C. D. Roberts, Phys. Lett. B **742**, 183 (2015).

[52] R. Williams, C. S. Fischer and W. Heupel, Phys. Rev. D **93**, 034026 (2016).

[53] D. Binosi, L. Chang, J. Papavassiliou, S.-X. Qin and C. D. Roberts, Phys. Rev. D **95**, 031501(R) (2017).

[54] D. Binosi, C. Mezrag, J. Papavassiliou, C. D. Roberts and J. Rodríguez-Quintero, Phys. Rev. D **96**, 054026 (2017).

[55] A. K. Cyrol, M. Mitter, J. M. Pawlowski and N. Strodthoff, Phys. Rev. D **97**, 054006 (2018).

[56] J. Rodríguez-Quintero *et al.*, Few Body Syst. **59**, 121 (2018).

[57] M. A. Ivanov, Yu. L. Kalinovsky and C. D. Roberts, Phys. Rev. D **60**, 034018 (1999).

[58] M. A. Ivanov, J. G. Körner, S. G. Kovalenko and C. D. Roberts, Phys. Rev. D **76**, 034018 (2007).

[59] S.-X. Qin, C. D. Roberts and S. M. Schmidt, Phys. Rev. D **97**, 114017 (2018).

[60] G. P. Lepage and S. J. Brodsky, Phys. Rev. D **22**, 2157 (1980).

[61] K. Raya *et al.*, Phys. Rev. D **93**, 074017 (2016).

[62] M. Oettel, G. Hellstern, R. Alkofer and H. Reinhardt, Phys. Rev. C **58**, 2459 (1998).

[63] I. C. Cloët, A. Krassnigg and C. D. Roberts, [arXiv:0710.5746 [nucl-th]].

[64] N. Suzuki *et al.*, Phys. Rev. Lett. **104**, 042302 (2010).

[65] M. Tanabashi *et al.*, Phys. Rev. D **98**, 030001 (2018), (*Particle Data Group*).

[66] C. D. Roberts, Few Body Syst. **58**, 5 (2017).

[67] C. Hanhart and S. Krewald, Phys. Lett. B **344**, 55-60 (1995).

[68] A. Buck and H. Reinhardt, Phys. Lett. B **356**, 168-174 (1995).

[69] M. Gell-Mann, "The Eightfold Way: A Theory of strong interaction symmetry", Tech. Rep. TID-12608; CTSL-20.

[70] S. Okubo, Prog. Theor. Phys. **27**, 949-966 (1962).

[71] S. Okubo, Prog. Theor. Phys. **28**, 24-32 (1962).

Towards a theory of baryon resonances

Ulf-G. Meißner[1,2,3,*]

[1]Helmholtz-Institut für Strahlen- und Kernphysik, Bethe Center for Theoretical Physics and Center for Science and Thought, Universität Bonn, D-53115 Bonn, Germany
[2]Institut für Kernphysik (Theorie), Institute for Advanced Simulation (IAS-4), and Jülich Center for Hadron Physics, Forschungszentrum Jülich, D-52425 Jülich, Germany
[3]Ivane Javakhishvili Tbilisi State University, 0186 Tbilisi, Georgia

Abstract. In this talk, I discuss methods that allow for a systematic and model-independent calculation of the hadron spectrum. These are lattice QCD and/or its corresponding Effective Field Theories. Assorted results are shown and I take the opportunity to discuss some misconceptions often found in the literature.

1 Introduction: QCD and excited states

QCD is a remarkable theory. Although its Lagrangian can be written in a line

$$\mathcal{L}_{\rm QCD} = -\frac{1}{4}G^a_{\mu\nu}G^{\mu\nu,a} + \sum_f \bar{q}_f(i\not{D} - \mathcal{M})q_f + \ldots\,, \quad (1)$$

its fundamental fields, the quarks and gluons, have never been observed in isolation but only appear as constituents of the strongly interacting particles, hadrons and nuclei. In Eq. (1), $D_\mu = \partial_\mu - igA^a_\mu\lambda^a/2$ is the gauge-covariant derivative, A^a_μ ($a = 1,\ldots,8$) the gluon field, $G^a_{\mu\nu} = \partial_\mu A^a_\nu - \partial_\nu A^a_\mu - ig[A^b_\mu, A^c_\nu]$ the gluon field strength tensor, g is the SU(3) gauge coupling, q_f a quark spinor of flavor f ($f = u,d,s,c,b,t$), $\mathcal{M}$ is the diagonal quark mass matrix and the ellipsis stand for gauge-fixing and CP-violating terms not considered here. The quarks come in two types, the light (u,d,s) and heavy (c,b,t) quark flavors, where light and heavy refers to the QCD scale $\Lambda_{\rm QCD} \simeq 210$ MeV (for $N_f = 5, \overline{MS}, \mu = 2\,$GeV). In the absence of the quark masses, $\Lambda_{\rm QCD}$ is the only dimensionful parameter in QCD that is generated by dimensional transmutation through the running of the strong coupling $\alpha_s = g^2/4\pi$.

The Lagrangian of QCD allows us to define two special limits, in which the theory can be analyzed in terms of appropriately formulated effective field theories (EFTs). In the light quark ($f = u,d,s$) sector, the effective Lagrangian can be written in terms of left- (q_L) and right-handed (q_R) quark fields, such that

$$\mathcal{L}_{\rm QCD} = \bar{q}_L\, i\not{D} q_L + \bar{q}_R\, i\not{D} q_R + \mathcal{O}(m_f/\Lambda_{\rm QCD})\,. \quad (2)$$

As can be seen, left- and right-handed quarks decouple, which is reflected in the chiral symmetry. It is explicitly broken by the finite but small quark masses m_f. Furthermore, chiral symmetry is spontaneously broken, leading to the eight pseudo-Goldstone bosons, the pions, the kaons and the eta. These are indeed the lightest hadrons. The pertinent EFT is chiral perturbation theory (CHPT).

Matters are very different for the heavy c and b quarks, where the leading order Lagrangian takes the form

$$\mathcal{L}_{\rm QCD} = \bar{Q}_f\, iv\cdot D\, Q_f + \mathcal{O}(\Lambda_{\rm QCD}/m_f)\,, \quad (3)$$

with v the four-velocity of the heavy quark and Q_f denotes a quark spinor of flavor f ($f = c,b$). Note that to leading order, this Lagrangian is independent of quark spin and flavor, which leads to *SU(2) spin* and *SU(2) flavor symmetries* (HQSS and HQFS, respectively). The pertinent EFT to analyze the consequences is heavy quark effective field theory (HQEFT), which comes in different manifestations. Finally, in heavy-light systems, where heavy quarks act as matter fields coupled to the light pions, one can combine CHPT and HQEFT.

There are various reasons to consider **excited states**. First, the spectrum of QCD is arguably its least understood feature. This is often phrased in terms of questions like: Why do we observe almost only qqq and $\bar{q}q$ states? What is the nature of the XYZ and other "exotic" states? Where are the glueballs predicted by QCD? Note that I put the word exotic in quotation marks, because this usually refers to states that can not be described within the (conventional) quark model. However, it is very obvious, but not accounted for by many, that the quark model is much too simple, particularly in the sector of the light quarks. E.g. it does not account for a whole class of important players in the hadron spectrum, the so-called hadronic molecules. These also provide the bridge to nuclear physics, because hadrons and nuclei are just different manifestations of structure formation in QCD which are intimately linked and should be considered together. It is also important that high-precision data for spectrum studies have been and will be produced with ELSA at Bonn, MAMI at Mainz, CEBAF at Jefferson Lab, the LHCb experiment at CERN, the BESIII experiment at the BEPCII,

*e-mail: meissner@hiskp.uni-bonn.de

GlueX at Jefferson Lab and in the future with PANDA at FAIR and other labs worldwide. These data clearly pose a challenge for any theoretical approach.

In what follows, I discuss theoretical approaches that will eventually unravel the physics behind the QCD spectrum. To give a wider perspective, I present results for mesons and baryons, although the center of attention of this workshop clearly is the excitation spectrum of the nucleon. Nevertheless, the QCD spectrum should be seen in a broader perspective. To be more precise, I only consider methods that are

- **model-independent**,
- **can be systematically improved**, and
- **allow for uncertainty estimates**.

If one of these conditions is not fulfilled, a given method will not be considered further. In particular, I eschew models here. So that leaves us with lattice QCD (LQCD) and EFTs or combinations thereof. LQCD can get ground-states and some excited states at (almost) physical pion masses, but the most distinctive feature of excited states are *decays*. These are only captured for very few states in lattice QCD, as discussed later.

This talk consists of a number of lessons. In Sec. 2 I give an answer to the question: What is a resonance? I also discuss briefly one example how resonance positions in the complex energy plane look and stress the two-pole structure of the $\Lambda(1405)$. Then, in Sec. 3, I discuss the case of a well-separated resonance in a finite volume, which brings us in contact with the Lüscher equation. This is extended to the multi-channel case in Sec. 4. I show in particular how chiral symmetry is important for the extrapolation into the complex plane, which leads us to the two-pole structure of the $D_0^\star(2400)$ from the analysis of high-precision lattice QCD data on the coupled channel $D\pi, D\eta, D_s\bar{K}$ scattering with isospin $I = 1/2$. I also point out that this two-pole structure is more common then so far believed. In Sec. 5, I discuss hadronic molecules, which are important players in the hadron spectrum. I address in particular the issues how to distinguish them from compact multi-quark states and the calculation of the production cross section in high-energy proton-(anti-)proton collisions. Then, in Sec. 6 I turn to the calculation of the width of the two lowest-lying baryon resonances, the $\Delta(1232)$ and the Roper $N^*(1440)$ using covariant baryon chiral perturbation theory and the complex-mass scheme. Sec. 7 contains a few remarks on the pion cloud of the nucleon and other hadrons, also often encountered at this workshop. I end with a short summary in terms of take-home messages.

2 Lesson 1: What is a resonance?

In the old times, resonances were searched for by looking at bumps in scattering cross sections. But as Moorhouse stated so eloquently: "Not every bump is a resonance and not every resonance is a bump" [1]. Now we know that resonances have **complex** properties, like their mass and

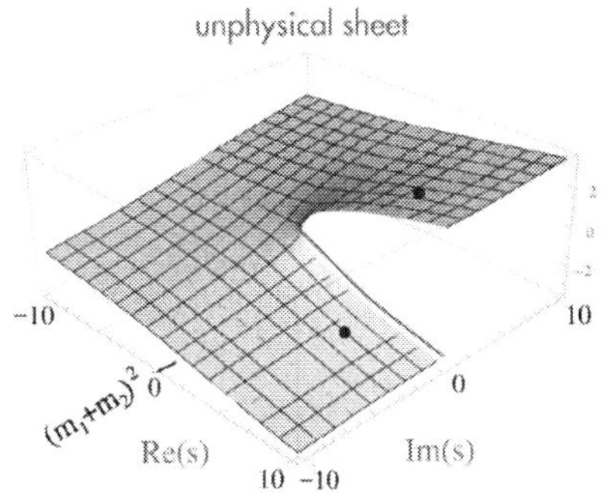

Figure 1. The imaginary part of a single-channel amplitude in the presence of a resonance. The solid dots indicate the allowed positions for resonance poles. Figure from [2].

width, their photo-couplings, etc. . In particular, these intrinsic properties do not depend on the experiment or theory (model). Most importantly, resonances correspond to **S-matrix poles on unphysical Riemann sheets**, as depicted in Fig. 1.

This is the most basic and *the only acceptable definition of a resonance* (with very few exceptions of well-isolated, single channel cases). So if somebody supplies you with the resonances parameters and has not looked for the corresponding pole in the complex plane, these numbers should be considered with suspicion or even be discarded. Especially in cases when a number of coupled channels is involved, the search in the complex energy plane is essentially the only viable method. A nice example of such a search in the multiple channel case (here: $\pi^0 p$, $\pi^+ n$, $p\eta$, ΛK^+, $\Sigma^0 K^+$ and $\Sigma^+ K^0$) are the two close-by poles corresponding to the two lowest S_{11} resonances in elastic pion-nucleon scattering with $J^P = 1/2^-$ as shown in Fig. 2 (from Ref. [3]). Another beautiful example is given by the two-pole structure of the $\Lambda(1405)$, first noted in Ref. [4]. Here, multiple channels need to be considered, namely $K^- p \to K^- p, \bar{K}^0 n, \Sigma^0\pi^0, \Sigma^+\pi^-, \Sigma^-\pi^+, \Lambda\pi^0, \Lambda\eta, \Xi^+ K^-, \Xi^0 K^0$. Only if one analyzes the pole structure in the complex energy plane, one finds these two poles, one being close to the $K^- p$ and the other closer to the $\pi\Sigma$ threshold, see Fig. 3. In fact, this two-pole structure has been verified by various groups world-wide, for a comparative study see [5] and later discussions.

Let us now consider the lattice, which corresponds to a finite, cubic box (keeping the time coordinate continuous for the moment). As it is well-known from quantum mechanics, the eigenstates of any Hamiltonian in a box are discrete energy levels. So does that mean that the excited states of QCD are not amenable to LQCD? Fortunately, as pointed out by Lüscher [6] and others, one can relate the volume dependence of the energy spectrum on the lattice to the continuum scattering phase shift. In fact, an isolated narrow resonance can be traced back to an avoided level crossing, see e.g. Ref. [7]. This, however, is not a practical method, so in the following sections I will discuss how to proceed in LQCD.

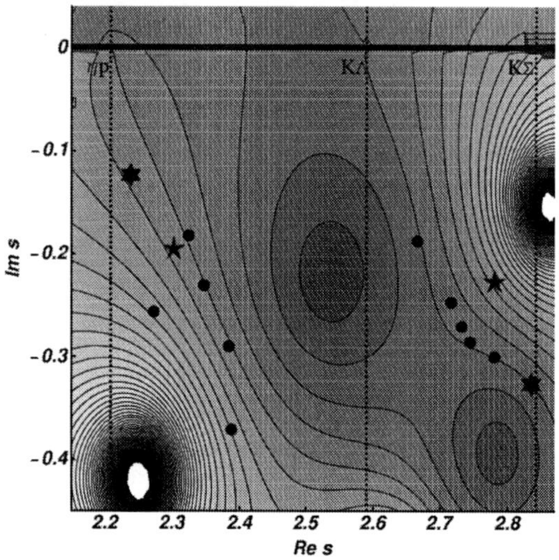

Figure 2. Modulus of the analytic continuation of the $J^P = 1/2^-$ πN scattering amplitude into the complex s-plane. Shown is the (- - - + + +) Riemann sheet. The two poles at $\sqrt{s}$ = $(1.51 - i0.14)$ GeV and $\sqrt{s} = (1.69 - i0.05)$ GeV, corresponding to the $S_{11}(1535)$ and the $S_{11}(1650)$, respectively, are clearly visible. The stars and black dots refer to other determinations as listed in [3]. For the definitions of the sheets, see e.g. Ref. [5].

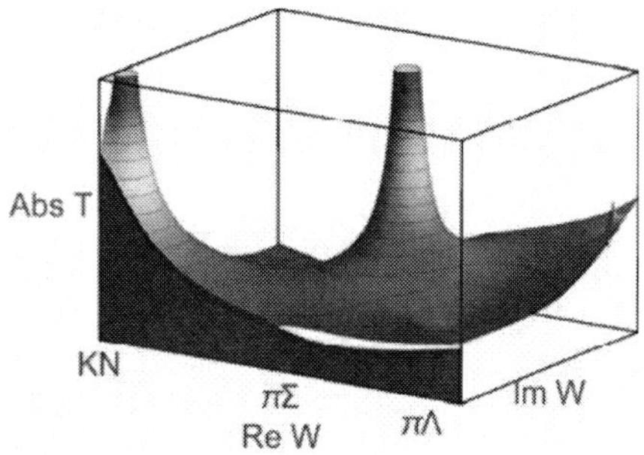

Figure 3. The S-wave amplitude f_{0+} for $K^-p \to K^-p$ for $I = 0$ in the complex energy plane clearly showing the two poles in the region of the Λ(1405). The red band indicates that the (+ + - - - - + + + +) Riemann sheet is only connected to the real energy axis between the $\bar{K}N$ and the $\pi\Sigma$ thresholds. For the definitions of the sheets, see e.g. Ref. [5]. Figure courtesy of Maxim Mai.

3 Lesson 2: Well separated resonances

Well separated resonances are, as already stated, more the exception than the rule. Still, they provide important benchmarks for any calculation of the spectrum. Let us consider an isolated resonance in a box. Before discussing the Lüscher formalism in three space dimensions, let me elaborate on the one-dimensional case as an instructive example. Consider the scattering of two particles on a line segment of length L (the lattice size), with a finite interaction range $R \ll L$. For simplicity, assume further an infinite time extent. As expected, outside the range of the interaction, the relative motion of the scattering particles should be described by a plane wave. Then, the interaction produces a finite phase shift $\delta(k)$ as known from elementary scattering theory. Let us impose periodic boundary conditions in the spatial dimension. Hence, for a plane wave with momentum k, one has

$$\exp(ikL + 2i\delta(k)) = \exp(ik0) = 1, \tag{4}$$

which gives

$$k_nL + 2\delta(k_n) = 2\pi n, \quad n \in \mathbb{N}, \tag{5}$$

for the quantization of the momenta. This result is remarkable, as it relates energies computed on the lattice to the continuum phase shift. Given an appropriate (lattice or continuum) dispersion relation which relates the lattice energy levels $E_n(L)$ to the momentum modes $k_n(L)$, the continuum phase shift at discrete values of the scattering momentum can be reconstructed from the energy levels on the lattice. Also, the non-interacting limit $\delta(k_n) = 0$ is recovered for $k_n = 2\pi n/L$. For non-relativistic particles with mass m, this method is applicable when $mL \gg 1$, so that further corrections are exponentially suppressed. Also, any inelasticity or coupling to other scattering channels introduces modifications. Let me now discuss the case of three space dimensions. The energy levels for two non-interacting identical particles of mass m are given by

$$E_n(L) = 2\sqrt{m^2 + \vec{k}^2}\,, \quad k_i = \frac{2\pi}{L}n_i\,, \quad n_i \in \mathbb{Z}\,. \tag{6}$$

Turning on the interaction, we obtain the continuum scattering phase $\delta(k)$ with the help of Lüscher's formula,

$$\begin{aligned} \delta(k) &= -\phi(q) \bmod \pi\,, \quad q = \frac{kL}{2\pi}\,, \\ \phi(q) &= -\arctan\frac{\pi^{3/2}q}{\mathcal{Z}_{00}(1;q^2)}\,, \\ \mathcal{Z}_{00}(1;q^2) &= \frac{1}{\sqrt{4\pi}}\sum_{\vec{n}\in\mathbb{Z}^3}\frac{1}{\vec{n}^2 - q^2}\,, \end{aligned} \tag{7}$$

for the S-wave. Note that the Lüscher ζ-function $\mathcal{Z}_{00}$ requires regularization, which can be done in different ways (a specific example is given in the next section). Generalizations to higher partial waves also exist. If we assume a resonance with mass $m_R > 2m$, we can also use the effective range expansion (ERE),

$$\tan\left(\delta - \frac{\pi}{2}\right) = \frac{E^2 - m_R^2}{m_R\Gamma_R}\,, \tag{8}$$

with m_R (Γ_R) the mass (width) of the resonance. Note, however, that this can only be used under very special circumstances (well isolated resonance, weak energy dependence of the background). In such a case, one can measure the phase shift in the resonance region and fit m_R, Γ_R, making also use of moving frames as pioneered in Ref. [8]. As an example, I show in Fig. 4 some pioneering (but unpublished) results for the Δ(1232), which is a well separated baryon resonance in the πN system, corresponding to the $l = 1, I = 3/2$ πN phase shift. These results were obtained for pion masses from 160 to 390 MeV in fairly large volumes [9]. More precisely, the calculation

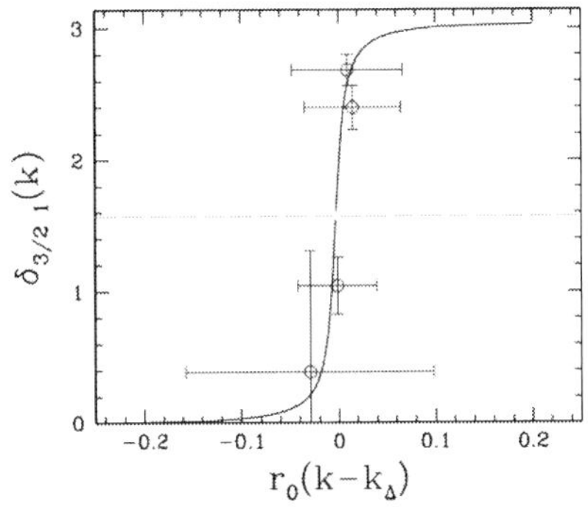

Figure 4. The πN phase shift of the Δ channel as a function of the center-of-mass momentum k. The solid line is the phase generated from the physical mass and width of the Δ-resonance. The dashed horizontal line indicates the 90° crossing of the phase.

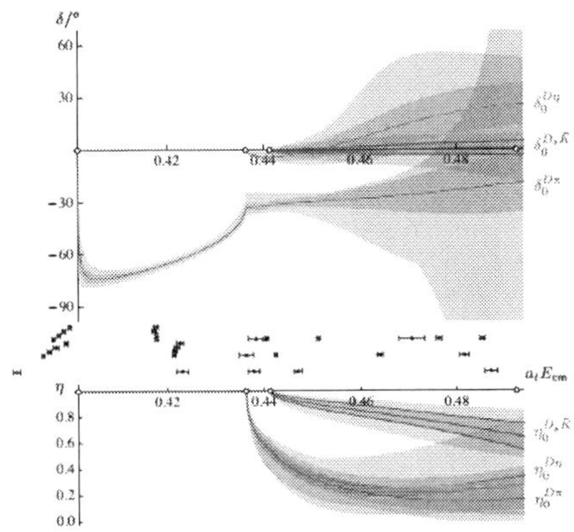

Figure 5. The upper (lower) panel shows the S-wave phase shifts (inelasticities) for the $D\pi$ (red), $D\eta$ (green) and $D_s\bar{K}$ (blue) channels, where the size of the bands incorporates all T-matrix parametrizations of the S-wave. The black points show the location of the finite-volume energy levels used to constrain the parametrizations. Figure from [26].

used a $N_f = 2$ clover action that consists of the plaquette gluon action together with nonperturbatively $O(a)$ improved Wilson (clover) fermions on $32^3 \times 64$ and $40^3 \times 64$ lattices at $\beta = 5.29, \kappa = 0.13632, a = 0.075$ fm and on $40^3 \times 64$ and $48^3 \times 64$ lattices at $\beta = 5.29, \kappa = 0.1364, a = 0.06$ fm. More recent results on the Δ are given in the talks by Morningstar [10] and by Petschlies [11]. For single channel calculations of elastic pion-nucleon scattering in the $J^P = 1/2^-$ channel, see Refs. [12, 13], and for the Roper in the $J^P = 1/2^+$ channel, see e.g. Ref. [14].

In the meson sector, the genuine case of an isolated resonance is the ρ that shows up in the $I = J = 1$ channel of pion-pion scattering. Here, the most recent calculation is due to the ETM Collaboration [15], and I refer to that paper for details.

4 Lesson 3: Coupled channels/thresholds

As said before, isolated (well-separated) resonances are the exception, in most cases we have to deal with coupled channel effects and/or close-by thresholds, e.g. for the scalar mesons $f_0(980)$ and $a_0(980)$ or the two poles in the $\Lambda(1405)$ region. There exist various extensions of Lüscher's approach to cope with coupled channels: 1) a purely quantum mechanical treatment, see e.g. [16], 2) the formulation within non-relativistic EFT (NREFT), see e.g. [17–20], 3) the use of finite-volume unitarized CHPT, see e.g. [21, 22] or 4) a relativistic version, see e.g. [23]. The main advantage of the third method is that it is based on an effective chiral Lagrangian that allows one to relate various processes. This will be exemplified below. So far, these methods have been mostly applied in the meson sector, with a few exceptions also for baryons [24, 25]. Before elaborating on one concrete example, I would like to issue a warning: Be aware of methods that can mislead you (K-matrix and alike). Some possible pitfalls will also be discussed in what follows.

The Hadron Spectrum Collaboration (HSC) has investigated the coupled channel $D\pi$, $D\eta$, $D_s\bar{K}$ scattering with isospin $I = 1/2$ [26] (for S-, P- and D-waves). They presented results for three lattice volumes, one spatial lattice spacing a_s and one temporal spacing a_t at $M_\pi \simeq 390$ MeV. Furthermore, they used various K-matrix type extrapolations (in fact, up to 11 variants) to look for poles in the complex plane. The resulting phase shifts and inelasticities are shown in Fig. 5. They report one S-wave pole at (2275.9 ± 0.9) MeV that is located very closely to the $D\pi$ threshold. The authors of Ref. [26] argue that this pole shares similarities with the $D_0^*(2400)$ of the Particle Data Group (PDG) [27] (note that the $D_0^*(2400)$ has been renamed as $D_0^*(2300)$ in the 2019 PDG update). However, in their T-matrix parametrizations, they ignored the important role of chiral symmetry. It is long known how to incorporate chiral symmetry in coupled-channel dynamics in the framework of unitarized CHPT (UCHPT), see e.g. the groundbreaking works in Refs. [4, 28–30]. To make contact to the work of the Hadron Spectrum Collaboration, let us consider the case of Goldstone bosons scattering off D-mesons ($D\phi$ scattering). The corresponding T-matrix in UCHPT takes the form

$$T^{-1}(s) = V^{-1}(s) - G(s)\,. \tag{9}$$

Here, the potential $V(s)$ can be obtained from the SU(3) heavy-light chiral effective Lagrangian, which up to next-to-leading order (NLO) contains 6 low-energy constants (LECs), as discussed below. Further, $G(s)$ is the 2-point scalar loop function, which is best written in a dispersion-theoretical representation and regularized with a subtraction constant $a(\mu)$, see e.g. [4]. Of course, T, V and G are matrices in the channel space, but for clarity I have suppressed the channel indices. Let us now discuss the NLO effective chiral Lagrangian for $D\phi$ coupled channel dynamics. Following Ref. [31], it takes the form

$$\begin{aligned}\mathcal{L}_{\rm eff} &= \mathcal{L}^{(1)} + \mathcal{L}^{(2)}\,,\\ \mathcal{L}^{(1)} &= \mathcal{D}_\mu D \mathcal{D}^\mu D^\dagger - M_D^2 D D^\dagger\,,\end{aligned}$$

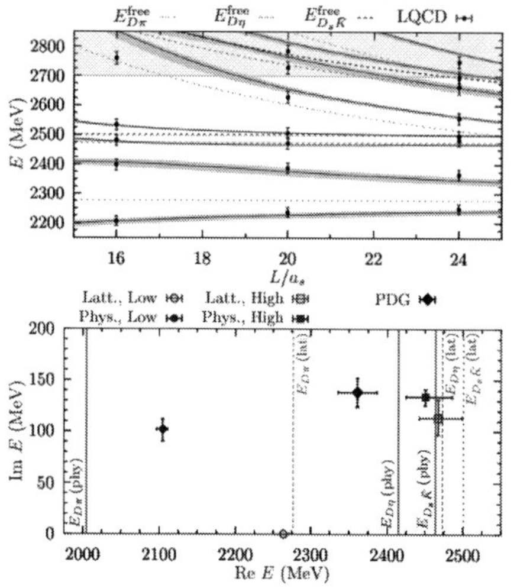

Figure 6. The upper panel shows the energy levels calculated in UCHPT with LECs determined in 2013 in comparison to the results of [26] in the $(S, I) = (0, 1/2)$ channel. Lower panel: Complex energy plane location of the two-pole-structure. Empty red (filled blue) symbols stand for the poles obtained when the LQCD [26] (physical) masses are used. The black diamond represents the isospin average of the PDG values for $D_0^\star(2400)^0$ and $D_0^\star(2400)^+$. Figures from [33].

$$\begin{aligned}\mathcal{L}^{(2)} &= D\left[-h_0\langle\chi_+\rangle - h_1\chi_+\right.\\ &+ h_2\langle u_\mu u^\mu\rangle - h_3 u_\mu u^\mu\left.\right]D^\dagger \\ &+ \mathcal{D}_\mu D\left[h_4\langle u^\mu u^\nu\rangle - h_5\{u^\mu, u^\nu\}\right]\mathcal{D}_\nu D^\dagger\ , \quad (10)\end{aligned}$$

with $D = (D^0, D^+, D_s^+)$, $\mathcal{D}_\mu$ is the chiral covariant derivative, M_D the D-meson mass (in the chiral limit) and the conventional chiral building blocks $u_\mu \sim \partial_\mu\phi$,$\chi_+ \sim \mathcal{M}$, etc. are used. The LECs can be determined as follows: h_0 can be fixed from the pion-mass dependence of the D and D_s masses and $h_1 = 0.42$ is given by the D_s-D splitting. Further, $h_{2,3,4,5}$ are obtained from a fit to lattice data ($D\pi \to D\pi, D\bar{K} \to D\bar{K}, ...$), see Ref. [32]. In that paper, 5 "simple" channels that do not contain disconnected diagrams where fitted and then UCHPT was used to predict the mass of the charm-strange scalar meson (the pole in the $(S, I) = (1, 0)$ channel) at 2315^{+18}_{-28} MeV, consistent with the PDG value for the $D_{s0}^\star(2317)$. To make contact to the results from [26], we need the appropriate formulation of UCHPT on a finite volume (FV). As noted before, in that case the momenta get quantized $\sim 1/L$ and the loop function $G(s)$ gets modified. Using the formalism of Ref. [21], we have (note the aforementioned regularization of the Lüscher function)

$$\tilde{G}(s, L) = \lim_{\Lambda\to\infty}\left[\frac{1}{L^3}\sum_{\vec{n}}^{|\vec{q}|<\Lambda} I(\vec{q}) - \int_0^\Lambda \frac{q^2 dq}{2\pi^2} I(\vec{q})\right]\ , \quad (11)$$

where $I(\vec{q})$ is the integrand of $G(s)$. The FV energy levels are obtained from the poles of $\tilde{T}(s, L)$ via $\tilde{T}^{-1}(s, L) = V^{-1}(s) - \tilde{G}(s, L)$. Note that on a torus the potential $V(s)$ is the same as in the continuum. Using this framework, the HSC data were re-analyzed in Ref. [33]. First, the energy levels were postdicted (no parameter needed to be fixed) to a high precision as shown in the upper panel of Fig. 6. Second, and even more stunning, was the observation of the two-pole structure of the $D_0^\star(2400)$, as depicted in the lower panel of Fig. 6, in complete analogy to the case of the $\Lambda(1405)$ [4]. This solves the enigma that the mass of the $D_0^\star(2400)$, which is made of a light (u, d) and a charm quark, is larger than the one of the $D_{s0}^\star(2317)$, which contains the heavier strange quark. The lower of the two poles is lying visibly below the $D_{s0}^\star(2317)$! This two-pole structure is easily understood from group theory. Consider the SU(3) limit, where all light and heavy mesons take common values, analogous to the study of the $\Lambda(1405)$ in [34]. Combining the D-meson anti-triplet with the Goldstone boson octet gives

$$\bar{3}\otimes 8 = \underbrace{\bar{3}\oplus 6}_{\text{attractive}} \oplus \overline{15}\ , \quad (12)$$

where the anti-triplet and the sextet are attractive, leading to two zero-width poles. Once the SU(3) breaking in the meson masses is switched on, these poles move to the positions in the complex energy plane shown in Fig. 6. We note that this two-pole structure had been observed in earlier calculations [35–38] but did not receive the proper attention then. In fact, as pointed out in Ref. [39], there is an easy lattice test for this scenario: The sextet pole becomes a bound state for $M_\phi > 575\,$MeV in the SU(3) limit. Such a calculation could be easily done and would be a fine test of the two-pole scenario. In fact, there is further phenomenological support of this picture. Analyzing the high-precision data on the decays $B \to D\phi\phi$ from LHCb [40–42] within the same framework (using the predetermined S-waves) leads to an excellent description of the so-called angular moments [39, 43] and a prediction of two cusps in the S-wave at the $D\eta$ and $D_s\bar{K}$ thresholds, as shown in Fig. 7. Shown are the so-called anchor points provided by LHCb, where the strength and the phase of

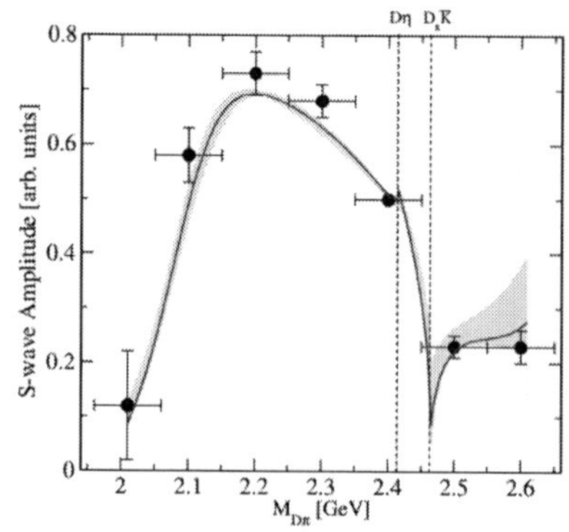

Figure 7. Comparison of the S-wave amplitude determined from the calculation of [39] with the S-wave anchor points from the experimental analysis, shown as the data points [42]. The red (solid) line gives the best fit results and the grey band quantifies the uncertainties. The dashed perpendicular lines indicate the location of the $D\eta$ and $D_s\bar{K}$ threshold, respectively. Figure from [39].

Table 1. Predictions of two states in various $I = 1/2$ channels in the heavy meson sector. Here $(M, \Gamma/2)$ denote the mass and the half-width, respectively, in units of [MeV]. From Ref. [39].

	Lower pole	Higher pole
$D_0^\star$	$\left(2105^{+6}_{-8}, 102^{+10}_{-11}\right)$	$\left(2451^{+36}_{-26}, 134^{+7}_{-8}\right)$
D_1	$\left(2247^{+5}_{-6}, 107^{+11}_{-10}\right)$	$\left(2555^{+47}_{-30}, 203^{+8}_{-9}\right)$
$B_0^\star$	$\left(5535^{+9}_{-11}, 113^{+15}_{-17}\right)$	$\left(5852^{+16}_{-19}, 36 \pm 5\right)$
B_1	$\left(5584^{+9}_{-11}, 119^{+14}_{-17}\right)$	$\left(5912^{+15}_{-18}, 42^{+5}_{-4}\right)$

the S-wave were extracted from the data and connected by a cubic spline in comparison to the UCHPT result of Ref. [39]. I point out that the higher mass pole at 2.45 GeV amplifies the predicted cusps. In fact, the two-pole scenario can be extended to the axial-vector states and the corresponding B-mesons using HQSS and HQFS, respectively. The emerging picture based on the calculations in Ref. [39] (and references therein) is summarized in Table 1. Note that the PDG lists one state for the $D_0^\star$ and for the D_1, these are located at $(2318 \pm 29, 134 \pm 20)$ MeV and $(2427 \pm 40, 192^{+65}_{-55})$ MeV, respectively. In view of the results obtained in UCHPT, time is ripe to change these entries in the PDG. The predicted two-pole scenarios for the $B_0^\star$ and the B_1 are a nice test to further validate this picture.

5 Lesson 4: Hadronic molecules

As already stated in the introduction, QCD offers yet another set of bound states, first seen in *nuclear physics*, namely **hadronic molecules**, which are bound states made of 2 or 3 hadrons. In what follows, I consider bound states of two hadrons in S-wave very close to a 2-particle threshold or between two close-by thresholds. Such molecular states are weakly bound, i.e. the binding energies are much smaller than the particle mass, and this weak binding also entails a large spatial extent. Further, hadronic molecules show particular decay patterns. The classical example is, of course, the deuteron, a bound state of a proton and a neutron. Its binding energy (BE) of about 2.22 MeV is much smaller than its mass, $M_D = m_p + m_n - E_B \simeq 1876$ MeV and its radius of 2.14 fm is much bigger than the proton radius of about 0.85 fm. Other examples are the two poles in the $\Lambda(1405)$ region, the $f_0(980)$, the $X(3872)$, and many others. For a recent review, see [2] and the discussion below.

Naturally the question arises how to distinguish these molecules from compact multi-quark states? This was originally answered by Weinberg [44] and then refined by various others, see e.g. [45–47]. For that, consider the wave function of a bound state $|\Psi\rangle$ with a compact component $|\psi_0\rangle$ and a two-hadron component $|h_1 h_2\rangle$ in S-wave:

$$|\Psi\rangle = \begin{pmatrix} \sqrt{Z}|\psi_0\rangle \\ \chi(\vec{k})|h_1 h_2\rangle \end{pmatrix}, \tag{13}$$

that is, such a bound state consists of a compact component with probability $\sqrt{Z}$ and a two-hadron component

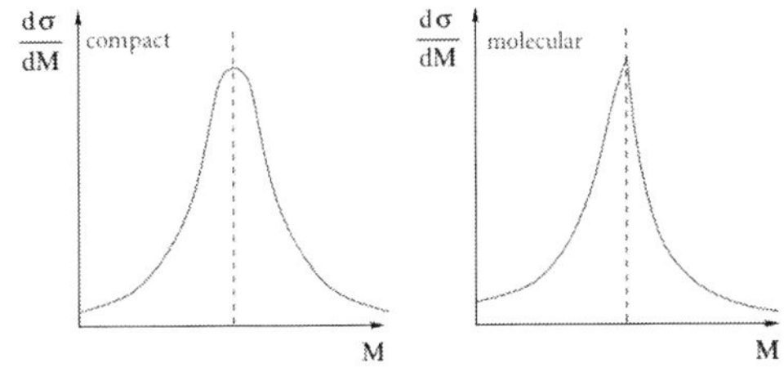

Figure 8. Typical near-threshold line shapes that emerge for compact (left panel) and molecular states (right panel). The dashed perpendicular lines indicate the location of the threshold. Note the cusp at the threshold for the molecular scenario. The x-axis shows $M = m_1 + m_2 + E$. Figure inspired from [2].

with a relative wave function $\chi(\vec{k})$ (which should be normalizable). Then, comparing the hadron-hadron scattering amplitude with the effective range expansion gives:

$$a = -2\frac{1-Z}{2-Z}\left(\frac{1}{\gamma}\right) + O\left(\frac{1}{\beta}\right),$$
$$r = -\frac{Z}{1-Z}\left(\frac{1}{\gamma}\right) + O\left(\frac{1}{\beta}\right), \tag{14}$$

with $\gamma = \sqrt{2\mu E_B}$ the binding momentum, μ the reduced mass, a the scattering length, r the effective range and β is the range of forces. For a pure molecule ($Z = 0$) we have the maximal scattering length, $a = -1/\gamma$, and a natural effective range $r = O(1/\beta)$. Very differently for a compact state ($Z = 1$), the scattering length $a = -O(1/\beta)$ is of natural size and the effective range diverges, which is cured by higher order effects not explicitly displayed in Eq. (14). As an example, consider the deuteron. Here, we have $\gamma = 45.7\,\text{MeV} = 0.23\,\text{fm}^{-1}$ and the range of forces is set by the one-pion-exchange, $1/\beta \sim 1/M_\pi \simeq 1.4$ fm. Setting $Z = 0$ in Eq. (14), we find $a_{\text{mol}} = -(4.3 \pm 1.4)$ fm and $r_{\text{mol}} = 1.4$ fm, consistent with the data, $a = -5.419(7)$ fm and $r = 1.764(8)$ fm.

Besides the deuteron all other (candidates for) hadronic molecules are unstable. Then the scattering T-matrix needs to be modified compared to the form discussed before. The corresponding modifications of the framework to cope with hadron resonances have been developed, see e.g. Refs. [47–50]. Assuming again closeness to a two-particle threshold, the scattering amplitude takes the form

$$\mathrm{T}(E) = \frac{g^2/2}{E - E_R + (g^2/2)(ik + \gamma) + i\Gamma_0/2}, \tag{15}$$

with g the coupling constant of the two-hadron system, $E = k^2/(2\mu)$, E_R the resonance energy and Γ_0 accounts for the inelasticities of other channels. This leads to very different *line shapes* for compact and molecular states. While for a compact state the k^2 term dominates leading to a symmetric line shape, for a molecular state the g^2 term dominates, leading to an asymmetric distribution exhibiting a cusp, cf. Fig. 8. Further generalizations to instable constituents, virtual poles, etc., have also been worked out. I refer again to the review [2] for details.

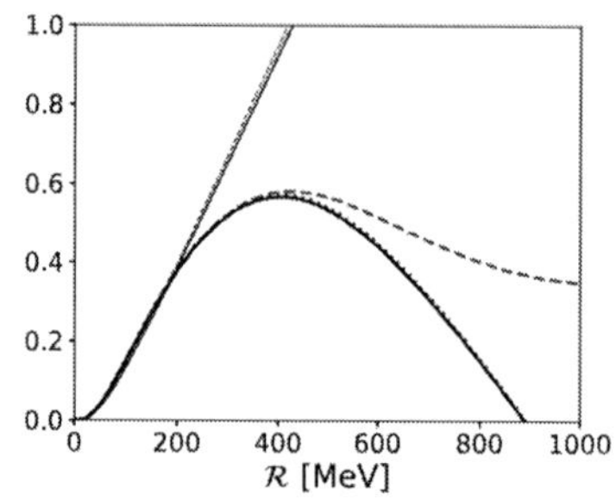

Figure 9. Estimate of $\bar{\Psi}_\lambda(\mathcal{R})$ for various deuteron wave functions: Results for $\lambda = 0.8, 1.6, 4$ GeV are shown as red (dashed), blue (dotted) and black (solid) curves, respectively. The thick (thin) lines depict the results with (without) OPE. Figure from [63].

Without going into details, let me list a few candidates for hadronic molecules (for details, see e.g. Ref. [2]). Prominent examples in the light quark sector are the scalar mesons $f_0(980)$, $a_0(980)$ as well as the already mentioned two $\Lambda(1405)$ states, here I refer to the talks by Mai [51] and Oset [52] for further discussion. There are also very prominent examples in the $c\bar{c}$ and $b\bar{b}$ spectrum, these are the $X(3872)$, $Z_c(3900)$, $Y(4260)$, $Y(4660)$, $Z_b(10610)$, $Z_b(10650)$, and others. Similarly, in heavy-light systems we have a number of candidates, especially the $D^\star_{s0}(2317), D_{s1}(2460)$, $D^\star_{s1}(2860)$, and others. The molecular nature of the charm-strange mesons was already discussed in Sect. 4. In the baryon sector, the LHCb pentaquarks [53, 54] are good candidates, especially the recently observed $P_c(4457)$, see e.g. Refs. [55–61].

I will take the opportunity to discuss a few misconceptions surrounding hadronic molecules. It is often claimed that molecules due to their large spatial extent can not be produced in high-energy collisions, say at the LHC, but this is not correct. An argument found in the literature is due to Ref. [62]. They consider the $X(3872)$ production at the Tevatron, and derived a bound on the cross section:

$$\begin{aligned}\sigma(\bar{p}p \to X) &\sim \left|\int d^3\mathbf{k}\langle X|D^0\bar{D}^{*0}(\mathbf{k})\rangle\langle D^0\bar{D}^{*0}(\mathbf{k})|\bar{p}p\rangle\right|^2 \\ &\simeq \left|\int_{\mathcal{R}} d^3\mathbf{k}\langle X|D^0\bar{D}^{*0}(\mathbf{k})\rangle\langle D^0\bar{D}^{*0}(\mathbf{k})|\bar{p}p\rangle\right|^2 \\ &\le \int_{\mathcal{R}} d^3\mathbf{k}\,|\Psi(\mathbf{k})|^2 \int_{\mathcal{R}} d^3\mathbf{k}\left|\langle D^0\bar{D}^{*0}(\mathbf{k})|\bar{p}p\rangle\right|^2 \\ &\le \int_{\mathcal{R}} d^3\mathbf{k}\left|\langle D^0\bar{D}^{*0}(\mathbf{k})|\bar{p}p\rangle\right|^2 , \end{aligned} \tag{16}$$

which depends crucially on the value of $\mathcal{R}$ which specifies the region where the bound state wave function "$\Psi(\mathbf{k})$ is significantly different from zero" [62]. These authors assume that $\mathcal{R}$ is of the order of the binding momentum γ, which amounts to $\mathcal{R} \simeq 35$ MeV for the then accepted BE of the $X(3872)$. From this they conclude that $\sigma(\bar{p}p \to X) \simeq 0.07$ nb, which is orders of magnitudes lower than the experimental bounds known then (updated numbers will be given below). So this leads them to conclude that the $X(3872)$ can not be a molecule. But I said that the $X(3872)$ is a premier candidate for a molecule in the charmonium spectrum, so what goes wrong with this argument? Let us take a step back and consider the best known molecule, the deuteron [63]. The relevant integral is

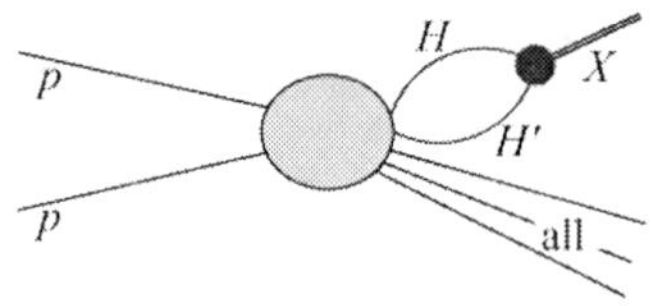

Figure 10. The mechanism for the inclusive production of the X as a HH' bound state in proton–proton collisions. Here, *all* denotes all the produced particles other than the H and H' in the collision. Figure from [66].

$$\bar{\Psi}_\lambda(\mathcal{R}) \equiv \int_{\mathcal{R}} d^3\mathbf{k}\,\Psi_\lambda(\mathbf{k})\,, \tag{17}$$

where λ specifies a regulator that needs to be introduced to render the wave function well defined. In the deuteron case, $\gamma \simeq 45$ MeV, and as in the case of the $X(3872)$ is of the order of the pion Compton wavelength $O(1/M_\pi)$. In Fig. 9, $\bar{\Psi}_\lambda(\mathcal{R})$ is shown for the deuteron with two sets of wave functions, one generated from a potential with a short-ranged term and one-pion exchange (OPE) and the other without OPE. For the former a Gaussian regulator is used and the small D-wave component is not shown. For the latter a sharp momentum space cut off is used for simplicity. One sees that $\bar{\Psi}_\lambda(\mathcal{R})$ is far from being saturated for $\mathcal{R} \simeq \gamma_d$ for all values of λ. A much larger value of $\mathcal{R}$ ~300 MeV~ $2M_\pi$ needs to be taken for the second line in Eq. (16) to be a good approximation of the first. This is, of course, related to the fact that two-pion exchange is required to bind the deuteron. The same should also be true for the $X(3872)$ as a $D\bar{D}^*$ molecule, since the range of forces is the same. A similar value was also favored in Ref. [64] based on rescattering arguments. In fact, Ref. [62] noted that with such a value of $\mathcal{R}$ the upper bound becomes consistent with the CDF result.

Let me briefly discuss how one can calculate the hadroproduction of the $X(3872)$ in an EFT if one realizes that this is a process involving short-distance physics, but still factorization is at work [64]. In such a scheme, event generators are used to calculate the collinear production of two hadrons and these then combine to form a molecule at large distances, see Fig. 10, which means that the molecule is not promptly produced. For high-energy processes like at the Tevatron or the LHC, the corresponding cross section of producing a molecule X from two hadrons H, H' can be calculated via [65, 66]:

$$\begin{aligned}\sigma[X] &= \frac{1}{4m_H m_{H'}} g^2 |G|^2 \left(\frac{d\sigma[HH'(k)]}{dk}\right)_{\rm MC} \frac{4\pi^2\mu}{k^2}\,, \\ G(E,\Lambda) &= -\frac{\mu}{\pi^2}\left[\sqrt{2\pi}\,\frac{\Lambda}{4} + \sqrt{\pi}\,\gamma D\left(\frac{\sqrt{2}\gamma}{\Lambda}\right) \right. \\ &\quad \left. -\frac{\pi}{2}\,\gamma\, e^{2\gamma^2/\Lambda^2}\right], \end{aligned} \tag{18}$$

Table 2. Integrated cross sections (in units of nb) for $pp/\bar{p} \to X(3872)$ compared to experimental measurements by CDF [69] and CMS [70].

$\sigma(pp/\bar{p}) \to X(3872))$	$\Lambda \in [0.5, 1.0]$ GeV	Exp.
Tevatron	5 - 29 [nb]	37 - 115 [nb]
LHC7	4 - 55 [nb]	13 - 39 [nb]

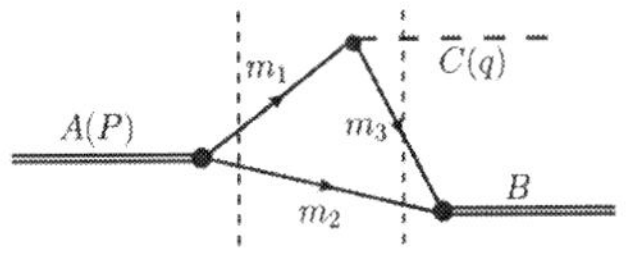

Figure 11. A triangle diagram illustrating the long-distance contribution to the transition between two heavy particles A and B with the emission of a light particle C. The two vertical dashed lines denote the two relevant cuts. Figure from [2].

with $(d\sigma[HH'(k)]/dk)_{\rm MC}$ the cross section of producing the hadrons H and H' in collinear geometry, G is the cut-off regularized two-meson (HH') loop function expressed in terms of the Dawson function, $D(x) = e^{x^2}\int_0^x e^{-y^2}\,dy$, $\gamma = \sqrt{-2\mu(E - m_H - m_{H'})}$ is the binding momentum and Λ is the cutoff. Following Ref. [67], a range of $[0.5, 1.0]$ GeV is used for the cutoff Λ. Typical results using the PYTHIA and HERWIG generators are collected in Tab. 2. We see that the calculations are not very precise, but they are perfectly consistent with the data, very different from the conclusions drawn in Ref. [62]. For a similar calculation concerning the hadroproduction of the charm-strange mesons, I refer to Ref. [68].

Finally, I mention that most candidates for hadronic molecules have been found through decays that often involve triangle diagrams with an anomalous singularity, as studied by Landau, Nambu and others in the 1950ties. There exist different EFTs that are used in the calculations of hadronic molecules and their decays. These are the already mentioned UCHPT as well as $\rm NREFT_I$ and $\rm NREFT_2$ when at least one heavy quark is involved. In $\rm NREFT_1$, all intermediate particles in the triangle diagram in Fig. 11 are close to their mass shell, so that one can expand in powers of the average velocity and external (small) momenta. This has e.g. been applied to systematic studies of a number of charmonium transitions. In $\rm NREFT_2$ one intermediate particle is further off its mass shell, so one integrates out this particle and then proceeds as before. This EFT was originally invented as XEFT for studies of the $X(3872)$ [71]. XEFT resembles much the pionless EFT of nuclear physics, see e.g. [72]. It has been used for systematic studies of processes involving the $X(3872)$ and the Z_b states, see e.g. [73].

6 The width of the lightest nucleon resonances from baryon CHPT

In this section, I will be concerned with calculating the width of the two lightest baryon resonances, the $\Delta(1232)$ and the Roper $N^*(1440)$. This might at first sight appear irritating, as imaginary parts are usually not precisely reproduced in CHPT. For that simple reason, one has to employ a complex-mass scheme and work to two loops. The complex-mass renormalization scheme is a method that was originally introduced for precision W, Z-physics, see e.g. [74, 75] and later transported to chiral EFT [76].

Let me give a brief outline of the complex-mass scheme (CMS), following Ref. [77]. Consider first an instable particle at tree level. The CMS amounts to treating the mass of this particle consistently as a complex quantity, defined as the location of the pole in the complex k^2-plane of the corresponding propagator with momentum k. It can be shown that this scheme is symmetry-preserving and leaves the corresponding Ward identities intact. Extending this to one loop, one splits the real bare masses into complex renormalized masses and complex counterterms. This is important, as only renormalized masses are observable. The corresponding Lagrangian yields Feynman rules with complex masses and counterterms, which allows for standard perturbative calculations. This is essentially a rearrangement of contributions that is not affected by double counting. The imaginary part of the particle mass appears in the propagator and is resummed in the Dyson series. In contrast to this, the imaginary part of the counterterm is not resummed. One can show that in such a case gauge invariance remains valid, and unitarity cancellations are respected order by order in the perturbative expansion. This also requires integrals with complex internal masses, as worked out in Ref. [78]. For further discussions of the method, the reader is referred to Ref. [77] (and references therein). In case of a chiral EFT, the perturbative expansion proceeds as usual in terms of small momenta and quark masses, with a proper treatment of the heavy particle mass in loop diagrams (like the heavy-baryon scheme or the so-called infrared-regularization or the extended-on-mass-scheme discussed below).

6.1 The width of the $\Delta(1232)$

Consider first the width of the Δ at two-loop order [79]. The pertinent effective Lagrangian contains, besides many other terms, the leading $\pi\Delta$ and $\pi N\Delta$ couplings, parametrized in terms of the LECs g_1 and h, respectively,

$$
\begin{aligned}
\mathcal{L}^{(1)}_{\pi\Delta} &= -\bar{\Psi}^i_\mu \xi^{\frac{3}{2}}_{ij}\Big\{\Big(i\not{D}^{jk} - m_\Delta\delta^{jk}\Big)g^{\mu\nu} \\
&- i\Big(\gamma^\mu D^{\nu,jk} + \gamma^\nu D^{\mu,jk}\Big) + i\gamma^\mu\not{D}^{jk}\gamma^\nu \\
&+ m_\Delta\delta^{jk}\gamma^\mu\gamma^\nu + g_1\frac{1}{2}\not{u}^{jk}\gamma_5 g^{\mu\nu} \\
&+ g_2\frac{1}{2}(\gamma^\mu u^{\nu,jk} + u^{\nu,jk}\gamma^\mu)\gamma_5 \\
&+ g_3\frac{1}{2}\gamma^\mu\not{u}^{jk}\gamma_5\gamma^\nu\Big\}\xi^{\frac{3}{2}}_{kl}\Psi^l_\nu\,, \\
\mathcal{L}^{(1)}_{\pi N\Delta} &= h\,\bar{\Psi}^i_\mu \xi^{\frac{3}{2}}_{ij}\Theta^{\mu\alpha}(z_1)\,\omega^j_\alpha \Psi_N + \text{h.c.}\,, \\
\mathcal{L}^{(2)}_{\pi N\Delta} &= \bar{\Psi}^i_\mu \xi^{\frac{3}{2}}_{ij}\Theta^{\mu\alpha}(z_2) \\
&\times \Big[i b_3 \omega^j_{\alpha\beta}\gamma^\beta + i\frac{b_8}{m}\omega^j_{\alpha\beta} i D^\beta\Big]\Psi_N + \text{h.c.}\,,
\end{aligned}
$$

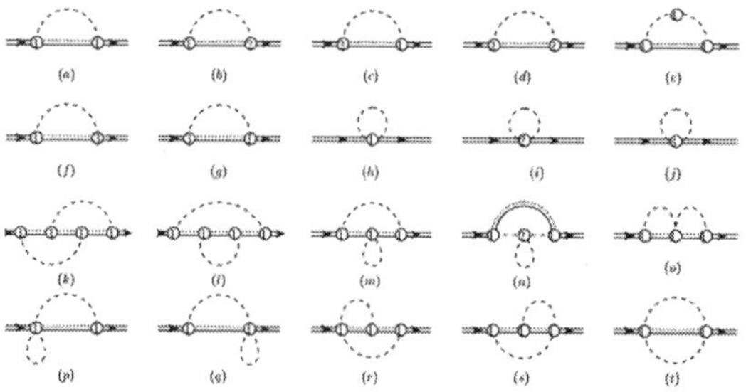

Figure 12. One and two-loop self-energy diagrams contributing to the width of the delta resonance up-to-and-including fifth order according to the standard power counting. The dashed and double solid lines represent the pions and the delta resonances, respectively. The double (solid-dotted) lines in the loops correspond to either nucleons or deltas. The numbers in the circles give the chiral orders of the vertices. Figure from [79].

$$\begin{aligned}\mathcal{L}^{(3)}_{\pi N\Delta} &= \bar{\Psi}^i_\mu \xi^{\frac{3}{2}}_{ij}\Theta^{\mu\nu}(z_3)\Big[\frac{f_1}{m}[D_\nu,\omega^j_{\alpha\beta}]\gamma^\alpha i D^\beta \\ &- \frac{f_2}{2m^2}[D_\nu,\omega^j_{\alpha\beta}]\{D^\alpha, D^\beta\} \\ &+ f_4\omega^j_\nu\langle\chi_+\rangle + f_5[D_\nu, i\chi^j_-]\Big]\Psi_N + \text{h.c.}, \quad (19)\end{aligned}$$

where Ψ_N and Ψ_ν are the isospin doublet field of the nucleon and the vector-spinor isovector-isospinor Rarita-Schwinger field of the Δ-resonance with bare masses m and $m_{\Delta 0}$, respectively. $\xi^{\frac{3}{2}}$ is the isospin-3/2 projector, $\omega^i_\alpha = \frac{1}{2}\langle\tau^i u_\alpha\rangle$ and $\Theta^{\mu\alpha}(z) = g^{\mu\alpha} + z\gamma^\mu\gamma^\alpha$. Using field redefinitions the off-shell parameters z can be absorbed in LECs of other terms of the effective Lagrangian and therefore they can be chosen arbitrarily [80, 81]. We fix the off-shell structure of the interactions with the delta by adopting $g_2 = g_3 = 0$ and $z_1 = z_2 = z_3 = 0$. For vanishing external sources, the covariant derivatives are given by

$$\begin{aligned}D_\mu\Psi_N &= \left(\partial_\mu + \Gamma_\mu\right)\Psi_N, \\ \Gamma_\mu &= \frac{1}{2}\left[u^\dagger\partial_\mu u + u\partial_\mu u^\dagger\right] = \tau_k\Gamma_{\mu,k}, \\ \left(D_\mu\Psi\right)_{\nu,i} &= \partial_\mu\Psi_{\nu,i} - 2\,i\,\epsilon_{ijk}\Gamma_{\mu,k}\Psi_{\nu,j} + \Gamma_\mu\Psi_{\nu,i}\,. \quad (20)\end{aligned}$$

The power counting rests on $m_\Delta - m_N$ being a small quantity. More precisely, the small parameters are the external momenta, the pion mass and the nucleon-Delta mass splitting, collectively denoted as q. However, there are so many LECs in Eq. (19), so how can one one possibly make a prediction? Let us evaluate the Δ self-energy on the complex pole,

$$z - m^0_\Delta - \Sigma(z) = 0 \quad \text{with} \quad z = m_\Delta - i\,\frac{\Gamma_\Delta}{2}\,. \quad (21)$$

The corresponding diagrams for the one- and two-loop self-energy contributing to the width of the delta resonance up to order q^5 are displayed in Fig. 12, where the counterterm diagrams are not shown. The one-loop diagrams are easily worked out. For the calculation of the two-loop graphs one uses the Cutkosky rules for instable particles, that relate the width to the pion-nucleon scattering amplitude, $\Gamma_\Delta \sim |A(\Delta \to N\pi)|^2$ [82]. One finds a remarkable

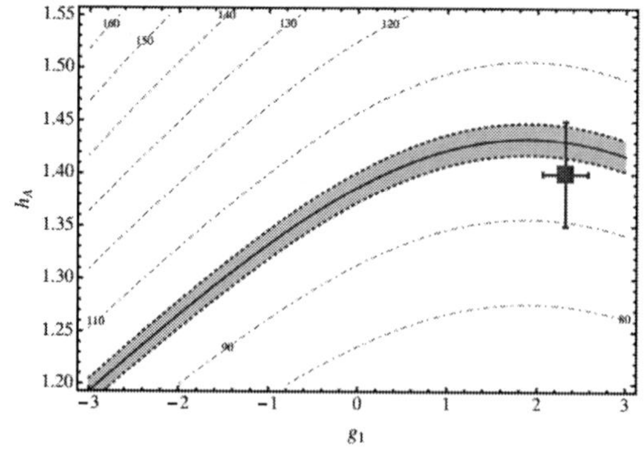

Figure 13. Correlation between the leading $\pi\Delta$ and $\pi N\Delta$ couplings. The central line corresponds to $\Gamma_\Delta = 100$ MeV while the band is obtained by varying Γ_Δ in the range of 98 to 102 MeV. The dot-dashed lines show the correlation for other values of the width of the Delta. The box with the error bars are the results from the analysis of Ref. [84]. Figure inspired from [79].

reduction of parameters that is reflected in the relation

$$\begin{aligned}h_A &= h - (b_3\Delta_{23} + b_8\,\Delta_{123}) \\ &- (f_1\Delta_{23} + f_2\,\Delta_{123})\,\Delta_{123} + 2(2f_4 - f_5)M_\pi^2\,, \\ \Delta_{23} &= m_N - m_\Delta\,, \\ \Delta_{123} &= \frac{M_\pi^2 + m_N^2 - m_\Delta^2}{2m_N}\,, \quad (22)\end{aligned}$$

which means that all of the LECs appearing in the $\pi N\Delta$ interaction at second and third order, the $b_i\,(i = 3, 8)$ and $f_i\,(i = 1, 2, 4, 5)$, respectively, merely lead to a renormalization of the LO $\pi N\Delta$ coupling h, and, consequently, one finds a very simple formula for the decay width $\Delta \to N\pi$,

$$\Gamma(\Delta \to N\pi) = (53.9\,h_A^2 + 0.9g_1^2h_A^2 - 3.3g_1h_A^2 - 1.0\,h_A^4)\,\text{MeV}\,. \quad (23)$$

This leads to a novel correlation that is independent of the number of colors, as N_c was not used as a parameter in the calculation. This correlation between h_A and g_1 is depicted in Fig. 13. It is obviously fulfilled by the analysis of Ref. [83], that showed that the inclusion of the Δ alleviates the tension between the threshold and subthreshold regions in the description of πN scattering found in baryon CHPT.

6.2 The width of the Roper resonance

Next, I consider the calculation of the width of the Roper $N^*(1440)$ at two-loop order [85]. A remarkable feature of the Roper is the fact that its decay width into a nucleon and a pion is similar to the width into a nucleon and two pions. Any model that is supposed to describe the Roper must account for this fact. In CHPT, consider the effective chiral Lagrangian of pions, nucleons and deltas coupled to the Roper [86–88],

$$\begin{aligned}\mathcal{L}_{\text{eff}} &= \mathcal{L}_{\pi\pi} + \mathcal{L}_{\pi N} + \mathcal{L}_{\pi\Delta} + \mathcal{L}_{\pi R} \\ &+ \mathcal{L}_{\pi N\Delta} + \mathcal{L}_{\pi NR} + \mathcal{L}_{\pi\Delta R}\,, \quad (24)\end{aligned}$$

with

$$
\begin{aligned}
\mathcal{L}_{\pi R}^{(1)} &= \bar{\Psi}_R \left\{ i\not{D} - m_R + \frac{1}{2} g_R \not{u} \gamma^5 \right\} \Psi_R \,, \\
\mathcal{L}_{\pi R}^{(2)} &= \bar{\Psi}_R \left\{ c_1^R \langle \chi_+ \rangle \right\} \Psi_R + \ldots \,, \\
\mathcal{L}_{\pi NR}^{(1)} &= \bar{\Psi}_R \left\{ \frac{1}{2} g_{\pi NR} \gamma^\mu \gamma_5 u_\mu \right\} \Psi_N + \text{h.c.} \,, \\
\mathcal{L}_{\pi \Delta R}^{(1)} &= h_R \, \bar{\Psi}_\mu^i \xi_{ij}^{\frac{3}{2}} \Theta^{\mu\alpha}(\tilde{z}) \, \omega_\alpha^j \Psi_R + \text{h.c.} \,, \qquad (25)
\end{aligned}
$$

where g_R, $g_{\pi NR}$ and h_R, respectively, are the leading Roper-pion, Roper-nucleon-pion and Delta-Roper-pion couplings. Here, Ψ_R denotes the Roper isospin doublet field and all other notations are as in the preceeding subsection and in [85].

In this case, the power counting is more complicated, but can be set up around the complex pole as (for more details, see [85]), assigning the following counting rules:

$$
m_R - m_N \sim \varepsilon \,, \quad m_R - m_\Delta \sim \varepsilon^2 \,, \quad m_\Delta - m_N \sim \varepsilon^2 \,, \quad M_\pi \sim \varepsilon^2 \,, \qquad (26)
$$

where ε denotes a small parameter. Again, let us calculate the self-energy to two loops at the complex pole $z_R = m_R - i\Gamma_R/2$. By applying the cutting rules to these self-energy diagrams, one obtains the graphs contributing to the decay amplitudes of the Roper resonance into the πN and $\pi\pi N$ systems, leading to the total width

$$
\Gamma_R = \Gamma_{R\to N\pi} + \Gamma_{R\to N\pi\pi} \,. \qquad (27)
$$

A somewhat lengthy calculation leads to:

$$
\Gamma(R \to N\pi) = 550(58)\, g_{\pi NR}^2 \,\text{MeV} \,, \qquad (28)
$$

and

$$
\begin{aligned}
\Gamma(R \to N\pi\pi) = \big(& 1.5(0.6)\, g_A^2\, g_{\pi NR}^2 \\
& -2.8(1.0)\, g_A\, g_{\pi NR}^2\, g_R \qquad (29) \\
& +1.5(0.6)\, g_{\pi NR}^2\, g_R^2 + 3.0(1.0)\, g_A\, g_{\pi NR}\, h_A h_R \\
& -3.8(1.4)\, g_{\pi NR}\, g_R\, h_A h_R + 9.9(5.5)\, h_A^2 h_R^2 \big) \,\text{MeV}.
\end{aligned}
$$

The total width thus depends on five LECs. The uncertainties in the round brackets are generated by the uncertainties in the LECs. We use $g_A = 1.27$ and $h_A = 1.42 \pm 0.02$. The latter value is the real part of this coupling taken from Ref. [84]. As for the other unknown parameters, the authors of [85] fixed $g_{\pi NR}$ so as to reproduce the width $\Gamma_{R\to\pi N} = (123.5 \pm 19.0)$ MeV from the PDG. This yields $g_{\pi NR} = \pm(0.47 \pm 0.04)$. In what follows, let us take the positive sign for our central value and use the negative one as part of the error budget. Further, assume $g_R = g_A$ and $h_R = h_A$, the so-called maximal mixing assumption [89]. Then, one can make a prediction for the two-pion decay width of the Roper,

$$
\Gamma(R \to N\pi\pi) = (41 \pm 22_{\text{LECs}} \pm 17_{\text{h.o.}}) \,\text{MeV} \,, \qquad (30)
$$

which is consistent with the PDG value of (67 ± 10) MeV. The error due to the neglect of the higher orders (h.o.) is simply given by multiplying the ε^5 result (central value) with $\varepsilon = (m_R - m_N)/m_N \simeq 0.43$. Clearly, to make further progress, we need an improved determination of the LECs g_R and h_R. This could be addressed within LQCD. Finally, I would like to mention that this scheme has also been used to consider the electromagnetic transition form factors of the Roper [90].

7 On the pion cloud of the nucleon and other hadrons

At this conference, one often encountered the situation that some observable (like e.g. a resonance transition form factor extracted from pion electroproduction experiments) is well described in terms of quark degrees of freedom at large momentum transfer, say Q^2 a few GeV^2, but then deviations between the model/theory show up when one approaches the photon point at $Q^2 = 0$. This is often cured by adding the pion cloud contribution from some hadronic model, see e.g. some examples in the opening talk by Burkert [91]. I would like to take the opportunity to issue a warning here. While the notion of the "pion cloud" is fairly intuitive, it is rather difficult to assign to it a quantitative measure. So let us look at that in more detail.

The so-called pion cloud is best analyzed in chiral perturbation theory. There is one very intriguing structural aspect generated by the (almost) massless Goldstone bosons of QCD, namely the so-called *chiral logarithms (logs)* or chiral singularities. Pions couple to themselves and to matter fields like the nucleons, thus generating a cloud that is Yukawa-suppressed at large distances as $\exp(-M_\pi r)/r$ for finite pion mass. In the chiral limit of vanishing quark and thus pion masses $M_\pi \to 0$, this Yukawa tail turns into a long-range Coulomb-like form. Consequently, S-matrix elements or transition currents can diverge in this limit. Famous examples are the pion vector radius or the nucleon isovector radius that scale as $\log(M_\pi)$, see e.g. Ref. [92], or the nucleon electromagnetic polarizabilities that are proportional to $1/M_\pi$ [93]. This is not a disaster but rather a natural consequence of having massless degrees of freedom, and it can serve as an important check for other calculational schemes, like e.g. the lattice formulation of QCD. Also, any model that is supposed to describe the pion or the nucleon structure at low energies should obey such constraints. Note also that such singularities are generated by loop graphs, so that in general there will also be a pion-mass-independent contribution from a counterterm (contact interaction) at the same order of the calculation. A typical example is the isovector charge radius of the pion, or any other observable that displays a chiral log. Since the argument of a log must be a number, it really depends on M_π/μ, with μ some regularization scale. Similarly, the corresponding LEC that appears must also depend on μ, thus making the observable scale-independent. This, however, also means that it is in general not possible to assign a definite value to the pion cloud contribution of an observable since shifts in the regularization scale allow one to shuffle strength from the long-range pion contribution to the shorter ranged contact term part. I refer the reader to Ref. [94], where the isovector charge radius of the proton is analyzed at one-loop order. It is shown that a modest change in the regularization scale from $\mu = 0.8$ GeV to $\mu = 1$ GeV (using conventional dimensional regularization) even leads to a sign change in the counterterm contribution. This explicitly shows that there is, of course, a contribution of the chiral (pion) physics to a given observable, but only the observable quantity (here: the isovector charge radius) is independent of the regularization scale. A more detailed discussion of this and related issues can be found in Refs. [95, 96].

8 Summary and outlook

The lessons learned and the take-home messages from this talk are:

- Resonances are defined as poles in the complex energy plane. To calculate their complex properties, one must locate the corresponding pole and then derive the resonance characteristics from suitably defined Laurent expansions around the pole position. Any other approach will in most cases lead to imprecise or even wrong results. Be aware of *how* many entries in the PDG tables have been obtained!
- The QCD spectrum is more than a collection of quark model states. The quark model should not be considered as a faithful representation of the QCD spectrum but rather as a simple approximation to a part of it. Note that the approximations underlying the conventional quark model are clearly better justifed for the heavy than for the light quarks.
- Structure formation in QCD ties nuclear and hadron physics together. This is most clearly seen through the appearance of molecular-type structures both in the nuclear as well as in the hadronic landscape. This was realized early by some researchers, see e.g. [97, 98].
- Lattice QCD is making progress in addressing complex resonance properties, even finding poles deep inside the complex plane, like for the lowest resonance in QCD, the $f_0(500)$, see e.g. [99, 100]. However, the extrapolation into the complex plane must respect chiral symmetry. Otherwise, one can easily miss a possible two-pole structure which was first observed for the $\Lambda(1405)$ but appears to play an even bigger role for states involving charm and bottom quarks.
- EFTs are of utmost importance in pushing this program forward. This has become apparent in the discussion of finding poles in the complex plane. The explicit calculations of finite-volume effects in hadron decays are another important playground for EFTs and this program is pursued vigorously, see e.g. the talk by Pang [101].

Acknowledgments

I thank all my collaborators for sharing their insights on the various topics discussed here. Special thanks to Maxim Mai for supplying Fig. 3, and to Feng-Kun Guo, Bernard Metsch and Akaki Rusetsky for a careful reading of the manuscript. This work is supported in part by the DFG (Grant No. TRR110) and the NSFC (Grant No. 11621131001) through the funds provided to the Sino-German CRC 110 "Symmetries and the Emergence of Structure in QCD", by the Chinese Academy of Sciences (CAS) President's International Fellowship Initiative (PIFI) (grant no. 2018DM0034) and by VolkswagenStiftung (grant no. 93562). Computational resources for this project were provided by the Jülich Supercomputing Centre (JSC) at the Forschungszentrum Jülich and by RWTH Aachen.

References

[1] R. G. Moorhouse, quote from the 1960ties.
[2] F. K. Guo *et al.*, Rev. Mod. Phys. **90**, 015004 (2018).
[3] P. C. Bruns *et al.*, Phys. Lett. B **697**, 254 (2011).
[4] J. A. Oller *et al.*, Phys. Lett. B **500**, 263 (2001).
[5] A. Cieplý *et al.*, Nucl. Phys. A **954**, 17 (2016).
[6] M. Lüscher, Commun. Math. Phys. **105**, 153 (1986).
[7] U. J. Wiese, Nucl. Phys. Proc. Suppl. **9**, 609 (1989).
[8] K. Rummukainen *et al.*, Nucl. Phys. B **450**, 397 (1995).
[9] Bonn-Jülich-QCDSF collaboration, unpublished (2010).
[10] C. Morningstar, *these proceedings*.
[11] M. Petschlies, *these proceedings*.
[12] C. B. Lang *et al.*, Phys. Rev. D **87**, 054502 (2013).
[13] V. Verduci *et al.*, PoS LATTICE **2014**, 121 (2014).
[14] L. Leskovec *et al.*, Few Body Syst. **59**, 95 (2018).
[15] M. Werner *et al.*, arXiv:1907.01237 [hep-lat].
[16] S. He *et al.*, JHEP **0507**, 011 (2005).
[17] M. Lage *et al.*, Phys. Lett. B **681**, 439 (2009).
[18] V. Bernard *et al.*, JHEP **1101**, 019 (2011).
[19] R. A. Briceño *et al.*, Phys. Rev. D **88**, 094507 (2013).
[20] R. A. Briceño *et al.*, Phys. Rev. D **88**, 034502 (2013).
[21] M. Döring *et al.*, Eur. Phys. J. A **47**, 139 (2011).
[22] M. Döring *et al.*, Eur. Phys. J. A **48**, 114 (2012).
[23] R. A. Briceño *et al.*, Phys. Rev. D **95**, 074510 (2017).
[24] J. M. M. Hall *et al.*, Phys. Rev. Lett. **114**, 132002 (2015).
[25] R. Molina *et al.*, Phys. Rev. D **94**, 056010 (2016) Addendum: [Phys. Rev. D **94**, 079901 (2016)].
[26] G. Moir *et al.*, JHEP **1610**, 011 (2016).
[27] see `http://pdg.lbl.gov/`.
[28] N. Kaiser *et al.*, Nucl. Phys. A **594**, 325 (1995).
[29] E. Oset *et al.*, Nucl. Phys. A **635**, 99 (1998).
[30] M. F. M. Lutz *et al.*, Nucl. Phys. A **700**, 193 (2002).
[31] F. K. Guo *et al.*, Phys. Lett. B **666**, 251 (2008).
[32] L. Liu *et al.*, Phys. Rev. D **87**, 014508 (2013).
[33] M. Albaladejo *et al.*, Phys. Lett. B **767**, 465 (2017).
[34] D. Jido *et al.*, Nucl. Phys. A **725**, 181 (2003).
[35] E. E. Kolomeitsev *et al.*, Phys. Lett. B **582**, 39 (2004).
[36] F. K. Guo *et al.*, Phys. Lett. B **641**, 278 (2006).
[37] F. K. Guo *et al.*, Eur. Phys. J. A **40**, 171 (2009).
[38] Z. H. Guo *et al.*, Phys. Rev. D **92**, 094008 (2015).
[39] M. L. Du *et al.*, Phys. Rev. D **98**, 094018 (2018).

[40] R. Aaij *et al.* [LHCb Collaboration], Phys. Rev. D **92**, 012012 (2015).
[41] R. Aaij *et al.* [LHCb Collaboration], Phys. Rev. D **91**, 092002 (2015) Erratum: [Phys. Rev. D **93**, 119901 (2016)].
[42] R. Aaij *et al.* [LHCb Collaboration], Phys. Rev. D **94**, 072001 (2016).
[43] M. L. Du *et al.*, Phys. Rev. D **99**, 114002 (2019).
[44] S. Weinberg, Phys. Rev. **137**, B672 (1965).
[45] D. Morgan *et al.*, Phys. Lett. B **258**, 444 (1991).
[46] N. A. Tornqvist, Phys. Rev. D **51**, 5312 (1995).
[47] V. Baru *et al.*, Phys. Lett. B **586**, 53 (2004).
[48] E. Braaten *et al.*, Phys. Rev. D **76**, 094028 (2007).
[49] F. Aceti *et al.*, Phys. Rev. D **86**, 014012 (2012).
[50] Z. H. Guo *et al.*, Phys. Rev. D **93**, 096001 (2016).
[51] M. Mai, *these proceedings*.
[52] E. Oset, *these proceedings*.
[53] R. Aaij *et al.* [LHCb Collaboration], Phys. Rev. Lett. **115** 072001 (2015).
[54] R. Aaij *et al.* [LHCb Collaboration], Phys. Rev. Lett. **122**, 222001 (2019).
[55] J. J. Wu *et al.*, Phys. Rev. Lett. **105**, 232001 (2010).
[56] R. Chen *et al.*, Phys. Rev. Lett. **115** 132002 (2015).
[57] T. J. Burns, Eur. Phys. J. A **51**, 152 (2015).
[58] L. Roca *et al.*, Phys. Rev. D **92**, 094003 (2015).
[59] R. Chen *et al.*, Phys. Rev. D **100**, 011502 (2019).
[60] F. K. Guo *et al.*, Phys. Rev. D **99** 091501 (2019).
[61] T. J. Burns *et al.*, arXiv:1908.03528 [hep-ph] (2019).
[62] C. Bignamini *et al.*, Phys. Rev. Lett. **103**, 162001 (2009).
[63] M. Albaladejo *et al.*, Chin. Phys. C **41**, 121001 (2017).
[64] P. Artoisenet *et al.*, Phys. Rev. D **81**, 114018 (2010).
[65] F. K. Guo *et al.*, Commun. Theor. Phys. **61**, 354 (2014).
[66] F. K. Guo *et al.*, Eur. Phys. J. C **74**, 3063 (2014).
[67] F. K. Guo *et al.*, Phys. Rev. D **88**, 054007 (2013).
[68] F. K. Guo *et al.*, JHEP **1405**, 138 (2014).
[69] G. Bauer [CDF Collaboration], Int. J. Mod. Phys. A **20**, 3765 (2005).
[70] S. Chatrchyan *et al.* [CMS Collaboration], JHEP **1304**, 154 (2013).
[71] S. Fleming *et al.*, Phys. Rev. D **76**, 034006 (2007).
[72] P. F. Bedaque *et al.*, Ann. Rev. Nucl. Part. Sci. **52**, 339 (2002).
[73] D. L. Canham *et al.*, Phys. Rev. D **80** 014009 (2009).
[74] R. G. Stuart, in "Z0 Physics", ed. J. Tran Thanh Van (Editions Frontieres, Gif-sur-Yvette, 1990), p. 41.
[75] A. Denner *et al.*, Nucl. Phys. B **560**, 33 (1999).
[76] D. Djukanovic *et al.*, Phys. Lett. B **680**, 235 (2009).
[77] A. Denner *et al.*, Nucl. Phys. Proc. Suppl. **160**, 22 (2006).
[78] W. Beenakker *et al.*, Nucl. Phys. B **338**, 349 (1990).
[79] J. Gegelia *et al.*, Phys. Lett. B **763**, 1 (2016).
[80] H. B. Tang *et al.*, Phys. Lett. B **387**, 9 (1996).
[81] H. Krebs *et al.*, Phys. Lett. B **683**, 222 (2010).
[82] M. J. G. Veltman, Physica **29**, 186 (1963).
[83] D. Siemens *et al.*, Phys. Lett. B **770**, 27 (2017).
[84] D. L. Yao *et al.*, JHEP **1605**, 038 (2016).
[85] J. Gegelia *et al.*, Phys. Lett. B **760**, 736 (2016.
[86] B. Borasoy *et al.*, Phys. Lett. B **641**, 294 (2006).
[87] D. Djukanovic *et al.*, Phys. Lett. B **690**, 123 (2010).
[88] B. Long *et al.*, Nucl. Phys. A **870-871**, 72 (2011).
[89] S. R. Beane *et al.*, J. Phys. G **31**, 921 (2005).
[90] M. Gelenava, Eur. Phys. J. A **54**, 88 (2018).
[91] V. Burkert, *these proceedings*.
[92] H. Pagels, Phys. Rept. **16**, 219 (1975).
[93] V. Bernard *et al.*, Phys. Rev. Lett. **67**, 1515 (1991).
[94] V. Bernard *et al.*, Nucl. Phys. A **635**, 121 (1998) [Erratum-ibid. A **642**, 563 (1998) 563].
[95] U.-G. Meißner, AIP Conf. Proc. **904**, 142 (2007).
[96] E. M. Henley *et al.*, *100 Years of Subatomic Physics* (World Scientific, Singapore, 2013) 199–229.
[97] M. B. Voloshin *et al.*, JETP Lett. **23**, 333 (1976) [Pisma Zh. Eksp. Teor. Fiz. **23**, 369 (1976)].
[98] N. A. Tornqvist, Z. Phys. C **61**, 525 (1994).
[99] R. A. Briceño, *et al.*, Phys. Rev. Lett. **118**, 022002 (2017).
[100] D. Guo *et al.*, Phys. Rev. D **98**, 014507 (2018).
[101] J.-Y. Pang, *these proceedings*.

Recent highlights with baryons from lattice QCD

Colin Morningstar[1,*]

[1]Department of Physics, Carnegie Mellon University, Pittsburgh, PA 15213, USA

Abstract. Highlights from recent computations in lattice QCD involving baryons are presented. Calculations of the proton mass and spin decompositions are discussed, a percent level determination of the nucleon axial coupling is described, and determinations of the proton and neutron electromagnetic form factors and light-cone parton distribution functions are outlined. Recent results applying the so-called Lüscher method to meson-baryon systems are presented. Key points emphasized are that much better precision with disconnected diagrams is being achieved, incorporating multi-hadron operators is now feasible, and more and more studies are being done with physical quark masses.

1 Introduction

Highlights from recent computations in lattice QCD involving baryons are presented in this talk. First, some introductory information about how baryons can be studied in lattice QCD is presented. Calculations of the proton mass and spin decompositions are then discussed, a percent level determination of the nucleon axial coupling is described, and determinations of the proton and neutron electromagnetic form factors and light-cone parton distribution functions are outlined. A current approach, involving the so-called Lüscher method, to confronting the challenge of studying baryon scattering and resonance properties in lattice QCD is then discussed. Some recent results applying the Lüscher method to meson-baryon systems are presented. A new calculation of the timelike pion form factor is highlighted to motivate future baryon form factor computations, especially the form factors for the Δ baryon which will be important to neutrino experiments such as DUNE. A new development with baryon-baryon interactions with the HAL QCD method is highlighted, and an H-dibaryon warm up calculation is presented. Key points emphasized are that much better precision with disconnected diagrams is being achieved, incorporating multi-hadron operators is now feasible, and more and more studies are being done with physical quark masses.

2 Studying baryons in lattice QCD

Finite-volume stationary-state energies are obtained in lattice QCD from Monte Carlo estimates of an $N\times N$ Hermitian correlation matrix

$$C_{ij}(t) = \langle 0|\, O_i(t+t_0)\, \overline{O}_j(t_0)\, |0\rangle. \tag{1}$$

Crucial to the success in extracting the energies is the use of judiciously designed operators $\overline{O}_j$ to create the states of interest

$$O_j(t) = O_j[\overline{\psi}(t), \psi(t), U(t)]. \tag{2}$$

Temporal correlator estimates are obtained from path integrals over the quark $\psi, \overline{\psi}$ and gluon U fields

$$C_{ij}(t) = \frac{\int \mathcal{D}(\overline{\psi},\psi,U)\;\, O_i(t+t_0)\,\overline{O}_j(t_0)\;\, \exp\left(-S[\overline{\psi},\psi,U]\right)}{\int \mathcal{D}(\overline{\psi},\psi,U)\;\exp\left(-S[\overline{\psi},\psi,U]\right)}$$

which involve the QCD action in imaginary time, formulated on a space-time lattice

$$S_{\rm QCD}[\overline{\psi},\psi,U] = \overline{\psi}\, K[U]\, \psi + S_G[U], \tag{3}$$

where $K[U]$ is the quark Dirac matrix and $S_G[U]$ is the gluon action. The integrals over the Grassmann-valued quark fields are done exactly, bringing the correlators into forms such as

$$C_{ij}(t) = \frac{\int \mathcal{D}U \det K[U] f(K^{-1}[U],\cdots,K^{-1}[U])e^{-S_G[U]}}{\int \mathcal{D}U\;\det K[U] e^{-S_G[U]}}$$

where $K^{-1}[U]$ are the quark propagators in the gluon field U. The remaining integrals over the gluon fields must be done with the Monte Carlo method, which uses a Markov chain to generate a sequence of gauge-field configurations $U_1, U_2, \ldots, U_N$. Including $\det K$ in the Monte Carlo updating and evaluating the K^{-1} in the numerator are the most computationally demanding parts of the calculation usually. The $\det K$ is handled by writing it as a multivariate Gaussian integral involving K^{-1}. A Metropolis method is employed with a sophisticated global updating proposal: Hybrid Monte Carlo (HMC) for the u, d quarks, and Rational Hybrid Monte Carlo (RHMC) for the s, c quarks.

Usually many different hadron operators are needed, so it is efficient to assemble them using basic building blocks, which we choose to be covariantly-displaced LapH-smeared[1] quark fields:

$$q^A_{a\alpha j} = D^{(j)}\widetilde{\psi}^{(A)}_{a\alpha}, \qquad \overline{q}^A_{a\alpha j} = \overline{\widetilde{\psi}}^{(A)}_{a\alpha}\,\gamma_4\, D^{(j)\dagger} \tag{4}$$

*e-mail: cmorning@andrew.cmu.edu

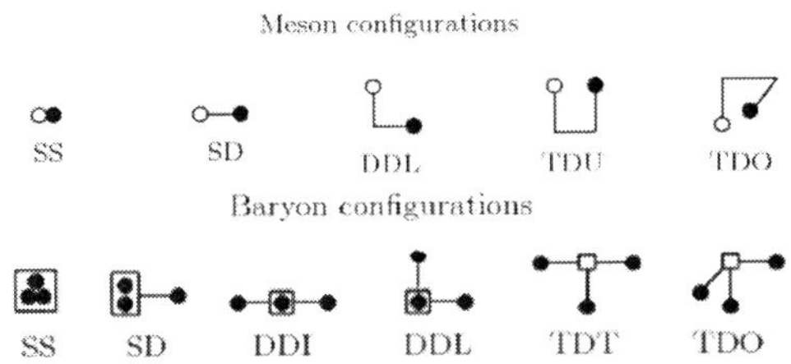

Figure 1. Spatial configurations of the meson and baryon operators in terms of displaced quark fields that we use. SS stands for single site, SD for singly displaced, DDL for doubly displaced in an L configuration, TDO for triply displaced orthogonal, and so on.

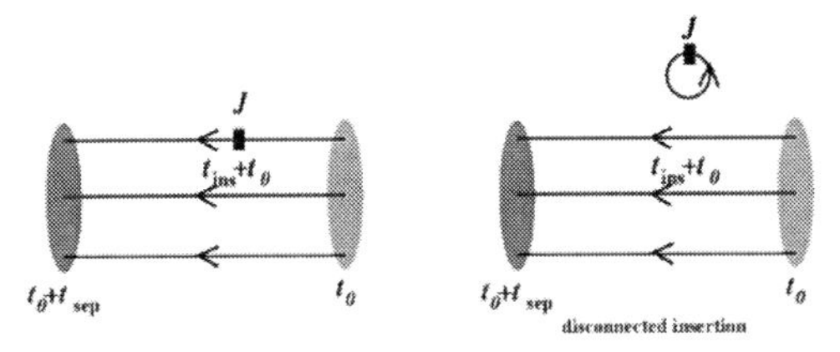

Figure 2. Diagrammatic representations of contributions to 3-point functions. (Left) The external current J occurs on one of the three quark lines connecting the baryon source (pink) to the baryon sink (blue). (Right) So-called disconnected contributions in which the insertion of the current J does not connect to the three quark lines between the baryon source and sink, but is involved in a separate sea quark loop.

where the LapH-smeared quark field is

$$\widetilde{\psi}_{a\alpha}(x) = \mathcal{S}_{ab}(x,y)\,\psi_{b\alpha}(y), \qquad \mathcal{S} = \Theta\left(\sigma_s^2 + \widetilde{\Delta}\right), \quad (5)$$

and each displacement $D^{(j)}$ is a product of smeared links:

$$D^{(j)}(x,x') = \widetilde{U}_{j_1}(x)\,\widetilde{U}_{j_2}(x{+}d_1)\ldots\widetilde{U}_{j_p}(x{+}d_{p-1})\delta_{x',\,x+d_p}. \quad (6)$$

Above, the a, b are color indices, α is a Dirac spin index, σ_s^2 is the smearing cutoff, and $\widetilde{\Delta}$ is the three-dimensional gauge-covariant Laplacian. The gauge-covariant displacements utilize stout link[2] variables $\widetilde{U}_k(x)$. To a good approximation, the LapH smearing operator is $\mathcal{S} = V_s V_s^\dagger$ where the columns of matrix V_s are the eigenvectors of $\widetilde{\Delta}$ on each time slice.

The quark displacements build up the orbital and radial structures of the mesons and baryons. The spatial configurations we use are shown in Fig. 1. So-called elemental meson and baryon operators are then given by

$$\overline{\Phi}^{AB}_{\alpha\beta}(\boldsymbol{p},t) = \textstyle\sum_{\boldsymbol{x}} e^{i\boldsymbol{p}\cdot(\boldsymbol{x}+\frac{1}{2}(\boldsymbol{d}_\alpha+\boldsymbol{d}_\beta))}\delta_{ab}\,\overline{q}^B_{b\beta}(\boldsymbol{x},t)\,q^A_{a\alpha}(\boldsymbol{x},t), \quad (7)$$

$$\overline{\Phi}^{ABC}_{\alpha\beta\gamma}(\boldsymbol{p},t) = \textstyle\sum_{\boldsymbol{x}} e^{i\boldsymbol{p}\cdot\boldsymbol{x}}\varepsilon_{abc}\,\overline{q}^C_{c\gamma}(\boldsymbol{x},t)\,\overline{q}^B_{b\beta}(\boldsymbol{x},t)\,\overline{q}^A_{a\alpha}(\boldsymbol{x},t). \quad (8)$$

Group-theory projections onto the irreps of the lattice symmetry group

$$\overline{M}_l(t) = c^{(l)*}_{\alpha\beta}\,\overline{\Phi}^{AB}_{\alpha\beta}(t), \qquad \overline{B}_l(t) = c^{(l)*}_{\alpha\beta\gamma}\,\overline{\Phi}^{ABC}_{\alpha\beta\gamma}(t), \quad (9)$$

then produce meson $\overline{M}_l(t)$ and baryon $\overline{B}_l(t)$ operators which create states of definite momentum $\boldsymbol{p}$ in the irreps of the little group of $\boldsymbol{p}$.

The low-lying QCD mass spectrum has been successfully determined (see, for example, Ref. [3]), with a level of precision such that isospin breaking is now relevant. The next challenge now is to evaluate scattering amplitudes and extract resonance information.

For matrix element calculations, the standard method requires evaluating 3-point functions, such as those shown in Fig. 2. A major issue is ensuring the removal of excited-state contamination by taking $t_{\rm sep}$, $t_{\rm ins}$, and $t_{\rm sep} - t_{\rm ins}$ large. In practice, this is difficult to achieve due to the signal-to-noise ratio which decreases with time separation. Computing so-called disconnected contributions, as shown on the right in Fig. 2, is much more difficult than the connected contributions, especially as the quark mass is taken to its physical value. Many previous studies simply neglected such contributions. However, a variety of new techniques are available now and such disconnected contributions are being reliably estimated with unprecedented precision. Our group uses the stochastic LapH method, whereas the results shown in Secs. 3-7 obtained by other groups use different methods, as mentioned in each case. Another issue with matrix element calculations is that the current operators require renormalization for comparison to $\overline{\rm MS}$. Such renormalizations can be evaluated using non-perturbative and/or perturbative methods.

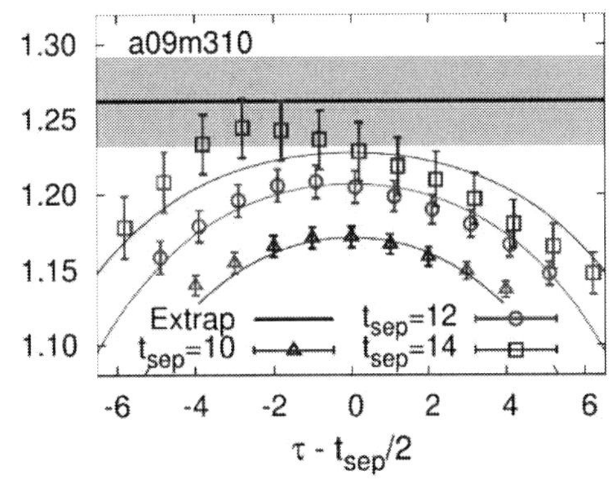

Figure 3. g_A^{u-d} extractions for a variety of $t_{\rm sep}$, the time separation between the source and sink nucleon, and $\tau = t_{\rm ins}$, the time where the current operator is inserted. Removal of excited-state contamination requires extrapolating to large $t_{\rm sep}$ and looking for insensitivity in τ around $t_{\rm sep}/2$. Results are from Ref. [8].

Difficulties in removing excited-state contamination in calculations related to g_A are illustrated in Fig. 3. The gray band shows the results of an extrapolation of the $t_{\rm sep} = 10, 12, 14$ results. It is also necessary to ensure insensitivity of the extrapolated results to the insertion time, which must be far from the source and sink.

3 Proton mass decomposition

A recent determination of the mass decomposition of the proton has been presented in Ref. [4]. The rest mass M of the proton is given by[5]

$$M = -\langle T_{44}\rangle = \langle H_m\rangle + \langle H_E\rangle(\mu) + \langle H_g\rangle(\mu) + \tfrac{1}{4}\langle H_a\rangle, \quad (10)$$

where $\langle T_{\mu\nu}\rangle$ is the expectation value of the energy momentum tensor in a hadron, and

$$H_m = \sum_{u,d,s\cdots}\int d^3x\, m\bar{\psi}\psi,$$
$$H_E = \sum_{u,d,s...}\int d^3x\, \bar{\psi}(\vec{D}\cdot\vec{\gamma})\psi,$$
$$H_g = \int d^3x\, \frac{1}{2}(B^2 - E^2),$$
$$H_a = \sum_{u,d,s\cdots}\int d^3x\, \gamma_m m\bar{\psi}\psi - \int d^3x\, \frac{\beta(g)}{g}(E^2+B^2),$$

where H_m is the quark condensate, H_E is the quark energy, H_g is the gluon field energy, and H_a is the anomaly term. Note that $\langle H_m\rangle$, $\langle H_a\rangle$, $\langle H_E + H_g\rangle$ are scale and scheme independent. This study obtained the quark and gluon energies from the renormalized quark and gluon momentum fractions

$$\langle H_g\rangle = \frac{3}{4}M\langle x\rangle_g, \qquad \langle H_E\rangle = \frac{3}{4}M\langle x\rangle_q - \frac{3}{4}\langle H_m\rangle,$$

and the anomaly term from $\langle H_a\rangle = M - \langle H_m\rangle$. The mass M was determined from the two-point correlator, as usual, and a previous determination of $\langle H_m\rangle$ by these authors was used. The momentum fractions were evaluated using

$$\langle x\rangle_{q,g} \equiv -\frac{\langle N|\frac{4}{3}\overline{T}_{44}^{q,g}|N\rangle}{M\langle N|N\rangle},$$
$$\overline{T}_{44}^q = \int d^3x\bar{\psi}(x)\frac{1}{2}(\gamma_4\overleftrightarrow{D}_4 - \frac{1}{4}\sum_{i=0,1,2,3}\gamma_i\overleftrightarrow{D}_i)\psi(x),$$
$$\overline{T}_{44}^g = \int d^3x\frac{1}{2}(E(x)^2 - B(x)^2),$$

taking renormalization into account:

$$\langle x\rangle^R_{u,d,s} = Z_{QQ}^{\overline{\text{MS}}}(\mu)\langle x\rangle_{u,d,s} + \delta Z_{QQ}^{\overline{\text{MS}}}(\mu)\sum_{q=u,d,s}\langle x\rangle_q + Z_{QG}^{\overline{\text{MS}}}(\mu)\langle x\rangle_g,$$
$$\langle x\rangle^R_g = Z_{GQ}^{\overline{\text{MS}}}(\mu)\sum_{q=u,d,s}\langle x\rangle_q + Z_{GG}^{\overline{\text{MS}}}\langle x\rangle_g.$$

Results were obtained on four ensembles using an $N_f = 2+1$ domain-wall fermion action with overlap valence propagators. The difficult disconnected insertions employed cluster-decomposition error reduction with all time slices looped over. Extrapolations to remove systematic errors were done with a global fit including finite-volume and finite-spacing corrections, as well as known chiral behavior. The impressive results are shown in Fig. 4 and can be summarized as quark energy 32(4)(4)%, glue energy 36(5)(4)%, quark condensate 9(2)(1)%, and trace anomaly 23(1)(1)%.

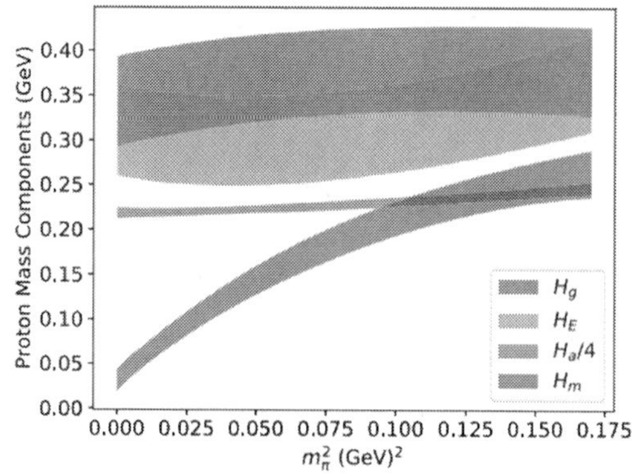

Figure 4. Proton mass decomposition from Ref. [4] against the pion mass squared. H_g is the gluon field energy, H_E is the quark energy, H_a is the anomaly term, and H_m is the quark condensate.

4 Nucleon spin decomposition

A recent determination of the spin and momentum fraction decomposition of the nucleon has been presented in Ref. [6]. From the Ji sum rule[7], the nucleon spin is given by

$$J_N = \sum_{q=u,d,s,c\cdots}\left(\tfrac{1}{2}\Delta\Sigma_q + L_q\right) + J_g = \textstyle\sum_q J_q + J_g, \quad (11)$$

where J_g is the gluon total angular momentum, L_q is the quark orbital angular momentum, and $\frac{1}{2}\Delta\Sigma_q$ is the contribution from the intrinsic quark spin. These quantities were obtained from the following nucleon matrix elements, with $Q=p'-p$ and $P=\frac{1}{2}(p'+p)$:

$$\langle N(p,s')|\bar{q}\gamma_\mu\gamma_5 q|N(p,s)\rangle = \bar{u}_N(p,s')\left[g_A^q\gamma^\mu\gamma_5\right]u_N(p,s),$$
$$\langle N(p',s')|\bar{q}\gamma^{\{\mu}\overleftrightarrow{D}^{\nu\}}q|N(p,s)\rangle = \bar{u}_N(p',s')\Lambda^q_{\mu\nu}(Q^2)u_N(p,s),$$
$$\Lambda^{\mu\nu}_{q(g)}(Q^2) = A_{20}^{q(g)}(Q^2)\gamma^{\{\mu}P^{\nu\}} + B_{20}^{q(g)}(Q^2)\frac{\sigma^{\{\mu\alpha}q_\alpha P^{\nu\}}}{2m} + C_{20}^{q(g)}(Q^2)\frac{1}{m}Q^{\{\mu}Q^{\nu\}}.$$

The quark(gluon) total angular momentum and quark momentum fraction and spin were extracted using

$$J_{q(g)} = \tfrac{1}{2}[A_{20}^{q(g)}(0) + B_{20}^{q(g)}(0)],$$
$$\langle x\rangle_q = A_{20}^q(0), \qquad \Delta\Sigma_q = g_A^q,$$

and the gluon momentum fraction was obtained from $O^g_{\mu\nu} = 2\text{Tr}[G_{\mu\sigma}G_{\nu\sigma}]$ with $\overline{O}^g \equiv O^g_{44} - \frac{1}{3}O^g_{jj}$, and

$$\langle N(p,s')|\overline{O}^g|N(p,s)\rangle = \left(-4E_N^2 - \frac{2}{3}\vec{p}^2\right)\langle x\rangle_g.$$

One ensemble at the physical point on a $48^3\times 96$ lattice using a twisted mass clover-improved action with lattice spacing $a = 0.0939(3)$ fm, set from the nucleon mass, was used. The u, d disconnected diagrams were estimated

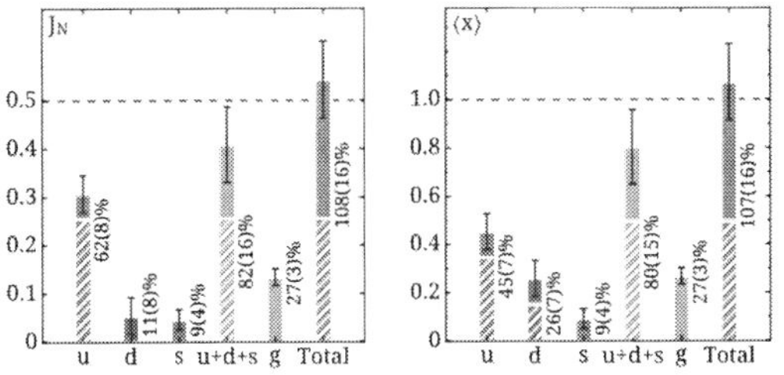

Figure 5. Nucleon spin (left) and momentum fraction (right) decompositions from Ref. [6] in terms of the contributions from each quark flavor and from gluons. Striped segments refer to valence quark contributions, while solid segments show sea quark and gluon contributions.

by exact deflation plus the one-end-trick, while the s disconnected diagrams were evaluated by a truncated solver method. Renormalization factors were determined non-perturbatively. Their final results are shown in Fig. 5.

5 Nucleon axial coupling

A remarkable percent level determination of the nucleon axial coupling g_A in lattice QCD recently appeared[8]. The use of a Feynman-Hellman method enabled a significant reduction in statistical errors. Their value was

$$g_A = 1.2711(103)^s(39)^\chi(15)^a(19)^V(04)^I(55)^M, \qquad (12)$$

where the errors, in order, are from the statistical estimation, chiral extrapolation, lattice spacing extrapolation, volume extrapolations, isospin corrections, and model selection. Extrapolations were done using several models and the final estimate is a model average. Sixteen ensembles from the MILC collaboration generated using a HISQ action with lattice spacings $a \sim 0.15$ fm, $a \sim 0.12$ fm, and $a \sim 0.09$ fm and pion masses ranging from near physical to 400 MeV were used. Domain-wall valence propagators were used, making this a mixed-action computation. Their main result is illustrated in Fig. 6, and a comparison of their estimate with other recent determinations is shown in Fig. 7.

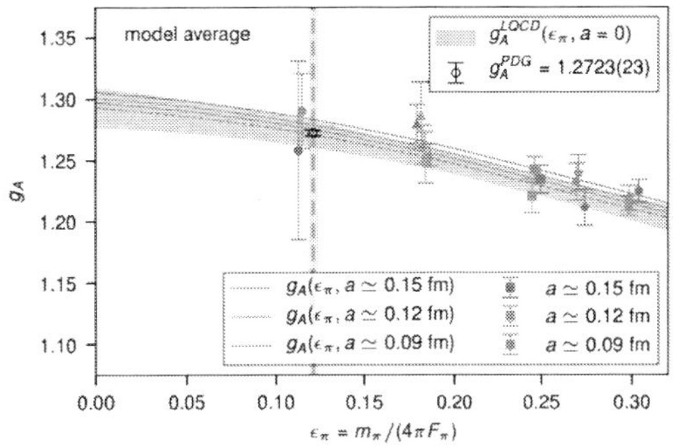

Figure 6. Extrapolation of a g_A determination in Ref. [8] to the physical point. Solid red, green, and blue curves are central values of g_A as a function of $\epsilon_\pi = m_\pi/(4\pi F_\pi)$ at fixed lattice spacing and infinite volume, and the black circle is the experimental value. Magenta band is the central 68% confidence band of the continuum and infinite volume extrapolations.

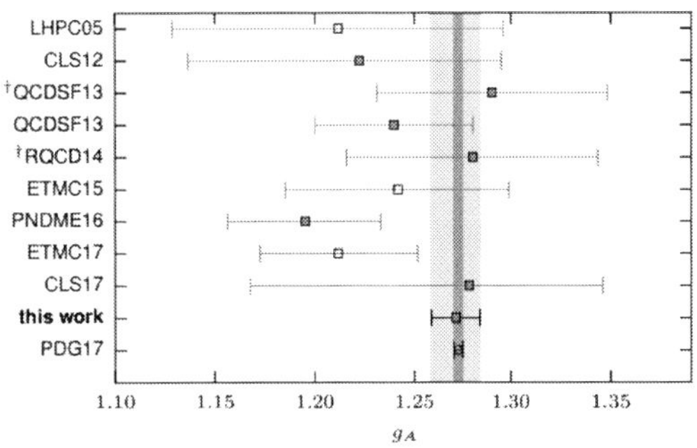

Figure 7. Comparison of the estimate of g_A from Ref. [8] with other recent determinations. See Ref. [8] for references to the other estimates. Results with closed symbols have included an extrapolation to the continuum limit, while results with open symbols have only included extrapolation to the physical pion mass. To guide the eye, the vertical magenta band is the full uncertainty from Ref. [8], while the vertical gray band is the experimental uncertainty.

6 Proton/neutron electromagnetic form factors

A recent study of the proton and neutron electromagnetic form factors in lattice QCD was presented in Ref. [9]. One ensemble using an $N_f = 2+1+1$ twisted mass action with $m_\pi = 130$ MeV, and two ensembles using an $N_f = 2$ twisted mass action with $m_\pi = 130$ MeV and two volumes $Lm_\pi \sim 3$ and $Lm_\pi \sim 4$ were utilized. An unprecedented precision of the disconnected diagram contributions was achieved using hierarchical probing, low mode deflation, and large numbers of smeared point sources to reduce gauge noise. The results demonstrated that the disconnected diagrams have nonnegligible effects. The study included a thorough investigation of excited-state contamination, but further study of finite-volume effects at low Q^2 is needed. Their $N_f = 2+1+1$ results are compared to experiment in Fig. 8.

7 Light-cone parton distribution function

A first determination of the unpolarized helicity parton distribution function (PDF) at the physical point with nonperturbative renormalization and large momenta treated was presented in Ref. [10]. Extracting PDFs from their moments is impractical, so these authors used a clever method proposed by Ji[11] with subsequent refinements. First, they computed spatial correlations between boosted nucleon states, then carried out Fourier transforms to produce quantities known as quasi-PDFs, then finally took the

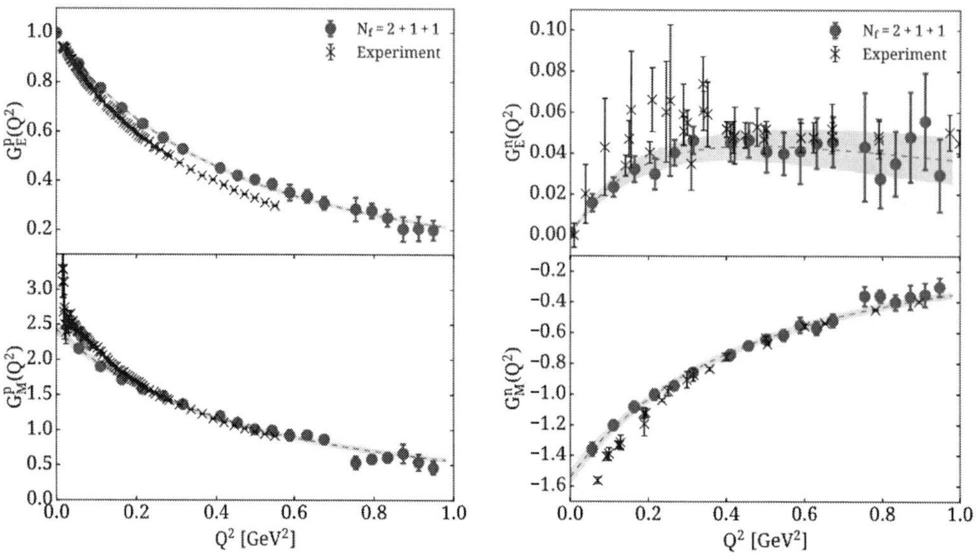

Figure 8. Electric (top) and magnetic (bottom) form factors of the proton (left) and neutron (right) from Ref. [9], compared to experiment.

infinite-momentum limit via a refined matching procedure. So-called target mass corrections were employed, as well as a renormalization scheme for the Wilson line operators.

Results were obtained on one ensemble using a $48^3 \times 96$ lattice with a twisted mass $N_f = 2$ action with lattice spacing $a = 0.0938(3)(2)$ fm and $m_\pi L = 2.98(1)$ at the physical point. Both unpolarized and polarized PDFs for three momenta were compared to some phenomenological curves, shown in Fig. 9. Improvement of these estimates of the PDFs is ongoing.

8 Excited baryon states

In finite volume, the stationary-state energies are discrete due to momentum quantization, so temporal correlation matrices have the spectral representation

$$C_{ij}(t) = \sum_n Z_i^{(n)} Z_j^{(n)*}\, e^{-E_n t}, \qquad Z_j^{(n)} = \langle 0|\, O_j\, |n\rangle, \quad (13)$$

neglecting wrap-around corrections. It is not practical to do fits using the above form, so diagonalization methods with single- and two-exponential fits are employed to extract some number of low-lying states.

To access excited baryon (and meson) states reliably, it is necessary to extract the energies of the multi-hadron states that are lower-lying than the excited baryons of interest. This requires evaluating correlators involving multi-hadron operators. Good multi-hadron operators involve combining good individual hadron operators which separately have well-defined momenta. With such operators, the usual point-to-all trick, which exploits translational invariance, cannot be utilized to drastically reduce the number of quark propagator sources that are needed, so an overly large number of Dirac matrix inversions are required. One solution to this is the stochastic LapH method[1] which estimates the entire matrix inverse, or a large portion of it, using an additional Monte Carlo calculation, exploiting the Laplacian-Heaviside quark smearing and various noise dilution projectors for variance reduction.

Excited Λ baryons were studied somewhat recently in Ref. [12]. A small sample of the results obtained is shown in Fig. 10.

9 Scattering amplitudes from lattice QCD

Finite-volume energies E in lattice QCD are related to the infinite-volume S matrix[13]. Utilizing such relations to obtain scattering amplitudes is often termed the Lüscher method. Introduce the K-matrix as usual,

$$S = (1 + iK)(1 - iK)^{-1} = (1 - iK)^{-1}(1 + iK). \quad (14)$$

In the $JLSa$ basis, for total angular momentum J, orbital angular momentum L, intrinsic spin S, and species channel a, introduce

$$K^{-1}_{L'S'a';\, LSa}(E) = q_{\mathrm{cm},a'}^{-L'-\frac{1}{2}}\; \widetilde{K}^{-1}_{L'S'a';\, LSa}(E_{\mathrm{cm}})\; q_{\mathrm{cm},a}^{-L-\frac{1}{2}}, \quad (15)$$

then below 3-particle thresholds, there is a quantization condition

$$\det(1 - B^{(P)}\widetilde{K}) = \det(1 - \widetilde{K}B^{(P)}) = 0 \quad (16)$$

or

$$\det(\widetilde{K}^{-1} - B^{(P)}) = 0 \quad (17)$$

where the Hermitian "box matrix" $B^{(P)}$ encodes the effects of the cubic finite-volume. Details about the box matrix may be found in Ref. [14].

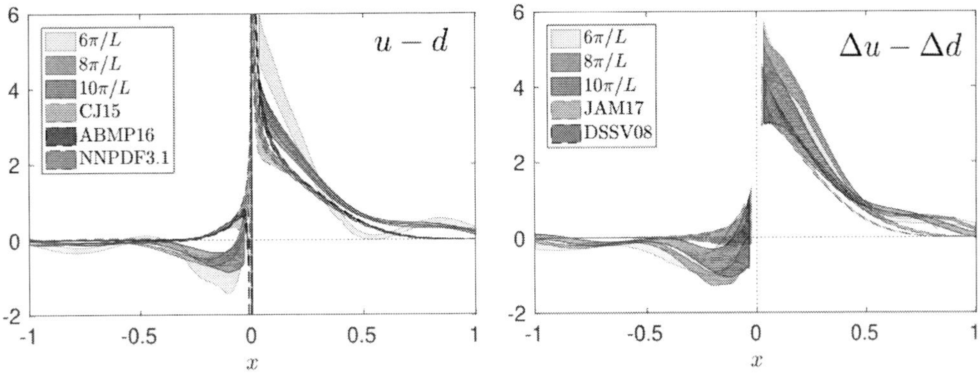

Figure 9. Unpolarized (left) and polarized (right) parton distribution functions for three momenta compared to some phenomenological curves. Results are from Ref. [10].

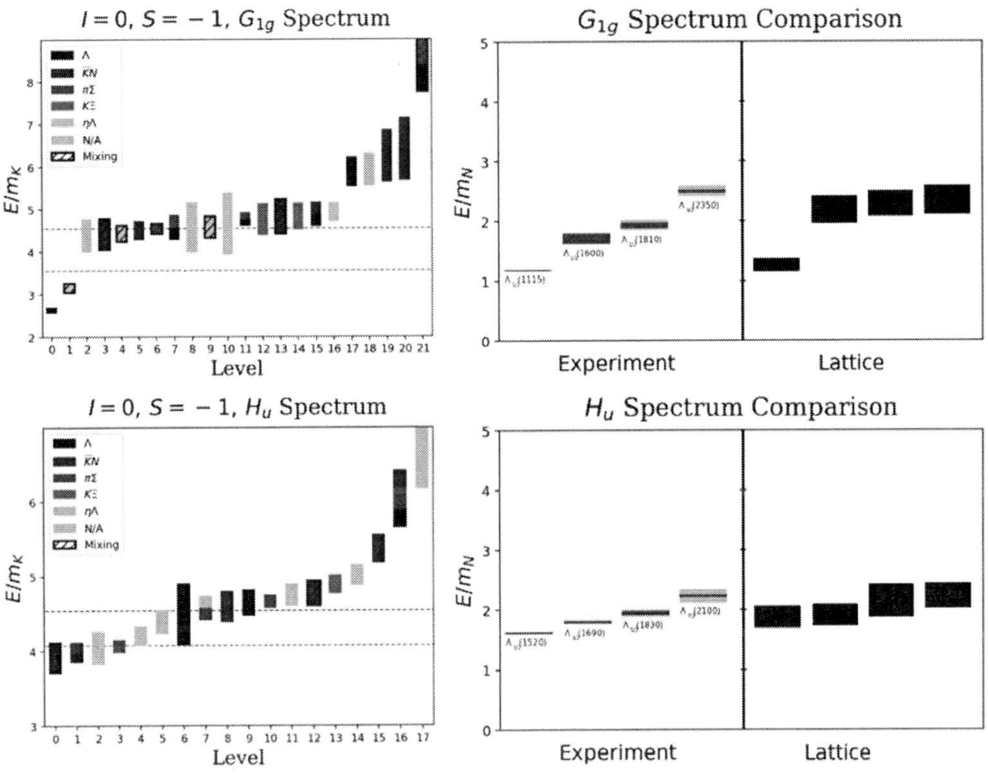

Figure 10. (Left) Stationary state energies in the $I = 0$, $S = -1$ baryonic flavor sector for a $32^3 \times 256$ anisotropic lattice with $m_\pi \sim 240$ MeV. The G_{1g} channel (even parity containing the spin-$\frac{1}{2}$ states) and the H_u channel (odd parity including the spin-$\frac{3}{2}$ states) are shown as ratios over the kaon mass m_K. Different colors indicate the different properties of the states as deduced by the fitted overlap factors $Z_j^{(n)}$. (Right) The single-baryon dominated levels are compared to experiment as ratios over the nucleon mass m_N. Results are from Ref. [12].

http://doi.org/10.1051/epjconf/202024102004

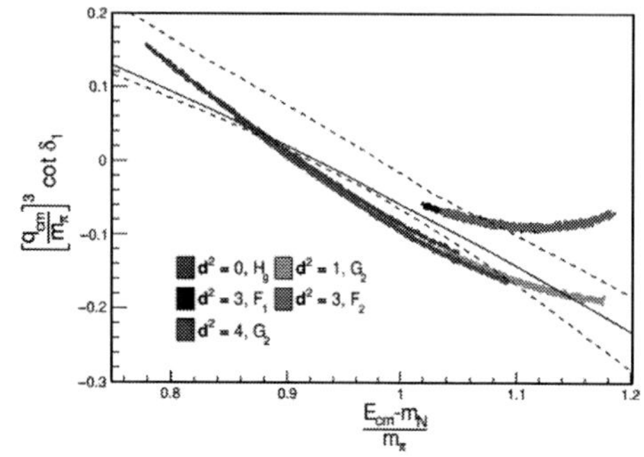

Figure 11. *P*-wave $N\pi$ scattering phase shift against center-of-mass energy showing the Δ resonance where the $\cot\delta_1$ crosses zero. Results are from Ref. [15].

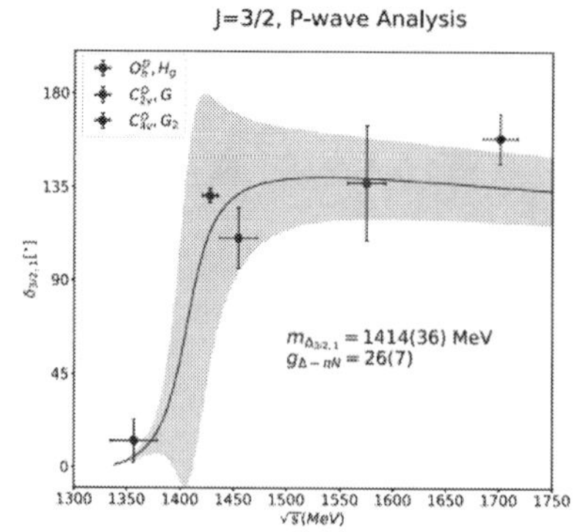

Figure 12. *P*-wave $N\pi$ scattering phase shift against the Mandelstam variable $\sqrt{s}$ showing the Δ resonance where the phase shift rises dramatically. Results are from Ref. [16].

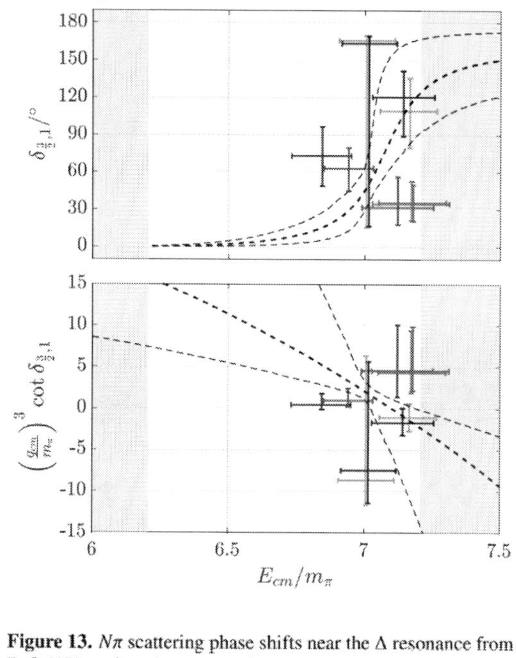

Figure 13. $N\pi$ scattering phase shifts near the Δ resonance from Ref. [17] against the center-of-mass energy as a ratio over the pion mass. The top plot shows the phase shift itself, and the bottom plot shows the cotangent of the phase shift multiplied by threshold factors.

The quantization condition relates a single energy E to the entire K-matrix, so one cannot solve for the K-matrix, except in a single channel for a single partial wave. Thus, we must approximate the K-matrix with functions depending on a handful of fit parameters. We then obtain estimates of the fit parameters using many different energies. The quantization condition involves an infinite-dimensional determinant. We make the condition practical by first transforming to a block-diagonal basis, then truncating in orbital angular momentum. Meson-meson scattering studies are becoming mature, whereas only a few meson-baryon scattering investigations have been attempted. Baryon-baryon scattering studies are currently gestating.

The decay $\Delta(1232) \to N\pi$ was recently studied in Ref. [15]. Only the $L = 1$ partial wave was included. Results were obtained on a large $48^3 \times 128$ isotropic lattice with $m_\pi \approx 280$ MeV and $a \sim 0.076$ fm. Their determination of the scattering phase shift is shown in Fig. 11. A Breit-Wigner fit gave $m_\Delta/m_\pi = 4.738(47)$ and $g_{\Delta N\pi} = 19.0(4.7)$. Note that experiment yields $g_{\Delta N\pi} \sim 16.9$.

Preliminary results for the Δ resonance in another recent study using a lattice with length $L = 2.8$ fm, spacing $a = 0.116$ fm and pion mass $m_\pi = 260$ MeV has appeared in Ref. [16]. No slice-to-slice propagators were used, three total momenta were studied, the ground and excited states were extracted, and their analysis included only a single partial wave. Their determination of the scattering phase shift is shown in Fig. 12.

Preliminary results from our most recent study[17] of the Δ resonance are shown in Fig. 13. The lattice length, lattice spacing, and pion mass are $L = 4.2$ fm, $a = 0.065$ fm, $m_\pi = 200$ MeV. Five total momenta have been used, and both ground and excited states were extracted. We expect to reduce statistical errors by a factor of 6 in our final results. The finite volume spectrum which produced these results is shown in Fig. 14. Fits included irreps which mix the S and P waves and relied on the automated determination of B-matrix elements from Ref. [14].

Results for $\Lambda(1405) \to \Sigma\pi$ will be presented in the near future in Ref. [18]. Very preliminary results on a lattice with $L = 3.2$ fm, $a = 0.065$ fm, $m_\pi = 280$ MeV are shown in Fig. 15.

10 Three last items

I finish this talk by reporting on three last items of interest.

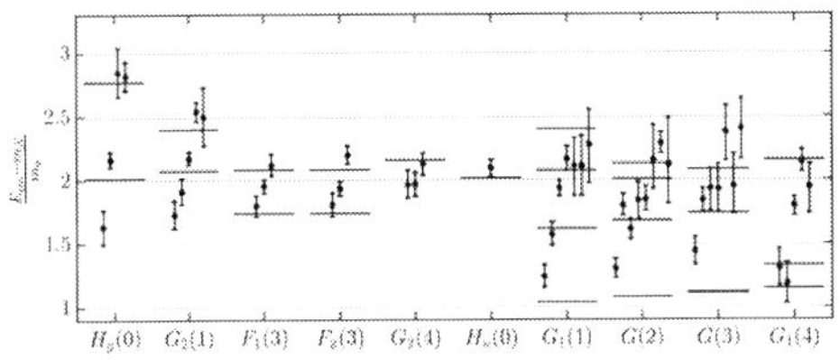

Figure 14. The energy spectrum of finite-volume states used in our most recent study in Ref. [17] of the Δ resonance. Differences of the center-of-mass energies from the nucleon mass as a ratio over the pion mass are shown for the different irreps and total momenta used. Horizontal lines show the non-interacting energies.

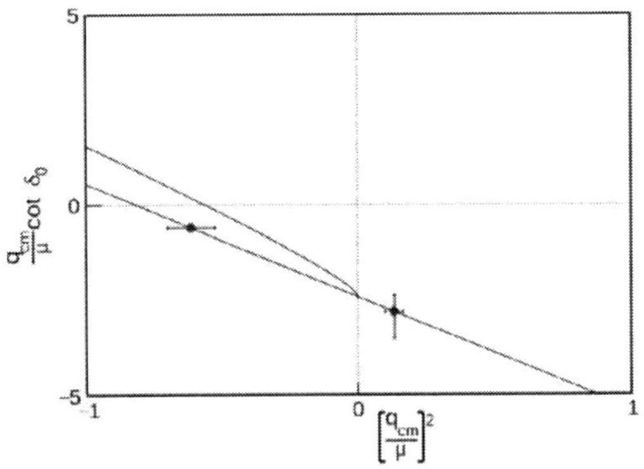

Figure 15. Very preliminary results for the $\Sigma\pi$ scattering phase shift near the $\Lambda(1405)$ resonance against the center-of-mass momentum as ratio with a reference energy μ. Result are from Ref. [18].

A recent determination of the time-like pion form factor appeared in Ref. [19]. It was extracted using

$$\begin{aligned}|F_\pi(E_{\rm cm})|^2 &= g_\Lambda(\gamma)\left(q_{\rm cm}\frac{\partial\delta_1}{\partial q_{\rm cm}} + u\frac{\partial\phi_1^{(d,\Lambda)}}{\partial u}\right)\\ &\quad\times\frac{3\pi E_{\rm cm}^2}{2q_{\rm cm}^5 L^3}|\langle 0|V^{(d,\Lambda)}|d\Lambda n\rangle|^2,\end{aligned}$$

where

$$\gamma = \frac{E}{E_{\rm cm}},\quad u = \frac{Lq_{\rm cm}}{2\pi},\quad g_\Lambda(\gamma) = \begin{cases}\gamma^{-1}, & \Lambda = A_1^+\\ \gamma, & \text{otherwise}\end{cases}$$

δ_1 is the physical phase shift, and the pseudophase $\phi_1^{(d,\Lambda)}$ is obtained from $B_{11}^{(d,\Lambda)} = (q_{\rm cm}/m_\pi)^3\cot\phi_1^{(d,\Lambda)}$. The matrix element

$$V^{(d,\Lambda)} = \sum_\mu b_\mu^{(d,\Lambda)} V_{R\mu},\qquad \sum_\mu b_\mu^{(d,\Lambda)*} b_\mu^{(d,\Lambda)} = 1,$$

with

$$\begin{aligned}V_{R\mu} &= Z_V(1 + ab_V m_l + a\bar{b}_V {\rm Tr}M_q)\, V_{I\mu},\\ V_{I\mu} &= V_\mu + ac_V\tilde{\partial}_\nu T_{\mu\nu},\end{aligned}$$

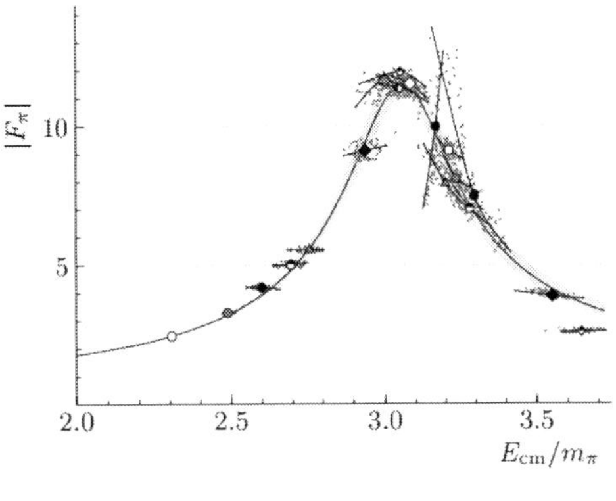

Figure 16. Timelike pion form factor $|F_\pi|$ again center-of-mass energy over pion mass $E_{\rm cm}/m_\pi$ using the CLS J303 ensemble on a $64^3 \times 192$ lattice with spacing $a = 0.050$ fm and $m_\pi = 260$ MeV. The curve is a fit with a thrice-subtracted dispersion relation. Results are from Ref. [19]. This shows the feasibility of future calculations of baryon form factors.

and

$$V_\mu^a = \tfrac{1}{2}\overline{\psi}\gamma_\mu\tau^a\psi,\qquad \tilde{\partial}_\nu T_{\mu\nu}^a = \tfrac{1}{2}i\tilde{\partial}_\nu\overline{\psi}\sigma_{\mu\nu}\tau^a\psi$$

was computed and used to determine the form factor. Results for CLS J303 ensemble on a $64^3 \times 192$ lattice with spacing $a = 0.050$ fm and $m_\pi = 260$ MeV are shown in Fig. 16. The success of this calculation paves the way for baryon form factor determinations. A similar method is now being used for Δ transition form factors needed by the Deep Underground Neutrino Experiment (DUNE).

The HAL QCD collaboration has extensively studied nucleon-nucleon interactions. Their method extracts observables from non-local kernels associated with tempo-spatial correlation functions. However, a controversy arose when disagreements of their results with so-called direct methods were found. A recent study[20] suggests that this discrepany arises from the misidentification of energies in the direct method. The $\Xi\Xi(^1S_0)$ temporal correlation function was studied in detail, with pertinent results shown in Fig. 17. In the right plot in this figure, the key point is the disagreement between the dashed line (the known result) and the apparent plateau of the effective mass points in blue. The shaded blue band is the expected behavior of this function for larger time separations. Hopefully this resolution will accelerate progress in baryon-baryon scattering.

A recent report on an ongoing study of the H-dibaryon was presented in Ref. [21]. These results were obtained at the $SU(3)$ flavor symmetric point using well-designed baryon-baryon operators since a previous study showed that a hexaquark operator would not saturate the signal. They found several finite-volume energies below the $\Lambda\Lambda$ threshold. Some of their results are shown in Fig. 18. A scattering amplitude analysis is needed to determine if the system is bound or a resonance. Due to the small lattices used and the very heavy pion, this must be viewed

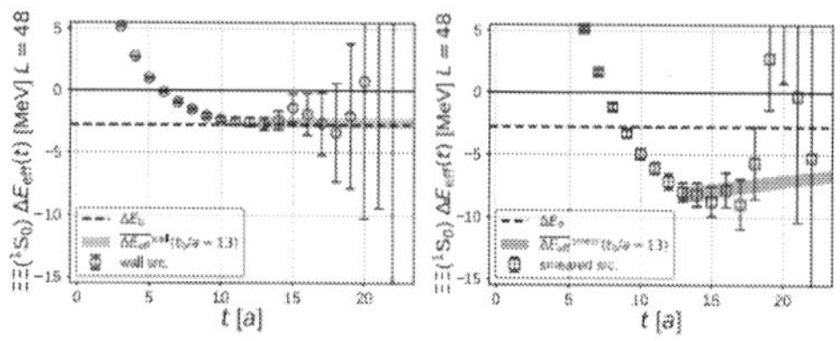

Figure 17. Effective masses associated with $\Xi\Xi(^1S_0)$ correlators using (left) a wall source and (right) a local smeared source. The dashed line indicates the asymptotic behavior using the wall source, and this result also agrees with the HAL QCD method. The key point, shown in the right plot, is the disagreement between the dashed line and the apparent plateau of the effective mass points in blue obtained using the local smeared source. The shaded blue band in the right plot is the expected large-time behavior. These results are from Ref. [20].

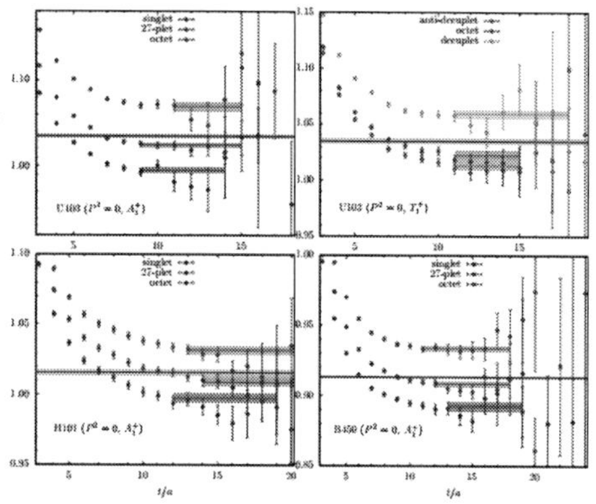

Figure 18. Effective masses for spin-0 and spin-1 dibaryon operators of different flavor irreps using 3 ensembles from Ref. [21]. Blue points show flavor singlet results, the red points show the flavor 27-plet, and the flavor octet is shown in green. Results on the CLS U103 ensemble in the A_1^+ spin-zero irrep are shown in the upper left. The upper right plot shows results on the U103 ensemble for the T_1^+ spin-one irrep. Results in the A_1^+ spin-zero irrep on the H101 and B450 ensembles are shown in the lower left and lower right plots, respectively. Horizontal black lines show the two-octet baryon thresholds. The $SU(3)$ flavor symmetric point is used, and the lattice volumes are small.

as a warm up exercise. Future work on larger lattices and lighter pions will involve the stochastic LapH method.

11 Conclusion

Highlights from recent computations in lattice QCD involving baryons were presented in this talk. How baryons can be studied in lattice QCD was first discussed, followed by results on the proton mass and spin decompositions, nucleon axial coupling, the proton and neutron electromagnetic form factors, and light-cone parton distribution functions. Recent works on meson-baryon scattering using the so-called Lüscher method were shown. Key points emphasized were that much better precision with disconnected diagrams is being achieved, incorporating multi-hadron operators is now feasible, and more and more studies are being done with physical quark masses.

12 Acknowledgements

Support from the U.S. National Science Foundation under award PHY-1613449 is gratefully acknowledged.

References

[1] C. Morningstar, J. Bulava, J. Foley, K.J. Juge, D. Lenkner, M. Peardon, C.H. Wong, Phys. Rev. D**83**, 114505 (2011).
[2] C. Morningstar and M. Peardon, Phys. Rev. D**69**, 054501 (2004).
[3] A. Kronfeld, Ann. Rev. Nucl.Part. Sci **62**, 265 (2012).
[4] Y. Yang, J. Liang, Y. Bi, Y. Chen, T. Draper, K. F. Liu, Z. Liu, Phys. Rev. Lett. **121**, 212001 (2018).
[5] X. Ji, Phys. Rev. Lett. **74**, 1071 (1995).
[6] C. Alexandrou, M. Constantinou, K. Hadjiyiannakou, K. Jansen, C. Kallidonis, G. Koutsou, A. V. Avilés-Casco, C. Wiese, Phys. Rev. Lett. **119**, 142002 (2017).
[7] X. Ji, Phys. Rev. Lett. **78**, 610 (1997).
[8] C.Chang *et al.*, Nature **558**, 91 (2018); arXiv:1805.12130.
[9] C.Alexandrou, S.Bacchio, M.Constantinou, J.Finkenrath, K.Hadjiyiannakou, K.Jansen, G.Koutsou, A.V.Aviles Casco, Phys. Rev. D **100**, 014509 (2019).
[10] C. Alexandrou, K. Cichy, M. Constantinou, K. Jansen, A. Scapellato, F. Steffens, Phys. Rev. Lett. **121**, 112001 (2018).
[11] X.Ji, Phys. Rev. Lett. **110**, 262002 (2013).
[12] J. Fallica, Ph.D. thesis, Carnegie Mellon University (2017).
[13] M. Lüscher, Nucl. Phys. B**354**, 531 (1991).
[14] C. Morningstar, J. Bulava, B. Singha, R. Brett, J. Fallica, A. Hanlon, B. Hörz, Nucl. Phys. B**924**, 477 (2017).
[15] C.W. Andersen, J. Bulava, B. Hörz, C. Morningstar, Phys. Rev. D **97**, 014506 (2018).
[16] S. Paul, G. Silvi, C. Alexandrou, G. Koutsou, S. Krieg, L. Leskovec, S. Meinel, J. Negele, M. Petschlies, A. Pochinsky, G. Rendon, S. Syritsyn, PoS(LATTICE2018)089.
[17] C. Andersen, B. Hörz, J. Bulava, C. Morningstar, in prep.
[18] B. Hörz, C. Andersen, J. Bulava, M. Hansen, D. Möhler, C. Morningstar, H. Wittig, in prep.
[19] C. Andersen, J. Bulava, B. Hörz, C. Morningstar, Nucl. Phys. B**939**, 145 (2019).
[20] T. Iritani, S. Aoki, T. Doi, T. Hatsuda, Y. Ikeda, T. Inoue, N. Ishii, H. Nemura, K. Sasaki, JHEP**03**, 007 (2019).
[21] A. Hanlon, A. Francis, J. Green, P. Junnarkar, H. Wittig, PoS LATTICE2018 (2018) 081.

Three-particle system in a finite volume: formalism, quantization condition, spectrum and energy shift

Jin-Yi Pang[1,2,*]

[1]College of Science, University of Shanghai for Science and Technology, Jungong Rd. 334, 200093 Shanghai, China
[2]Helmholtz-Institut für Strahlen- und Kernphysik and Bethe Center for Theoretical Physics, Universität Bonn, Nußallee 14-16, 53115 Bonn, Germany

Abstract. Lattice QCD calculations provide an ab initio access to hadronic process. These calculations are usually performed in a small cubic volume with periodic boundary conditions. The infinite volume extrapolations for three-body systems are indispensable to understand many systems of high current interest. We derive the three-body quantization condition in a finite volume using an effective field theory in the particle-dimer picture. Our work shows a powerful and transparent method to read off three-body physical observables from lattice simulations. In this paper, we review the formalism, quantization condition, spectrum analysis and energy shifts calculation both for 3-body bound states and scattering states.

1 Introduction

A rapid progress is developed aiming to analyze of the lattice data in the three-particle sector [1–29] (for the recent review, see Ref. [30]). This is also the request of lattice simulation which has touched the three-body kinematics region. Among these approaches, our works [17, 18, 20, 29] has showed its transparent and powerful to the problem via the non-relativistic effective theory and the particle-dimer picture in a finite volume (a very similar approach was proposed in Refs. [14, 21]). It is demonstrated that effective field theory is very important to connect lattice simulation results with physical experiment data [18]. Three-body force, the critical point of the problem is fit to the lattice spectrum [17]. Efimov states, as the typical bound states are reproduced and interpreted in a finite volume within our formalism [20]. We also show that three-body threshold amplitude can be extracted from energy shifts of scattering states [29]. The equivalence to other approaches in the literature has been shown [18] as well.

The paper is arranged as followed. We show the formalism describing 3-body system in Sec. 2. The formalism is put into a finite volume and quantization condition is derived and projected in Sec. 3. We apply the quantization condition to analyze the 3-body bound states and scattering states in Sec. 4 and 5. Sec. 6 gives the conclusion and outlook.

2 Three-body formalism

2.1 Three-body effective theory

We have the effective Lagrangian including kinematics terms, 2-body operators and 3-body operators.

$$\mathcal{L} = \mathcal{L}_0 + \mathcal{L}_I^{(2\text{-body})} + \mathcal{L}_I^{(3\text{-body})}. \tag{1}$$

Non-relativistic kinematics for scalar particles is considered,

$$\mathcal{L}_0 = \psi^\dagger\Big(i\partial_0 + \frac{\nabla^2}{2m}\Big)\psi \tag{2}$$

and the interaction operators are arranged in order of pertinent power counting,

$$\mathcal{L}_I^{(2\text{-body})} = C_0\,\psi^\dagger\psi^\dagger\psi\psi + C_2\,(\psi^\dagger\overset{\leftrightarrow}{\nabla}{}^2\,\psi^\dagger\psi\psi + \text{h.c.}) + \cdots \tag{3}$$

and

$$\mathcal{L}_I^{(3\text{-body})} = D_0\,\psi^\dagger\psi^\dagger\psi^\dagger\psi\psi\psi + D_2\,(\psi^\dagger\psi^\dagger\,\overset{\leftrightarrow}{\nabla}{}^2\,\psi^\dagger\,\psi\psi\psi + \text{h.c.}) + \cdots. \tag{4}$$

Here we use $\overset{\leftrightarrow}{\nabla} = (\nabla - \overset{\leftarrow}{\nabla})/2$ to produce relative momentum of two particles which preserves Galilean transformation invariance.

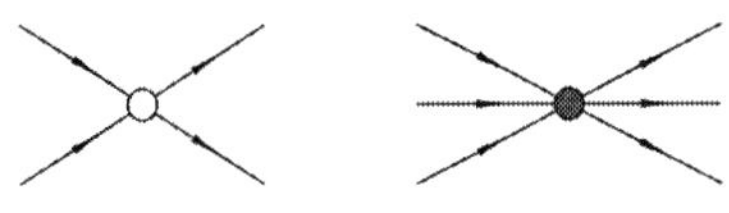

Figure 1: 2-body operator and 3-body operator

*e-mail: pang@hiskp.uni-bonn.de

2.2 Particle-dimer formalism

We introduce particle-dimer formalism to make the 3-body effective theory more compact and clear. Dimer field is an auxiliary degree of freedom which describes a 2-body subsystem. For simplicity, only the scalar dimer is considered[1]. The kinematics of dimer field T is added by,

$$\mathcal{L}_0 = \psi^\dagger\left(i\partial_0 + \frac{\nabla^2}{2m}\right)\psi + \sigma T^\dagger T. \quad (5)$$

Dimer has no internal structure at tree level. The dynamics of dimer comes from its interaction with particles.

$$\mathcal{L}_I^{(2\text{-body})} = T^\dagger\psi[f(-i\overleftrightarrow{\nabla})]\psi + \text{h.c.} \quad (6)$$

where

$$\psi[f(-i\overleftrightarrow{\nabla})]\psi = f_0\psi\psi + f_2\psi(-i\overleftrightarrow{\nabla})^2\psi + \cdots, \quad (7)$$

Also the 3-body force is expressed by the form of,

$$\mathcal{L}_I^{(3\text{-body})} = h_0 T^\dagger T\psi^\dagger\psi + h_2 T^\dagger T(\psi^\dagger\nabla^2\psi + \nabla^2\psi^\dagger\psi) + \cdots. \quad (8)$$

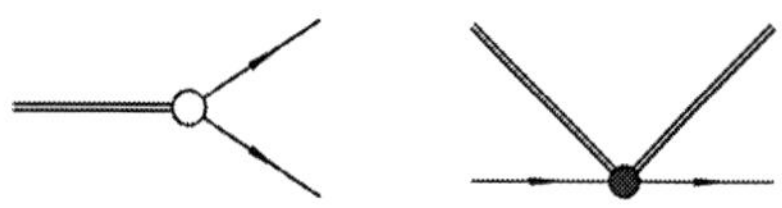

Figure 2: 2-body operator and 3-body operator in particle-dimer formalism

The particle-dimer formalism is totally equivalent to 3-body effective theory without any approximation. It can be seen in the path integral representation . Since dimer field is auxiliary, the complete 3-body effective theory can be reproduced by integrating dimer field out from particle-dimer formalism [18].

2.3 Two-body sector

The parameters in 2-body sector, *i.e.*, $f_0, f_2, \cdots$ in eq.(7) should be in principal matched to 2-body physical observables, such as 2-body scattering phase shift. However, since we are focused in 3-body dynamics, these 2-body parameters can be taken account as input, either from experiment data or 2-body lattice simulation. We have demonstrated that the 2-body low-energy constants (LECs) can be encoded in phase shift (or its analytic continuation) for both particle-dimer scattering equation and 3-body quantization condition [18].

Explicitly speaking, we can write down the dimer propagator[2] in terms of,

$$\tau(\mathbf{k}; E) = \frac{1}{k^*\cot\delta(k^*) + \sqrt{(k^*)^2}} \quad (9)$$

where the momentum k^* is on-shell momentum[3],

$$(k^*)^2 = \frac{3}{4}\mathbf{k}^2 - mE - i\epsilon. \quad (10)$$

At the same time, the hopping potential (see Fig. 3) is expressed by

$$z(\mathbf{p}, \mathbf{q}; E) = \frac{1}{p^2 + q^2 + \mathbf{pq} - mE}. \quad (11)$$

All the 2-body parameters have been hidden. The price is that we have to input 2-body physical information, *i.e.*, phase shift.

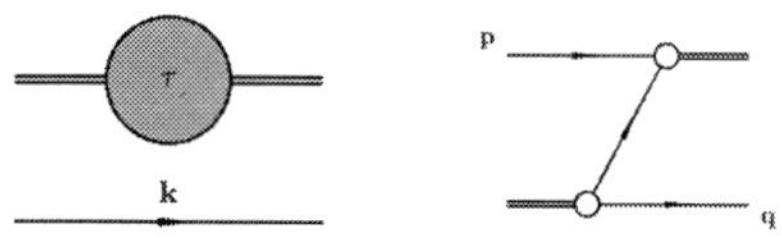

Figure 3: The left plane: dimer-propagator. The right plane: hopping potential.

2.4 Three-body force

The only one kind of parameters left is *three-body force*, *i.e.*, $h_0, h_2, \cdots$ in eq.(8). They appear in particle-dimer scattering equation or quantization condition. Our main aim is to determine the 3-body force from lattice simulation. The total particle-dimer potential reads,

$$Z(\mathbf{p}, \mathbf{q}; E) = \frac{1}{p^2 + q^2 + \mathbf{pq} - mE} + \frac{H(\Lambda)}{\Lambda^2}. \quad (12)$$

Here we use dimensionless parameter H instead of $h_0, h_2, \cdots$ and Λ is three-momentum cut-off[4]. Although the 3-body force H is cut-off dependent, it does not lead to any cut-off dependence of physical result, for example, binding energy of 3-body bound state. In a finite volume, it turns out that all the energy levels are cut-off independent as well.

2.5 Particle-dimer scattering equation

The particle-dimer scattering equation is written down as,

$$\mathcal{M}(\mathbf{p},\mathbf{q}; E) = Z(\mathbf{p}, \mathbf{q}; E) + 8\pi\int^\Lambda \frac{d^3k}{(2\pi)^3} Z(\mathbf{p}, \mathbf{k}; E)\tau(\mathbf{k}; E)\mathcal{M}(\mathbf{k}, \mathbf{q}; E). \quad (13)$$

As long as 3-body force is determined, we can solve the scattering equation. The analytic structure of particle-dimer scattering amplitude $\mathcal{M}$ then encodes all the 3-body dynamics.

In the infinite volume, rotational symmetry is respected. So the scattering equation can be projected to the specific angular momentum quantum number.

[1] The dimer with arbitrary spin can be checked in our paper [18].

[2] Strictly speaking, since we are using the spectator momentum $\mathbf{k}$, it is actually spectator-dimer propagator.

[3] For 3-body system with total momentum, $P = (E, 0)$, one particle is on-shell and has momentum $\mathbf{k}$. If another on-shell particle has momentum $|\mathbf{p}| = k^*$, the third particle must be on-shell as well.

[4] We adopt cut-off regularization to deal with ultra-violet divergence here. Other regularization schemes such as dimension regularization are available as well in the infinite volume. However, in a finite volume, cut-off scheme is most clear and natural.

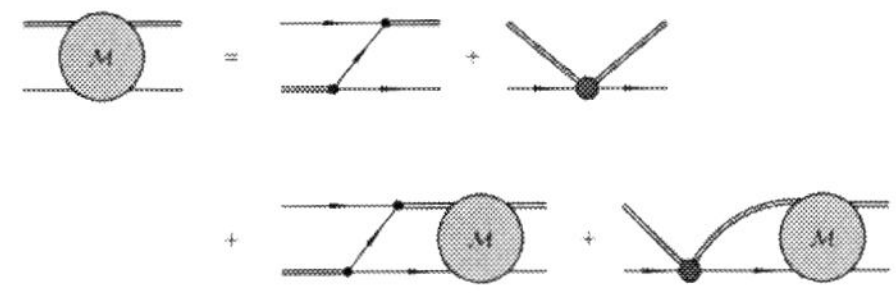

Figure 4: Particle-dimer scattering equation

3 Projected quantization condition

3.1 Particle-dimer scattering equation in a finite volume

Imposing periodical boundary condition for lattice configuration makes the momentum of the system discrete. Whereas the local LECs is non-sensitive to the volume effect. Therefore both 2-body input and 3-body force are preserved. The finite volume effects are reflected by changing loop integrals to loop sums. There are two corrections in the finite volume scattering equation, iteration part and dimer propagator.

$$\mathcal{M}_L(\mathbf{p},\mathbf{q};E) = Z(\mathbf{p},\mathbf{q};E) + \frac{8\pi}{L^3}\sum_{\mathbf{k}}^{\Lambda} Z(\mathbf{p},\mathbf{k};E)\tau_L(\mathbf{k};E)\mathcal{M}_L(\mathbf{k},\mathbf{q};E)\,, \quad (14)$$

where $\mathbf{k} = \frac{2\pi}{L}\mathbf{n}$, $\mathbf{n} \in \mathbb{Z}^3$ is a quantized three-momentum in a finite volume and

$$\tau_L^{-1}(\mathbf{k};E) = k^* \cot\delta(k^*) + S(\mathbf{k},E). \quad (15)$$

The finite volume correction for dimer propagator is

$$S(\mathbf{k},E) = -\frac{4\pi}{L^3}\sum_{\mathbf{l}} \frac{1}{\mathbf{k}^2 + \mathbf{l}^2 + \mathbf{kl} - mE}\,. \quad (16)$$

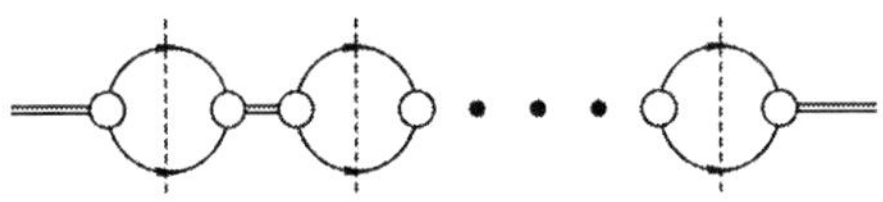

Figure 5: Finite volume correction for dimer propagator

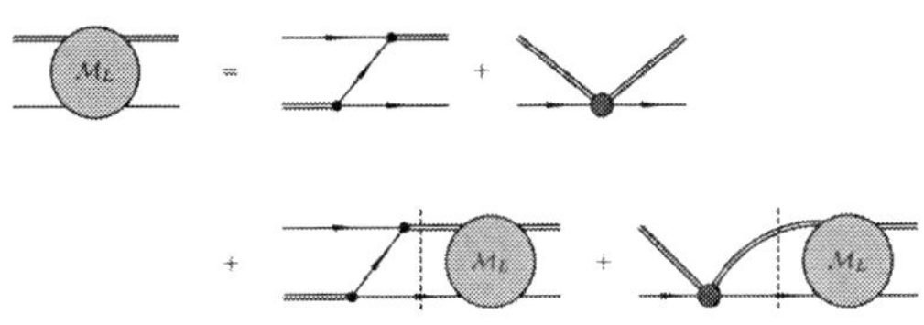

Figure 6: Finite volume scattering equation

3.2 Quantization condition

In a finite volume, the spectrum is discrete both below and above threshold. All the energy levels are determined by the poles of finite volume scattering amplitude $\mathcal{M}_L$. Based on the finite volume scattering equation, eq.(14), these poles are obtained from quantization condition,

$$\det(\tau_L^{-1} - Z) = 0\,. \quad (17)$$

Now we have written down the 3-body quantization condition. It gives the finite volume spectrum of 3-body system. Inversely, 3-body force can be fixed by plugging lattice simulation results into quantization condition. With 3-body force and pertinent 2-body input, we can solve the scattering equation in the infinite volume, *i.e.*, eq.(13). Particle-dimer scattering amplitude $\mathcal{M}$ encodes all the 3-body dynamics information.

3.3 Octahedral group, shell structure and irreducible representation expansion

The cubic periodical boundary condition breaks rotational symmetry in the infinite volume. The remaining symmetry for the discrete momentum configuration ($\mathbf{p} = \frac{2\pi}{L}\mathbf{n}$, $\mathbf{n} \in \mathbb{Z}^3$) is octahedral group, $\mathcal{G} = O_h$. Combined with inversion operator, there are 48 group elements. O_h group has 10 irreducible representations, $A_1^\pm, A_2^\pm, E^\pm, T_1^\pm, T_2^\pm$.

One the other hand, with respect to the octahedral group, we define the shell structure for the discrete momenta. *Shell* is the set of momenta with the same magnitude. The arbitrary momentum $\mathbf{p}$ in a shell can be obtained by O_h transformation acted on *reference momentum* denoted by $\mathbf{p}_0$, *i.e.*, $\mathbf{p} = g\,\mathbf{p}_0$, $g \in \mathcal{G}$. Notice that $\mathbf{n} = (3,0,0)$ and $\mathbf{n}' = (2,2,1)$ which have the same magnitude, although, cannot be related by any O_h transformation. Thus they belong to two different shells. We will specify the reference momentum for each shell at the beginning. For example, shell-1 has 6 momenta, $(1,0,0), (0,1,0), \cdots$. We choose $\mathbf{p}_0 = \frac{2\pi}{L}(1,0,0)$. Finally, nothing depends on the choice of reference. We also define *multiplicity* $\vartheta(r)$ of the shell r, for example, $\vartheta(1) = 6$.

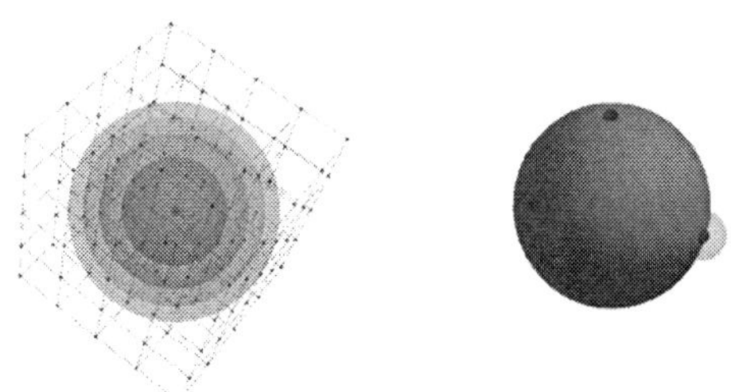

Figure 7: The left plane: Shell structure of cubic momentum configuration. The right plane: Shell-1.

Similar to the partial wave expansion corresponding to rotational symmetry, we develop the irreducible represen-

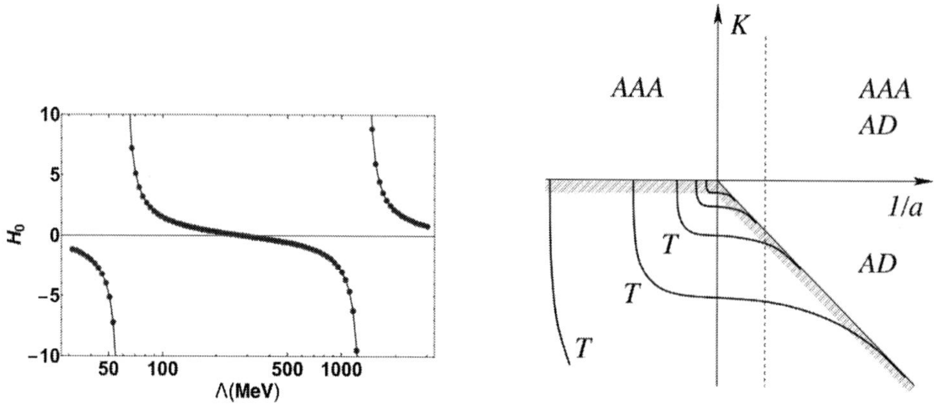

Figure 8: The left plane: 3-body force H_0 is running as momentum cut-off Λ in the log-periodical manner. The right plane: Efimov states, a is 2-body scattering length ($1/a = 0$ corresponding to unitary limit), K is 3-body energy, AAA means 3-atom state, AD is the state of a spectating atom and a 2-atom bound state, T means 3-body bound state, *i.e.*, Efimov state. The vertical dashed line shows the system we are concerning.

tation expansion for the cubic symmetry[5].

$$f(\mathbf{p}) = f(g\mathbf{p}_0) = \sum_{\Gamma}\sum_{\rho\sigma} T^{\Gamma}_{\sigma\rho}(g) f^{\Gamma}_{\rho\sigma}(\mathbf{p}_0) \qquad (18)$$

where $\Gamma = A_1^{\pm}, A_2^{\pm}, E^{\pm}, T_1^{\pm}, T_2^{\pm}$, and $T^{\Gamma}_{\sigma\rho}(g)$ are the matrices of the irreducible representations [20]. Inversely, we have

$$f^{\Gamma}_{\rho\sigma}(\mathbf{p}_0) = \frac{s_\Gamma}{G}\sum_{g\in\mathcal{G}} (T^{\Gamma}_{\sigma\rho}(g))^* f(g\mathbf{p}_0) \qquad (19)$$

where $G = 48$ is the total number of elements in the group $\mathcal{G}$ and s_Γ is the dimension of the representation Γ.

For the finite volume scattering equation, the potential is projected as

$$\begin{aligned} Z(\mathbf{p},\mathbf{q};E) &= Z(g\mathbf{p}_0(s), g'\mathbf{q}_0(r); E) \\ &= \sum_{\Gamma,\rho\sigma,\lambda} \frac{s_\Gamma}{G} T^{(\Gamma)}_{\rho\sigma}(g) Z^{(\Gamma)}_{\sigma\lambda}(s,r;E) T^{(\Gamma)*}_{\rho\lambda}(g') \end{aligned} \qquad (20)$$

and

$$Z^{\Gamma}_{\sigma\lambda}(s,r;E) = \sum_{g\in\mathcal{G}} (T^{\Gamma}_{\lambda\sigma}(g))^* Z(\mathbf{p}_0(s), g\mathbf{q}_0(r); E). \qquad (21)$$

The dimer propagator is projected automatically,

$$\tau_L(s;E) = \tau_L(\mathbf{p}_0(s);E) = \tau_L(\mathbf{p}(s);E). \qquad (22)$$

Finally, we arrive at projected quantization condition,

$$\det\left(\tau_L(s)^{-1}\delta_{rs}\delta_{\sigma\rho} - \frac{8\pi\vartheta(s)}{GL^3} Z^{\Gamma}_{\sigma\rho}(r,s)\right) = 0\,. \qquad (23)$$

This equation determines the energy levels in the irreducible representation Γ.

[5]The expansion is complete based on

$$\sum_{g\in\mathcal{G}} (T^{\Gamma}_{\rho\sigma}(g))^* T^{\Gamma'}_{\rho'\sigma'}(g) = \delta_{\Gamma\Gamma'}\delta_{\rho\rho'}\delta_{\sigma\sigma'}\frac{G}{s_\Gamma}$$

and

$$\sum_{\Gamma,\rho\sigma} \frac{s_\Gamma}{G} (T^{\Gamma}_{\rho\sigma}(g))^* T^{\Gamma}_{\rho\sigma}(g') = \delta_{gg'}\,.$$

4 Three-body bound states

4.1 Efimov states

We apply 3-body quantization condition on the system where Efimov states exist. The 3-body force plays an essential role for the forming of Efimov states which are the typical 3-body bound states.

The 2-body physical information as input is encoded in phase shift giving the following effective range expansion,

$$p\cot\delta = -\frac{1}{a} + \frac{1}{2}rp^2 + \cdots, \qquad a \gg r. \qquad (24)$$

The pertinent 3-body force generates 3-body bound state whatever 2-body interaction is attractive or repulsive. By solving eq.(13), we produce Efimov states via a log-periodical 3-body force H_0 (see Fig. 8).

In the system with positive 2-body scattering length, *i.e.*, $a > 0$, there exists a 2-body bound state with binding momentum[6] $\sim 1/a$. It induces an additional threshold above which AD states (the state of a spectating particle and a 2-body bound state) emerge. The threshold is an effective 2-body threshold (AD threshold) contrast to 3-body threshold above which AAA states (the state of three particles) emerge. We find two bound states in the 3-body sector. The shallow one is close to AD threshold from below while the deeper one is far away relatively (see Fig. 8). All the physics are encoded in scattering equation in the infinite volume, *i.e.*, eq.(13).

4.2 Efimov states in a finite volume

In a finite volume, we apply 3-body quantization condition (23) by inputting the 3-body force which has been fixed in the infinite volume. The two 3-body bound states in a finite

[6]Here we define binding momentum $\kappa = \sqrt{m|E_B|}$, E_B is the energy of the bound state.

volume are both reproduced (see Fig. 9). They are in the irreducible representation A_1^+ corresponding to S-wave in the infinite volume.

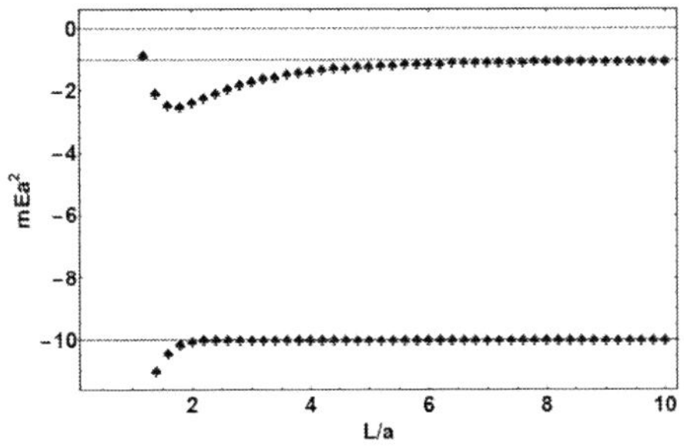

Figure 9: The two Efimov states in a finite volume

4.3 Energy shifts of bound states

There are two different scenarios for the 3-body bound states in the infinite volume.

- Effective 2-body limit: $\kappa^2 - a^{-2} \ll \kappa^2$;
- 3-body limit: $\kappa a \gg 1$.

Here κ is 3-body binding momentum defined by $\kappa = \sqrt{m|E_B|}$ and E_B is bound state pole. These two cases have different energy shifts in a finite volume, respectively.

We firstly relate the finite volume amplitude $\mathcal{M}_L$ in eq.(14) to the infinite volume amplitude $\mathcal{M}$ in eq.(13) by

$$\begin{aligned}\mathcal{M}_L(\mathbf{p},\mathbf{q};E) &= \mathcal{M}(\mathbf{p},\mathbf{q};E)\\ &+ 8\pi\int\frac{d^3k}{(2\pi)^3}\mathcal{M}(\mathbf{p},\mathbf{k};E)\,\delta\tau(\mathbf{k};\mathbf{E})\mathcal{M}_L(\mathbf{k},\mathbf{q};E)\end{aligned} \quad (25)$$

where

$$\delta\tau(\mathbf{k};\mathbf{E}) = \sum_{\mathbf{n}\neq 0} e^{i\mathbf{n}\mathbf{k}L}\tau(\mathbf{k};E) + \left(\text{high order in } \frac{1}{L}\right). \quad (26)$$

The infinite volume amplitude $\mathcal{M}$ can be expanded around the bound state pole E_B,

$$\mathcal{M}(\mathbf{p},\mathbf{q};E) = \frac{\psi(\mathbf{p})\psi^\dagger(\mathbf{q})}{E-E_B} + \text{reg.}. \quad (27)$$

The residue at the pole gives the wave-function $\psi(\mathbf{p})$ of Efimov state as an effective 2-body system of dimer field and the spectator. The energy shift, therefore, can be calculated by

$$\begin{aligned}\Delta E &= E_L - E_B\\ &= 8\pi\sum_{\mathbf{n}\neq 0}\int\frac{d^3k}{(2\pi)^3}\psi^\dagger(\mathbf{k})\tau(\mathbf{k};E_B)e^{i\mathbf{n}\mathbf{k}L}\psi(\mathbf{k}).\end{aligned} \quad (28)$$

In the effective 2-body limit, eq.(28) gives that

$$\frac{\Delta E}{|E_B|} \sim \frac{C_2}{\kappa L}e^{-\frac{2}{\sqrt{3}}\sqrt{\kappa^2-a^{-2}}L} \quad (29)$$

which is consistent with 2-body Lüscher formula. While in the 3-body limit, we have energy shift,

$$\frac{\Delta E}{|E_B|} \sim \frac{C_3}{(\kappa L)^{3/2}}e^{-\frac{2}{\sqrt{3}}\kappa L}. \quad (30)$$

C_3 is proportional to asymptotic normalization coefficient $|\mathcal{A}|^2$. It turns out that in the unitary limit, $a \to \infty$, the ratio $C_3/|\mathcal{A}|^2 = -96.351$ is universal [2].

The general 3-body bound state is the mixing of the two limits. By analyzing the 3-body contribution and effective 2-body contribution in the energy shifts of two bound states (see Fig. 10), we find that:

- The shallow bound state is dominated by effective 2-body scenario.
- 3-body picture and effective 2-body picture are both important in the deeper bound state.

5 Three-body scattering states

5.1 Avoided level crossing

Since there are two thresholds in the system, AD threshold and 3-body threshold, the scattering states in a finite volume[7] tend to these two thresholds in the infinite volume limit, respectively. Consequently, we can find both effective 2-body states and 3-body states above threshold in the lattice spectrum (see Fig. 11). All the states are determined by quantization condition (23) which is projected on A_1^+-irrep.

Avoided level crossing happens between the two kinds of scattering states. In 2-body lattice spectrum, avoided level crossing usually implies the existence of resonance. Reproducing this behavior in 3-body spectrum is a good sign for us to interpret 3-body resonance in the future simulation.

5.2 Energy shifts of scattering states

The scattering states distribute around free lines which are the non-interacting energy levels. The energy shifts of 3-body states above threshold can be expressed in the series of $1/L$, *i.e.*,

$$\begin{aligned}mE_L &= N\left(\frac{2\pi}{L}\right)^2 + \frac{\#}{L^3} + \frac{\#}{L^4} + \frac{\#}{L^5}\\ &+ \frac{\#}{L^5}\log\left(\frac{mL}{2\pi}\right) + \frac{\#}{L^6} + \cdots\end{aligned} \quad (31)$$

where $N(2\pi/L)^2$ is the free energy level. The high order coefficients can be obtained by solving quantization condition (23) perturbatively.

The expansions around free lines also indicate that the symmetry representations of these spectra are related to the specific momentum shell. For example, the lowest energy level is related to shell-0, *i.e.*, $\mathbf{p} = 0$. Shell-0 contains only A_1^+-irrep which means the ground state must be in A_1^+. The 1st excited states are related to shell-1 including

[7] In the infinite volume, the states above threshold are continuous. Whereas in a finite volume, they are discrete.

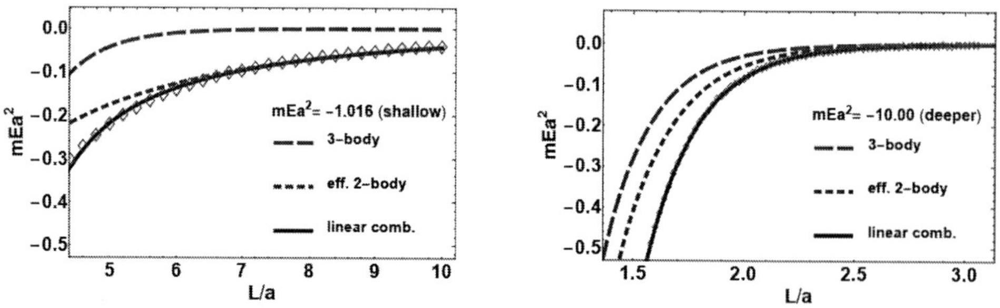

Figure 10: The left plane: energy shift of the shallow bound state. The right plane: energy shift of the deeper bound state.

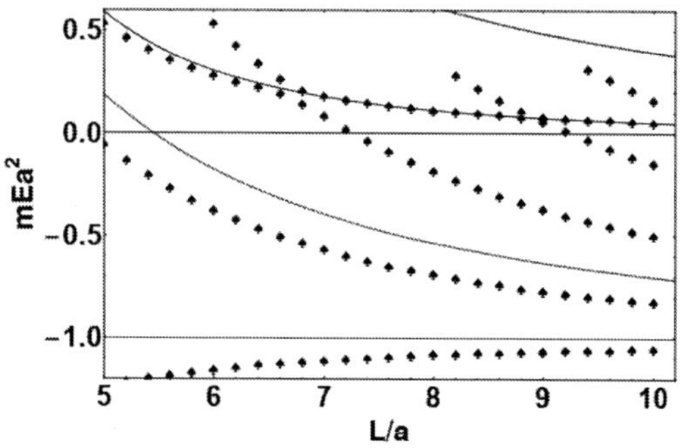

Figure 11: The 3-body lattice spectrum in A_1^+. AD threshold is located at -1.0 while 3-body threshold at 0. Red dashed line indicates the spectrum for free state of a 2-body bound state and a spectator. Black dashed line is the free line for the 3 free particles. Black solid line is the theoretical calculation for the 3-body ground state.

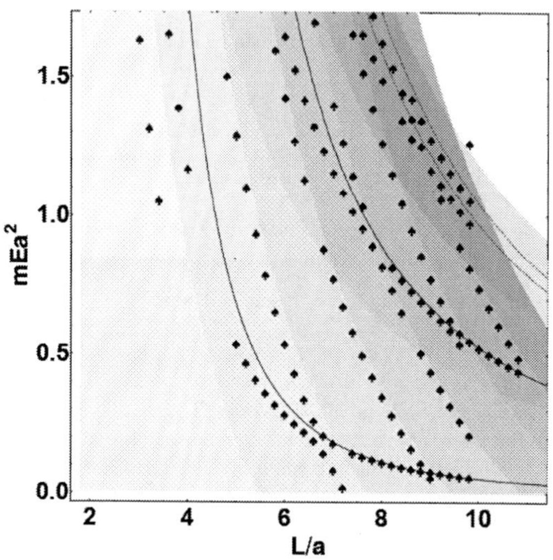

Figure 12: The lower blue line is the theoretical calculation for energy shifts of ground states up to $O(1/L^6)$. The upper blue line is for the excited states in A_1^+ up to $O(1/L^6)$ as well. The two red lines are the calculation for higher excited states at the next to leading order, that is $O(1/L^3)$.

A_1^+, E^+ and T_1^--irrep. Therefore we are supposed to find three spectra around shell-1. But due to the identical particles principal, T_1^- spectrum does not exist actually.

The quantization condition

$$\det(\tau^{-1} - Z) = 0 \tag{32}$$

can be solved perturbatively around shell-r in the form of

$$(\tau^{-1} - Z)_{rr} = \sum_{k\neq r} Z_{rk}\tau_k Z_{kr} + \sum_{k,l\neq r} Z_{rk}\tau_k Z_{kl}\tau_l Z_{lr} + \cdots \tag{33}$$

where k, l are both shell numbers. They also denote the matrix indices of τ, Z and so on. Both sides of eq.(33) are expanded with respect to $1/L$, so the energy shifts can be calculated order by order.

The energy shift of the ground state (which resides in the A_1^+-irrep) is:

$$mE_L = \left(\frac{2\pi}{L}\right)^2 \frac{g_0}{L}\left(1 + \frac{g_1}{L} + \frac{g_2}{L^2} + \frac{g_3}{L^3}\log\frac{mL}{2\pi} + \frac{g_4}{L^3} + \cdots\right) \tag{34}$$

where

$$\begin{aligned} g_0 &= \frac{3}{\pi}a\,, \\ g_1 &= 2.837297480\,a\,, \\ g_2 &= 9.725330808\,a^2\,, \\ g_3 &= 8\pi\left(\frac{2\sqrt{3}}{\pi} - \frac{8}{3}\right)a^3\,, \\ g_4 &= \left(-5.159159617 + 6\pi\left(\frac{r}{a}\right) - 8\pi\left(\frac{\hat{\mathcal{M}}}{a^2}\right)\right)a^3\,. \end{aligned} \tag{35}$$

Here a is 2-body scattering length and r is 2-body scattering effective range. In g_4, the 3-body threshold amplitude $\hat{\mathcal{M}}$ emerges. This 3-body physical quantity can be fit to lattice spectrum up to the order of $1/L^6$. $\hat{\mathcal{M}}$ also appears in the energy shift of the 1st excited state in the A_1^+-irrep,

$$mE_L = \left(\frac{2\pi}{L}\right)^2\left(1 + \frac{h_0}{L}\left(1 + \frac{h_1}{L} + \frac{h_2}{L^2} + \frac{h_3}{L^3}\ln\frac{mL}{2\pi} + \frac{h_4}{L^3} + \cdots\right)\right) \tag{36}$$

with

$$\begin{aligned} h_0 &= \frac{10}{\pi}a, \\ h_1 &= 0.279070\, a, \\ h_2 &= \left(8.494802 + \frac{7\pi^2}{5}\left(\frac{r}{a}\right)\right)a^2, \\ h_3 &= \frac{27}{5} \times 8\pi\left(\frac{2\sqrt{3}}{\pi} - \frac{8}{3}\right)a^3, \\ h_4 &= \left(-172.001650 + 83.745841\left(\frac{r}{a}\right) - \frac{27}{5} \times 8\pi\left(\frac{\hat{\mathcal{M}}}{a^2}\right)\right)a^3. \end{aligned} \tag{37}$$

However, $\hat{\mathcal{M}}$-dependence of the energy shift of the 1st excited state in the E^+-irrep disappears because E^+ corresponds to high derivative force while $\hat{\mathcal{M}}$ is non-derivative quantity.

$$mE_L = \left(\frac{2\pi}{L}\right)^2\left(1 + \frac{h_0'}{L}\left(1 + \frac{h_1'}{L} + \frac{h_2'}{L^2} + \frac{h_4'}{L^3} + \cdots\right)\right), \tag{38}$$

with

$$\begin{aligned} h_0' &= \frac{4}{\pi}a, \\ h_1' &= 2.984094\, a, \\ h_2' &= \left(3.001706 + \frac{\pi^2}{2}\left(\frac{r}{a}\right)\right)a^2, \\ h_4' &= \left(-28.89478538 + 54.584571\left(\frac{r}{a}\right)\right)a^3. \end{aligned} \tag{39}$$

6 Conclusion

We have reviewed the effective field theory describing 3-body system. Particle-dimer formalism shows its powerful to analyze the 3-body problems both in the infinite volume and a finite volume. 3-body quantization condition is derived. Considering the cubic symmetry for the lattice configuration, we also develop the irreducible representation expansion to project quantization condition on specific quantum number. The finite volume spectrum in the corresponding irreducible representation is determined consequently. The physical pictures of 3-body bound states in the infinite volume can be interpreted via energy shifts in a finite volume. For 3-body scattering states, avoided level crossing is observed and energy shifts are calculated perturbatively by expanding the quantization condition. 3-body threshold amplitude emerging in the energy shifts up to $O(1/L^6)$ can be extracted in principal from lattice simulation.

Our formalism needs further development including introducing twisted boundary condition, relativistic kinematics and so on. The application on realistic system, such as triton and Roper resonance is on-going as well.

Acknowledgments: The work is done in collaboration with H.-W. Hammer and A. Rusetsky. The authors would like to thank M. Döring, M. Mai, Ulf.-G. Meißner and J.-J. Wu for the collaboration on the most of works. We would also like to thank S. Beane, W. Detmold, M. Hansen, F. Romero-López, M. Savage, S. Sharpe, C. Thomas and C. Urbach for interesting discussions. JYP acknowledges the support from the DFG (CRC 110 "Symmetries and the Emergence of Structure in QCD").

References

[1] K. Polejaeva and A. Rusetsky, Eur. Phys. J. A **48** (2012) 67, arXiv:1203.1241 [hep-lat].

[2] U.-G. Meißner, G.-Rios and A. Rusetsky, Phys. Rev. Lett. **114** (2015) 091602 Erratum: Phys. Rev. Lett. **117** (2016) 069902, arXiv:1412.4969 [hep-lat].

[3] P. Guo, Phys. Rev. D **95** (2017) 054508, arXiv:1607.03184 [hep-lat].

[4] P. Guo and V. Gasparian, Phys. Lett. B **774** (2017) 441, arXiv:1701.00438 [hep-lat].

[5] R. A. Briceño and Z. Davoudi, Phys. Rev. D **87** (2013) 094507, arXiv:1212.3398 [hep-lat].

[6] M. T. Hansen and S. R. Sharpe, Phys. Rev. D **90** (2014) 116003, arXiv:1408.5933 [hep-lat].

[7] M. T. Hansen and S. R. Sharpe, Phys. Rev. D **92** (2015) 114509, arXiv:1504.04248 [hep-lat].

[8] M. T. Hansen and S. R. Sharpe, Phys. Rev. D **93** (2016) 014506, arXiv:1509.07929 [hep-lat].

[9] M. T. Hansen and S. R. Sharpe, Phys. Rev. D **93** (2016) 096006, arXiv:1602.00324 [hep-lat].

[10] M. T. Hansen and S. R. Sharpe, Phys. Rev. D **95** (2017) 034501, arXiv:1609.04317 [hep-lat].

[11] R. A. Briceño, M. T. Hansen and S. R. Sharpe, Phys. Rev. D **95** (2017) 074510, arXiv:1701.07465 [hep-lat].

[12] S. Kreuzer and H.-W. Hammer, Phys. Lett. B **694** (2011) 424, arXiv:1008.4499 [hep-lat].

[13] S. Kreuzer and H.-W. Hammer, Eur. Phys. J. A **43** (2010) 229, arXiv:0910.2191 [nucl-th].

[14] S. Kreuzer and H.-W. Hammer, Phys. Lett. B **673** (2009) 260, arXiv:0811.0159 [nucl-th].

[15] S. Kreuzer and H.-W. Grießhammer, Eur. Phys. J. A **48** (2012) 93, arXiv:1205.0277 [nucl-th].

[16] S. R. Sharpe, Phys. Rev. D **96** (2017) 054515, arXiv:1707.04279 [hep-lat].

[17] H.-W. Hammer, J.-Y. Pang and A. Rusetsky, JHEP **1709** (2017) 109, arXiv:1706.07700 [hep-lat].

[18] H.-W. Hammer, J.-Y. Pang and A. Rusetsky, JHEP **1710** (2017) 115, arXiv:1707.02176 [hep-lat].

[19] Y. Meng, C. Liu, U.-G. Meißner and A. Rusetsky, Phys. Rev. D **98** (2018) 014508, arXiv:1712.08464 [hep-lat].

[20] M. Döring, H.-W. Hammer, M. Mai, J.-Y. Pang, A. Rusetsky and J. Wu, Phys. Rev. D **97** (2018) no.11, 114508, arXiv:1802.03362 [hep-lat].

[21] M. Mai and M. Döring, Eur. Phys. J. A **53** (2017) 240, arXiv:1709.08222 [hep-lat].

[22] R. A. Briceño, M. T. Hansen and S. R. Sharpe, arXiv:1803.04169 [hep-lat].

[23] F. Romero-López, A. Rusetsky and C. Urbach, Eur. Phys. J. C **78** (2018) no.10, 846, arXiv:1806.02367 [hep-lat].

[24] R. A. Briceño, M. T. Hansen and S. R. Sharpe, arXiv:1810.01429 [hep-lat].

[25] M. Mai and M. Döring, arXiv:1807.04746 [hep-lat].

[26] P. Guo and T. Morris, arXiv:1808.07397 [hep-lat].

[27] P. Guo, M. Döring and A. P. Szczepaniak, Phys. Rev. D **98** (2018) 094502, arXiv:1810.01261 [hep-lat].

[28] T. D. Blanton, F. Romero-López and S. R. Sharpe, arXiv:1901.07095 [hep-lat].

[29] J.-Y. Pang, J.J. Wu, H.-W. Hammer, Ulf.-G. Meißner and A. Rusetsky Phys. Rev. D **99** (2019) no.7, 074513, arXiv:1902.01111 [hep-lat].

[30] M. T. Hansen and S. R. Sharpe, arXiv:1901.00483 [hep-lat].

[31] M. Lüscher, Commun. Math. Phys. **105** (1986) 153.

[32] K. Huang and C. N. Yang, Phys. Rev. **105** (1957) 767.

[33] T. T. Wu, Phys. Rev. **115** (1959) 1390.

[34] S. Tan, Phys. Rev. A **78** (2008) 013636, arXiv:0709.2530 [cond-mat.stat-mech].

[35] S. R. Beane, W. Detmold and M. J. Savage, Phys. Rev. D **76** (2007) 074507, arXiv:0707.1670 [hep-lat].

[36] W. Detmold and M. J. Savage, Phys. Rev. D **77** (2008) 057502 arXiv:0801.0763 [hep-lat].

[37] E. Braaten and H.-W. Hammer, Phys. Rept. **428** (2006) 259 [cond-mat/0410417].

[38] V. Bernard, M. Lage, U.-G. Meißner and A. Rusetsky, JHEP **0808** (2008) 024, arXiv:0806.4495 [hep-lat].

[39] M. Göckeler, R. Horsley, M. Lage, U.-G. Meißner, P. E. L. Rakow, A. Rusetsky, G. Schierholz and J. M. Zanotti, Phys. Rev. D **86** (2012) 094513, arXiv:1206.4141 [hep-lat].

[40] F. Romero-López, A. Rusetsky and C. Urbach, Phys. Rev. D **98** (2018) 014503, arXiv:1802.03458 [hep-lat].

[41] H. W. Hammer and T. Mehen, Phys. Lett. B **516** (2001) 353, ,nucl-th/0105072.

[42] C. Ji, D. R. Phillips and L. Platter, Annals Phys. **327** (2012) 1803, arXiv:1106.3837 [nucl-th]

Low-energy pion-nucleon scattering and the Δ resonance in lattice QCD

Constantia Alexandrou[1,2,*], *Giannis* Koutsou[1,**], *Stefan* Krieg[3,4,***], *Luka* Leskovec[5,****], *Stefan* Meinel[6,†], *John W.* Negele[7,‡], *Srijit* Paul[1,3,§], *Marcus* Petschlies[8,9,¶], *Andrew* Pochinsky[7,||], *Gumaro* Rendon[6,**], *Giorgio* Silvi[4,††], and *Sergey* Syritsyn[10,‡‡]

[1]Computation-based Science and Technology Research Center, The Cyprus Institute
[2]Department of Physics, University of Cyprus, POB 20537, 1678 Nicosia, Cyprus
[3]Faculty of Mathematics und Natural Sciences, University of Wuppertal
[4]Institute for Advanced Simulation, Forschungszentrum Jülich GmbH, Jülich 52425, Germany
[5]Theory Center, Jefferson Lab, Newport News, VA 23606, USA
[6]Department of Physics, University of Arizona, Tucson, AZ 85721, USA
[7]Center for Theoretical Physics, Massachusetts Institute of Technology
[8]Helmholtz-Institut für Strahlen- und Kernphysik
[9]Bethe-Center for Theoretical Physics
[10]Department of Physics and Astronomy, Stony Brook University, Stony Brook, NY 11794, USA

Abstract. We report on our investigation of low-energy pion-nucleon scattering from lattice QCD with Wilson-type fermions. Our focus is on the isospin $I = 3/2$ channel, which couples to the Delta resonance. We discuss our calculations aiming at the extraction of the Delta resonance mass, width and effective coupling at pion mass 250 MeV, using the Lüscher finite-volume method to extract the phase shift $\delta_{J=3/2,l=1}$.

1 Introduction

In this proceedings contribution we give an account of our extraction of low-energy pion-nucleon scattering properties in the channel $I\left(J^P\right) = \frac{3}{2}\left(\frac{3}{2}^+\right)$, which features the Δ resonance, from lattice QCD data at pion mass 250 MeV. We will focus on the well-known parametrization in terms of Breit-Wigner parameters with resonance mass m_Δ and (energy-dependent) decay width $\Gamma_{\Delta-\pi N}(s)$. For the latter we will use the parametrization in leading-order effective theory of pions and nucleons, which defines the effective coupling $g_{\Delta-\pi N}$.

The Δ resonance is the lowest-lying meson-baryon resonance, which is accessible within purely elastic scattering, decaying with branching ratio approximately 100% to the QCD-stable (= lattice QCD ground) states of pion and nucleon. It is thus a natural starting point for a full-fledged scattering analysis based on Lüscher's finite-volume method [1], including its generalizations to moving frames, particles of non-equal mass, multiple channels [2–5] and extensions to particles of arbitrary spin [6, 7].

Even beyond in of itself the lattice QCD scattering analysis for Δ is of great interest for and a prequel to the calculation of pion-nucleon transition amplitudes via electro-weak currents, in the Delta resonance region. These include in particular the matrix elements for so-called $2 \rightarrow 1$ and $2 \rightarrow 2$ transitions as depicted diagrammatically in Fig. 1. Two very prominent examples

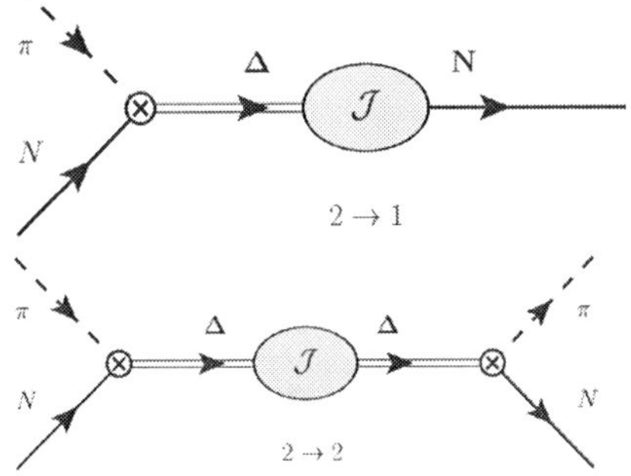

Figure 1. Diagrammatic illustration of $2 \rightarrow 1$ and $2 \rightarrow 2$ transition matrix elements of the Δ

of foremost interest to us for the $2 \rightarrow 1$ case are the transition via the electromagnetic current $\mathcal{J} = V_\mu$ and the

*e-mail: alexand@ucy.ac.cy
**e-mail: koutsou@cyi.ac.cy
***e-mail: s.krieg@fz-juelich.de
****e-mail: leskovec@jlab.org
†e-mail: smeinel@email.arizona.edu
‡e-mail: negele@mit.edu
§e-mail: s.paul@hpc-leap.eu
¶e-mail: marcus.petschlies@hiskp.uni-bonn.de
||e-mail: avp@mit.edu
**e-mail: jgrs@email.arizona.edu
††e-mail: g.silvi@fz-juelich.de
‡‡e-mail: ssyritsyn@quark.phy.bnl.gov

axial vector current $\mathcal{J} = A_\mu$. The former gives rise to the transition amplitude and thus cross-section for resonant pion-photoproduction off the nucleon, which can be compared directly to experiment. The lattice calculation of the radiative decay together with the strong decay, which is under consideration in this study, then gives in particular a complete ab-initio theory characterization of the decay structure of the Δ resonance.
The axial $N \to \Delta$ transition amplitude is important input to neutrino reaction cross sections like neutrino-nucleus scattering, which are investigated in contemporary experiments such as DUNE and MiniBooNE [8].
Common to all these analyses is the requirement to convert finite-volume lattice amplitudes to their infinite volume counterpart. This normalization based on the Lellouch-Lüscher factor [9] or its generalizations [10–14] in turn requires knowledge of the scattering matrix in the Δ resonance region, which is the goal our calculation presented here.

The pion-nucleon elastic scattering in the Δ channel has been the subject of several lattice QCD calculations. Among those are the analysis using the Michael/McNeile method [15, 16], requiring the lattice Hamiltonian to exhibit a Δ-like state perturbatively close and below the pion-nucleon threshold and determining the resonance coupling based on the leading-order effective pion-nucleon interaction. The Lüscher method (to all orders in the effective πN interaction) has been initially explored in several lattice QCD works [17–19]. However, the dedicated analysis of partial wave mixing is still to be achieved.

Here we discuss the first steps of our analysis of pion-nucleon scattering in the Δ channel [20] from lattice QCD focusing on the irreps with leading p-wave contribution and neglecting partial wave mixing.

2 Calculation of Lattice Correlation functions

We consider QCD with up, down and strange quark on a Euclidean space-time lattice of spatial and temporal size $L = 2.8\,\mathrm{fm}$ and $T = 2\,L$. The gauge field ensemble features $N_f = 2+1$ tree-level clover-improved Wilson quarks and the tree-level $O\left(a^2\right)$ Symanzik-improved gauge action [21] with two levels of HEX smearing of the gauge links coupling to the quarks.

For the specific ensemble A7 investigated here, the pion mass is $m_\pi \approx 258\,\mathrm{MeV}$ and with an ensuing $m_\pi L \approx 3.6$, residual exponential finite volume effects are expected to be sufficiently suppressed for the pertinent polynomial finite-volume effects underlying the Lüscher method to be dominant. For convenience all relevant ensemble parameters are collected in Table 1.

We perform the spectral analysis for the isospin channel $I = 3/2$, $I_3 = +3/2$ based on the generalized eigenvalue problem [22] for a matrix of correlation functions with single-hadron operators of Δ-type and 2-hadron operators of πN type. In detail, we list the operators in Eqs.. (1) to (6) below.

Label	$N_s^3 \times N_t$	L (fm)	a (fm)	m_π (MeV)	$m_\pi L$
A7	$24^3 \times 48$	2.8	0.116	258	3.6

Table 1. Parameters of the ensemble A7 used in this work [].

For the single-hadron interpolators we use standard quark-antiquark and three-quark operators with two spin-structures for nucleon and Delta each.

$$\pi(t,\vec{p}) = \sum_{\vec{x}} \bar{d}(t,\vec{x})\,\gamma_5\, u(t,\vec{x})\, e^{i\vec{p}\cdot\vec{x}} \tag{1}$$

$$N_\mu^{(1)}(t,\vec{p}) = \sum_{\vec{x}} \epsilon_{abc}\,(u_a(t,\vec{x}))_\mu\,(u_b^T(t,\vec{x})\,C\gamma_5\, d_c(t,\vec{x}))\, e^{i\vec{p}\cdot\vec{x}} \tag{2}$$

$$N_\mu^{(2)}(t,\vec{p}) = \sum_{\vec{x}} \epsilon_{abc}\,(u_a(t,\vec{x}))_\mu\,(u_b^T(t,\vec{x})\,C\gamma_0\gamma_5\, d_c(t,\vec{x}))\, e^{i\vec{p}\cdot\vec{x}} \tag{3}$$

$$\Delta_{\mu i}^{(1)}(t,\vec{p}) = \sum_{\vec{x}} \epsilon_{abc}\,(u_a(t,\vec{x}))_\mu\,(u_b^T(t,\vec{x})\,C\gamma_i\, u_c(t,\vec{x}))\, e^{i\vec{p}\cdot\vec{x}} \tag{4}$$

$$\Delta_{\mu i}^{(2)}(t,\vec{p}) = \sum_{\vec{x}} \epsilon_{abc}\,(u_a(t,\vec{x}))_\mu\,(u_b^T(t,\vec{x})\,C\gamma_i\gamma_0\, u_c(t,\vec{x}))\, e^{i\vec{p}\cdot\vec{x}} \tag{5}$$

The two-hadron interpolators of $N\pi$-type are the product of the above pion and nucleon operators

$$(N\pi)^{(i)}(t,\vec{p}_N,\vec{p}_\pi) = N^{(i)}(t,\vec{p}_N)\,\pi(t,\vec{p}_\pi)\,,$$
$$\vec{p}_N + \vec{p}_\pi = \vec{P}\,,\ i = 1,2 \tag{6}$$

with specified total momentum $\vec{P}$. The individual hadron momenta range in $0 \le |\vec{p}|/(2\pi/L) \le \sqrt{3}$.

From the operator basis in Eqs. (1) to (6) above we build all (cross-)correlation functions $\pi - \pi$ for the pion, $N^{(i)} - N^{(k)}$ for the nucleon and $\Delta^{(i)} - \Delta^{(k)}$, $(N\pi)^{(i)} - \Delta^{(k)}$, $(N\pi)^{(i)} - (N\pi)^{(k)}$ for the Delta channel. The corresponding types of Wick contractions for the latter are depicted in the diagrams of Fig. 2, where each diagram-type stands as a placeholder for 2 to 4 distinct Wick contractions. To compute the diagrams we follow our "sequential + stochastic + smearing" method, previously successfully applied in the calculation of elastic $\pi\pi$ and $K\pi$ scattering amplitudes [23, 24]: contractions are based on point-to-all, sequential and stochastic timeslice-to-all propagators as well propagator products from usage of the one-end-trick. The propagators receive Wuppertal source and sink smearing [25] at each meson and baryon vertex, using the APE smeared [26] gauge field in the smearing operator.

3 Lattice spectrum determination and Lüscher analysis

The finite-volume and discretization of the lattice box lead to the breaking of the continuum $SU(2)\times$ parity rotation-reflection symmetry group down to O_h^D (superscript D denoting double cover) for a system of hadrons

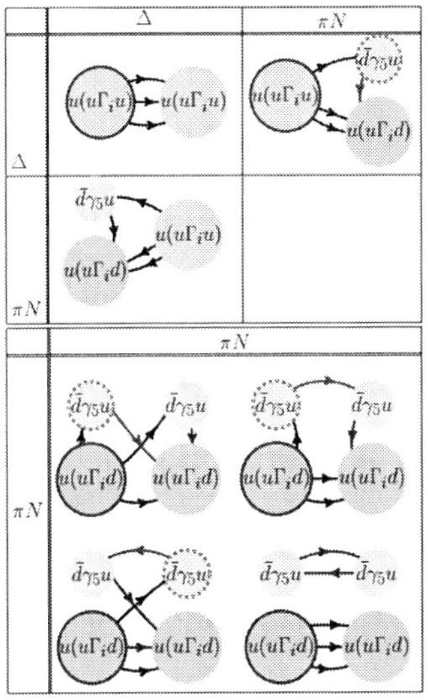

Figure 2. Topology-types of Wick contractions for $\Delta/N\pi$ correlation functions

$\frac{L}{2\pi}\vec{P}_{ref}[N_{dir}]$	Group	N_{elem}	$\Lambda(J) : \pi(0^-)$
(0, 0, 0) [1]	O_h^D	96	$A_{1u}(0, 4, ...)$
(0, 0, 1) [6]	C_{4v}^D	16	$A_2(0, 1, ...)$
(0, 1, 1) [12]	C_{2v}^D	8	$A_2(0, 1, ...)$
(1, 1, 1) [8]	C_{3v}^D	12	$A_2(0, 1, ...)$

Table 2. Little group, lattice irreps and angular momentum content per moving frame relevant for isospin 1 π operator.

$\frac{L}{2\pi}\vec{P}_{ref}[N_{dir}]$	Group	N_{elem}	$\Lambda(J) : N(\frac{1}{2}^+)$
(0, 0, 0) [1]	O_h^D	96	$G_{1g}(\frac{1}{2}, \frac{7}{2}, ...) \oplus G_{1u}(\frac{1}{2}, \frac{7}{2}, ...)$
(0, 0, 1) [6]	C_{4v}^D	16	$G_1(\frac{1}{2}, \frac{3}{2}, ...)$
(0, 1, 1) [12]	C_{2v}^D	8	$G(\frac{1}{2}, \frac{3}{2}, ...)$
(1, 1, 1) [8]	C_{3v}^D	12	$G(\frac{1}{2}, \frac{3}{2}, ...)$

Table 3. Little group, lattice irreps and angular momentum content per moving frame relevant for isospin 1/2 nucleon-type operators.

at rest. A boosted system with non-zero total momentum $\vec{P}$ will exhibit an even further reduced symmetry group depending on the boost vector $\vec{P}$, denoted as the little group $LG(\vec{P}) \subset O_h^D$. The interpolators in eqs.(1) to (6) are thus no longer classified according to continuum $SU(2)$ irreps of orbital and total angular momentum l, $J = l \oplus 1/2$, but according to the lattice irreducible representations (irreps) of $LG(\vec{P})$. In our analysis we consider the rest frame $\vec{P} = 0$ as well the first three moving frames $\vec{P}/(2\pi/L) = (0, 0, 1), (0, 1, 1), (1, 1, 1)$ (including permutations and sign changes for non-zero elements). The mixing pattern of the different l and spin degrees of freedom for these frames under subduction to lattice irreps is shown in Tables 2 to 4. To project our basis operators (or correlation functions) to the target lattice irreps, we use the

$\frac{L}{2\pi}\vec{P}_{ref}[N_{dir}]$	Group	N_{elem}	$\Lambda(J) : \Delta(\frac{3}{2}^+)$
(0, 0, 0) [1]	O_h^D	96	$H_g(\frac{3}{2}, \frac{5}{2}, ...) \oplus H_u(\frac{3}{2}, \frac{5}{2}, ...)$
(0, 0, 1) [6]	C_{4v}^D	16	$G_1(\frac{1}{2}, \frac{3}{2}, ...) \oplus G_2(\frac{3}{2}, \frac{5}{2}, ...)$
(0, 1, 1) [12]	C_{2v}^D	8	$G(\frac{1}{2}, \frac{3}{2}, ...)$
(1, 1, 1) [8]	C_{3v}^D	12	$G(\frac{1}{2}, \frac{3}{2}, ...) \oplus F_1(\frac{3}{2}, \frac{5}{2}, ...) \oplus F_2(\frac{3}{2}, \frac{5}{2}, ...)$

Table 4. Little group, lattice irreps and angular momentum content per moving frame relevant for isospin 3/2 Δ-type operators.

group theory master formula

$$O^{\Lambda,r,n}(t, \vec{p}) = \frac{d_\Lambda}{g_{LG(\vec{P})}} \sum_{R \in LG(\vec{P})} \Gamma^\Lambda_{r,r}(R)\, U_R\, O(t, \vec{p})\, U_R^\dagger \quad (7)$$

In eq. (7) $O(\vec{p})$ stands for either single- or two-hadron operators (with appropriate 3-momentum arguments), Λ denotes an irrep of $LG(\vec{P})$ with dimension d_Λ, $r \in \{1, \dots, d_\Lambda\}$ one of its rows and g gives the number of elements in the little group. U_R is the unitary representation of the rotation on the operator algebra. For completeness, we list the action of rotations and reflections on the individual single-hadron interpolating fields for π, N, Δ in eqs. (8), (9) below.

$$\begin{aligned} U_R\, \pi(t, \vec{x})\, U_R^{-1} &= \pi(t, R^{-1}\vec{x}) \\ U_R\, N(t, \vec{x})\, U_R^{-1} &= S(R)\, N(t, R^{-1}\vec{x}) \\ U_R\, \Delta(t, \vec{x})_k^\alpha\, U_R^{-1} &= A(R)_{kk'}\, S(R)\, \Delta_{k'}(t, R^{-1}\vec{x}) \end{aligned} \quad (8)$$

$$\begin{aligned} I\, \pi(t, \vec{x})\, I^{-1} &= -\pi(t, -\vec{x}) \\ I\, N(t, \vec{x})\, I^{-1} &= \gamma_t\, N(t, -\vec{x}) \\ I\, \Delta(t, \vec{x})\, I^{-1} &= \gamma_t\, \Delta(t, -\vec{x}) \end{aligned} \quad (9)$$

In eq. (8) $S(R)$ is the 2-dimensional spin-1/2 representation of $SU(2)$ and $A(R)$ the 3-dimensional vector representation.

Given the many-to-one mapping of $SU(2)$ spin and vector degrees of freedom to lattice irrep states, an individual irrep Λ can appear multiple times for a specific little group. The multiplicity is counted using the characters of the representation matrices [27, 28]

$$m = \frac{1}{g_{LG(\vec{P})}} \sum_{R \in LG(\vec{P})} \chi^{\Gamma^\Lambda(R)}\, \chi^{M(R)}, \quad (10)$$

where M denotes the spin-1/2 bispinor representation for the nucleon Dirac-spinor and the vector-bispinor product representation for the Δ. The multiplicity pattern in the center-of-mass frame and the 3 moving frames used in this calculation is shown in the upper panel of Fig. 3 for the nucleon case and the lower panel for the Δ.

From the set of irrep-projected single- and two-hadron interpolators with all included momentum configurations we draw a linearly independent set using the Gram-Schmidt procedure and thus arrive at a sum total of 1720 Δ- and $N\pi$ operators, across all frames and irreps, from

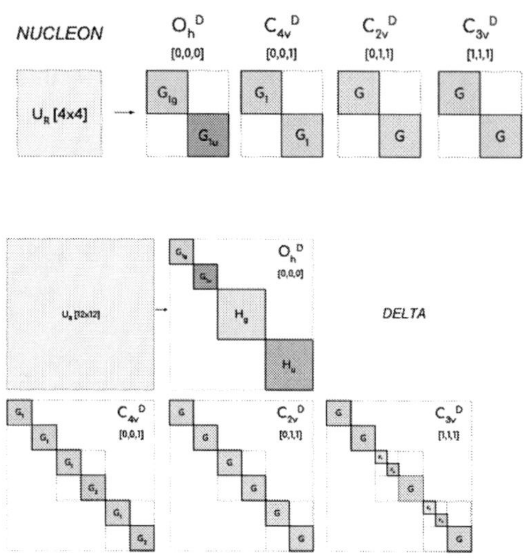

Figure 3. Multiplicities of irreps in frames with $\vec{P}/(2\pi/L)$ = $(0, 0, 0)$, $(0, 0, 1)$, $(0, 1, 1)$, $(1, 1, 1)$ for the nucleon (upper) and Delta.

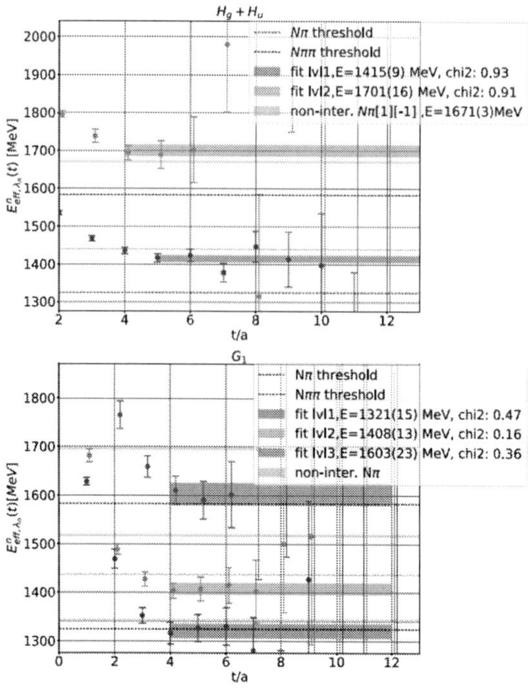

Figure 4. Examples of effective energy plateaus for combined $H_{u,g}$ in cmf and G_1 of the first moving frame.

which we then build the correlation matrices for the generalized eigenvalue problem. Total momentum $\vec{P}$ and the $LG(\vec{P})$ irrep are good quantum numbers for the finite-volume lattice, so the GEVP is solved per frame and per lattice irrep. In Fig. 4 we give two examples for the effective energy obtained from single exponential decay fits in the range $[t_{min}, t_{max}]$ to the generalized eigenvalues. In the top panel we show the effective energy plateau for the two lowest-lying levels of the combined H_g and H_u (time-forward and backward propagation) of the center-of-mass frame, $\vec{P} = 0$, $LG(\vec{P}) = O_h^D$. The bottom panel shows the same for the lowest 3 levels for the first moving frame and irrep G_1. The error bands with same color as the data points give the values from the nominal single exponential fits to the corresponding eigenvalues. The dashed lines give the $N\pi$ 2-particle elastic and the $N\pi\pi$ 3-particle inelastic threshold. Moreover, the light green bars locate the non-interacting $N\pi$ energy levels to appreciate the finite-volume induced shift to the interacting levels.

In Fig. 5 we demonstrate the stability of fit results for the same levels as in Fig. 4 under variation of the fit range $[t_{min}, t_{max}]$ of the single exponential fits to the lowest-lying eigenvalues, in particular of the lower limit t_{min}, which controls the impact of excited states. At each data point marking the fit result, the observed value for the fit's χ^2/dof is given. We find, that in general our extracted spectrum is stable within uncertainties for $t_{min}/a \geq 4$.

4 Scattering phase shift and resonance parameters

At this stage in our analysis we disregard partial wave mixing and thus have a 1:1 relation between between the lattice energy levels obtained in Sec. 3 and the infinite volume pion-nucleon P-wave elastic scattering phase shift $\delta_{J=3/2,l=1}$ from the Lüscher Quantization Condition (LQC) in eq. (11) [6].

$$\det\left[M^{\vec{P}}_{Jlm,J'l'm'}(s) - \delta_{JJ'}\,\delta_{ll'}\,\delta_{mm'}\,\cot(\delta_{Jl}(s))\right]_{\sqrt{s}=E_k^{\Lambda,\vec{P}}} = 0\,, \quad (11)$$

to be fulfilled for the kth energy level obtained from the frame with total momentum $\vec{P}$ and irrep Λ.
The LQC in eq. (11) is taken in the space of angular momentum states $|J, l, m\rangle$, for a given frame (total momentum $\vec{P}$). However, with irrep Λ, its row r and occurrence n good quantum numbers in the finite volume lattice calculation the Lüscher matrix can be block-diagonalized by a change to basis states $|\Lambda, r, n; J, l\rangle$ [6, 29]. An example of the outcome of this transformation is depicted in Fig. 6 for the case of $\vec{P} = (0, 0, 1)(2\pi/L)$ with $LG(\vec{P}) = C_{4v}^D$. Thus we obtain a corresponding factorized quantization condition, where individual factors solve the quantization of energy levels for individual irreps,

$$\det\left[M^{\Lambda,\vec{P}}_{Jln,J'l'n'}(s) - \delta_{JJ'}\,\delta_{ll'}\,\delta_{nn'}\,\cot(\delta_{Jl}(s))\right]_{\sqrt{s}=E_k^{\Lambda,\vec{P}}} = 0\,. \quad (12)$$

To extract resonance parameters for the Δ we use the Breit-Wigner parametrization of the scattering amplitude

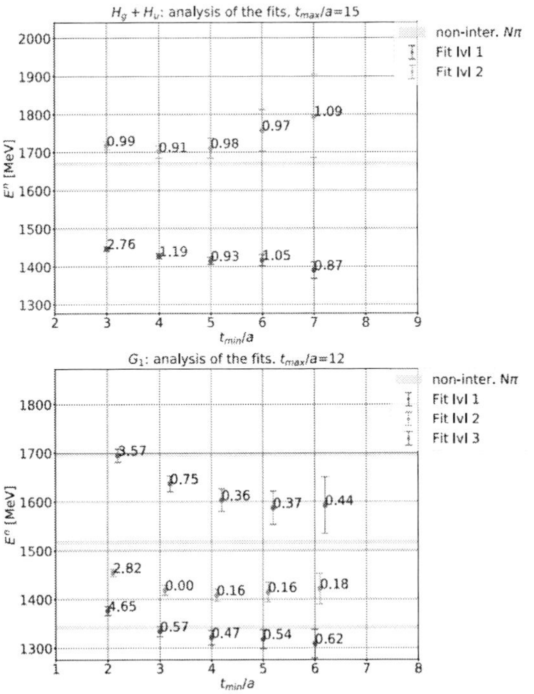

Figure 5. Check of stability of the GEVP energy levels by variation of lower end of the single exponential fit range; top: cmf first two levels; bottom: first three levels in G_1 of the first moving frame.

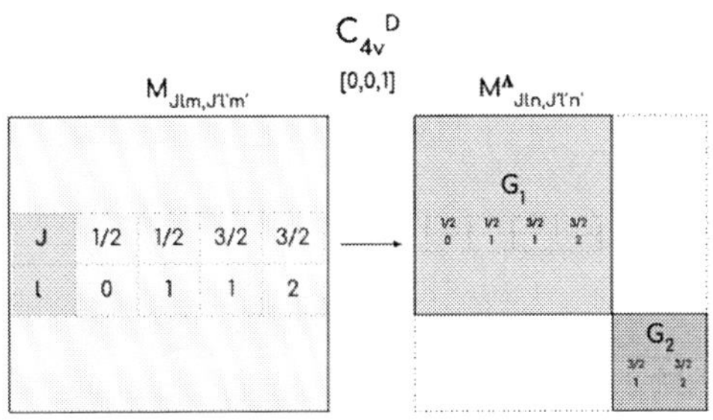

Figure 6. Block-diagonalization of the Lüscher matrix by change to irrep basis

(with contribution from a single partial wave at this point)

$$t_l(s) = \frac{\sqrt{s}\,\Gamma(s)}{m_R^2 - s - i\sqrt{s}\,\Gamma(s)}, \tag{13}$$

which incorporates the pole for the isolated, narrow Delta resonance.

Moreover, the s-dependence of the decay width Γ is described using the result from leading-order effective theory of the pion-nucleon system [30], which introduces the

Particle	J^P	$\Gamma_{N\pi}[MeV]$
$\Delta(1232)$	$3/2^+$	112.4(5)
$\Delta(1600)$	$3/2^+$	18(4)
$\Delta(1620)$	$1/2^-$	37(2)
$\Delta(1700)$	$3/2^-$	36(2)
...	...	...

Table 5. Experimentally observed resonances in the isospin-3/2 channel.

effective coupling $g_{\Delta-\pi N}$

$$\Gamma^{LO}_{EFT}(s) = \frac{g^2_{\Delta-\pi N}}{48\pi}\frac{E_N + m_N}{E_N + E_\pi}\frac{p^{*3}}{m_N^2} \tag{14}$$

To extract the target quantities m_R and $g_{\Delta-\pi N}$ we employ the inverse Lüscher method: with the Breit-Wigner parametrization of the scattering amplitude in Eq. (13) and together with the energy dependence of the decay width in Eq. (14) the quantization condition in Eq. (12) allows for the prediction of the expected lattice energy levels (disregarding the effect of non-zero lattice spacing artefacts). We compare the predicted values ("model") with our observed energy levels ("avg") and fit the resonance parameters to minimize the cost function in eq. (15).

$$\chi^2 = \sum_{\vec{P},\Lambda,n}\sum_{\vec{P}',\Lambda',n'}\left(\sqrt{s_n^{\Lambda,\vec{P}}[\text{avg}]} - \sqrt{s_n^{\Lambda,\vec{P}}[\text{model}]}\right)\times$$
$$\times\left[C^{-1}\right]^{\vec{P}',\Lambda',n'}_{\vec{P},\Lambda,n}\left(\sqrt{s_{n'}^{\Lambda',\vec{P}'}[\text{avg}]} - \sqrt{s_{n'}^{\Lambda',\vec{P}'}[\text{model}]}\right) \tag{15}$$

From the GEVP analysis in the nucleon channel we obtain our estimate for the nucleon mass given in Eq. (16) together with the pion mass for the ensemble studied here.

$$m_\pi = 258\,(1.1)\,\text{MeV}, \quad m_N = 1066.4\,(2.7)\,\text{MeV} \tag{16}$$

Our preliminary estimate of the Delta resonance mass and effective coupling to the pion-nucleon state is based on the evaluation of data from three frames and one irrep each, $\vec{P} = 0$, $H_{g/u}$, $\vec{P} = (0,0,1)\,2\pi/L$, G_2 and $\vec{P} = (0,1,1)\,2\pi/L$, G (cf. Tab. 4). The first two cases have our target continuum partial wave $J = 3/2$ as the leading contribution (with contamination from only higher partial waves, $J \geq 5/2$). The latter case on the other hand, has a contribution from $J = 1/2$. However, considering the experimentally observed excitations [31] collected in Tab. 5 (at physical pion mass), we expect the dominant resonant features to be only mildly influenced by higher-lying resonance contributions, largely to the upper end of the energy spectrum considered in our study. From the ensuing fit we then obtain the parameters

$$m_\Delta = 1414\,(128)\,\text{MeV}, \quad g_{\Delta-\pi N} = 26\,(4) \tag{17}$$

with fit quality $\chi^2/\text{dof} = 3.15$.

5 Summary and outlook

We present our data points for energy levels and phase shift $\delta_{J=3/2,l=1}$, which enter our fit, in Fig. 7. The green

Collaboration	Ref.	Methodology
Verduci (2014)	[18]	Dist., Lüscher
Alexandrou et.al. (2013)	[15]	Michael & McNeile
Alexandrou et.al. (2015)	[16]	Michael & McNeile
Andersen et.al. (2017)	[19]	Stoch. Dist., Lüscher
Our study (preliminary)	[20]	Smeared src., Lüscher
Physical value (PDG)	[32]	pheno, K-matrix

Table 6. Collection of contemporary results of studies of the Delta resonance mass and $g_{\Delta-\pi N}$ in lattice QCD and from analysis of experimental data (bottom line): analysis methods used.

Ref.	m_π [MeV]	m_Δ [MeV]	$g_{\Delta-\pi N}$
[18]	266	1396 (19)	19.90 (83)
[15]	360	1535 (25)	26.7 (1.5)
[16]	180	1350 (50)	23.7 (1.3)
[19]	280	1344 (20)	37.1 (9.2)
[20]	258	1414 (128)	26 (4)
[32]	140	1232(1)	29.4 (3), 28.6 (3)

Table 7. Collection of contemporary results of studies of the Delta resonance mass and $g_{\Delta-\pi N}$: numerical values obtained for the analyses listed in Tab. 6.

line shows the evaluation of the fit function with the best fit parameters and the shaded region gives the uncertainty obtained from the error propagation with the variance-covariance matrix of the fit parameters. We find in our

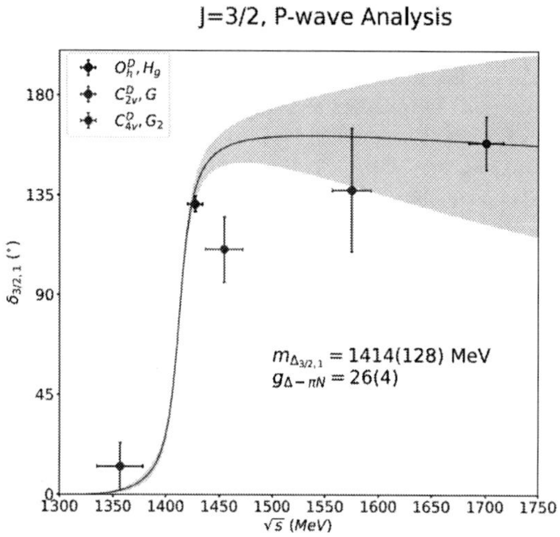

Figure 7. Our current result for the $I\left(J^P\right) = \frac{3}{2}\left(\frac{3}{2}^+\right)$ phase shift using the spectrum from little groups O_h^D, C_{4v}^D and C_{2v}^D.

data the generically expected features of resonant behavior in the phase shift with a steep rise from $\delta_{3/2,1} = 0$ to ≈ 180 degrees. However, at present the levels from C_{2v}^D/G give rise to some tension with the model prediction, leading to a χ^2/dof-value of about 3 in our fit. Their role deserves further investigation, in particular since right for this case the angular momentum content of the irrep starts at $J = 1/2$.

Finally, we compare the preliminary results of our study presented here with other determinations of Delta resonance parameters in lattice QCD calculations in Tabs. 6, 7. The comparison can only be a superficial one at this point, as the results listed stem from very different calculations, without commonalities like pion mass or a continuum extrapolation. Nevertheless, we find rough agreement of m_Δ for comparable pion mass. The effective coupling is already in the region of the value extracted from phenomenological and K-matrix analysis of experimental data (last line in both tables). The latter tendency is consistent for the dimensionless quantity $g_{\Delta-\pi N}$, which is expected to display only a mild quark mass dependence.

At the time of this report, we continue to build up the gauge statistics for the ensemble `A7`, which is the basis for the data shown here. Moreover, two immediate follow-up simulations at same pion mass but larger volume and at smaller pion mass of approximately 200 MeV are in preparation.

6 Acknowledgements

This research used resources of the National Energy Research Scientific Computing Center (NERSC), a U.S. Department of Energy Office of Science User Facility operated under Contract No. DE-AC02-05CH11231. SM and GR are supported by the U.S. Department of Energy, Office of Science, Office of High Energy Physics under Award Number DE-SC0009913. SK is supported by the Deutsche Forschungsgemeinschaft grant SFB-TRR 55. SK and GS were partially funded by the IVF of the HGF. SM and SS also thank the RIKEN BNL Research Center for support. JN was supported in part by the DOE Office of Nuclear Physics under grant DE-SC-0011090. AP was supported in part by the U.S. Department of Energy Office of Nuclear Physics under grant DE-FC02-06ER41444. LL was supported by the U.S. Department of Energy, Office of Science, Office of Nuclear Physics under contract DE-AC05-06OR23177. SP was supported by the Horizon 2020 of the European Commission research and innovation programme under the Marie Sklodowska-Curie grant agreement No. 642069. MP gratefully acknowledges support by the Sino-German collaborative research center CRC-110. We acknowledge the use of the USQCD software QLUA [33] for the calculation of the correlators.

References

[1] M. Lüscher, U. Wolff, Nuclear Physics B **339**, 222 (1990)

[2] K. Rummukainen, S.A. Gottlieb, Nucl. Phys. **B450**, 397 (1995), `hep-lat/9503028`

[3] C.H. Kim, C.T. Sachrajda, S.R. Sharpe, Nucl. Phys. **B727**, 218 (2005), `hep-lat/0507006`

[4] L. Leskovec, S. Prelovsek, Phys. Rev. **D85**, 114507 (2012), 1202.2145
[5] R.A. Briceño, Phys. Rev. **D89**, 074507 (2014), 1401.3312
[6] M. Gockeler, R. Horsley, M. Lage, U.G. Meissner, P.E.L. Rakow, A. Rusetsky, G. Schierholz, J.M. Zanotti, Phys. Rev. **D86**, 094513 (2012), 1206.4141
[7] R.A. Briceño, M.T. Hansen, S.R. Sharpe, Phys. Rev. **D95**, 074510 (2017), 1701.07465
[8] L. Alvarez-Ruso et al., Prog. Part. Nucl. Phys. **100**, 1 (2018), 1706.03621
[9] L. Lellouch, M. Lüscher, Commun.Math.Phys. **219**, 31 (2001)
[10] M.T. Hansen, S.R. Sharpe, Phys. Rev. **D86**, 016007 (2012), 1204.0826
[11] R.A. Briceño, M.T. Hansen, A. Walker-Loud, Phys. Rev. **D91**, 034501 (2015), 1406.5965
[12] A. Agadjanov, V. Bernard, U.G. Meißner, A. Rusetsky, Nuclear Physics B **886**, 1199 (2014)
[13] R.A. Briceño, M.T. Hansen, Phys. Rev. **D94**, 013008 (2016), 1509.08507
[14] R.A. Briceño, M.T. Hansen, Phys. Rev. **D92**, 074509 (2015), 1502.04314
[15] C. Alexandrou, J.W. Negele, M. Petschlies, A. Strelchenko, A. Tsapalis, Phys. Rev. **D88**, 031501 (2013), 1305.6081
[16] C. Alexandrou, J.W. Negele, M. Petschlies, A.V. Pochinsky, S.N. Syritsyn, Phys. Rev. **D93**, 114515 (2016), 1507.02724
[17] U.G. Meissner, J. Phys. Conf. Ser. **295**, 012001 (2011), 1012.0924
[18] V. Verduci, Ph.D. thesis, Graz U. (2014), http://unipub.uni-graz.at/obvugrhs/content/titleinfo/252105
[19] C.W. Andersen, J. Bulava, B. Hörz, C. Morningstar, Phys. Rev. **D97**, 014506 (2018), 1710.01557
[20] S. Paul et al., PoS **LATTICE2018**, 089 (2018), 1812.01059
[21] S. Durr, Z. Fodor, C. Hoelbling, S.D. Katz, S. Krieg, T. Kurth, L. Lellouch, T. Lippert, K.K. Szabo, G. Vulvert, JHEP **08**, 148 (2011), 1011.2711
[22] B. Blossier, M. Della Morte, G. von Hippel, T. Mendes, R. Sommer, JHEP **04**, 094 (2009), 0902.1265
[23] C. Alexandrou, L. Leskovec, S. Meinel, J. Negele, S. Paul, M. Petschlies, A. Pochinsky, G. Rendon, S. Syritsyn, Phys. Rev. **D96**, 034525 (2017), 1704.05439
[24] G. Rendon, L. Leskovec, S. Meinel, J. Negele, S. Paul, M. Petschlies, A. Pochinsky, G. Silvi, S. Syritsyn, PoS **LATTICE2018**, 073 (2019), 1811.10750
[25] S. Güsken, U. Low, K.H. Mutter, R. Sommer, A. Patel, K. Schilling, Phys. Lett. **B227**, 266 (1989)
[26] M. Albanese, F. Costantini, G. Fiorentini, F. Flore, M. Lombardo, R. Tripiccione, P. Bacilieri, L. Fonti, P. Giacomelli, E. Remiddi et al., Physics Letters B **192**, 163 (1987)
[27] D.C. Moore, G.T. Fleming, Phys. Rev. **D74**, 054504 (2006), hep-lat/0607004
[28] F.A. Cotton, *Chemical applications of group theory* (John Wiley & Sons, 2003)
[29] V. Bernard, M. Lage, U.G. Meissner, A. Rusetsky, JHEP **08**, 024 (2008), 0806.4495
[30] V. Pascalutsa, M. Vanderhaeghen, Phys. Rev. **D73**, 034003 (2006), hep-ph/0512244
[31] K. Olive, Chinese Physics C **40**, 100001 (2016)
[32] C. Patrignani et al. (Particle Data Group), Chin. Phys. **C40**, 100001 (2016)
[33] *USQCD software Qlua package*, https://usqcd.lns.mit.edu/w/index.php/QLUA

Quark model calculations of transition form factors at high photon virtualities

G. Ramalho[1]

[1]Laboratório de Física Teórica e Computacional – LFTC, Universidade Cruzeiro do Sul/Universidade Cidade de São Paulo, 01506-000, São Paulo, SP, Brazil

Abstract. We present calculations of $\gamma^* N \to N^*$ transition form factors, where N is the nucleon and N^* is a nucleon resonance, based on a covariant quark model. Our main focus is at high photon virtualities (large Q^2) where the valence quark degrees of freedom dominate the contributions to the transition form factors and helicity amplitudes. In that regime, the quark model estimates can be compared with the available data, particularly with the Jefferson Lab data at intermediate and large momentum transfer ($Q^2 > 2$ GeV^2). The main focus is on the $\Delta(1232)3/2^+$, $N(1440)1/2^+$, $N(1535)1/2^-$ and $N(1520)3/2^-$ resonances, but estimates for other higher mass resonances are also discussed.

1 Introduction

In the last two decades important information about the electromagnetic structure of the nucleon and the nucleon resonances have been collected in modern facilities such as Jefferson Lab (JLab), MAMI, MIT-Bates and ELSA, among others [1–5]. Helicity amplitudes and form factors associated with the $\gamma^* N \to N^*$ transitions have been measured for resonances N^* with masses up to 2 GeV for values of Q^2 in the range $Q^2 = 0$– 8 GeV^2 [1, 6], where $Q^2 = -q^2$, and q is the momentum transfer. With the JLab-12 GeV upgrade, we expect to probe even larger values of Q^2 [1, 5].

To interpret the new data above $Q^2 = 2$ GeV^2, theoretical models based on relativity are fundamental. Models which include valence quark degrees of freedom are preferable, since those are the expected dominant degrees of freedom at large Q^2 [1–4]. These models can be used to make predictions for transition form factors at large Q^2, and may also be used to guide future experiments.

In the present work, we discuss mainly estimates from the covariant spectator quark model [5, 7, 8]. The covariant spectator quark model is a model based on constituent quarks where the quark electromagnetic structure is parametrized in order to describe the nucleon electromagnetic structure. The wave functions of the baryons are ruled by the spin-flavor-radial symmetries [9], with radial wave functions determined phenomenologically with the assistance of empirical data, lattice data or estimates of the quark core contributions [1, 5]. One can then use parametrizations of a few resonances N^* to make predictions for other states based on the symmetries. The model is covariant by construction and can then be used in applications at large Q^2. In some cases the model can be extended with the inclusion of effective descriptions of the meson cloud effects, which can have significant contributions at small Q^2 [10–19].

In the next section we describe the formalism associated with the covariant spectator quark model. In Sect. 3, we present estimates for $\gamma^* N \to N^*$ transition form factors at large Q^2 for the states: $\Delta(1232)$, $N(1440)$, $N(1520)$, $N(1535)$, $N(1880)$, $N(1700)$ and $\Delta(1700)$. In Sect. 4 we present estimates for the $\Delta(1232)$ quadrupole form factors at low Q^2. Our summary and conclusions are presented in Sect. 5.

2 Covariant spectator quark model

The covariant spectator quark model [5, 7, 8] is based on the three main ingredients:

1. the baryon wave function (including the nucleon) is represented in terms of the spin-flavor structure of the individual quarks based on the $SU_S(2) \times SU_F(3)$ spin-flavor symmetry, rearranged as an active quark and a spectator quark-pair [7, 8, 20];

2. within the impulse approximation, the three-quark system is reduced to a quark-diquark system, parametrized by a radial wave function ψ_B, after the integration into the quark-pair degrees of freedom [7, 8, 20];

3. the electromagnetic structure of the quark is parametrized by quark isoscalar/isovector and strange quark form factors $f_{i\pm}(Q^2)$ and $f_{is}(Q^2)$ ($i = 1$ for Dirac, and $i = 2$ for Pauli), which simulate the substructure associated with the gluons and quark-antiquark effects, represented according to the vector meson mechanism [7, 8, 12, 21].

When the nucleon wave function (Ψ_N) and the resonance wave function (Ψ_R) are both expressed in terms of the single quark and quark-pair states, the transition current in impulse approximation can be written as [7, 8, 20]

$$J^\mu = 3 \sum_\Gamma \int_k \bar{\Psi}_R(P_R, k) j_q^\mu \Psi_N(P_N, k), \qquad (1)$$

where P_R, P_N, and k are the resonance, the nucleon, and the diquark momenta, respectively. In the previous equation the index Γ labels the intermediate diquark polarization states, the factor 3 takes account of the contributions from the other quark pairs by the symmetry, and the integration symbol represents the covariant integration over the diquark on-mass-shell momentum. In the study of the inelastic transitions, we use the Landau prescription to ensure the current conservation [14, 22, 23].

Using Eq. (1), we can express the transition current in terms of the quark electromagnetic form factors $f_{i\pm}$ ($i = 1, 2$) and the radial wave functions ψ_N and ψ_R for resonances with no strange quarks and determine the transition form factors [7, 14, 23, 24]. Taking advantage of the quark form factor structure, based on vector meson dominance, the model has been extended to the lattice QCD regime (heavy pions and reduced meson cloud) [8, 12, 21], to the nuclear medium [25, 26] and to the timelike regime ($Q^2 < 0$) [15, 17, 27, 28].

The present description of the covariant spectator quark model takes into account only the effects associated with the valence quark degrees of freedom. There are however some processes, such as the meson exchanged between the different quarks inside the baryon, which cannot be reduced to processes associated with the dressing of a single quark. Those processes can be regarded as a consequence of the meson exchanged between the different quarks inside the baryon, and can be classified as meson cloud corrections to the hadronic reactions [14, 15, 18, 19].

The study of the role of the meson cloud effects on the $\gamma^* N \to N^*$ transition can be done also in the context of the dynamical coupled-channel reaction models (dynamical models) [4, 29–32]. Those models use baryon-meson states to describe the photo- and electro-production of mesons by nucleons, taking into account the meson dressing of propagators and vertices, and are fitted to the data. The bare core contribution to the data can be determined by removing the effect of the meson-baryon dressing of propagators and vertices [31]. Those estimates of the bare core can be used to test the limits of models based on valence quarks or even to calibrate parameters of the quark models.

The separation of the transition form factors into the valence quark and meson cloud contributions is model dependent [33, 34], and the identification of the bare baryon states depends of the calibration of the background [4, 32]. One can, however, reduce the impact of the model dependence comparing our estimates of the valence quark contributions with lattice QCD simulations for large pion masses. Although quenched and unquenched lattice QCD simulations include some meson cloud effects, those effects are reduced for large pion masses [35, 36].

To compare our model with the lattice QCD simulations, we use our calibration of the radial wave functions for the baryon lattice QCD masses and the quark current, j_q^μ, expressed in terms of the lattice QCD vector meson masses [8, 12, 21]. The free parameters of the radial wave functions and quark current are the same in the physical and lattice QCD regimes. The extrapolation of the results for the physical limit is obtained when we replace the dependence on the lattice QCD masses by the physical masses.

3 Transition form factors at large Q^2

We present next our estimates for the transition form factors for the $\Delta(1232)$, radial excitations of the nucleon and some negative parity states, including the states $N(1520)$ and $N(1535)$.

We use M_N and M_R to represent the nucleon and the resonance masses, respectively. In the case of the $\Delta(1232)$, we use M_Δ, exceptionally.

3.1 $\gamma^* N \to \Delta(1232)$ transition

The $\gamma^* N \to \Delta(1232)$ transition is known to be dominated by the magnetic transition [1, 37], as a consequence of a quark spin flip in the nucleon to form a $\Delta(1232)$ with spin $\frac{3}{2}$ (Δ^+ or Δ^0). Quark models, however, underestimate the magnitude of the magnetic form factor, G_M^*, near $Q^2 = 0$. This property can be naturally explained in the context of the covariant spectator quark model, when we consider that the nucleon and the $\Delta(1232)$ systems are mainly composed by s-wave quark-diquark states [10]. In that case, one concludes, based on the normalization of the radial wave functions and the Cauchy-Schwarz inequality for integrals that $G_M^*(0) \leq 2.07$, well below the experimental value $G_M^*(0) \simeq 3$ [5, 10, 18].

The missing strength in quark model estimates of G_M^* can be understood when we consider the effect of the pion cloud dressing in addition to the valence quark effects. Dynamical models estimate that the contribution of the pion cloud effects can be about 30-45% of G_M^* at low Q^2 [4, 29–31]. The natural conclusion is that quark models cannot describe the $\gamma^* N \to \Delta(1232)$ completely, and therefore the $\Delta(1232)$ radial wave function ψ_Δ cannot be extracted directly from the data [5, 10, 12].

One needs, then to rely on indirect methods to estimate ψ_Δ. There are two main options: i) use some independent estimate of the contribution of the valence quark effects to G_M^*; ii) use lattice QCD simulations of the form factors to determine the function ψ_Δ, assuming that the pion cloud effects are small due to the large pion masses.

The first option can be implemented using the EBAC/Argonne-Osaka [31] estimate for the bare core contributions, obtained when the baryon-meson couplings are turned off [30, 31]. In this limit, we obtain the blue points from Fig. 1 in the left and central panels. One can see that those points (bare contribution) are well described by the fits used to determine ψ_Δ (dashed lines in the left and central panels). At large Q^2, we can see in the central panel of Fig. 1, that the valence quark contributions are dominant and describe well the data.

As mentioned, ψ_Δ, can also be estimated by a fit of the model parameters to a lattice QCD data with

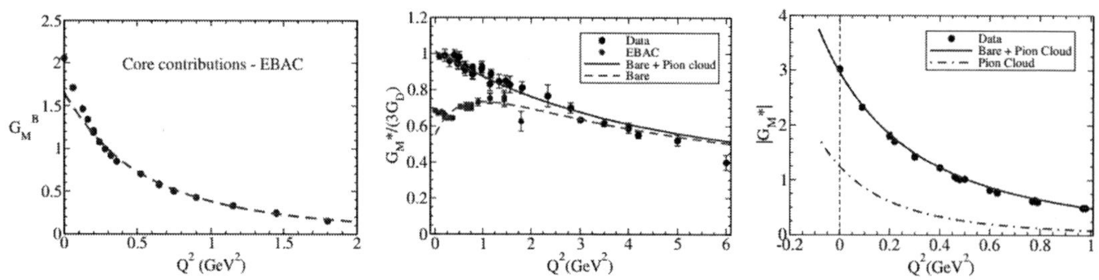

Figure 1. $\gamma^* N \to \Delta(1232)$ magnetic form factor. **Left:** Comparison of bare contribution with EBAC estimate from [31]. **Center:** Form factor normalized by $3G_D$, where $G_D = (1 + Q^2/\Lambda_D^2)^{-2}$, with $\Lambda_D^2 = 0.71$ GeV^2. **Right:** Re-parametrization of $|G_M^*|$ based on expression (2) for the pion cloud with $\Lambda_D^2 = 0.9$ GeV^2 [28]. Extension to the timelike region. Database from [10, 11].

$m_\pi \simeq 400$–600 MeV [12, 36]. In this case, the pion cloud effects are reduced and one obtains a cleaner estimate of the valence quark effects. Based on the extension of the model to the lattice QCD regime [12, 21], we conclude that the parametrization derived from the EBAC estimate is also consistent with the lattice QCD data from [36], as shown in [12]. The main conclusion is then that ψ_Δ can be determined either by lattice QCD data or by the EBAC estimate of the bare core.

To describe the G_M^* data, we need to include some estimate of the pion cloud effects (G_M^π). Since our framework is based on valence quarks, we choose to simulate the pion cloud effects by a simple parametrization: $G_M^\pi = 3\lambda_\pi G_D \left(\frac{\Lambda_\pi^2}{\Lambda_\pi^2+Q^2}\right)^2$, where $\lambda_\pi \simeq 0.448$ and $\Lambda_\pi^2 \simeq 1.53$ GeV^2 [11]. The final result for G_M^* is represented by the solid line in the central and right panels of Fig. 1. In the more recent applications of the model, which include the $\gamma^* N \to \Delta(1232)$ transition in the timelike region [27, 28], we consider an alternative parametrization based on the connection with the microscopic pion cloud mechanisms [18, 28]. We consider in particular, the parametrization

$$G_M^\pi = \frac{3}{2}\lambda_\pi F_\pi(q^2)\left(\frac{\Lambda_\pi^2}{\Lambda_\pi^2 - q^2}\right)^2 + \frac{3}{2}\lambda_\pi \frac{\Lambda_D^4}{(\Lambda_D^2 - q^2)^2 + \Lambda_D^2[\Gamma_D(q^2)]^2}, \quad (2)$$

where the first term is associated with the photon-pion coupling, proportional to the pion electromagnetic form factor $F_\pi(q^2)$, and the second term is associated with the photon coupling with the intermediate baryon states. This last coupling is simulated by an effective dipole function G_D with $\Lambda_D^2 = 0.90$ GeV^2, and include an effective width Γ_D in order to avoid singularities in the region $q^2 = -Q^2 > 0$. The new parametrization is presented in the right panel of Fig. 1 (dashed-dotted line). The advantage of the last parametrization of G_M^π is that the model can be naturally extended to the timelike region, above $Q^2 = -(M_\Delta - M_N)^2$. These estimates are useful for the study of the Δ Dalitz decay ($\Delta \to e^+e^- N$) measured at HADES. The comparison of our calculations with the HADES $\Delta(1232)$ Dalitz decay cross-sections can be found in [38]. Those results are also discussed in the presentations of B. Ramstain and P. Salabura [39, 40].

The present discussion of the $\gamma^* N \to \Delta(1232)$ transition is simplified because we ignore the contributions of the $\Delta(1232)$ d-wave states [11, 12, 41, 42]. The procedure is justified because the admixture terms are of the order of 0.7% for both states. Those states are, however, important for the sub-leading transition form factors, the electric (G_E^*) and Coulomb (G_C^*) quadrupole form factors. Those form factors are described in the Sect. 4, at low Q^2.

3.2 Radial excitations

The covariant spectator quark model can also be applied to the radial excitations of the nucleon. The first radial excitation, $N(1440)$ or Roper, can be defined as a system which differs from the nucleon only by the radial structure, and has a zero in the radial wave function [43, 44]. The $N(1440)$ state can then be determined defining a radial wave function ψ_R which is orthogonal to the nucleon and has a single node. The transition form factors are then calculated without any adjustable parameters. The results (solid line) are presented in Fig. 2 in comparison with the CLAS data [45–47].

From the graphs, we conclude that the model describe well the data above 1.5 GeV^2 for both form factors. Those results are an indication that $N(1440)$ is in fact the first radial excitation of the nucleon. The disagreement at low Q^2 can be a consequence of not taking into account possible meson cloud effects [43, 44].

The $\gamma^* N \to N(1440)$ form factors have been also estimated within the AdS/QCD framework [48, 49]. In the leading twist approximation, where the baryons correspond to three-quark systems, the nucleon and the Roper transition form factors depend on three independent couplings, which can be fixed by the nucleon elastic form factor data, within certain limits. The results for the Roper are then determined without any additional fit. Those are also represented in Fig. 2 by the red band between the upper and lower limits. The results of the Pauli form factor F_2^* are in excellent agreement with the data. Overall, it is interesting to observe that the two estimates converge to the same result for large values of Q^2, which can be seen as a consequence of the dominance of the valence quark degrees of freedom at large Q^2.

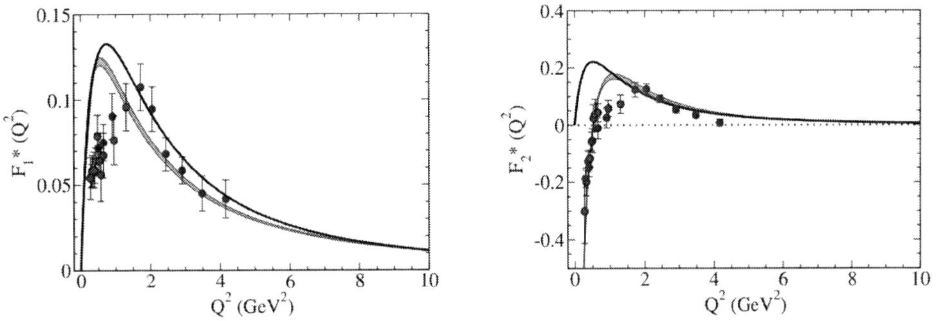

Figure 2. $\gamma^* N \to N(1440)$ transition form factors. The solid line represent the estimate of the covariant quark model [43]. The red band represent the estimated from AdS/QCD taking into account the uncertainty of the couplings [48]. Data from CLAS [45–47].

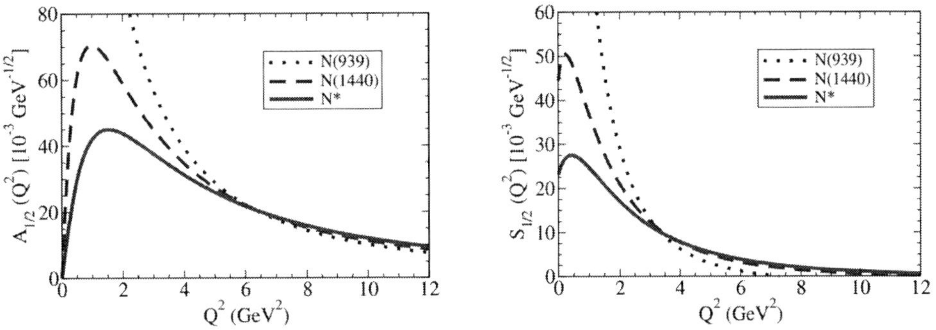

Figure 3. $\gamma^* N \to N^*$ transition amplitudes for the first and the second (N^*) radial excitations of the nucleon. $N(1930)$ represent the nucleon's equivalent amplitudes (description in the main text).

The covariant spectator quark model can also be used to estimate the second radial excitation of the nucleon. In a previous work [44], we assumed tentatively that the state could be the $N(1710)\frac{1}{2}^+$, but the assumption was discarded by recent measurements from CLAS [50]. The next candidate is then the state $N(1880)\frac{1}{2}^+$. In this case, we compare the results directly to the helicity amplitudes $A_{1/2} \propto (F_{1p}^* + F_{2p}^*)$ and $S_{1/2} \propto (F_{1p}^* - \tau F_{2p}^*)$, where $\tau = \frac{Q^2}{(M_N+M_R)^2}$. The results for the Roper and the next radial excitation (N^*) are presented in Fig. 3. In the graphs, we include also the equivalent amplitudes of the nucleon $A_{1/2} \propto G_M$ and $S_{1/2} \propto G_E$ (the conversion factors are calculated using the mass of the Roper) [44]. Note that, the helicity amplitude representation is particularly useful, because it emphasizes the similarity of the amplitudes associated with the different states at large Q^2, particularly to $A_{1/2}$. The corollary of this result is that, one can use the experimental results of the magnetic form factor of the proton, G_{Mp}, to predict the amplitude $A_{1/2}$ for the Roper and $N(1880)\frac{1}{2}^+$ states in the range $Q^2 = 10$–30 GeV^2. Hopefully our predictions for $A_{1/2} \propto G_{Mp}$ can be tested in a near future.

Similarly to the case of the nucleon, we can use the formalism to estimate the first radial excitation of the $\Delta(1232)$, the state $\Delta(1600)$, based on our model for the $\Delta(1232)$. Our estimates are valid for large Q^2 and differ from the present measurements at $Q^2 = 0$, although the estimates approach the experimental data when we take into account the meson cloud dressing based on a simplified model derived from the $\gamma^* N \to \Delta(1232)$ pion cloud parametrization [13]. Our estimates for the $\gamma^* N \to \Delta(1600)$ form factors may be tested in a near future, since data are under analysis and results are expected soon [51].

3.3 $\gamma^* N \to N(1520)$ and $\gamma^* N \to N(1535)$ transitions

The $N(1520)$ and $N(1535)$ are negative parity states. Applications of the covariant spectator quark model to the $\gamma^* N \to N(1535)$ and $\gamma^* N \to N(1520)$ transitions are discussed in Refs. [14, 15, 24] based on different prescriptions to the radial wave functions.

In particular the $\gamma^* N \to N(1520)$ transition was extended to the timelike region [15] and to the study of the HADES $N(1520)$ Dalitz decays. Some preliminary results are discussed in the B. Ramstein presentation [39].

Recently the $\gamma^* N \to N(1535)$ and $\gamma^* N \to N(1520)$ transitions have been studied within the semirelativistic approximation. The semirelativistic approximation is characterized by the following properties [22]: i) the mass difference (M_R and M_N) is neglected in a first approximation; ii) the radial wave function of the resonance ψ_R has the same form as the nucleon radial wave function.

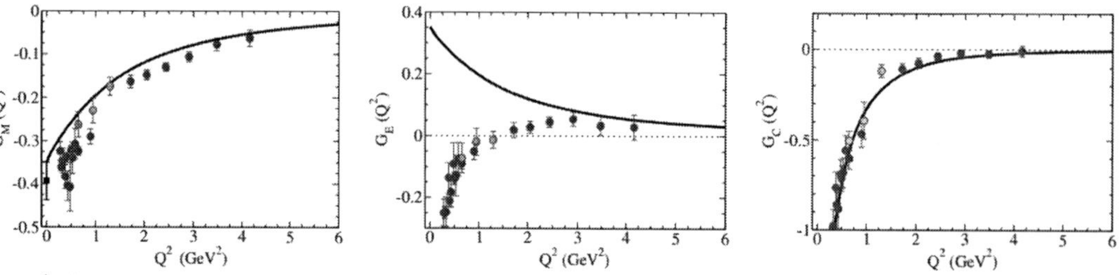

Figure 4. $\gamma^* N \to N(1520)$ transition form factors. The solid line is the result of the semirelativistic approximation. Data from JLab (circles) [6, 45, 47] and PDG (square) [6].

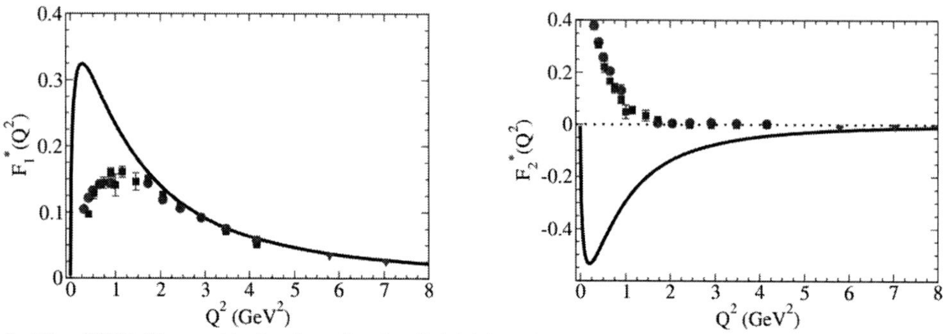

Figure 5. $\gamma^* N \to N(1535)$ Dirac transition form factor. Data from JLab (circles and triangles) [6, 45, 54] and MAID (squares) [3].

Under these conditions the orthogonality between the states is naturally ensured, and the transition form factors are calculated without the introduction of any additional parameters [22]. Those estimates are therefore true predictions. All model parameters are determined by the original parametrization of the nucleon system. The semirelativistic estimates for the $\gamma^* N \to N(1520)$ and $\gamma^* N \to N(1535)$ form factors are presented in Figs. 4 and 5, respectively.

In Fig. 4 we compare our estimates for the $\gamma^* N \to N(1520)$ transition form factors: G_M (magnetic), G_E (electric) and G_C (Coulomb) with the available data. One can notice that one obtain a good description of the data for G_M and G_C, particularly for $Q^2 > 1$ GeV^2. As for G_E the estimate fails for $Q^2 < 2.5$ GeV^2, which seems to be a consequence of the omission of the meson cloud effects. In the context of the covariant spectator quark model the result for G_E is a consequence of the estimate $A_{3/2} \equiv 0$, indicating that the amplitude $A_{3/2}$ is dominated by meson cloud effects [14, 15].

In Fig. 5, we compare the estimates of the Dirac (F_1^*) and Pauli (F_2^*) $\gamma^* N \to N(1535)$ with the available data. From the graph for F_1^*, we conclude that the model describe well the data above 1.5 GeV^2. From the graph for F_2^*, we can notice that the model fails completely the description of the data (wrong sign). These two discrepancies may be an indication of the impact of the meson cloud effects, not included in the present calculations. There are evidences that the inclusion of meson cloud effects can in fact improve the description of the data [16, 17, 23, 24, 52, 53].

The results for F_2^* in particular raises the possibility that there are cancellations between the valence quark effects (estimated by our model) and the meson cloud effects (not included in our estimate). In that case, we can explain the experimental result $F_2^* \approx 0$ for $Q^2 > 1.5$ GeV^2. The implication of the previous result is that the $\gamma^* N \to N(1535)$ amplitudes are related by [52]

$$S_{1/2} = -\frac{\sqrt{1+\tau}}{\sqrt{2}} \frac{M_R^2 - M_N^2}{2M_R Q} A_{1/2}, \tag{3}$$

for $Q^2 > 1.5$ GeV^2, where $\tau = \frac{Q^2}{(M_R+M_N)^2}$ and $Q = \sqrt{Q^2}$.

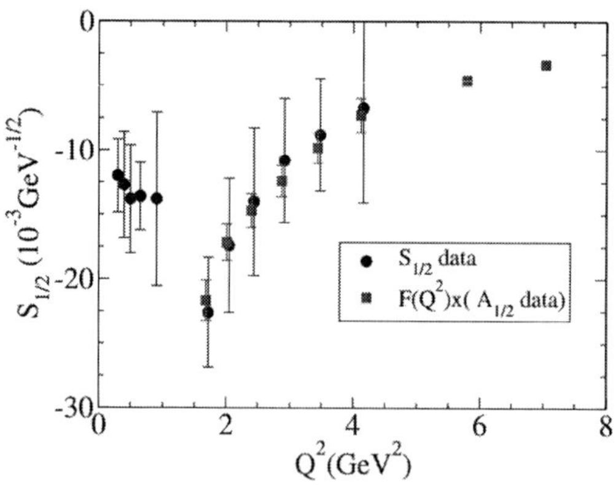

Figure 6. $\gamma^* N \to N(1535)$ amplitude $S_{1/2}$. Data compared with estimate from Eq. (3). Data from CLAS [45, 54].

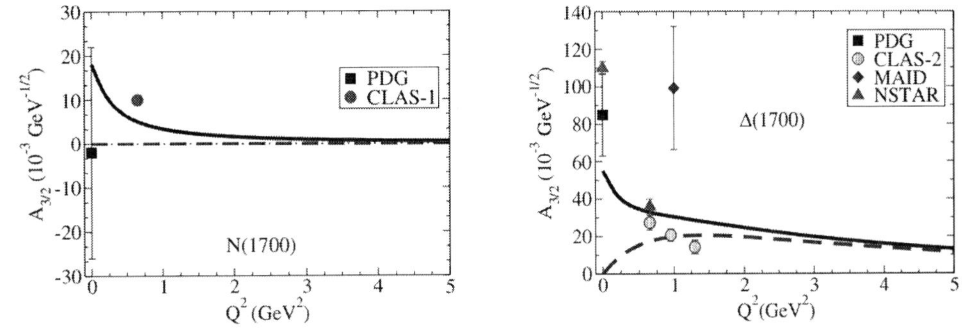

Figure 7. Examples of estimates of amplitudes associated with negative parity states based on the single quark transition model [23]. The solid line represent a model where the meson cloud effects are taken into account. Check [23] for the description of the data.

The relation (3) is tested in Fig. 6 based on the data for the amplitude $A_{1/2}$. The last two point present a prediction for the $S_{1/2}$ data. New large Q^2 data are needed in order to test the relation (3) for higher values of Q^2.

The results for the $\gamma^* N \to N(1535)$ transition motivated the use of the valence quark model to calibrate the meson cloud contributions and the extension of the results to the timelike region and the $N(1535)$ Dalitz decay [17].

Overall, we can conclude that the semirelativistic approximation provides a good description of the $\gamma^* N \to N(1520)$ and $\gamma^* N \to N(1535)$ transition form factors for $Q^2 > 1.5$ GeV^2. The exceptions are the Pauli form factor for $N(1535)$ and the electric form factor for $N(1520)$, the two cases for which there are indications that meson cloud effects may be more significant. We emphasize that all the calculations include no free parameters, apart the calibration of the nucleon system. A more detailed discussion of the results for the $\gamma^* N \to N(1535)$ and $\gamma^* N \to N(1520)$ transitions can be found in [14, 16, 17, 23, 24, 52, 53].

3.4 Single Quark Transition Model

The covariant spectator quark model can also be used in combination with single quark transition model (SQTM) [23, 55], since the two models are based on similar principles: electromagnetic interaction in impulse approximation and wave functions determined by the $SU(6) \otimes O(3)$ structure [9].

Of particular interest is the $[70, 1^-]$ supermultiplet, whose transverse amplitudes, $A_{1/2}$ and $A_{3/2}$, can be determined by three independent functions of Q^2: A, B and C, according with the SQTM. In these circumstances, one can determine the transverse amplitudes of the states $N(1650)\frac{1}{2}^-$, $\Delta(1620)\frac{1}{2}^-$, $N(1700)\frac{3}{2}^-$ and $\Delta(1700)\frac{3}{2}^-$, once the functions A, B and C are known. These three functions can be calculated using the covariant spectator quark models for the $N(1520)\frac{3}{2}^-$ and $N(1535)\frac{1}{2}^-$ transverse amplitudes [5, 14, 23, 24].

The estimates for the $N(1650)\frac{1}{2}^-$ and $\Delta(1620)\frac{1}{2}^-$ amplitudes are consistent with available data [3, 6] for large Q^2 [23]. For the remaining cases, there are no available data for $Q^2 > 2$ GeV^2. As an example, we present in Fig. 7, the results for the amplitude $A_{3/2}$ associated with the states $N(1700)$ and $\Delta(1700)$. Note in both cases the lack of data for $Q^2 > 1.5$ GeV^2. Our estimates can be tested, then only when new large Q^2 data from the JLab-12 GeV upgrade became available [1].

4 Transition form factors at low Q^2

One can also use the covariant spectator quark model to estimate transition form factors at low Q^2. In general the estimates cannot be compared directly with the data because the meson cloud contributions can be significant and camouflage the valence quark contributions.

In the case of the $\gamma^* N \to \Delta(1232)$ there is the possibility of estimating the valence quark effects since there are lattice QCD data available [36]. We obtain non-zero contributions to the electric (G_E) and Coulomb (G_C) quadrupole form factors when we consider d-wave contributions to the $\Delta(1232)$ wave function [11, 12, 41, 42]. The contributions of those states for the transition form factors are small (about one order of magnitude) when compared with the experimental data, but their magnitude is comparable with the lattice QCD data [11, 12]. One can then use lattice QCD data to estimate the free parameters of the quark model performing fits to the data [12, 21]. Once extrapolated to the physical limit, one obtains the valence quark contributions to the quadrupole form factors [12].

The experimental data for G_E and G_C are represented in Fig. 8. For convenience of the discussion the results for G_C are multiplied by $\kappa = \frac{M_\Delta - M_N}{2M_\Delta}$. The valence quark contributions discussed above are represented by the thin lines (solid line for G_E and dashed line for κG_C). Notice, in particular, the difference of magnitude in comparison with the final results (thick lines). The thick lines are discussed next.

To describe the physical data, we need to take into account the pion cloud contributions to the electric and Coulomb quadrupole form factors. There are in the literature estimates of the pion cloud contributions to the quadrupole form factors based on the large-N_c limit [56, 57]. Those estimates are, however, inconsistent with the long wavelength limit also known as Siegert's theorem, which states that in the limit $Q^2 = -(M_\Delta - M_N)^2$, the two

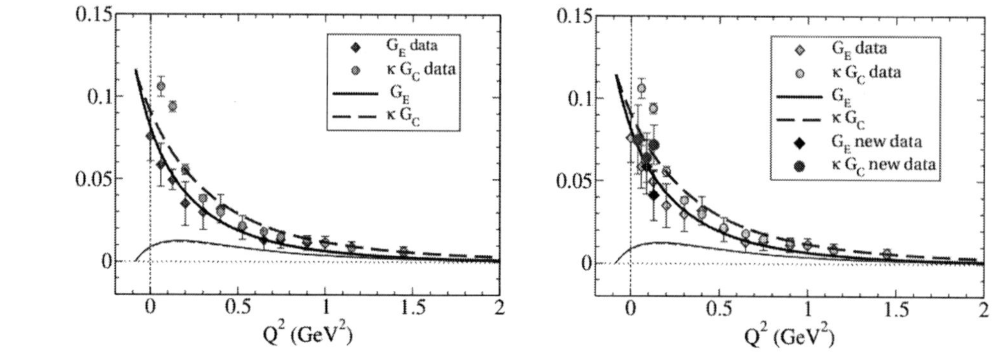

Figure 8. $\gamma^* N \to \Delta(1232)$ quadrupole form factors. The Coulomb form factor is normalized by $\kappa = \frac{M_\Delta - M_N}{2M_\Delta}$ for convenience. The model combines valence quark and pion cloud contributions. Data from [6, 64]. **Left:** Estimate with pion cloud contribution from [62]. **Right:** Estimate with pion cloud contribution from [63]. New low-Q^2 data from [66].

quadrupole form factors are related by [37, 58–65]

$$G_E = \frac{M_\Delta - M_N}{2M_\Delta} G_C. \tag{4}$$

The point $Q^2 = -(M_\Delta - M_N)^2$ is also known as pseudothreshold.

The original large-N_c estimates of the pion cloud contributions [56, 57, 62], can be improved with a correction of the order $1/N_c^2$ in G_E in order to be consistent with Siegert's theorem (4) [62–64]. Combining the valence quark contributions with an improved large-N_c estimate of the pion cloud for G_E [62], we obtain a good description of the quadrupole form factor data. This can be observed in the left panel of the Fig. 8. The parametrization of [62] violates Siegert's theorem only a term in $1/N_c^4$. The estimate fails only for the first two points for G_C.

On the right panel of Fig. 8, we present a new calculation where the pion cloud parametrization is further improved in order to satisfy (4) exactly. In the graph, the low-Q^2 data for G_E and G_C are replaced by new data which corrects the previous analysis [66]. The new estimate is in perfect agreement with the new data for both form factors.

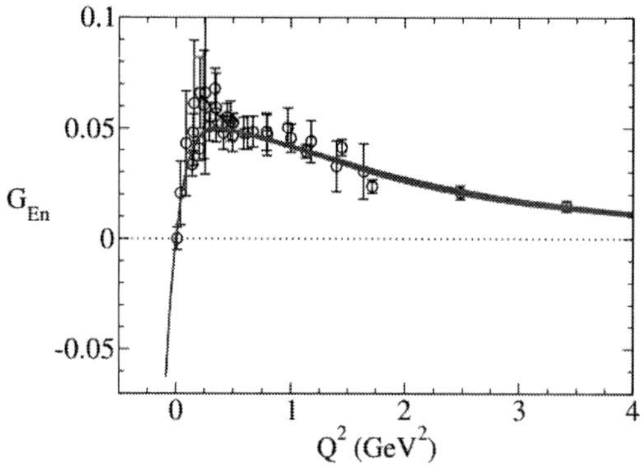

Figure 9. Neutron electric form factor G_{En} determined by a global fit to the G_{En} and G_E and G_C quadrupole form factors [64].

These results, demonstrate that the $\gamma^* N \to \Delta(1232)$ quadrupole form factors can be completely described by a combination of valence quark and pion cloud contributions. The pion cloud parametrizations include no free parameters, and the valence quark contributions are determined exclusively by the lattice QCD data [12]. There are, therefore, no fits to the physical data in the estimates from Fig. 8.

The pion cloud parametrizations for G_E and G_C are based on relations $G_E \propto G_{En}$ and $G_C \propto G_{En}$, valid at low Q^2 [56, 57, 62, 63], where G_{En} is the neutron electric form factor. Inspired by the previous results, we use the relation between the $\gamma^* N \to \Delta(1232)$ G_E and G_C form factors and G_{En} to obtain a more accurate parametrization of G_{En} [64]. Our final parametrization for G_{En} is presented in Fig. 9. Also estimated are the second moment of the neutron electric form factor: $r_4^4 \simeq -0.4$ fm^4, and the electric and Coulomb square radii: $r_{E2}^2 = 2.2 \pm 0.2$ fm^2 and $r_{C2}^2 = 1.8 \pm 0.1$ fm^2.

5 Summary and Conclusions

We present covariant estimates of the $\gamma^* N \to N^*$ transition form factors for large Q^2 for the N^* states: $\Delta(1232)$, $N(1440)$, $N(1520)$, $N(1535)$, $N(1880)$, $N(1700)$ and $\Delta(1700)$. There are also estimates for the states $N(1650)$, $\Delta(1600)$ and $\Delta(1620)$, not discussed in detail due to the lack of data for $Q^2 > 2$ GeV^2.

Our estimates are valid in principle for $Q^2 > 2$ GeV^2, since the valence quark effects are expected to dominate over the meson cloud effects when Q^2 increases. In some cases, however, we conclude that the valence quark contributions are insufficient to explain the data in the range $Q^2 = 1$–4 GeV^2, implying that the meson cloud effects may be significant in that range. Examples of those cases are the $\gamma^* N \to N(1535)$ Pauli form factor and the $\gamma^* N \to \Delta(1232)$ quadrupole form factors.

Our estimates are based mainly on the calibrations for the nucleon and $\Delta(1232)$ systems. Future large Q^2 data, such as the data from the JLab-12 GeV upgrade will be fundamental to test our estimates at large Q^2, or to refine our calibrations, if the new data for the nucleon and the $\Delta(1232)$ deviates from the present trend.

Future data at low Q^2, particularly in the range $Q^2 = 0$–0.3 GeV^2, will be also very important to establish the shape of the transition form factors near $Q^2 = 0$. The N^* contributions to the nucleon Compton scattering are particularly sensitive to those form factors [67].

Acknowledgments

G. R. was supported by the Fundação de Amparo à Pesquisa do Estado de São Paulo (FAPESP): Project No. 2017/02684-5, Grant No. 2017/17020-BCO-JP.

References

[1] I. G. Aznauryan *et al.*, Int. J. Mod. Phys. E **22**, 1330015 (2013).
[2] I. G. Aznauryan and V. D. Burkert, Prog. Part. Nucl. Phys. **67**, 1 (2012).
[3] D. Drechsel, S. S. Kamalov and L. Tiator, Eur. Phys. J. A **34**, 69 (2007).
[4] V. D. Burkert and T. S. H. Lee, Int. J. Mod. Phys. E **13**, 1035 (2004).
[5] G. Ramalho, Few Body Syst. **59**, 92 (2018).
[6] V. I. Mokeev, https://userweb.jlab.org/~mokeev/resonance_electrocouplings/
[7] F. Gross, G. Ramalho and M. T. Peña, Phys. Rev. C **77**, 015202 (2008).
[8] G. Ramalho, K. Tsushima and F. Gross, Phys. Rev. D **80**, 033004 (2009).
[9] S. Capstick and W. Roberts, Prog. Part. Nucl. Phys. **45**, S241 (2000).
[10] G. Ramalho, M. T. Peña and F. Gross, Eur. Phys. J. A **36**, 329 (2008).
[11] G. Ramalho, M. T. Peña and F. Gross, Phys. Rev. D **78**, 114017 (2008).
[12] G. Ramalho and M. T. Peña, Phys. Rev. D **80**, 013008 (2009).
[13] G. Ramalho and K. Tsushima, Phys. Rev. D **82**, 073007 (2010).
[14] G. Ramalho and M. T. Peña, Phys. Rev. D **89**, 094016 (2014).
[15] G. Ramalho and M. T. Peña, Phys. Rev. D **95**, 014003 (2017).
[16] G. Ramalho, D. Jido and K. Tsushima, Phys. Rev. D **85**, 093014 (2012).
[17] G. Ramalho and M. T. Peña, Phys. Rev. D **101**, 114008 (2020).
[18] G. Ramalho and K. Tsushima, Phys. Rev. D **87**, 093011 (2013).
[19] G. Ramalho and K. Tsushima, Phys. Rev. D **88**, 053002 (2013).
[20] F. Gross, G. Ramalho and M. T. Peña, Phys. Rev. D **85**, 093005 (2012).
[21] G. Ramalho and M. T. Peña, J. Phys. G **36**, 115011 (2009).
[22] G. Ramalho, Phys. Rev. D **95**, 054008 (2017).
[23] G. Ramalho, Phys. Rev. D **90**, 033010 (2014).
[24] G. Ramalho and M. T. Peña, Phys. Rev. D **84**, 033007 (2011).
[25] G. Ramalho, K. Tsushima and A. W. Thomas, J. Phys. G **40**, 015102 (2013).
[26] G. Ramalho, J. P. B. C. De Melo and K. Tsushima, Phys. Rev. D **100**, 014030 (2019).
[27] G. Ramalho and M. T. Peña, Phys. Rev. D **85**, 113014 (2012).
[28] G. Ramalho, M. T. Peña, J. Weil, H. van Hees and U. Mosel, Phys. Rev. D **93**, 033004 (2016).
[29] S. S. Kamalov and S. N. Yang, Phys. Rev. Lett. **83**, 4494 (1999).
[30] T. Sato and T. S. H. Lee, Phys. Rev. C **63**, 055201 (2001).
[31] B. Julia-Diaz, H. Kamano, T.-S. H. Lee, A. Matsuyama, T. Sato and N. Suzuki, Phys. Rev. C **80**, 025207 (2009).
[32] D. Ronchen *et al.*, Eur. Phys. J. A **49**, 44 (2013).
[33] H. W. Hammer, D. Drechsel and U. G. Meissner, Phys. Lett. B **586**, 291 (2004).
[34] U. G. Meissner, AIP Conf. Proc. **904**, 142 (2007).
[35] V. Pascalutsa, M. Vanderhaeghen and S. N. Yang, Phys. Rept. **437**, 125 (2007).
[36] C. Alexandrou, G. Koutsou, H. Neff, J. W. Negele, W. Schroers and A. Tsapalis, Phys. Rev. D **77**, 085012 (2008).
[37] H. F. Jones and M. D. Scadron, Annals Phys. **81**, 1 (1973).
[38] J. Adamczewski-Musch *et al.* [HADES Collaboration], Phys. Rev. C **95**, 065205 (2017).
[39] B. Ramstein, these proceedings.
[40] P. Salabura, these proceedings.
[41] G. Ramalho, M. T. Peña and F. Gross, Phys. Rev. D **81**, 113011 (2010).
[42] G. Ramalho, M. T. Peña and A. Stadler, Phys. Rev. D **86**, 093022 (2012).
[43] G. Ramalho and K. Tsushima, Phys. Rev. D **81**, 074020 (2010).
[44] G. Ramalho and K. Tsushima, Phys. Rev. D **89**, 073010 (2014).
[45] I. G. Aznauryan *et al.* [CLAS Collaboration], Phys. Rev. C **80**, 055203 (2009).
[46] V. I. Mokeev *et al.* [CLAS Collaboration], Phys. Rev. C **86**, 035203 (2012).
[47] V. I. Mokeev *et al.*, Phys. Rev. C **93**, 025206 (2016).
[48] G. Ramalho and D. Melnikov, Phys. Rev. D **97**, 034037 (2018).
[49] G. Ramalho, Phys. Rev. D **96**, 054021 (2017).
[50] K. Park *et al.* [CLAS Collaboration], Phys. Rev. C **91**, 045203 (2015).
[51] R. Gothe, these proceedings.
[52] G. Ramalho and K. Tsushima, Phys. Rev. D **84**, 051301 (2011).
[53] D. Jido, M. Doering and E. Oset, Phys. Rev. C **77**, 065207 (2008).
[54] M. M. Dalton *et al.*, Phys. Rev. C **80**, 015205 (2009).

[55] V. D. Burkert, R. De Vita, M. Battaglieri, M. Ripani and V. Mokeev, Phys. Rev. C **67**, 035204 (2003).
[56] V. Pascalutsa and M. Vanderhaeghen, Phys. Rev. D **76**, 111501 (2007).
[57] A. J. Buchmann, Can. J. Phys. **87**, 773 (2009).
[58] R. C. E. Devenish, T. S. Eisenschitz and J. G. Korner, Phys. Rev. D **14**, 3063 (1976).
[59] A. J. Buchmann, U. Meyer, A. Faessler and E. Hernandez, Phys. Rev. C **58**, 2478 (1998).
[60] G. Ramalho, Phys. Lett. B **759**, 126 (2016).
[61] G. Ramalho, Phys. Rev. D **93**, 113012 (2016).
[62] G. Ramalho, Phys. Rev. D **94**, 114001 (2016).
[63] G. Ramalho, Eur. Phys. J. A **54**, 75 (2018).
[64] G. Ramalho, Eur. Phys. J. A **55**, 32 (2019).
[65] G. Ramalho, Phys. Rev. D **100**, 114014 (2019).
[66] A. Blomberg *et al.*, Phys. Lett. B **760**, 267 (2016).
[67] G. Eichmann and G. Ramalho, Phys. Rev. D **98**, 093007 (2018).

Resonance Electroproduction and the Origin of Mass

Craig D. Roberts[1,2,*]

[1]School of Physics, Nanjing University, Nanjing, Jiangsu 210093, China
[2]Institute for Nonperturbative Physics, Nanjing University, Nanjing, Jiangsu 210093, China

Abstract. One of the greatest challenges within the Standard Model is to discover the source of visible mass. Indeed, this is the focus of a "Millennium Problem", posed by the Clay Mathematics Institute. The answer is hidden within quantum chromodynamics (QCD); and it is probable that revealing the origin of mass will also explain the nature of confinement. In connection with these issues, this perspective will describe insights that have recently been drawn using contemporary methods for solving the continuum bound-state problem in relativistic quantum field theory and how they have been informed and enabled by modern experiments on nucleon-resonance electroproduction.

1 Emergence of Mass

The natural energy-scale for strong interactions is characterised by the proton mass:

$$m_p \approx 1\,\text{GeV} \approx 2000\, m_e\,, \tag{1}$$

where m_e is the mass of the electron. In the Standard Model, m_e is rightly attributed to the Higgs boson; but what is the source of the enormous enhancement to produce m_p? This is the crux: the source of the vast majority of visible mass in the Universe is unknown. Followed logically to its origin, this question leads to an appreciation that the existence of our Universe depends critically on, *inter alia*, the following empirical facts: (*i*) the proton is massive, *i.e.* the mass-scale for strong interactions is vastly different to that of electromagnetism; (*ii*) the proton is absolutely stable, despite being a composite object constituted from three valence-quarks; and (*iii*) the pion, responsible for long-range interactions between nucleons, is unnaturally light (not massless), possessing a lepton-like mass despite being a strongly interacting composite object built from a valence-quark and valence-antiquark. These are basic emergent features of Nature; and QCD must somehow explain them and many other high-level phenomena with enormous apparent complexity.

The Lagrangian of chromodynamics is simple:

$$\mathcal{L}_{\rm QCD} = \sum_{j=u,d,s,\ldots} \bar{q}_j [i\gamma^\mu D_\mu - m_j] q_j - \tfrac{1}{4} G^a_{\mu\nu} G^{a\mu\nu}, \tag{2a}$$

$$D_\mu = \partial_\mu + i g \tfrac{1}{2}\lambda^a A^a_\mu\,, \tag{2b}$$

$$G^a_{\mu\nu} = \partial_\mu A^a_\nu + \partial_\nu A^a_\mu + \underline{i g f^{abc} A^b_\mu A^c_\nu}. \tag{2c}$$

Here: $\{q_j\}$ are the quark fields, with j their flavor label and m_j their Higgs-generated current-quark masses; and $\{A^a_\mu, a = 1,\ldots,8\}$ are the gluon fields, with $\{\frac{1}{2}\lambda^a\}$ the generators of the SU(3) (color/chromo) gauge-group in the fundamental representation. In comparison with quantum electrodynamics (QED), the *solitary* difference is the term describing gluon self-interactions, marked as the underlined piece in Eq. (2c). Somehow, $\mathcal{L}_{\rm QCD}$ – this one line, along with two definitions – is responsible for the origin, mass, and size of almost all visible matter in the Universe. That being true, then QCD is quite possibly the most remarkable fundamental theory ever invented.

The only apparent mass-scales in Eq. (2) are the current-quark masses, generated by the Higgs boson; but focusing on the valence quarks which define nucleons, *i.e.* u, d, this scale is more-than 100-times smaller than m_p. No amount of "staring" at $\mathcal{L}_{\rm QCD}$ can reveal the source of that enormous amount of "missing mass"; yet, it is there. This contrasts starkly with QED wherein, *e.g.* the scale in the spectrum of the hydrogen atom is set by m_e, a prominent feature of $\mathcal{L}_{\rm QED}$ that is generated by the Higgs boson.

The emergence of mass is seen even more keenly by considering that even treated as a classical theory, chromodynamics is a non-Abelian local gauge field theory. As with all such theories formulated in four spacetime dimensions, no energy-scale exists in the absence of Lagrangian masses for the matter fields. (The absence of such masses defines the chiral limit.) There is no dynamics in a scale-invariant theory, only kinematics: the theory looks the same at all length-scales; hence, there can be no clumps of anything. Bound states are therefore impossible and, accordingly, our Universe cannot exist.

A *spontaneous* breaking of symmetry, as realised via the Higgs mechanism, does not solve this problem: the masses of the neutron and proton, the kernels of all visible matter, are roughly 100-times larger than the Higgs-generated current-masses of the light u- and d-quarks, the main building blocks of protons and neutrons.

Consequently, the questions of how does a mass-scale appear and why does it have the value we observe are in-

*e-mail: cdroberts@nju.edu.cn

separable from the question of how did the Universe come into being.

Modern quantum field theories are built upon Poincaré invariance, so consider the energy-momentum tensor in classical chromodynamics, $T_{\mu\nu}$. Conservation of energy and momentum is a consequence of spacetime translational invariance, elements in the set of Poincaré transformations; hence, $\partial_\mu T_{\mu\nu} = 0$. Suppose now that one performs a global scale transformation on the coordinates and fields in classical chromodynamics. In the absence of fermion masses, *viz.* suppressing the Higgs mechanism, the classical action is invariant under the scale transformation and the dilation current is conserved:

$$\mathcal{D}_\mu = T_{\mu\nu}x_\nu \mid \partial_\mu \mathcal{D}_\mu = 0 = [\partial_\mu T_{\mu\nu}]x_\nu + T_{\mu\nu}\delta_{\mu\nu} = T_{\mu\mu} \,. \quad (3)$$

Evidently, the energy-momentum tensor must be traceless in a scale invariant theory.

Classical chromodynamics is not a meaningful framework for many reasons; amongst them the fact that strong interactions are empirically known to be characterised by a large mass-scale, $m_p \approx 1\,$GeV. In quantising the theory, regularisation and renormalisation of ultraviolet divergences introduces a mass-scale. This is "dimensional transmutation": all quantities, including the field operators themselves, become dependent on a mass-scale. It entails violation of Eq. (3), *i.e.* the appearance of the chiral-limit "trace anomaly":

$$T_{\mu\mu} = \beta(\alpha(\zeta))\tfrac{1}{4}G^a_{\mu\nu}G^a_{\mu\nu} =: \Theta_0 \,, \quad (4)$$

where $\beta(\alpha(\zeta))$ is QCD's β-function, $\alpha(\zeta)$ is the associated running-coupling, and ζ is the renormalisation scale. Eq. (4) indicates that a mass-scale related to the resolving power of a given measurement is introduced via quantisation, *viz.* the scale *emerges* as an integral part of the theory's quantum definition.

There is another aspect of chromodynamics that should be highlighted; namely, the classical Lagrangian still defines a non-Abelian local gauge theory. Accordingly, the concept of local gauge invariance persists; but without a mass-scale, there is no notion of confinement. For example, three quarks can be prepared in a colour-singlet combination and colour rotations will keep the three-body system neutral; but the quarks involved need not have any proximity to one another. Indeed, proximity is meaningless because all lengths are equivalent in a scale invariant theory. Hence, the question of "Whence a mass-scale?" is equivalent to "Whence a confinement scale?".

Knowing a trace anomaly exists does not actually deliver a great deal: it only indicates that there is a mass-scale. The crucial issue is whether one can compute and/or understand the magnitude of that scale.

One can certainly measure the size of the scale anomaly, for consider the expectation value of the energy-momentum tensor in the proton:

$$\langle p(P)|T_{\mu\nu}|p(P)\rangle = -P_\mu P_\nu \,, \quad (5)$$

where the right-hand-side follows from the equations-of-motion for a one-particle proton state. In the chiral limit,

$$\langle p(P)|T_{\mu\mu}|p(P)\rangle = -P^2 = m_p^2 = \langle p(P)|\Theta_0|p(P)\rangle \,; \quad (6)$$

namely, there is a clear sense in which it is possible to conclude that the entirety of the proton mass is produced by gluons. The trace anomaly is measurably large; and that property must logically owe to gluon self-interactions, which are also responsible for asymptotic freedom.

This is a valid conclusion. After all, what else could be responsible for a mass-scale in QCD? QCD is all about gluon self-interactions; and it's gluon self-interactions that (potentially) enable one to rigorously (nonperturbatively) define the expectation value in Eq. (6). On the other hand, it's only a sensible conclusion when the operator and the wave function are defined at a resolving-scale $\zeta \gg m_p$, *viz.* when one employs a parton-model basis.

There is also another issue, which can be exposed by returning to Eq. (5) and replacing the proton by the pion:

$$\langle \pi(q)|T_{\mu\nu}|\pi(q)\rangle = -q_\mu q_\nu \overset{\text{chiral limit}}{\Rightarrow} \langle \pi(q)|\Theta_0|\pi(q)\rangle = 0 \quad (7)$$

because the pion is a massless Nambu-Goldstone (NG) mode. Equation (7) could mean that the scale anomaly vanishes trivially in the pion state, *viz.* that gluons and their self-interactions have no impact within a pion because each term in the practical computation of the operator expectation value vanishes when evaluated in the pion. However, that is a difficult way to achieve Eq. (7). It is easier to imagine that Eq. (7) owes to cancellations between different operator-component contributions. Of course, such precise cancellation should not be an accident. It could only arise naturally because of some symmetry and/or symmetry-breaking pattern.

Equations (6) and (7) present a quandary, which highlights that no understanding of the source of the proton's mass can be complete unless it simultaneously explains the meaning of Eq. (7). Moreover, any discussion of confinement, fundamental to the proton's absolute stability, is impossible before this conundrum is resolved. The explanation of these features of Nature must lie in the dynamics responsible for the emergence of m_p as the natural mass-scale for nuclear physics; and one of the most important goals in modern science is to explain and elucidate the entire array of empirical consequences of this dynamics.

2 Gluons are Massive

Gluons must be the key; after all, their self interactions separate QCD from QED. Gluons are supposed to be massless. This is true in perturbation theory; but it is a feature that is not preserved nonperturbatively. Beginning with a pioneering effort almost forty years ago [1], continuum and lattice studies of QCD's gauge sector have been increasing in sophistication and reliability; and today it is known that the gluon propagator saturates at infrared momenta [2–10]:

$$\Delta(k^2 \simeq 0) = 1/m_g^2 \,. \quad (8)$$

Thus, the long-range propagation characteristics of gluons are dramatically affected by their self-interactions. Importantly, one may associate a renormalisation-group-invariant gluon mass-scale with this effect: $m_0 \approx 0.5\,\text{GeV} \approx m_p/2$, and summarise a large body of work by

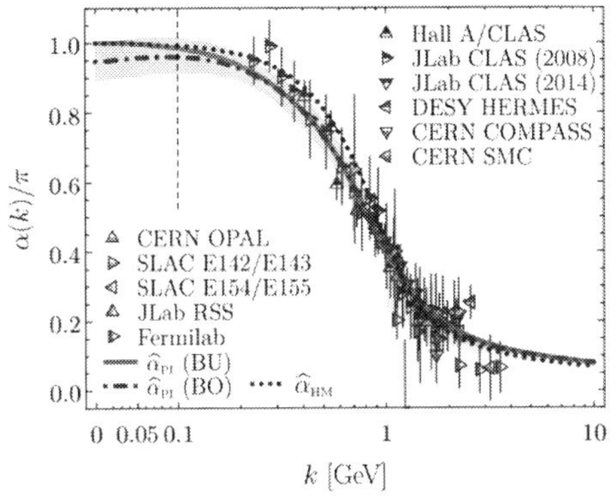

Figure 1. Dot-dashed [blue] curve: process-independent running-coupling $\alpha_{PI}(k^2)$ [6]: the shaded [blue] band bracketing this curve combines a 95% confidence-level window based on existing lQCD results for the gluon two-point function with an error of 10% in the continuum analysis of relevant ghost-gluon dynamics. Solid [black] curve, updated result [9]. Also depicted, world's data on the process-dependent coupling α_{g_1}, defined via the Bjorken sum rule, the sources of which are listed elsewhere [6]. The shaded [yellow] band on $k > 1$ GeV represents α_{g_1} obtained from the Bjorken sum by using QCD evolution to extrapolate high-k^2 data into the depicted region [14]; and, for additional context, the dashed [red] curve is the effective charge obtained in a light-front holographic model [15]. The k-axis scale is linear to the left of the vertical dashed line and logarithmic otherwise.

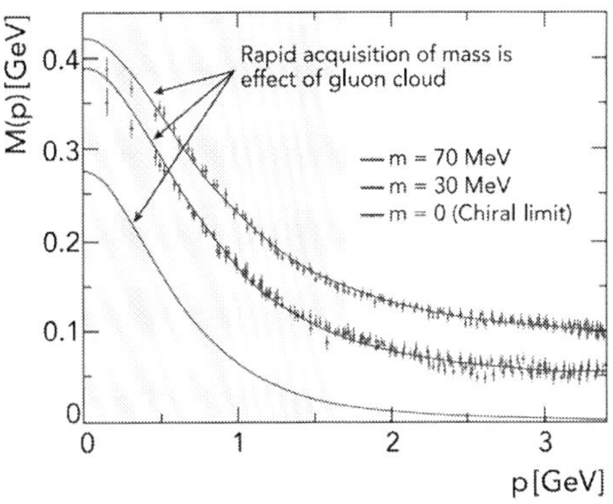

Figure 2. Renormalisation-group-invariant dressed-quark mass function, $M(p)$ in Eq. (9): *solid curves* – continuum nonperturbative results [16, 17]; "data" – numerical simulations of lQCD [18]. The current-quark of perturbative QCD evolves into a constituent-quark as its momentum becomes smaller. The constituent-quark mass arises from a cloud of low-momentum gluons attaching themselves to the current-quark. This is DCSB, the essentially nonperturbative effect that generates a quark *mass from nothing*; namely, it occurs even in the absence of a Higgs mechanism. The size of $M(0)$ is a measure of the magnitude of the QCD scale anomaly in $n = 1$-point Schwinger functions [19]. Moreover, experiments on $Q^2 \in [0, 12]\,\text{GeV}^2$ at the modern Thomas Jefferson National Accelerator Facility (JLab) will be sensitive to the momentum dependence of $M(p)$ within a domain that is here indicated approximately by the shaded region.

stating that gluons, although acting as massless degrees-of-freedom on the perturbative domain, actually possess a running mass, whose value at infrared momenta is characterised by m_0.

Asymptotic freedom ensures that QCD's ultraviolet behaviour is controllable; but the emergence of a gluon mass reveals a new frontier within the Standard Model because the existence of a running gluon mass, large at infrared momenta, has an impact on all analyses of the bound-state problem. For instance, it could be a harbinger of gluon saturation [11, 12]. Furthermore, $m_0 > 0$ entails that QCD dynamically generates its own infrared cutoff, so that gluons with wavelengths $\lambda \gtrsim \sigma := 1/m_0 \approx 0.5$ fm decouple from the strong interaction, hinting at a dynamical realisation of confinement [13].

3 Process-Independent Effective Charge

There are many other consequences of the intricate nonperturbative nature of QCD's gauge-sector dynamics. Important amongst them is the generation of a process-independent running coupling, $\alpha_{PI}(k^2)$ – see Refs. [6, 9]. Depicted as the solid [black] curve in Fig. 1, this is a new type of effective charge, which is an analogue of the Gell-Mann–Low effective coupling in QED because it is completely determined by the gauge-boson propagator. The result in Fig. 1 is a parameter-free prediction, capitalising on analyses of QCD's gauge sector undertaken using continuum methods and informed by numerical simulations of lattice-QCD (lQCD).

As a unique process-independent effective charge, α_{PI} appears in every one of QCD's dynamical equations of motion, including the gap equation, setting the strength of all interactions. α_{PI} therefore plays a crucial role in understanding the dynamical origin of light-quark masses in the Standard Model even in the absence of a Higgs coupling, as described in the next section.

4 Dynamical Chiral Symmetry Breaking

The emergence of a gluon mass-scale in the Standard Model drives an enormous array of phenomena. Crucial amongst them is dynamical chiral symmetry breaking (DCSB), which is most readily apparent in the dressed-quark propagator:

$$S(p) = 1/[i\gamma \cdot pA(p^2) + B(p^2)] = Z(p^2)/[i\gamma \cdot p + M(p^2)]\,. \tag{9}$$

$S(p)$ can be predicted in QCD using nonperturbative continuum and lattice techniques. The quantity $M(p^2)$ in Eq. (9) is the dressed-quark mass-function, whose computed behaviour is depicted and explained in Fig. 2. Evidently, even in the absence of a Higgs mechanism, when current-masses vanish, quarks acquire a running mass whose value in the infrared is approximately $m_p/3$. This is the scale required to support the constituent quark model and all its successes. It follows that DCSB can be identified as the source for more than 98% of the visible mass in the Universe, *viz.* DCSB is fundamentally connected with the *origin of mass from nothing*.

One must insist that chiral symmetry breaking in the absence of a Higgs mechanism is "dynamical," as distinct from spontaneous, because nothing is added to QCD in order to effect this remarkable outcome and there is no change of variables in $\mathcal{L}_{\rm QCD}$ that will make it apparent. Instead, through the act of quantising the classical chromodynamics of massless gluons and quarks, a large mass-scale is generated in both the gauge and matter sectors.

DCSB is empirically revealed very clearly in properties of the pion. In fact [19], the key to understanding Eq. (7) is a set of Goldberger-Treiman-like (GT) relations [20, 21], the best known of which states:

$$m \simeq 0 \; \Big| \; f_\pi E_\pi(k;0) = B(k^2)\,, \qquad (10)$$

where E_π is the leading piece of the pion's Bethe-Salpeter amplitude (wave function), and B is the scalar piece of the dressed-quark self-energy, Eq. (9). This equation is exact in chiral QCD and expresses the fact that the Nambu-Goldstone theorem is fundamentally an expression of equivalence between the quark one-body problem and the two-body bound-state problem in QCD's color-singlet flavor-nonsinglet pseudoscalar channel. Consequently and enigmatically:

The properties of the nearly-massless pion are the cleanest expression of the mechanism that is responsible for (almost) all the visible mass in the Universe.

The manner by which this can be exploited at modern and planned facilities, in order to chart, *inter alia*, the origin and distribution of mass, is canvassed elsewhere [22–24].

With the GT relations in hand, one can construct an algebraic proof [20, 21], that at any and each order in a symmetry-preserving truncation of those equations in quantum field theory necessary to describe a pseudoscalar bound state, there is a precise cancellation between the mass-generating effect of dressing the valence-quark and -antiquark which constitute the system and the attraction generated by the interactions between them, *i.e.*

$$M^{\rm dressed}_{\rm quark} + M^{\rm dressed}_{\rm antiquark} + U^{\rm dressed}_{\rm quark-antiquark\ interaction} \overset{\rm chiral\ limit}{\equiv} 0\,. \qquad (11)$$

This guarantees the "disappearance" of the scale anomaly in the chiral-limit pion. An analogy with quantum mechanics thus arises: the mass of a QCD bound-state is the sum of the mass-scales characteristic of the constituents plus a (negative and sometimes large) binding energy.

Since QCD's interactions are universal, similar cancellations must take place within the proton. However, in the proton channel there is no symmetry that requires the cancellations to be complete. Hence, the proton's mass has a value that is typical of the magnitude of scale breaking in QCD's one body sectors, *viz*. the dressed-gluon and -quark mass-scales.

The perspective just described may be called the "DCSB paradigm". It provides a basis for understanding why the mass-scale for strong interactions is vastly different to that of electromagnetism, why the proton mass expresses that scale, and why the pion is nevertheless unnaturally light. In this picture, no significant mass-scale is possible in QCD unless one of commensurate size is expressed in the dressed-propagators of gluons and quarks.

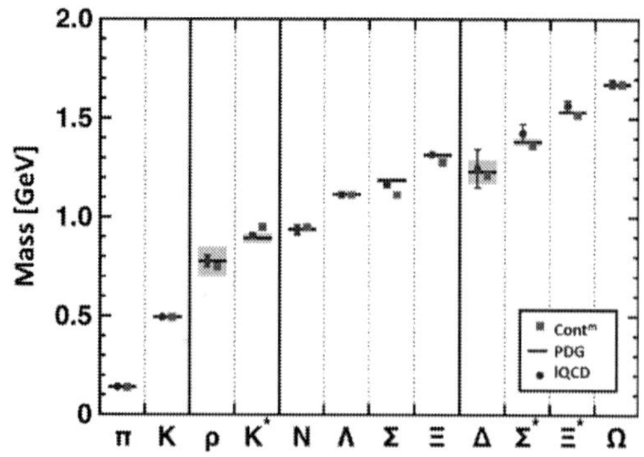

Figure 3. Masses of pseudoscalar and vector mesons, and ground-state positive-parity octet and decuplet baryons calculated using continuum (Cont^{m} – squares, red) [33] and lattice [34] methods in QCD compared with experiment (PDG – black bars, with decay-widths of unstable states shaded in grey) [35]. The continuum study did not include isospin symmetry breaking effects, which are evidently small, as highlighted by the empirically determined Σ-Λ mass difference.

5 Faddeev Equation for Baryons

The Faddeev equation was introduced almost sixty years ago [25]. It treats the quantum mechanical problem of three-bodies interacting via pairwise potentials by reducing it to a sum of three terms, each of which describes a solvable scattering problem in distinct two-body subsystems. An analogous approach to the three-valence-quark (baryon) bound-state problem in quantum chromodynamics (QCD) is explained in Refs. [26–28]. In this case, owing to DCSB and the importance of symmetries [21], a Poincaré-covariant quantum field theory generalisation of the Faddeev equation is required. Like the Bethe-Salpeter equation for mesons, it is natural to consider such a Faddeev equation as one of the tower of QCD's Dyson-Schwinger equations (DSEs), which are being used to develop a systematic continuum approach to the strong-interaction bound-state problem [29, 30].

The first direct treatment of the Poincaré-covariant Faddeev equation for the nucleon is described in Ref. [31]. Following that approach, Refs. [32, 33] calculated the spectrum of ground-state $J = 1/2^+, 3/2^+$ $(qq'q'')$-baryons, where $q, q', q'' \in \{u, d, s, c, b\}$, their first positive-parity excitations and parity partners. Introducing two parameters, to compensate for deficiencies of the leading-order truncation when used for excited hadrons, a description of the known spectrum of 39 such states was obtained, with a mean-absolute-relative-difference between calculation and experiment of 3.6 (2.7)%. This is exemplified in Fig. 3. The framework was subsequently used to predict the masses of 90 states not yet seen empirically.

Since solving the Faddeev equation yields Poincaré-covariant wave functions for all these systems, information about their internal structure is also revealed. For instance, considering $J = 3/2^+$ states, each has a complicated rest-frame angular-momentum profile. The ground-states are primarily S-wave in character, but each possesses P-, D- and F-wave components: the P-wave is large in $\{u, d, s\}$-quark baryons. Turning to the first positive-parity excitations, each has a large D-wave component, which grows with increasing current-quark mass, but they otherwise exhibit features consistent with radial excitations.

Whilst a complete treatment of the Poincaré-covariant Faddeev equation is now possible using modern hardware, it remains a complex task. Hence, for the development of insights in a wide array of baryon problems, it remains common to treat the equation in a quark-diquark approximation, where the diquark correlations are non-pointlike and dynamical [36]. This amounts to a simplified treatment of the scattering problem in the two-body subchannels (as explained, *e.g.* in Ref. [37, Sec. II.A.2]), which is founded on an observation that the same interaction which describes colour-singlet mesons also generates diquark correlations in the colour-antitriplet ($\bar{3}$) channel [38–40]. Whilst the diquarks do not survive as asymptotic states, *viz.* they are absent from the strong interaction spectrum [41, 42], the attraction between the quarks in the $\bar{3}$ channel sustains a system in which two quarks are always correlated as a colour-$\bar{3}$ pseudoparticle, and binding within the baryon is effected by the iterated exchange of roles between the bystander and diquark-participant quarks. Contemporary studies indicate that diquark correlations are an important component of all baryons; and owing to the dynamical character of the diquarks, it is typically the lightest allowed diquark correlation which defines the most important component of a baryon's Faddeev amplitude [43].

6 Δ Baryons

The quark+diquark picture of baryons has yielded a unified picture of the four lightest $(I, J^P) = (1/2, 1/2^+)$ baryon isospin doublets [44]. Turning to $(3/2, 3/2^+)$ states with no net strangeness, the $\Delta^{+,0}$ baryons are conventionally viewed as the lightest isospin- and spin-flip excitations of the proton and neutron (nucleons, N), respectively. Hence, since nucleons are the basic elements of all nuclei, developing a detailed understanding of the Δ-baryons is of fundamental importance. Without this, hadron physics remains at a level akin to atomic physics based only on knowledge of the hydrogen atom's ground state.

In this connection, given the simplicity of electromagnetic probes, it is worth using virtual photons to chart Δ-resonance structure; and using intense, energetic electron-beams at the JLab, $\gamma^* p \to \Delta^+$ data are now available for $0 \leq Q^2 \lesssim 8\,\text{GeV}^2$ [45–47]. These data have stimulated much theoretical analysis and speculation about, *inter alia*: the role that resonance electroproduction experiments can play in exposing nonperturbative aspects of QCD, such as the nature of confinement and DCSB; the relevance of perturbative QCD to processes involving moderate momentum transfers; and hadron shape deformation.

Deformation can be characterised by a given hadron's spectroscopic quadrupole moment:

$$Q = \frac{3J_z^2 - J(J+1)}{(J+1)(2J+3)} Q_0, \tag{12}$$

where Q_0, the state's intrinsic deformation, is a measure of the D-wave component of its rest-frame-projected Faddeev wave function. The difference between Q and Q_0 expressed in Eq. (12) represents the averaging of the non-spherical charge distribution due to its rotational motion as seen in the laboratory/rest-frame. Plainly, $Q = 0$ when $J = 1/2$, *i.e.* there is no measurable quadrupole moment for an isolated $J = 1/2$ bound-state.

It is worth remarking here that just above the Δ-baryon level lies the nucleon's first positive-parity excitation, *viz.* the Roper resonance, $N(1440)\,1/2^+$. Discovered in 1963 [48], its characteristics were long the source of puzzlement because, *e.g.* constituent-quark potential models typically (and erroneously) produce a spectrum in which this excitation lies above the first negative-parity state $N(1535)\,1/2^-$ [49–51]. This has now changed following: acquisition and analysis of high-precision proton-target exclusive electroproduction data with single- and double-pion final states, on a large energy domain and with momentum-transfers out to $Q^2 \approx 5\,\text{GeV}^2$; development of a dynamical reaction theory capable of simultaneously describing all partial waves extracted from available, reliable data; and formulation and application of a Poincaré covariant approach to the continuum bound-state problem in relativistic quantum field theory. Today, it is widely accepted that the Roper is, at heart, the first radial excitation of the nucleon, consisting of a well-defined dressed-quark core that is augmented by a meson cloud, which both reduces the Roper's core mass by approximately 20% and contributes materially to the electroproduction form factors at low-Q^2 [52, 53].

Notably, a similar pattern of energy levels is found in the spectrum of Δ-baryons, *viz.* contradicting quark-model predictions, the first positive-parity excitation, $\Delta(1600)\,3/2^+$, lies below the negative parity $\Delta(1700)\,3/2^-$, with the splitting being roughly the same as that in the nucleon sector. This being so and given the Roper-resonance example, it is likely that elucidating the nature of the Δ(1600)-baryon will require both (*i*) data on its electroproduction form factors which extends well beyond the meson-cloud domain and (*ii*) predictions for these form factors to compare with that data. The data exist [54, 55]; and can be analysed with this aim understood. This is especially important now that theoretical predictions are available [56].

The Poincaré-covariant wave functions for the Δ(1232) and Δ(1600) are depicted in Ref. [56, Fig. 2], from which it is plain that, in their rest-frames, both the ground-state and first positive-parity excitation are primarily S-wave in character. Notably, the Δ(1232) mass is almost insensitive to non-S-wave components; and its quadrupole moment is large in magnitude and negative, indicating oblate intrinsic deformation. Turning to the Δ(1600) and considering its mass, P-wave components generate noticeable repulsion, D-waves produce some attraction, and F-waves can be neglected. Interestingly, in this quark+diquark approach, too, when comparing the ground state with the first positive-parity excitation, some of the S-wave strength is shifted into P- and D-wave contributions just as in the direct Faddeev equation solution described in Sec. 5. The intrinsic quadrupole moment of the $\Delta(1600)$ is just 45% of that of the $\Delta(1232)$. By this measure, the $\Delta(1600)$ is still oblate, but possesses less quadrupole deformation than the $\Delta(1232)$.

7 $\gamma^* + \mathrm{p} \to \Delta(1232), \Delta(1600)$

Regarding the $\gamma^* p \to \Delta^+(1232)$ transition, on $Q^2 \gtrsim 0.5 m_p^2$, *i.e.* outside the meson cloud domain for this process, the magnetic dipole and Coulomb quadrupole form factors reported in Ref. [56] agree well with available data. Consistent with the data, too, the electric quadrupole form factor is very small in magnitude; hence, it is particularly sensitive to the diquark content and quark-diquark angular-momentum structure of the baryons involved, and also to meson-baryon final-state-interactions (MB FSIs) on a larger domain than the other form factors. These remarks are supported by the following observations: the role played by higher partial waves in the wave functions increases with momentum transfer (something also observed in meson form factors), here generating destructive interference; agreement with data on G_M^* is impossible without the higher partial waves; and the effect of such components is very large in G_E^*, with the complete result for G_E^* exhibiting a zero at $Q^2 \approx 4m_p^2$, which is absent in the S-wave-only result(s).

Drawn from Ref. [56], predictions for the $\gamma^* p \to \Delta^+(1600)$ transition form factors are displayed in Fig. 4. Empirical results are here only available at the real-photon point: $G_M^*(Q^2 = 0)$, $G_E^*(Q^2 = 0)$. Evidently, the quark model results – [shaded grey band] [58], dot-dashed (brown) curve [59] and dot-dot-dashed [orange] curve [60]) – are very sensitive to the wave functions employed for the initial and final states. Furthermore, inclusion of relativistic effects has a sizeable impact on transitions to positive-parity excited states [58].

The quark+diquark Faddeev equation prediction is the solid [black] curve in each panel of Fig. 4. In this instance, every transition form factor is of unique sign on the domain displayed. Notably, the mismatches with the empirical results for $G_M^*(0)$, $G_E^*(0)$ are commensurate in relative sizes with those in the $\Delta(1232)$ case, suggesting that MB FSIs are of similar importance in both channels.

Axial-vector diquark contributions interfere constructively with MB FSIs [37]; hence, regarding form factors, one can mimic some meson-cloud effects by modifying the axial-vector diquark content of the participating hadrons. Accordingly, to illustrate the potential impact of MB FSIs, the transition form factors were also computed using an enhanced axial-vector diquark content in the proton. This was achieved by setting $m_{1^+} = m_{0^+} = 0.85\,\mathrm{GeV}$, values with which the proton's mass is practically unchanged. The procedure produced the dotted [green] curves in Fig. 4; better aligning the $x \simeq 0$ results with experiment and suggesting thereby that MB FSIs will improve the Faddeev equation predictions.

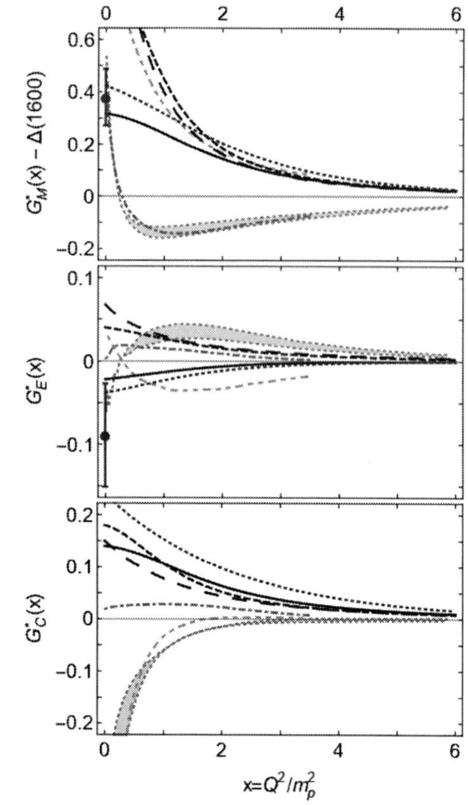

Figure 4. *Top panel* – Magnetic dipole $\gamma^* p \to \Delta^+(1600)$ transition form factor; *middle* – electric quadrupole; and *bottom*: Coulomb quadrupole. Data from Ref. [35]; and the conventions of Ref. [57] are employed. All panels: solid [black] curve, complete result; long-dashed [blue] curve, result obtained when $\Delta(1600)$ is reduced to S-wave state; dashed [blue] curve, both the proton and $\Delta(1600)$ are reduced to S-wave states; dotted [green] curve, obtained by enhancing proton's axial-vector diquark content; shaded [grey] band, light-front relativistic Hamiltonian dynamics (LFRHD) [58]; dot-dashed [brown] curve, light-front relativistic quark model (LFRQM) with unmixed wave functions [59]; and dot-dot-dashed [orange] curve, LFRQM with configuration mixing [60].

The short-dashed [blue] curve in Fig. 4 is the result obtained when only rest-frame S-wave components are retained in the wave functions of the proton and $\Delta(1600)$; and the long-dashed [blue] curve is that computed with a complete proton wave function and a S-wave-projected $\Delta(1600)$. Once again, the higher partial-waves have a visible impact on all form factors, with G_E^* being most affected: the higher waves produce a change in sign. This reemphasises one of the conclusions from the quark model studies, *viz.* data on the $\gamma^* p \to \Delta^+(1600)$ transition form factors will be sensitive to the structure of the $\Delta^+(1600)$.

It is interesting to observe that whilst all $\gamma^* p \to \Delta^+(1232)$ transition form factors are larger in magnitude than those for $\gamma^* p \to \Delta^+(1600)$ in some neighbourhood of $Q^2 = 0$, this ordering is reversed on $Q^2 \gtrsim 2m_p^2$. One can thus argue that the $\gamma^* p \to \Delta^+(1600)$ transition is more localised in configuration space.

It is also notable that $R_{SM} \sim G_C^*/G_M^*$ is qualitatively similar for both $\gamma^* p \to \Delta^+$ transitions; but $R_{EM} \sim G_E^*/G_M^*$ is markedly different, being of opposite sign on $Q^2 \lesssim 4m_p^2$ and uniformly larger in magnitude for the $\Delta(1600)$. These observations again highlight the sensitivity of the electric quadrupole form factor to the degree of intrinsic deformation of the Δ-baryons.

8 Summary and Outlook

8.1 Particulars

Poincaré-covariance entails that no realistic hadron wave function can be purely S-wave in character. Hence, all $J \neq 0$ hadrons possess intrinsic deformation; and for baryons, such deformation can be exposed via measurements of nucleon-to-resonance transition form factors. Of course, in relativistic quantum field theory, the magnitude and mixing between orbital angular momentum and spin is frame and renormalisation-scale dependent. This is also true of the interpretations of these quantities because the natural degrees of freedom evolve with scale. Consequently, even though Poincaré-invariant form factors are the same for all observers, their interpretation via subsystem degrees-of-freedom can change with frame and scale. All these remarks extend to deformation. A close connection with quantum mechanics notions of probability distributions can be made via light-front projections of Poincaré-covariant wave functions; but unless the studied quantity is protected by symmetry, even conclusions drawn from such projections are scale dependent.

8.2 Broader View

N^* physics has made great strides in this millennium. For instance, continuum and lattice QCD approaches now appear to qualitatively confirm the constituent quark model spectrum; at least, they produce a spectrum of baryons that possesses a richness which cannot be explained by a two-body model; and the associated "missing resonance" problem is being addressed, with empirical gaps being filled.

Novel images of N^* structure have been drawn through synergistic efforts between experiment and theory, with the availability of electroproduction data on a large Q^2 domain playing an especially important role. A signature example here is the emerging understanding of the Roper resonance as the first radial excitation of the nucleon's dressed-quark core augmented by a meson cloud whose presence has a material impact on its low-Q^2 properties. With this picture, the Roper becomes "normal", being much like the Δ-baryon in many ways.

Such experiment-theory interplay is proving critical in validating modern QCD theory predictions for the momentum-dependence of the running-coupling, -masses, and other basic Schwinger functions that are computable using continuum and lattice methods.

The next decade will see completion of the analysis of existing electroproduction data available on $Q^2 \in [0, 5]\,\text{GeV}^2$ and the gathering of new results out to $Q^2 \approx 12\,\text{GeV}^2$. The new large-$Q^2$ data will probe deeply into the quark-core domain of N^* structure, enabling empirical flavour separation of transition form factor contributions and, thereby, stringent tests of theory predictions regarding the role of diquark correlations within baryons. Furthermore, completed and new analyses of data on electroexcitation amplitudes of baryon parity partners will enable validation of predictions concerning the role of DCSB (emergent mass) in the spectrum and structure of nucleon resonances and help resolve questions relating to chiral symmetry restoration in highly excited systems.

Studies of these types, both ongoing and anticipated, will reveal facts about the running-coupling and -masses in QCD at length-scales associated with more-than 98% of the observable mass of a typical hadron. They are therefore crucial in developing answers to some of the Standard Model's most fundamental questions; namely, what is mass, what is confinement, and how are they related?

Acknowledgements

I am grateful to R. W. Gothe for convening the session in which this perspective was presented. The material contained herein is the product of collaborations with many people, to all of whom I am greatly indebted. Related discussions can be found in contributions to this volume from, *e.g.* V. Burkert, C. Chen, G. Eichmann, C. S. Fischer, R. W. Gothe, V. Mokeev and J. Rodríguez-Quintero. Work partly supported by Jiangsu Province *Hundred Talents Plan for Professionals*.

References

[1] J.M. Cornwall, Phys. Rev. D **26**, 1453 (1982)
[2] P. Boucaud, J.P. Leroy, A. Le-Yaouanc, J. Micheli, O. Pene, J. Rodríguez-Quintero, Few Body Syst. **53**, 387 (2012)
[3] S. Strauss, C.S. Fischer, C. Kellermann, Phys. Rev. Lett. **109**, 252001 (2012)
[4] D. Binosi, L. Chang, J. Papavassiliou, C.D. Roberts, Phys. Lett. B **742**, 183 (2015)
[5] A.C. Aguilar, D. Binosi, J. Papavassiliou, Front. Phys. China **11**, 111203 (2016)
[6] D. Binosi, C. Mezrag, J. Papavassiliou, C.D. Roberts, J. Rodríguez-Quintero, Phys. Rev. D **96**, 054026 (2017)
[7] F. Gao, S.X. Qin, C.D. Roberts, J. Rodríguez-Quintero, Phys. Rev. D **97**, 034010 (2018)
[8] A.K. Cyrol, M. Mitter, J.M. Pawlowski, N. Strodthoff, Phys. Rev. D **97**, 054006 (2018)
[9] J. Rodríguez-Quintero, D. Binosi, C. Mezrag, J. Papavassiliou, C.D. Roberts, Few Body Syst. **59**, 121 (2018)
[10] D. Binosi, R.A. Tripolt (arXiv:1904.08172 [hep-ph]), *Spectral functions of confined particles*
[11] A. Accardi et al., Eur. Phys. J. A **52**, 268 (2016)
[12] US National Academies, *An Assessment of U.S.-Based Electron-Ion Collider Science* (The National

Academies Press, Washington, DC, 2018), ISBN 978-0-309-47856-4
[13] S.J. Brodsky, A.L. Deshpande, H. Gao, R.D. McKeown, C.A. Meyer, Z.E. Meziani, R.G. Milner, J.W. Qiu, D.G. Richards, C.D. Roberts (aXiv:1502.05728 [hep-ph]), *QCD and Hadron Physics*
[14] A. Deur, V. Burkert, J.P. Chen, W. Korsch, Phys. Lett. B **665**, 349 (2008)
[15] A. Deur, S.J. Brodsky, G.F. de Teramond, Prog. Part. Nucl. Phys. **90**, 1 (2016)
[16] M.S. Bhagwat, M.A. Pichowsky, C.D. Roberts, P.C. Tandy, Phys. Rev. C **68**, 015203 (2003)
[17] M.S. Bhagwat, P.C. Tandy, AIP Conf. Proc. **842**, 225 (2006)
[18] P.O. Bowman et al., Phys. Rev. D **71**, 054507 (2005)
[19] C.D. Roberts, Few Body Syst. **58**, 5 (2017)
[20] S.X. Qin, C.D. Roberts, S.M. Schmidt, Phys. Lett. B **733**, 202 (2014)
[21] D. Binosi, L. Chang, S.X. Qin, J. Papavassiliou, C.D. Roberts, Phys. Rev. D **93**, 096010 (2016)
[22] T. Horn, C.D. Roberts, J. Phys. G. **43**, 073001 (2016)
[23] O. Denisov et al. (arXiv:1808.00848 [hep-ex]), *Letter of Intent (Draft 2.0): A New QCD facility at the M2 beam line of the CERN SPS*
[24] A.C. Aguilar et al., Eur. Phys. J. A *in press* (2019), *Pion and Kaon Structure at the Electron-Ion Collider* - arXiv:1907.08218 [nucl-ex]
[25] L.D. Faddeev, Sov. Phys. JETP **12**, 1014 (1961), [Zh. Eksp. Teor. Fiz. **39** (1960) 1459]
[26] R.T. Cahill, C.D. Roberts, J. Praschifka, Austral. J. Phys. **42**, 129 (1989)
[27] H. Reinhardt, Phys. Lett. B **244**, 316 (1990)
[28] G.V. Efimov, M.A. Ivanov, V.E. Lyubovitskij, Z. Phys. C **47**, 583 (1990)
[29] C.D. Roberts, J. Phys. Conf. Ser. **706**, 022003 (2016)
[30] G. Eichmann, H. Sanchis-Alepuz, R. Williams, R. Alkofer, C.S. Fischer, Prog. Part. Nucl. Phys. **91**, 1 (2016)
[31] G. Eichmann, R. Alkofer, A. Krassnigg, D. Nicmorus, Phys. Rev. Lett. **104**, 201601 (2010)
[32] S.X. Qin, C.D. Roberts, S.M. Schmidt, Phys. Rev. D **97**, 114017 (2018)
[33] S.X. Qin, C.D. Roberts, S.M. Schmidt, Few Body Syst. **60**, 26 (2019)
[34] S. Durr et al., Science **322**, 1224 (2008)
[35] M. Tanabashi et al., Phys. Rev. D **98**, 030001 (2018), (*Particle Data Group*)
[36] J. Segovia, C.D. Roberts, S.M. Schmidt, Phys. Lett. B **750**, 100 (2015)
[37] M.B. Hecht, C.D. Roberts, M. Oettel, A.W. Thomas, S.M. Schmidt, P.C. Tandy, Phys. Rev. C **65**, 055204 (2002)
[38] R.T. Cahill, C.D. Roberts, J. Praschifka, Phys. Rev. D **36**, 2804 (1987)
[39] P. Maris, Few Body Syst. **32**, 41 (2002)
[40] Y. Bi, H. Cai, Y. Chen, M. Gong, Z. Liu, H.X. Qiao, Y.B. Yang, Chin. Phys. C **40**, 073106 (2016)
[41] A. Bender, C.D. Roberts, L. von Smekal, Phys. Lett. B **380**, 7 (1996)
[42] M.S. Bhagwat, A. Höll, A. Krassnigg, C.D. Roberts, P.C. Tandy, Phys. Rev. C **70**, 035205 (2004)
[43] P.L. Yin, C. Chen, G. Krein, C.D. Roberts, J. Segovia, S.S. Xu, Phys. Rev. D **100**, 034008 (2019)
[44] C. Chen, B. El-Bennich, C.D. Roberts, S.M. Schmidt, J. Segovia, S. Wan, Phys. Rev. D **97**, 034016 (2018)
[45] I. Aznauryan, V. Burkert, T.S. Lee, V. Mokeev, J. Phys. Conf. Ser. **299**, 012008 (2011)
[46] I. Aznauryan, V. Burkert, Prog. Part. Nucl. Phys. **67**, 1 (2012)
[47] I. Aznauryan, A. Bashir, V. Braun, S. Brodsky, V. Burkert et al., Int. J. Mod. Phys. E **22**, 1330015 (2013)
[48] L.D. Roper, Phys. Rev. Lett. **12**, 340 (1964)
[49] S. Capstick, W. Roberts, Prog. Part. Nucl. Phys. **45**, S241 (2000)
[50] V. Crede, W. Roberts, Rept. Prog. Phys. **76**, 076301 (2013)
[51] M.M. Giannini, E. Santopinto, Chin. J. Phys. **53**, 020301 (2015), 1501.03722
[52] V.D. Burkert, C.D. Roberts, Rev. Mod. Phys. **91**, 011003 (2019)
[53] C. Chen, Y. Lu, D. Binosi, C.D. Roberts, J. Rodríguez-Quintero, J. Segovia, Phys. Rev. D **99**, 034013 (2019)
[54] A. Trivedi, Few Body Syst. **60**, 5 (2019)
[55] V.D. Burkert, V.I. Mokeev, B.S. Ishkhanov (arXiv:1901.09709 [nucl-ex]), *The nucleon resonance structure from exclusive $\pi^+\pi^-p$ photo-/electroproduction off protons*
[56] Y. Lu, C. Chen, Z.F. Cui, C.D. Roberts, S.M. Schmidt, J. Segovia, H.S. Zong, Phys. Rev. D **100**, 034001 (2019)
[57] H.F. Jones, M.D. Scadron, Annals Phys. **81**, 1 (1973)
[58] S. Capstick, B.D. Keister, Phys. Rev. D **51**, 3598 (1995)
[59] I.G. Aznauryan, V.D. Burkert, Phys. Rev. C **92**, 035211 (2015)
[60] I.G. Aznauryan, V.D. Burkert (arXiv:1603.06692 [hep-ph]), Configuration mixings and light-front relativistic quark model predictions for the electroexcitation of the $\Delta(1232)3/2^+$, $N(1440)1/2^+$, and $\Delta(1600)3/2^+$

Form factors for the Nucleon-to-Roper electromagnetic transition at large-Q^2

José Rodríguez-Quintero[1,*], *Daniele* Binosi[2], *Chen* Chen[3], *Ya* Lu[4,5], *Craig D.* Roberts[4,5], and *Jorge* Segovia[6]

[1]Department of Integrated Sciences, University of Huelva, E-21071 Huelva, Spain
[2]European Centre for Theoretical Studies in Nuclear Physics and Related Areas (ECT*) and Fondazione Bruno Kessler Villa Tambosi, Strada delle Tabarelle 286, I-38123 Villazzano (TN) Italy
[3]Justus-Liebig-Universität Giessen, Institut für Theoretische Physik, Heinrich-Buff-Ring 16, 35392 Giessen, Germany
[4]School of Physics, Nanjing University, Nanjing, Jiangsu 210093, China
[5]Institute for Nonperturbative Physics, Nanjing University, Nanjing, Jiangsu 210093, China
[6]Departamento de Sistemas Físicos, Químicos y Naturales, Universidad Pablo de Olavide, E-41013 Sevilla, Spain

Abstract. We report on a recent calculation of all Roper-related electromagnetic transtions form factors, covering the range of energies that next-to-come planned experiments are expected to map. Direct reliable calculations were performed, within a Poincaré covariant approach of the three-body bound-state problem, up to Q^2/m_N^2=6; approximated then by applying the Schlessinger point method and the results eventually extended up to $Q^2/m_N^2 \simeq 12$ *via* analytic continuation.

1 Introduction

Albeit there is no doubt about the nature of nucleons as bound states made of three valence-quarks, that of their first excited states – $N(1440)\,1/2^+$, $N(1535)\,1/2^-$ – is less certain. The parity-positive one, discovered in 1963 [1–5] and dubbed "Roper resonance" since then, puzzled immediately the nuclear physicists because, *e.g.* a second positive-parity state lying above the first negative-parity one in the baryon spectrum had been predicted by a wide array of constituent-quark potential models (*e.g.* see [6]). During the last 20 years, a combined effort for the acquisition of a vast amount of high-precision proton-target exclusive electroproduction data, for their analysis with sophisticated tools from dynamical reaction theory, and for the formulation of wide-ranging application of a Poincaré covariant approach to the continuum bound-state problem in quantum field theory led to the wide acceptance about the Roper being the first radial excitation of the nucleon [7]. And about its being sensibly described as a well-defined dressed-quark core, augmented by a meson cloud which both reduces the Roper's core mass by approximately 20% and contributes materially to the electroproduction transition form factors at low-Q^2.

The high-Q^2 electroproduction data were crucial to reaching this understanding of the Roper and, with a new era of experiments beginning at the upgraded JLab12 in the near future, are expected to sharpen the contemporary picture for the Roper resonance by fostering a unified explanation of elastic and transtion form factors for nucleons, Δ-baryon and Roper resonances. To pursue this goal, given that JLab 12 will deliver results for the Q^2-dependence of $R^{0,+}$ electrocouplings, reaching to $Q^2 \approx 12\,m_N^2$ (m_N for the nucleon mass) [8–10], we have recently reported [11] calculations of all Roper-related transition form factors covering the Q^2-domain that planned experiments expect to map. We will herein sketch these calculations, grounded on the framework of many previous Poincaré-covariant continuum analyses of the three valence-quark bound-state problems associated with the nucleon, Δ-baryon and Roper resonance [12–18].

*e-mail: jose.rodriguez@dfaie.uhu.es

2 Nucleon and Roper Structure

The structure of baryons, treated as three–valence-body bound-states, should be revealed by the solutions of Poincaré-covariant Faddeev equation [19–23], which sums all possible exchanges and interactions that can take place between the three valence dressed-quarks. Notwithstanding this, soft (nonpointlike) diquark correlations have been predicted to exist within baryons, owing to the use of a realistic quark-quark interaction [24–26] and appropriate implementation of chiral symmetry breaking (DCSB) [27]. This strongly supports a truncation of the three-body Faddeev equation such that the problem of determining the properties of the baryon's dressed-quark core transforms into that of solving the linear, homogeneous matrix equation depicted in Fig. 1.

Then, following Ref. [14], the Faddeev equation can be solved to obtain masses and Poincaré-covariant wave functions for the nucleon and its first radial excitation. The masses are found to be, respectively, 1.18 and 1.73 GeV,[1]

[1]It is worth to mention herein that the mass of the nucleon's parity partner, *i.e.* $N(1535)\frac{1}{2}^-$, has been computed within the same formalism in Ref. [16], and its value is 1.83 GeV. Therefore, our formalism produces, on one hand, correct mass order of the $N(1440)\frac{1}{2}^+$ and $N(1535)1/2^-$ states and, on the other hand, right mass splittings with re-

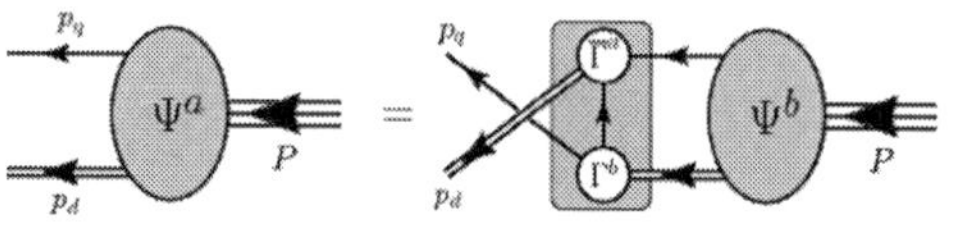

Figure 1. Poincaré covariant Faddeev equation: a linear integral equation for the matrix-valued function Ψ, being the Faddeev amplitude for a baryon of total momentum $P = p_q + p_d$, which expresses the relative momentum correlation between the dressed-quarks and -diquarks within the baryon. The shaded rectangle demarcates the kernel of the Faddeev equation: *single line*, dressed-quark propagator; Γ, diquark correlation amplitude; and *double line*, diquark propagator.

which are predicted to be constituted only from isoscalar-scalar and isovector-pseudovector diquarks, with negligible impact from correlations in the pseudoscalar and vector channels. The masses correspond to the locations of the two lowest-magnitude $I = 1/2$, $J^P = 1/2^+$ poles in the three dressed-quark scattering problem, while the associated residues are the canonically-normalized Faddeev wave functions, depending upon $(\ell^2, \ell \cdot P)$, where ℓ is the quark-diquark relative momentum and P is the baryon's total momentum.

Then, with the wave function in hand, the zeroth Chebyshev moments of all its S-wave components,

$$\mathcal{W}(\ell^2; P^2) = \frac{2}{\pi} \int_{-1}^{1} du\, \sqrt{1-u^2}\, \mathcal{W}(\ell^2, u; P^2), \qquad (1)$$

where $u = \ell \cdot P / \sqrt{\ell^2 P^2}$, can be computed and put under examination; and they can be seen to be either positive- or negative-definite for the ground-state nucleon, while a single zero is exhibitted for the first excited state (See, *e.g.* Ref. [14], Fig. 2, or Ref. [16], Fig. 4.), strongly indicating that this state is a radial excitation of the quark-diquark system.

Concerning the masses, they can be confronted to the empirical values of the pole locations for the first two states in the nucleon channel, which are: 0.939 GeV for the nucleon; and two poles for the Roper, $1.357 - i\,0.076$, $1.364 - i\,0.105$ GeV, [28]. They differ at first glance, but the discrepancies can be well understood: The solutions of the Faddeev equation with the kernel in Fig. 1 should be interpreted as the *dressed-quark core* of the bound-state, instead of representing the fully-dressed observable object, because the kernel omits [29, 30] all those resonant contributions which may be associated with the meson-baryon-cloud finite-state-interactions (MB FSIs) resummed in DCC models [28, 31–36]. Therefore, the critical comparison is not between the computed dressed-quark core masses and empirical values of the pole-positions but between the former and the values determined for the meson-undressed bare bound-state; *e.g.* for the Roper, *viz.* (in GeV):

	$R_{\rm core}^{[14,\,16]}$	$R_{\rm core}^{[37]}$	$R_{\rm core}^{[38]}$	$R_{\rm DCC\,bare}^{[28]}$
mass	1.73	1.82	1.72	1.76

; (2)

spect to nucleon's mass, being their masses higher than experiment for the reasons that shall be explained below. We refer the interested reader to the Ref. [16] for further details.

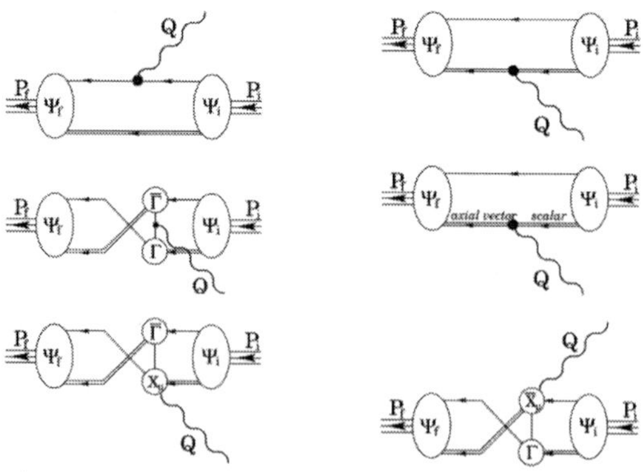

Figure 2. Vertex that ensures a conserved current for on-shell baryons that are described by the Faddeev amplitudes produced by the equation depicted in Fig. 1: *single line*, dressed-quark propagator; *undulating line*, photon; Γ, diquark correlation amplitude; and *double line*, diquark propagator. Diagram 1 is the top-left image; the top-right is Diagram 2; and so on, with Diagram 6 being the bottom-right image. (Details are provided in Ref. [12], Appendix C.)

The particular DCC bare-Roper mass shown above agreeing remarkably well with the quark core results obtained using both a QCD-kindred interaction [14] and refined treatments of a strictly-implemented vector$\otimes$vector contact-interaction [37, 38]. An alternative test comes with clothing nucleon's dressed-quark core by including resonant contributions to the kernel. One thus produces a physical nucleon and Δ-baryon masses which are ≈ 0.2 GeV lower than those of the core [31, 39, 40]. Consistently, our estimate for the nucleon is 0.2 GeV greater than the empirical value.

3 Electromagnetic Currents

The next brick for the desired computation of elastic and transition form factors is the electromagnetic current carrying the interaction between the participating states. When the initial and final states are $I = 1/2$, $J = 1/2^+$ baryons, that current is completely specified by two form factors, *viz.*

$$\bar{u}_f(P_f)[\gamma_\mu^T F_1^{fi}(Q^2) + \frac{1}{m_{fi}} \sigma_{\mu\nu} Q_\nu F_2^{fi}(Q^2)] u_i(P_i), \qquad (3)$$

where: u_i, $\bar{u}_f$ are, respectively, Dirac spinors describing the incoming/outgoing baryons, with four-momenta $P_{i,f}$ and masses $m_{i,f}$ so that $P_{i,f}^2 = -m_{i,f}^2$; $Q = P_f - P_i$; $m_{fi} = (m_f + m_i)$; and $\gamma^T \cdot Q = 0$.

The interaction of a photon with a baryon generated by the Faddeev equation in Fig. 1 can be sufficiently expressed by a vertex including the six terms depicted in Fig. 2 [12, 41], with the photon separately probing the quarks and diquarks in various ways. Then, elastic and transition electromagnetic form factors involving the nucleon and Roper may be dissected in two separate ways, each of which can be considered as a sum of three distinct terms, namely:

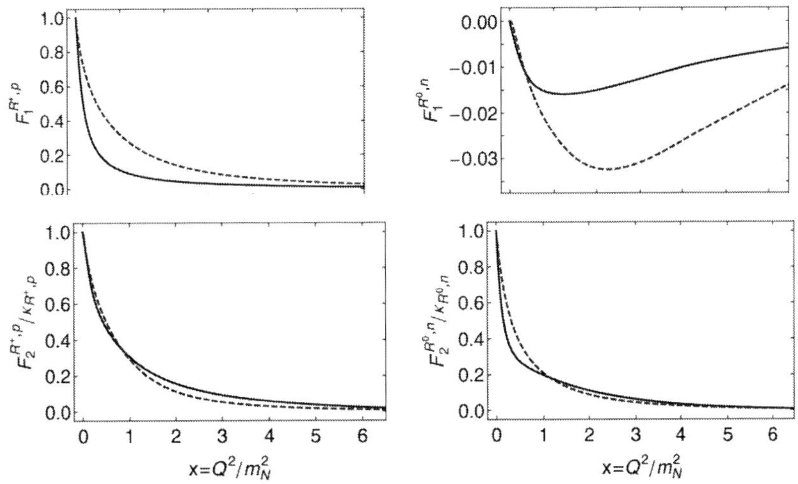

Figure 3. Solid (black) curves – Dirac (upper panels) and Pauli (lower) elastic electromagnetic form factors associated with the dressed-quark cores of the charged (left) and neutral (right) Roper systems. Dashed (blue) curves – analogous results for proton and neutron. ($\kappa_{R,N} = F_2^{R,N}(x = 0)$; $x = Q^2/m_N^2$, where $M_N = 1.18$ GeV is the nucleon' dressed-quark core mass.)

DD=diquark dissection: [*DD1*] scalar diquark, $[ud]$, in both the initial- and final-state baryon; [*DD2*] pseudovector diquark, $\{qq\}$, in both the initial- and final-state baryon; and [*DD3*] a different diquark in the initial- and final-state baryon.

DS = scatterer dissection: [*DS1*] photon strikes a bystander dressed-quark (Diagram 1 in Fig. 2); [*DS2*] photon interacts with a diquark, elastically or causing a transition scalar $\leftrightarrow$ pseudovector (Diagrams 2 and 4 in Fig. 2); and [*DS3*] photon strikes a dressed-quark in-flight, as one diquark breaks up and another is formed (Diagram 3 in Fig. 2), or appears in one of the two associated "seagull" terms (Diagrams 5 and 6).

4 Elastic Form Factors

The analogous elastic form factors for the states involved, $F_1^{f=i}(Q^2 = 0)$, are needed to fix the normalisation of the transition. The elastic Dirac and Pauli form factors associated with the dressed-quark core of the charged and neutral Roper are depicted in Fig. 3 and compared with those for the proton and neutron from Ref. [12]. Evidently, there are qualitative similarities and quantitative differences, discussed at length in Ref. [11].

Table 1. Static properties derived from the elastic form factors depicted in Fig. 3, see Eq. (5) and following text. ($M_N = 1.18$ GeV is the nucleon dressed-quark core mass.)

	R^+	p	R^0	n
$r_E M_N$	6.23	3.65	$0.93i$	$1.67i$
$r_M M_N$	4.49	3.17	4.15	4.19
μ	2.67	2.50	−1.24	−1.83

Defining (Sachs) electric and magnetic form factors:

$$G_E = F_1 - \frac{Q^2}{4m_B^2} F_2\,, \quad G_M = F_1 + F_2\,, \tag{4}$$

where m_B is the baryon's mass, the $Q^2 = 0$ values and slopes of the form factors in Fig. 3 yield the static properties listed in Table 1, where the radii are defined via

$$r^2 = -\frac{6}{n}\frac{d}{dQ^2} G(Q^2)\bigg|_{Q^2=0}\,, \tag{5}$$

with $n = G(Q^2 = 0)$ when this quantity is nonzero, $n = 1$ otherwise, and the anomalous magnetic moment $\mu = G_M(0)$. The electromagnetic radii of the charged-Roper core are larger than those of the proton core, but the magnetic moments are similar; and this pattern is reversed in the neutral-Roper/neutron comparison.

Figs. 4, 5 display the contrast of diquark and scatterer dissections of the Dirac and Pauli form factors for the proton and R^+. It is apparent from Fig. 4 that every contribution to the R^+ elastic Dirac form factor falls more rapidly than its analogue in the proton; while, on the other hand, the relative importance of each is typically the same within the proton and R^+. The former observation highlights that the relative strengths of the various diquarks in both the nucleon and Roper are almost identical. Concerning Fig. 5, DD1×DS1 is dominant in both columns, implying that the primary contribution to both proton and charged Roper comes from a photon striking a bystander quark in association with a $[ud]$-diquark; while differences appear for the subleading terms.

Diquark and scatterer dissections for the Dirac and Pauli form factors of the neutral-Roper, which can be contrasted to $\gamma^* n \to R^0$ transition form factors, appear also displayed in Fig. 6.

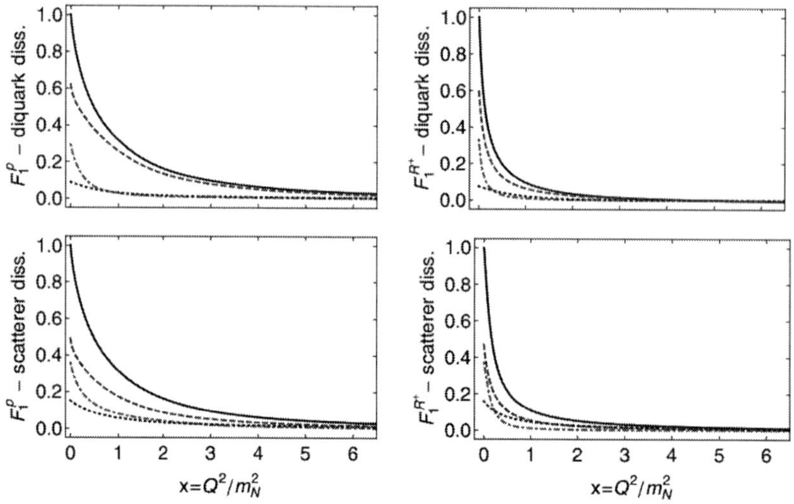

Figure 4. Dirac form factors of the proton (left) and charged-Roper (right). *Upper panels* – diquark breakdown: *DD1* (dashed red), scalar diquark in initial and final baryon; *DD2* (dot-dashed green), pseudovector diquark in both initial and final states; *DD3* (dotted blue), scalar diquark in incoming baryon, pseudovector diquark in outgoing baryon, and vice versa. *Lower panels* – scatterer breakdown: *DS1* (red dashed), photon strikes an uncorrelated dressed quark; *DS2* (dot-dashed green), photon strikes a diquark; and *DS3* (dotted blue), diquark breakup contributions, including photon striking exchanged dressed-quark.

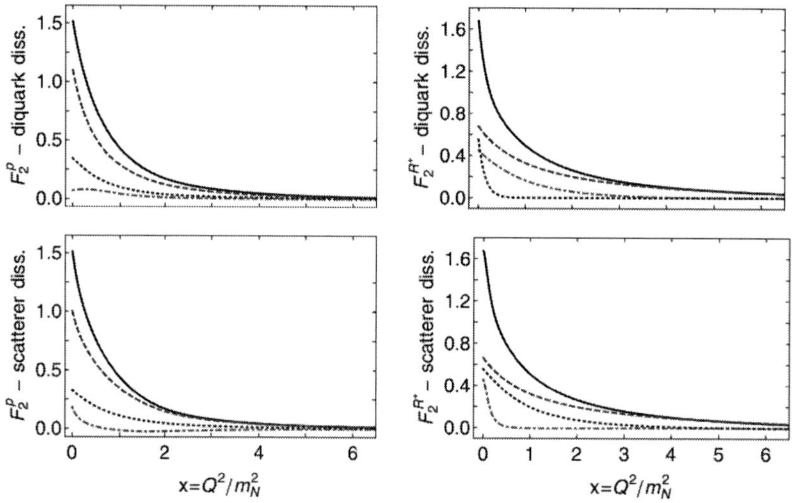

Figure 5. Pauli form factors of the proton and charged-Roper. *Upper panels* – diquark breakdown: *DD1* (dashed red), scalar diquark in initial and final baryon; *DD2* (dot-dashed green), pseudovector diquark in both initial and final states; *DD3* (dotted blue), scalar diquark in incoming baryon, pseudovector diquark in outgoing baryon, and vice versa. *Lower panels* – scatterer breakdown: *DS1* (red dashed), photon strikes an uncorrelated dressed quark; *DS2* (dot-dashed green), photon strikes a diquark; and *DS3* (dotted blue), diquark breakup contributions, including photon striking exchanged dressed-quark.

5 Nucleon-to-Roper Transition

The Dirac transition form factors for $\gamma^* N \to R$, drawn in Fig. 7, show qualitative similarities to the elastic Dirac form factors for the charged channels depicted in Fig. 4, dominated by the contribution in which a photon strikes a bystander dressed-quark in association with a $[ud]$-diquark, with lesser but non-negligible contributions from other processes. The same process for a photon striking a bystander dressed-quark partenered by $[ud]$ primarily contributes, too, to the transition in the neutral channel. Comparisons with the the left panels in Fig. 6, for R^0 elastic Dirac form factor, are however less transparent because,

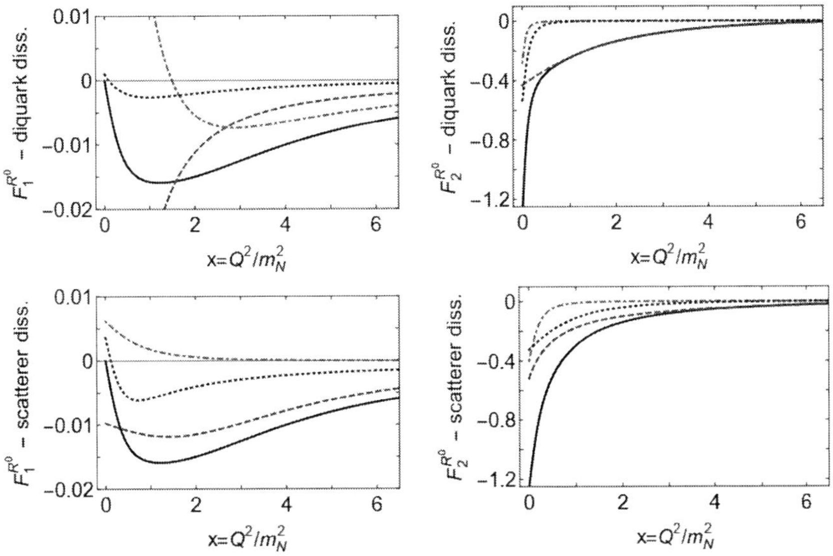

Figure 6. Dirac (left) and Pauli (right) elastic form factors of neutral-Roper. *Upper panels* – diquark breakdown: *DD1* (dashed red), scalar diquark in initial and final baryon; *DD2* (dot-dashed green), pseudovector diquark in both initial and final states; *DD3* (dotted blue), scalar diquark in incoming baryon, pseudovector diquark in outgoing baryon, and vice versa. *Lower panels* – scatterer breakdown: *DS1* (red dashed), photon strikes an uncorrelated dressed quark; *DS2* (dot-dashed green), photon strikes a diquark; and *DS3* (dotted blue), diquark breakup contributions, including photon striking exchanged dressed-quark.

both $F_1^{R^0}$ and $F_{1,n}^*$ vainishing at the origin, in the latter case each contribution does separately, owing to state othogonality, while in the former charge neutrality only ensures that all terms sum to zero [11].

Owing to the QCD-derived momentum-dependence of the propagators and vertices employed in solving the bound-state and scattering problems, $F_{1,p}^*(x)$ agrees quantitatively in magnitude and trend with the data above $x \simeq 2$. Below this value, the mismatch between prediction and data can be plausibly attributed to MB FSIs, as described in Sec. 5 of Ref. [45]. (See also Ref. [36].)

A computation of the associated light-front-transverse transition charge-density is [46],

$$\rho^{pR}(|\vec{b}|) := \int \frac{d^2\vec{q}_\perp}{(2\pi)^2}\, e^{i\vec{q}_\perp \cdot \vec{b}} F_1^*(|\vec{q}_\perp|^2)\,, \tag{6}$$

the frame defined by $Q = (\vec{q}_\perp = (Q_1, Q_2), Q_3 = 0, Q_4 = 0)$, may be found elsewhere [18], along with a related analysis of the impact of MB FSIs, which have been found to introduce significant attraction, screening the long negative tail of the quark-core contribution to $\rho^{pR}(|\vec{b}|)$ (consistently with their role in reducing the nucleon and Roper quark-core masses) and thereby compressing the transition domain in transverse space. Concerning the neutral transition, owing to its being uniformly small, $F_{1,n}^*(x)$ can be plausibly impacted by MB FSIs on a larger Q^2-domain, as it appears to be the case, *e.g.* with the electric quadrupole form factor in the $\gamma^* N \to \Delta$ transition [12].

Pauli transition form factors for $\gamma^* N \to R$ are drawn in Fig. 8: They are all nonzero at $x = 0$ and each crosses zero at roughly the same location, *viz.* $x \approx 0.2$. As happens with F_2^{p,R^+} and F_2^{n,R^0} in Fig. 3, $F_{2,p}^*$ and $F_{2,n}^*$ are similar in magnitude and Q^2-dependence and, remarkably, the ratio $F_{2,p}^*(0)/F_{2,n}^*(0) \approx -3/2$ is consistent with available data [48]. Furthermore, MB FSIs also apply to Pauli form factors as to Dirac but, although affecting its precise location, do not spoil the existence of a zero in F_2^*, which can be confidently consisered as a robust prediction.

6 Transitions at Larger Q^2

We have recently reported [11] projections for all transitions from factors on $x \in [0, 12]$, aiming at anticipating the deliverance of data on the Roper-resonance electroproduction form factors out to $Q^2 \approx 12\, m_N^2$ in both the charged and neutral channels by the CLAS12 detector at JLab 12. Let us sketch them here.

Direct calculations of all the contributions to our transitions form factors in Fig. 2 imply eight-dimensional integrals to be evaluated (Diagrams 3, 5, 6), *e.g.* by applying Monte-Carlo methods which are imprecise when they are required to deliver a small result and not all contributions share the same sign (it is the case with form factors at large photon virtuality). We then circumvented this difficulty in Ref. [11] by using the Schlessinger point method (SPM) [49–51] to construct analytic approximations on $x \in [0, 6]$, and then defining the results on $x \in [6, 12]$ via the analytic continuation of those approximations.

Details for the practical implementation of SPM in our case can be found in Ref. [11], suffice it to indicate here

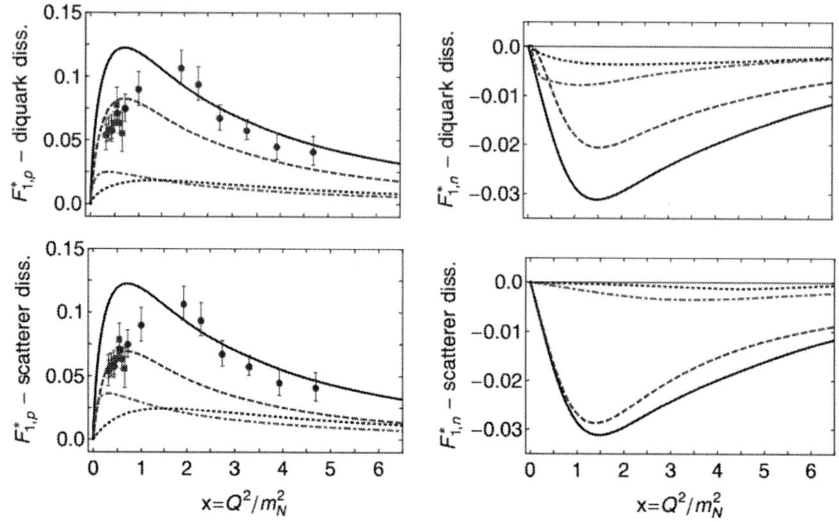

Figure 7. Computed Dirac transition form factor, F_1^*, for the charged reaction $\gamma^* p \to R^+$ (left panels) and the neutral reaction $\gamma^* n \to R^0$ (right panels): solid (black) curve in each panel. Data, left panels: circles (blue) [42], and squares (purple) [43, 44]. *Upper panels* – diquark breakdown: *DD1* (dashed red), scalar diquark in both nucleon and Roper; *DD2* (dot-dashed green), pseudovector diquark in both nucleon and Roper; *DD3* (dotted blue), scalar diquark in nucleon, pseudovector diquark in Roper, and vice versa. *Lower panels* – scatterer breakdown: *DS1* (red dashed), photon strikes an uncorrelated dressed quark; *DS2* (dot-dashed green), photon strikes a diquark; and *DS3* (dotted blue), diquark breakup contributions, including photon striking exchanged dressed-quark.

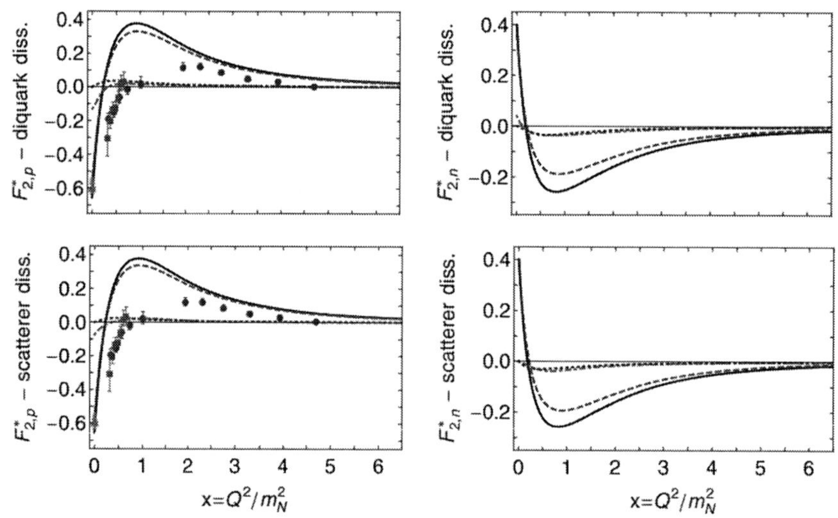

Figure 8. Computed Pauli transition form factor, F_2^*, for the charged reaction $\gamma^* p \to R^+$ (left panels) and the neutral reaction $\gamma^* n \to R^0$ (right panels): solid (black) curve in each panel. Data: circles (blue) [42], squares (purple) [43, 44], triangle (gold) [47], and star (green) [48]. *Upper panels* – diquark breakdown: *DD1* (dashed red), scalar diquark in both nucleon and Roper; *DD2* (dot-dashed green), pseudovector diquark in both nucleon and Roper; *DD3* (dotted blue), scalar diquark in nucleon, pseudovector diquark in Roper, and vice versa. *Lower panels* – scatterer breakdown: *DS1* (red dashed), photon strikes an uncorrelated dressed quark; *DS2* (dot-dashed green), photon strikes a diquark; and *DS3* (dotted blue), diquark breakup contributions, including photon striking exchanged dressed-quark.

that we are eventually left with a band of extrapolated curves whose collective reliability at any $Q^2 > Q^2_{\rm max}$ is expressed by the width of the band at that point, which is itself determined by the SPM's extrapolation error and the precision of the original output on $Q^2 \leq Q^2_{\rm max}$; with $Q^2_{\rm max}$ being the upper bound for the domain upon which direct

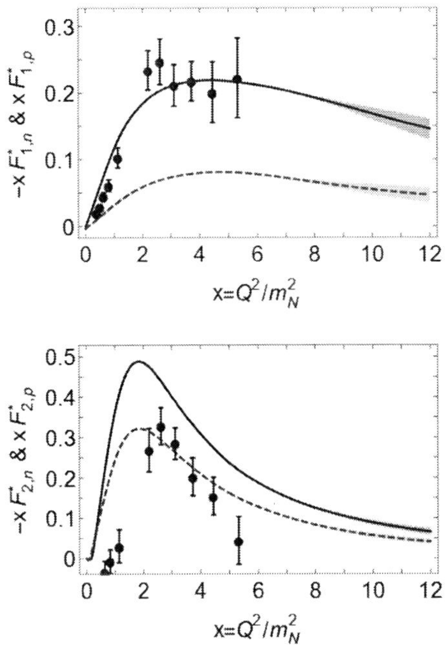

Figure 9. Computed x-weighted Dirac (upper panel) and Pauli (lower panel) transition form factors for the reactions $\gamma^* p \to R^+$ (solid blue curves) and $\gamma^* n \to R^0$ (dashed green curves). In all cases, the results on $x \in [6, 12]$ are projections, obtained via extrapolation of analytic approximations to our results on $x \in [0, 6]$: at each x, the width of the band associated with a given curve indicates our confidence in the extrapolated value. (See text for details.) Data in both panels are for the charged channel transitions, $F^*_{1,p}$ and $F^*_{2,p}$: circles (blue) [42]. No data currently exist for the neutral channel.

computations have been confidently performed. Results for the x-weighted Dirac and Pauli transition form factors for the reactions $\gamma^* p \to R^+$, $\gamma^* n \to R^0$ appear displayed in Fig. 9 on the domain $0 < x < 12$. The precision of our SPM projections can be exemplified by quoting the form factor values at the upper bound of the extrapolation domain, $x_{12} = 12$:

$$F^*_{1,p} = 0.0121(14)\,, \qquad -F^*_{1,n} = 0.0039(10)\,, \quad \text{(7a)}$$
$$x_{12} F^*_{1,p} = 0.145(17)\,, \qquad -x_{12} F^*_{1,n} = 0.046(11)\,, \quad \text{(7b)}$$

$$F^*_{2,p} = 0.0055(8)\,, \qquad -F^*_{2,n} = 0.0034(7)\,, \quad \text{(8a)}$$
$$x_{12} F^*_{2,p} = 0.066(10)\,, \qquad -x_{12} F^*_{2,n} = 0.041(9)\,. \quad \text{(8b)}$$

The x-weighted results accentuate, without overmagnifying, the larger-x behaviour of the form factors. On the domain depicted, there is no indication of the scaling behaviour expected of the transition form factors: $F^*_1 \sim 1/x^2$, $F^*_2 \sim 1/x^3$. One would anyhow expect such behaviour becoming manifest on $x \gtrsim 20$, since each dressed-quark in the baryons must roughly share the impulse momentum, Q.

7 Conclusions

The existence of strong nonpointlike quark-quark (diquark) correlations within baryons has long been argued. On the ground of this, we used a quark-diquark approximation to the Poincaré-covariant three-body bound-state problem to compute all form factors relevant to the $\gamma^* N \to R$ transitions in Ref. [11] and have now reported here the results. We have found that both scalar and pseudovector diquarks are essential for a description of existing data, but correlations in other diquark channels can be neglected. The same holds true for nucleon elastic and $\gamma^* N \to \Delta(1232)$ transition form factors.

Focusing on $\gamma^* N \to R$, precise measurements in the charged channel already exist [42, 52–56], novel experiments are approved at JLab 12 and elsewhere, and others are either planned or under consideration as part of an international effort to measure transition electrocouplings of all prominent nucleon resonances [8–10, 53, 57]. Hence, our predictions, including those in the neutral channel, are likely to be tested in the foreseeable future.

The empirical information that can be delivered by such experiments has the potential to address a wide range of issues. In particular, it should help to confirm whether, as happens for the nucleon, Δ-baryon and Roper resonance, the quark-quark correlations play an essential role in the structure of all baryons. And, underlying this, one could be able to inquire about whether the expression of DCSB is the same in each baryon. Indeed, the nucleon, Δ-baryon and Roper resonance are merely a small collection of closely-related positive-parity baryons; and, hence, consistency with available data may not be seen as conclusive. Even less conclusive because there can be emerging evidences indicating that pseudoscalar and vector diquark correlations also play a material role in low-lying negative-parity baryons [16, 37, 58], that excited states of the Δ-baryon possess unexpectedly complicated wave functions [59], and very little is known about the Poincaré-covariant wave functions of $I = 1/2$, $J = 3/2$ baryons. These systems are the focus of forthcoming analyses using the methods herein described.

acknowledgements

We are grateful to R. W. Gothe for convening the session in which this perspective was presented. Work partly supported by Spanish ministry project PPA2017-86380-P and by Jiangsu Province Hundred Talents Plan for Professionals.

References

[1] L.D. Roper, Phys. Rev. Lett. **12**, 340 (1964)
[2] P. Bareyre, C. Bricman, G. Valladas, G. Villet, J. Bizard, J. Seguinot, Phys. Lett. **8**, 137 (1964)
[3] P. Auvil, C. Lovelace, A. Donnachie, A.T. Lea, Phys. Lett. **12**, 76 (1964)
[4] S.L. Adelman, Phys. Rev. Lett. **13**, 555 (1964)
[5] L.D. Roper, R.M. Wright, B.T. Feld, Phys. Rev. **138**, B190 (1965)

[6] V. Crede, W. Roberts, Rept. Prog. Phys. **76**, 076301 (2013), 1302.7299
[7] V.D. Burkert, C.D. Roberts, Rev. Mod. Phys. **91**, 011003 (2019), 1710.02549
[8] V.I. Mokeev (CLAS), Few Body Syst. **59**, 46 (2018), 1801.09750
[9] D.S. Carman, Few Body Syst. **59**, 82 (2018)
[10] P. Cole, B. Ramstein, A. Sarantsev, Few Body Syst. **59**, 144 (2018)
[11] C. Chen, Y. Lu, D. Binosi, C.D. Roberts, J. Rodríguez-Quintero, J. Segovia, Phys. Rev. **D99**, 034013 (2019), 1811.08440
[12] J. Segovia, I.C. Cloet, C.D. Roberts, S.M. Schmidt, Few Body Syst. **55**, 1185 (2014), 1408.2919
[13] C.D. Roberts, J. Phys. Conf. Ser. **630**, 012051 (2015), 1501.06581
[14] J. Segovia, B. El-Bennich, E. Rojas, I.C. Cloet, C.D. Roberts, S.S. Xu, H.S. Zong, Phys. Rev. Lett. **115**, 171801 (2015), 1504.04386
[15] J. Segovia, C.D. Roberts, Phys. Rev. **C94**, 042201 (2016), 1607.04405
[16] C. Chen, B. El-Bennich, C.D. Roberts, S.M. Schmidt, J. Segovia, S. Wan, Phys. Rev. **D97**, 034016 (2018), 1711.03142
[17] C. Mezrag, J. Segovia, L. Chang, C.D. Roberts, Phys. Lett. **B783**, 263 (2018), 1711.09101
[18] C.D. Roberts, Few Body Syst. **59**, 72 (2018), 1801.08562
[19] R.T. Cahill, C.D. Roberts, J. Praschifka, Austral. J. Phys. **42**, 129 (1989)
[20] C.J. Burden, R.T. Cahill, J. Praschifka, Austral. J. Phys. **42**, 147 (1989)
[21] R.T. Cahill, Austral. J. Phys. **42**, 171 (1989)
[22] H. Reinhardt, Phys. Lett. **B244**, 316 (1990)
[23] G.V. Efimov, M.A. Ivanov, V.E. Lyubovitskij, Z. Phys. **C47**, 583 (1990)
[24] D. Binosi, L. Chang, J. Papavassiliou, C.D. Roberts, Phys. Lett. **B742**, 183 (2015), 1412.4782
[25] D. Binosi, C. Mezrag, J. Papavassiliou, C.D. Roberts, J. Rodriguez-Quintero, Phys. Rev. **D96**, 054026 (2017), 1612.04835
[26] J. Rodríguez-Quintero, D. Binosi, C. Mezrag, J. Papavassiliou, C.D. Roberts, Few Body Syst. **59**, 121 (2018), 1801.10164
[27] J. Segovia, C.D. Roberts, S.M. Schmidt, Phys. Lett. **B750**, 100 (2015), 1506.05112
[28] N. Suzuki, B. Julia-Diaz, H. Kamano, T.S.H. Lee, A. Matsuyama, T. Sato, Phys. Rev. Lett. **104**, 042302 (2010), 0909.1356
[29] G. Eichmann, R. Alkofer, I.C. Cloet, A. Krassnigg, C.D. Roberts, Phys. Rev. **C77**, 042202 (2008), 0802.1948
[30] G. Eichmann, I.C. Cloet, R. Alkofer, A. Krassnigg, C.D. Roberts, Phys. Rev. **C79**, 012202 (2009), 0810.1222
[31] B. Julia-Diaz, T.S.H. Lee, A. Matsuyama, T. Sato, Phys. Rev. **C76**, 065201 (2007), 0704.1615
[32] H. Kamano, S.X. Nakamura, T.S.H. Lee, T. Sato, Phys. Rev. **C81**, 065207 (2010), 1001.5083
[33] D. Ronchen, M. Doring, F. Huang, H. Haberzettl, J. Haidenbauer, C. Hanhart, S. Krewald, U.G. Meissner, K. Nakayama, Eur. Phys. J. **A49**, 44 (2013), 1211.6998
[34] H. Kamano, S.X. Nakamura, T.S.H. Lee, T. Sato, Phys. Rev. **C88**, 035209 (2013), 1305.4351
[35] M. Doring, Int. J. Mod. Phys. Conf. Ser. **26**, 1460054 (2014)
[36] H. Kamano, Few Body Syst. **59**, 24 (2018)
[37] Y. Lu, C. Chen, C.D. Roberts, J. Segovia, S.S. Xu, H.S. Zong, Phys. Rev. **C96**, 015208 (2017), 1705.03988
[38] D.J. Wilson, I.C. Cloet, L. Chang, C.D. Roberts, Phys. Rev. **C85**, 025205 (2012), 1112.2212
[39] N. Ishii, Phys. Lett. **B431**, 1 (1998)
[40] M.B. Hecht, M. Oettel, C.D. Roberts, S.M. Schmidt, P.C. Tandy, A.W. Thomas, Phys. Rev. **C65**, 055204 (2002), nucl-th/0201084
[41] M. Oettel, M. Pichowsky, L. von Smekal, Eur. Phys. J. **A8**, 251 (2000), nucl-th/9909082
[42] I.G. Aznauryan et al. (CLAS), Phys. Rev. **C80**, 055203 (2009), 0909.2349
[43] V.I. Mokeev et al. (CLAS), Phys. Rev. **C86**, 035203 (2012), 1205.3948
[44] V.I. Mokeev et al., Phys. Rev. **C93**, 025206 (2016), 1509.05460
[45] C.D. Roberts, J. Segovia, Few Body Syst. **57**, 1067 (2016), 1603.02722
[46] L. Tiator, M. Vanderhaeghen, Phys. Lett. **B672**, 344 (2009), 0811.2285
[47] M. Dugger et al. (CLAS), Phys. Rev. **C79**, 065206 (2009), 0903.1110
[48] M. Tanabashi et al. (Particle Data Group), Phys. Rev. **D98**, 030001 (2018)
[49] L. Schlessinger, C. Schwartz, Phys. Rev. Lett. **16**, 1173 (1966)
[50] L. Schlessinger, Phys. Rev. **171**, 1523 (1968)
[51] R.A. Tripolt, I. Haritan, J. Wambach, N. Moiseyev, Phys. Lett. **B774**, 411 (2017), 1610.03252
[52] I.G. Aznauryan, V.D. Burkert, Phys. Rev. **C85**, 055202 (2012), 1201.5759
[53] I.G. Aznauryan et al., Int. J. Mod. Phys. **E22**, 1330015 (2013), 1212.4891
[54] K. Park et al. (CLAS), Phys. Rev. **C91**, 045203 (2015), 1412.0274
[55] E.L. Isupov et al. (CLAS), Phys. Rev. **C96**, 025209 (2017), 1705.01901
[56] G.V. Fedotov et al. (CLAS), Phys. Rev. **C98**, 025203 (2018), 1804.05136
[57] B. Ramstein, Few Body Syst. **59**, 143 (2018)
[58] G. Eichmann, C.S. Fischer, H. Sanchis-Alepuz, Phys. Rev. **D94**, 094033 (2016), 1607.05748
[59] S.X. Qin, C.D. Roberts, S.M. Schmidt, Phys. Rev. **D97**, 114017 (2018), 1801.09697

Role of $N^*(1535)$ in the $\Lambda_c^+ \to \bar{K}^0\eta p$ decay and the possible ϕp state in the $\Lambda_c^+ \to \pi^0\phi p$ decay

Ju-Jun Xie[1,2,3,*], *Li-Sheng* Geng[4,3], and *Feng-Kun* Guo[5,2]

[1]Institute of Modern Physics, Chinese Academy of Sciences, Lanzhou 730000, China
[2]School of Nuclear Science and Technology, University of Chinese Academy of Sciences, Beijing 100049, China
[3]School of Physics and Microelectronics, Zhengzhou University, Zhengzhou, Henan 450001, China
[4]School of Physics & Beijing Advanced Innovation Center for Big Data-based Precision Medicine, Beihang University, Beijing 100191, China
[5]CAS Key Laboratory of Theoretical Physics, Institute of Theoretical Physics, Chinese Academy of Sciences, Zhong Guan Cun East Street 55, Beijing 100190, China

Abstract. The nonleptonic weak decays of $\Lambda_c^+ \to \bar{K}^0\eta p$ and $\Lambda_c^+ \to \pi^0\phi p$ are investigated from the viewpoint of probing the $N^*(1535)$ resonance and the possible ϕp state. For the $\Lambda_c^+ \to \bar{K}^0\eta p$ decay, we study the invariant mass distribution of ηp with both the chiral unitary approach and an effective Lagrangian model. Within the chiral unitary approach, the $N^*(1535)$ resonance is dynamically generated from the final state interaction of mesons and baryons in coupled channels. While for the effective Lagrangian model, we take a Breit-Wigner formula for the $N^*(1535)$ resonance. We found that the behavior of the $N^*(1535)$ resonance in the $\Lambda_c^+ \to \bar{K}^0 N^*(1535) \to \bar{K}^0\eta p$ decay within the two approaches is different. For the $\Lambda_c^+ \to \pi^0\phi p$ decay, we consider a triangle singularity mechanism, where the Λ_c^+ decays into the $K^*\Sigma^*(1385)$, the $\Sigma^*(1385)$ decays into the $\pi^0\Sigma/\Lambda$, and then the $K^*\Sigma/\Lambda$ merge to produce the ϕp in the final state. This mechanism produces a peak structure around 2020 MeV. In addition, the possibility that there is a hidden-strange pentaquark-like state is also considered by taking into account the final state interactions of $K^*\Lambda$, $K^*\Sigma$, and ϕp. We conclude that it is difficult to search for the hidden-strange state in this decay. However, we do expect nontrivial behavior in the ϕp invariant mass distribution. The proposed Λ_c^+ decay mechanism here can provide valuable information on the properties of these nuclear resonances and can in principle be tested by experiments such as BESIII, LHCb and Belle-II.

1 Introduction

The nonleptonic weak decays of charmed hadrons provide a useful platform to study hadronic resonances, some of which are subjects of intense debate about their nature [1–3]. For instance, the $\Lambda_c^+ \to \pi^+ MB$ weak decays are studied in Refs. [4, 5] from the viewpoint of probing the $\Lambda(1405)$ and $\Lambda(1670)$ resonances and extracting the $\pi\Sigma$ scattering lengths, where M and B stand for mesons and baryons. In Ref. [6], the $\Lambda(1670)$ and $a_0(980)$ resonances are investigated in the $\Lambda_c^+ \to \pi^+\eta\Lambda$ decay. The role of the possible $\Sigma^*(1380)$ state with $J^P = 1/2^-$ in the $\Lambda_c^+ \to \eta\pi^+\Lambda$ decay is also studied in Ref. [7] where the color-suppressed W-exchange diagram is considered for the production of the $\Sigma^*(1385)$ with $J^P = 3/2^+$. Experimentally, in Ref. [8], the decays $\Lambda_c^+ \to \eta\pi^+\Lambda$ and $\eta\Sigma^{*+}(1385)$ was investigated by the BESIII collaboration. On the other hand, in Ref. [9] the role of the exclusive Λ_c^+ decays into a neutron in testing the flavor symmetry and final state interaction was proposed. It was shown that the three body nonleptonic decays of Λ_c^+ are of great interest to explore the final state interactions.

Analogous to the hidden-charm pentaquark states that were observed in the $J/\psi p$ invariant mass spectrum via the $\Lambda_b^0 \to K^- J/\psi p$ decay by the LHCb collaboration [10, 11], one may consider the possible existence of hidden-strange pentaquarks $P_s = uuds\bar{s}$, in which the $c\bar{c}$ pair is replaced by $s\bar{s}$ [12]. In fact, in the light flavor sector below 2 GeV, understanding the nature of the $N^*(1535)$ resonance with spin parity $J^P = 1/2^-$ is very challenging [13, 14]. One peculiar property of the $N^*(1535)$ is that it couples strongly to the channels with strangeness, which is difficult to understand in the classical quark models. However, the strange decay properties of the $N^*(1535)$ resonance can be easily understood if it contains large five-quark components [15–18]. Within this pentaquark picture, the $N^*(1535)$ resonance could be the lowest $L = 1$ orbitally excited uud state with a large admixture of $[ud][us]\bar{s}$ pentaquark component. This makes the $N^*(1535)$ heavier than the $N^*(1440)$ and also gives a natural explanation of its large couplings to the channels with strangeness [19]. The large $N^*(1535)$ coupling to the $K\Lambda$ channel obtained in Refs. [15, 20] from the analysis of $J/\psi \to \bar{p}K^+\Lambda$ and $J/\psi \to \bar{p}p\eta$ reactions has been challenged in Ref. [21].

At around 2 GeV, a ϕN bound state is predicted in several models [22–24]. Such a ϕN states can be viewed as a P_s pentaquark state. The forward-direction enhancement

*e-mail: xiejujun@impcas.ac.cn

at around invariant mass $W = 2.1$ GeV in the $\gamma p \to p\phi$ reaction [25, 26] may indicate the s-channel contributions from nucleon excited state around 2.1 GeV [27].

In Ref. [28], the role of the $N^*(1535)$ resonance in the $\Lambda_c^+ \to \bar{K}^0\eta p$ decay was investigated by taking the advantage of the strong coupling of the $N^*(1535)$ to the ηN channel and its large $uuds\bar{s}$ component. While for the possible P_s state, it was studied in Ref. [29] in the singly Cabibbo suppressed process $\Lambda_c^+ \to \pi^0\phi p$, as shown in Fig. 1 in the quark level. However, the main difficulty is that the decay $\Lambda_c^+ \to \pi^0\phi p$ has a very limited phase space [12].

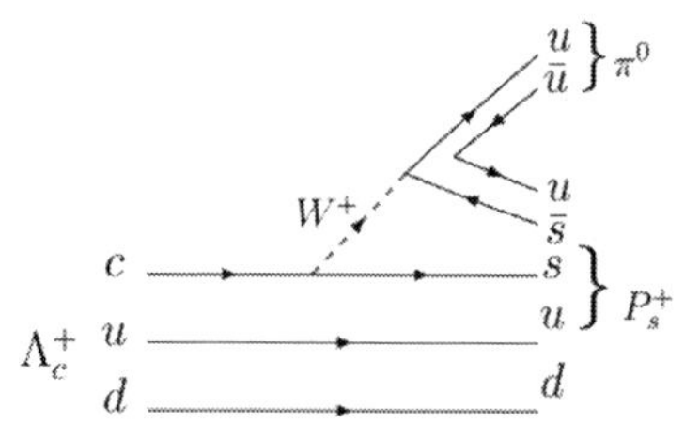

Figure 1. Quark level diagram for the $\Lambda_c^+ \to \pi^0 P_s^+$ decay.

Furthermore, in Ref. [29], a contribution from triangle singularities close to the physical region was also considered to the $\Lambda_c^+ \to \pi^0\phi p$ decay, where the Λ_c^+ decays into $K^*\Sigma^*(1385)$, the $\Sigma^*(1385)$ ($\equiv \Sigma^*$) decays to the $\pi^0\Sigma$ (or Λ) and the $K^*\Sigma$ (or Λ) rescatter into ϕp in the final state. In addition to the effects of the triangle mechanism, the final state interaction of $K^*\Lambda \to \phi p$ and $K^*\Sigma \to \phi p$ are also taken into account. If there was a P_s resonance, it must couple to both the ϕp and $K^*\Sigma/\Lambda$ and thus may be manifest in the ϕp invariant mass distribution. Yet, because of the small phase space and depending on the mass and width of such a P_s state, it could be difficult to search for it. Indeed, searching for the decay of $\Lambda_c^+ \to \pi^0 p\phi$ was done by the Belle Collaboration, and no significant signal was observed with an upper limit on the branching fraction of $\mathcal{B}(\Lambda_c^+ \to \pi^0 p\phi) < 15.3 \times 10^{-5}$ at the 90% confidence level [30].

2 The $\Lambda_c^+ \to \bar{K}^0\eta p$ decay

The quark content of Λ_c^+ is *cud*, where the up and down quarks are in a state of spin zero and isospin zero. Thus, these two light quarks are flavor antisymmetric, and the following simplified notation can be used for the Λ_c^+:

$$\Lambda_c^+ = \frac{1}{\sqrt{2}}|c(ud - du)> . \tag{1}$$

In the Λ_c^+ weak decay, the charm quark first decays into a strange quark by emitting a W^+ boson, then the W^+ decays into a pair of $\bar{d}$ and u quarks, which is the most dominated process for the charmed baryon decays. In Fig. 2 we show the mechanism for the decay of Λ_c^+ to produce the $\bar{K}^0$ from the $s\bar{d}$ pair and the insertion of a new $\bar{q}q$ pair with the quantum numbers of the vacuum, $\bar{u}u + \bar{d}d + \bar{s}s$, to construct the intermediate meson-baryon state MB from the *uud* cluster with the assumption that the u and d quarks in the Λ_c^+ are spectators in the weak decay. Thus, after the hadronization these u and d quarks in the Λ_c^+ are part of the baryon, and the u quark originated from the weak decay forms the meson. In this process, the ud diquark in Λ_c^+ is the spectator, and the uud cluster is combined into a pure $I = 1/2$ state.

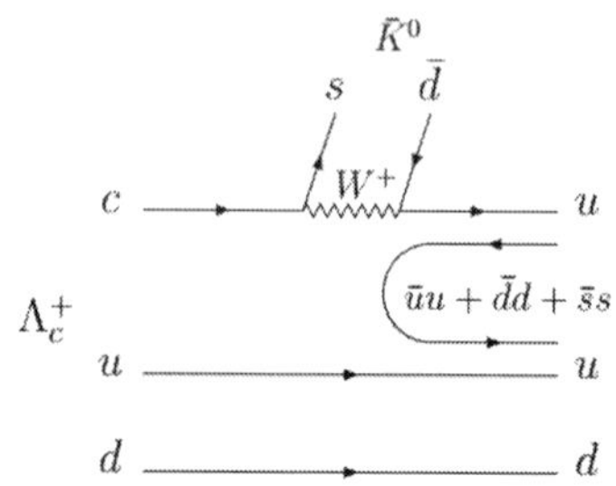

Figure 2. Quark level diagram for the $\Lambda_c^+ \to \bar{K}^0 MB$ decay with the $\bar{K}^0$ emission from the $s\bar{d}$ pair.

Following the procedure of Refs. [31–33], one can straightforwardly obtain the meson-baryon states after the $\bar{q}q$ pair production as

$$|MB\rangle = \frac{\sqrt{3}}{3}|\eta p\rangle + \frac{\sqrt{2}}{2}|\pi^0 p\rangle + |\pi^+ n\rangle - \frac{\sqrt{6}}{3}|K^+\Lambda\rangle, \tag{2}$$

where we have omitted the $\eta' p$ term because of its large mass threshold compared to other channels that we considered.

After the production of a meson-baryon pair, the final-state interaction between them takes place, which can be parameterized by the re-scattering shown in Fig. 3 at the hadronic level for the $\Lambda_c^+ \to \bar{K}^0\eta p$ decay. Finally, the final-state interaction of MB, in $I = 1/2$, will lead to the dynamical generation of the $N^*(1535)$ resonance [34, 35].

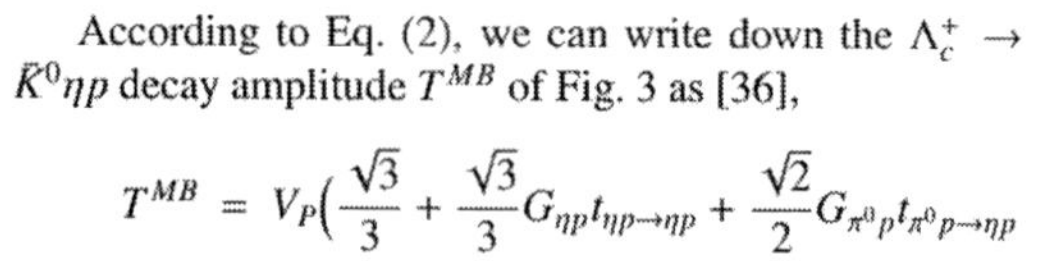

Figure 3. Diagrams for the $\Lambda_c^+ \to \bar{K}^0\eta p$ decay: (a) direct $\bar{K}^0\eta p$ vertex at tree level, (b) final-state interaction of the ηp.

According to Eq. (2), we can write down the $\Lambda_c^+ \to \bar{K}^0\eta p$ decay amplitude T^{MB} of Fig. 3 as [36],

$$T^{MB} = V_P\Big(\frac{\sqrt{3}}{3} + \frac{\sqrt{3}}{3}G_{\eta p}t_{\eta p\to\eta p} + \frac{\sqrt{2}}{2}G_{\pi^0 p}t_{\pi^0 p\to\eta p} + G_{\pi^+ n}t_{\pi^+ n\to\eta p} - \frac{\sqrt{6}}{3}G_{K^+\Lambda}t_{K^+\Lambda\to\eta p}\Big), \tag{3}$$

where V_P expresses the weak and hadronization strength, which is assumed to be a constant and independent of the final state interaction. The G_{MB} denotes the one-meson-one-baryon loop function, which depends on the invariant mass of the final ηp system, $M_{\eta p}$. The meson-baryon scattering amplitudes $t_{MB\to\eta p}$ are those obtained in the chiral unitary approach, which depend also on $M_{\eta p}$.

On the other hand, because the $N^*(1535)$ has a large $uuds\bar{s}$ component, it can also be produced via the process shown in Fig. 4 (a). After the $N^*(1535)$ is formed with $uuds\bar{s}$, it decays into ηp. The hadron level diagram for the decay of $\Lambda_c^+ \to \bar{K}^0 N^*(1535) \to \bar{K}^0 \eta p$ is also shown in Fig. 4 (b). However, we should note that the strangeness component of $N^*(1535)$ can not be guaranteed from the decay process shown in Fig. 4. Indeed, the $N^*(1535)$ can also be produced from the process shown in Fig. 2, where the $s\bar{d}$ forms the $\bar{K}^0$, while the $N^*(1535)$ is constructed from the uud cluster and then it decays into ηp because of its large coupling to this channel.

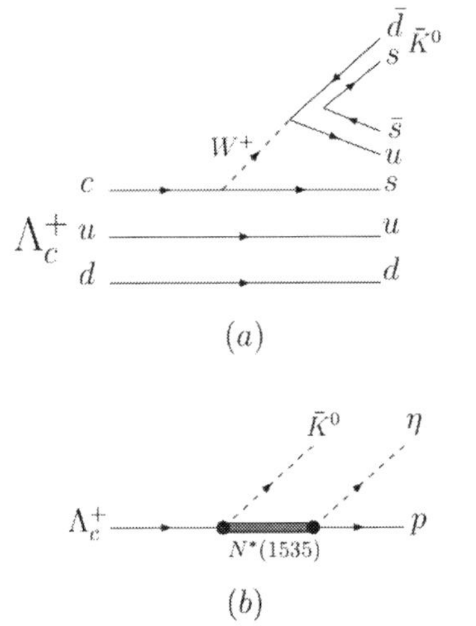

Figure 4. Quark level diagram for $\Lambda_c^+ \to \bar{K}^0 N^*(1535)$ (a) and hadron level diagram for $\Lambda_c^+ \to \bar{K}^0 \eta p$ decay (b).

The general decay amplitudes for $\Lambda_c^+ \to \bar{K}^0 N^*(1535)$, shown in Fig. 4, can be decomposed into two different structures with the corresponding coefficients C_1 and C_2,

$$\mathcal{M} = i\bar{u}(q)(C_1 + C_2\gamma_5)u(p), \tag{4}$$

where q and p are the momentum of the $N^*(1535)$ and Λ_c^+, respectively. The coefficients C_1 and C_2 for charmed baryons decaying into ground meson and baryon states, in general, can be calculated in the framework of the pole model [37] or within the perturbative QCD approach [38]. In the present case, because the $N^*(1535)$ resonance is not well understood in the classical quark model, the values of C_1 and C_2 in Eq. (4) are very difficult to be pined down, and we have to determine them with future experimental data. In this work, we take $C_1 = C_2$.

To get the whole decay amplitude of $\Lambda_c^+ \to \bar{K}^0 N^*(1535) \to \bar{K}^0 \eta p$ as shown in Fig. 4 (b), we use the effective interaction Lagrangian as in Ref. [39] for the $N^*(1535)N\eta$ vertex,

$$\mathcal{L}_{N^*N\eta} = -ig_{N^*N\eta}\bar{N}\eta N^* + \text{h.c.}. \tag{5}$$

Then the invariant decay amplitude of the $\Lambda_c^+ \to \bar{K}^0 N^*(1535) \to \bar{K}^0 \eta p$ decay can be written as

$$T^{N^*} = ig_{N^*N\eta}\bar{u}(p_3, s_p)G_{N^*}(q)(C_1 + C_2\gamma_5)u(p, s_{\Lambda_c^+}), \tag{6}$$

where p_3 is the momentum of the final proton. The s_p and $s_{\Lambda_c^+}$ are the spin polarization variables for the proton and Λ_c^+ baryon, respectively. The $G_{N^*}(q)$ is the propagator of the $N^*(1535)$, which is given by a Breit-Wigner (BW) form as,

$$G_{N^*}(q) = i\frac{\not{q} + M_{N^*}}{q^2 - M_{N^*}^2 + iM_{N^*}\Gamma_{N^*}(q^2)}, \tag{7}$$

with M_{N^*} and $\Gamma_{N^*}(q^2)$ the mass and total decay width of the $N^*(1535)$, respectively. For $\Gamma_{N^*}(q^2)$, since the dominant decay channels for the $N^*(1535)$ resonance are πN and ηN [40], we take the following form as used in Refs. [41, 42]

$$\Gamma_{N^*}(q^2) = \Gamma_{N^*\to\pi N}(q^2) + \Gamma_{N^*\to\eta N}(q^2) + \Gamma_0, \tag{8}$$

with

$$\Gamma_{N^*\to\pi N}(q^2) = \frac{3g_{N^*N\pi}^2}{4\pi}\frac{E_1 + m_p}{\sqrt{q^2}}\sqrt{E_1^2 - m_p^2}, \tag{9}$$

$$\Gamma_{N^*\to\eta N}(q^2) = \frac{g_{N^*N\eta}^2}{4\pi}\frac{E_2 + m_p}{\sqrt{q^2}}\sqrt{E_2^2 - m_p^2}. \tag{10}$$

Here

$$E_1 = \frac{q^2 + m_p^2 - m_\pi^2}{2\sqrt{q^2}}, \quad E_2 = \frac{q^2 + m_p^2 - m_\eta^2}{2\sqrt{q^2}}. \tag{11}$$

We take $g_{N^*N\pi}^2/4\pi = 0.037$ and $g_{N^*N\eta}^2/4\pi = 0.28$ as used in Ref. [43]. With these values we can get $\Gamma_{N^*\to N\pi} = 67.5$ MeV and $\Gamma_{N^*\to N\eta} = 63$ MeV if we take $\sqrt{q^2} = 1535$ MeV. In this work, we choose $\Gamma_0 = 19.5$ MeV for $\Gamma_{N^*}(\sqrt{q^2} = 1535\text{ MeV}) = 150$ MeV.

With all the ingredients obtained in the previous subsection, one can write down the invariant ηp mass distribution of the $\Lambda_c^+ \to \bar{K}^0 \eta p$ decay as:

$$\frac{d\Gamma}{dM_{\eta p}} = \frac{1}{16\pi^3}\frac{m_p p_{\bar{K}^0} p_\eta^*}{M_{\Lambda_c^+}}|T|^2, \tag{12}$$

where T is the total decay amplitude. The $p_{\bar{K}^0}$ and p_η^* are the three-momenta of the outgoing $\bar{K}^0$ meson in the Λ_c^+ rest frame and the outgoing η meson in the center of mass frame of the final ηp system, respectively.

In Fig. 5, we show the numerical results for the $d\Gamma/dM_{\eta p}$ with two models: Model I takes $T = T^{MB}$, while Model II takes $T = T^{N^*}$, $M_{N^*} = 1535$ MeV and Γ_{N^*} is energy dependent as in Eq. (8). The results of Model I are obtained with $V_P = 1$ MeV^{-1}. The results of Model II are obtained with $C_1 = C_2 = 47.2$, which are normalized to the peak of Model I.

For Model I, there is a clear peak around 1524 MeV corresponding to the $N^*(1535)$ resonance. The peak position of Model II is very close to the central value, 1535

http://doi.org/10.1051/epjconf/202024102010

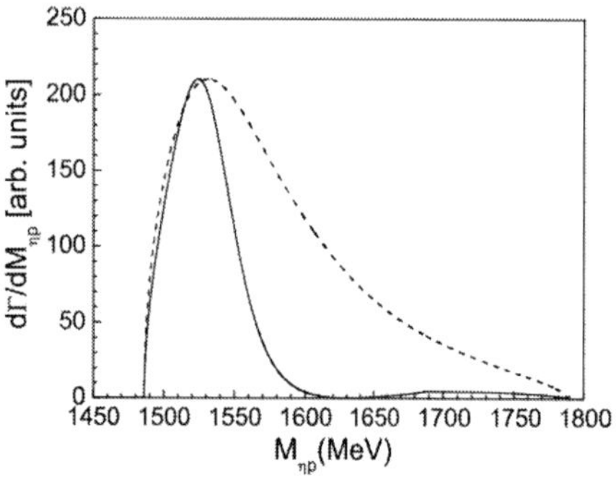

Figure 5. Invariant ηp mass distribution for the $\Lambda_c^+ \to \bar{K}^0\eta p$ decay. The solid and dashed curves represent the results obtained in Model I and II, respectively.

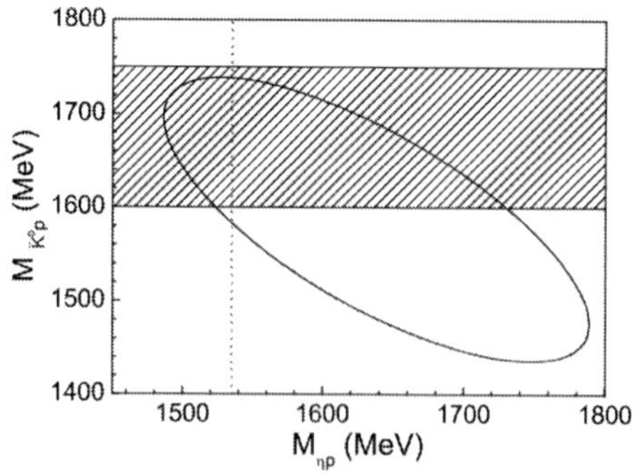

Figure 6. Dalitz plot for $M_{\eta p}$ and $M_{\bar{K}^0 p}$ in the $\Lambda_c^+ \to \bar{K}^0\eta p$ decay. The $N^*(1535)$ energy is shown by the vertical dotted line, which the horizontal band represents the masses of Σ^* states from 1600 to 1800 MeV.

MeV, estimated in the PDG [40] for the $N^*(1535)$. The peak position of Model I is also close to the value 1535 MeV, but with a narrow width. From the results of Model I and II shown in Fig. 5, we see that these two different descriptions of the $N^*(1535)$ resonance give different line shapes for the invariant ηp mass distributions. The findings here are similar to that obtained in Refs. [44, 45]. For the $N^*(1535)$, the amplitude square obtained with the chiral unitary approach does not behave like an usual BW resonance, even at the peak position. Note that in the chiral unitary approach, only the meson-baryon components of $N^*(1535)$ are included. However, in the works of Refs. [46, 47], it was shown that the $N^*(1535)$ contains a mixture of a genuine quark state apart from the meson-baryon components.

Up to now, we have considered only the contribution from $N^*(1535)$, while the contributions from other baryon resonances, such as, $N^*(1650)\frac{1}{2}^-$, $N^*(1710)\frac{1}{2}^+$, $N^*(1720)\frac{3}{2}^+$, $\Sigma^*(1660)\frac{1}{2}^+$, $\Sigma^*(1670)\frac{3}{2}^-$, and $\Sigma^*(1750)\frac{1}{2}^-$, are not taken into account. In Fig. 6, we show the Dalitz plot for the $\Lambda_c^+ \to \bar{K}^0\eta p$ decay, where we see that the the decay of $\Lambda_c^+ \to \bar{K}^0 N^*(1710)$ and $\Lambda_c^+ \to \bar{K}^0 N^*(1720)$ have very limited phase space, and $N^*(1710)$ and $N^*(1720)$ decay into ηp in p-wave, hence, their contributions should be much suppressed. On the other hand, in the $N^*(1535)$ energy region, the Dalitz plot overlaps with these Σ^* resonances from 1600 to 1800 MeV in the $\bar{K}^0 p$ channel, which may make the analysis of $N^*(1535)$ difficult. Fortunately, the $\Sigma^*(1660)\frac{1}{2}^+$ and $\Sigma^*(1670)\frac{3}{2}^-$ decay into $\bar{K}^0 p$ in p-wave and D-wave, respectively. These contributions will be suppressed because of the higher partial waves involved. For the $\Sigma^*(1750)\frac{1}{2}^-$, it decays into $\bar{K}^0 p$ in s-wave. However, it lies in the kinematic end-point region and therefore the decay of $\Lambda_c^+ \to \eta\Sigma^*(1750)$ has a relative small phase space.

It is worthy to mention that both the $N^*(1535)$ and $N^*(1650)$ are dynamically generated from the analysis of the s-wave πN scattering [48–50]. Indeed, the effect of the $N^*(1535)$ and the $N^*(1650)$ resonances are studied in Ref. [51], where both effects of the $N^*(1535)$ and the $N^*(1650)$ are seen in the mass distributions, but, their manifestation is different. Thus, the contributions from other N^* and Σ^* resonances might be small compared with the contribution from the $N^*(1535)$, and we expect that their contributions will not change much the model predictions here. If future experimental measurements provide enough data to disentangle the contributions from these resonances, one can also study them.

3 The $\Lambda_c^+ \to \pi^0\phi p$ decay

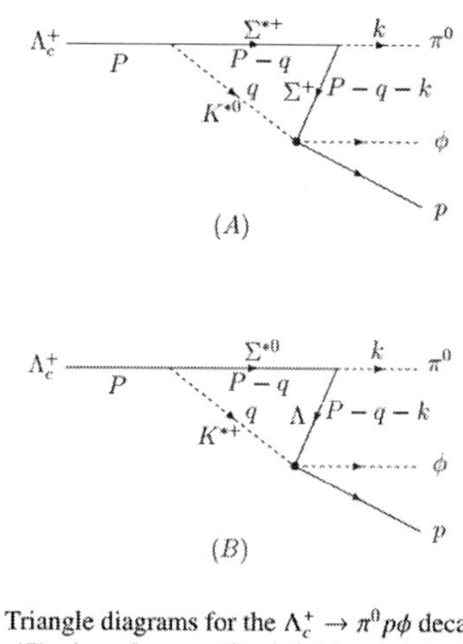

Figure 7. Triangle diagrams for the $\Lambda_c^+ \to \pi^0 p\phi$ decay. (*A*): Σ^+-exchange. (*B*): Λ-exchange. The definitions of the kinematical variables (P, q, k) are also shown.

In addition to the diagram shown in Fig. 1, the $\Lambda_c^+ \to \pi^0 p\phi$ decay can proceed through the triangle diagrams depicted in Fig. 7, where the decay of $\Lambda_c^+ \to (\Sigma^*(1385)K^*)^+$ can proceed by W-exchange diagram [52, 53]: $(cd)u \to (du)u$, and the duu are hadronized, together with a $s\bar{s}$

pair with the vacuum quantum numbers, into the $\Sigma^*(1385)$ ($\equiv \Sigma^*$) and K^*. Then the $\Sigma^*(1385)$ decays into $\pi^0\Sigma/\Lambda$ and, after the production of the $K^{*0}\Sigma^+$ and $K^{*+}\Lambda$, they re-scatter into the ϕp in the final state, as shown in Fig. 7.

In Fig. 8, we show explicitly the W-exchange diagram at quark level for the decays $\Lambda_c^+ \to \Sigma^{*0}(1385)K^{*+}$ and $\Lambda_c^+ \to \Sigma^{*+}(1385)K^{*0}$.

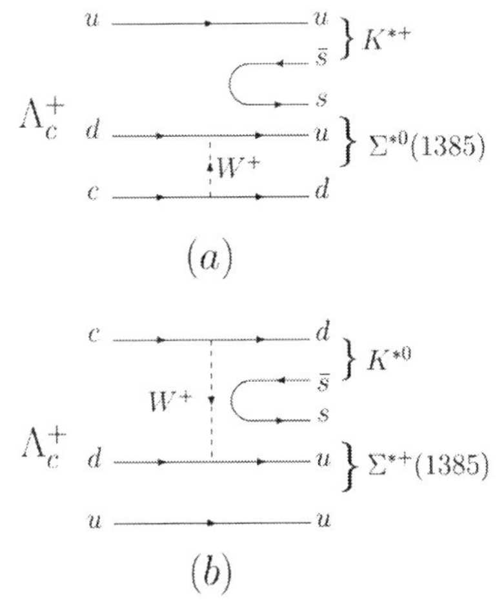

Figure 8. Quark level W-exchange diagram for the $\Lambda_c^+ \to (\Sigma^*(1385)K^*)^+$ decay. (a): $\Lambda_c^+ \to \Sigma^{*0}(1385)K^{*+}$. (B): $\Lambda_c^+ \to \Sigma^{*+}(1385)K^{*0}$.

Next, to get the total decay amplitude of the decay $\Lambda_c^+ \to \pi^0\phi p$. We first need the effective interaction for the $\Lambda_c^+ \to (\Sigma^*K^*)^+$ vertex. Because the Σ^*K^* threshold is very close to the mass of Λ_c^+, we consider only the S-wave coupling, which can be written as

$$t_{\Lambda_c^+\to\Sigma^*K^*} = f_I g_{\Lambda_c\Sigma^*K^*}\bar{u}^\mu(P-q)u(P)\varepsilon_\mu(q), \quad (13)$$

where f_I is the isospin factor with $f_I = \sqrt{6}/3$ for the $\Sigma^{*+}K^{*0}$ and $-\sqrt{3}/3$ for the $\Sigma^{*0}K^{*+}$, and $g_{\Lambda_c\Sigma^*K^*}$ is an effective coupling constant which can be estimated, in general, from the branching faction of $\mathrm{Br}[\Lambda_c^+ \to \Sigma^*(1385)K^*]$.

Second, we consider the $\Sigma^*\pi\Sigma$ and $\Sigma^*\pi\Lambda$ vertexes. Since he decays of $\Sigma^* \to \pi\Sigma$ and $\pi\Lambda$ are in P waves, we can write

$$t_{\Sigma^{*+}\to\pi^0\Sigma^+} = \frac{g}{m_\pi}\bar{u}(P-q-k)u^\mu(P-q)k_\mu, \quad (14)$$

$$t_{\Sigma^{*0}\to\pi^0\Lambda} = \sqrt{3}\frac{g}{m_\pi}\bar{u}(P-q-k)u^\mu(P-q)k_\mu, \quad (15)$$

with $g = 0.69$ obtained from the total decay width $\Gamma_{\Sigma^*} = 37.13$ MeV and the branching fraction of $\mathrm{Br}(\Sigma^* \to \pi\Sigma) = 11.7\%$ [40].

Then the total decay amplitude for the processes shown in Fig. 7 can be written as

$$t = \frac{g_{\Lambda_c\Sigma^*K^*}\,g}{m_\pi}\vec{\epsilon}_\phi\cdot\vec{k}\sum_{i=\Sigma,\Lambda}C_i\int\frac{d^4q}{(2\pi)^4}\times \frac{i2m_{\Sigma^*}}{(P-q)^2-m_{\Sigma^*}^2+im_{\Sigma^*}\Gamma_{\Sigma^*}}\frac{i}{q^2-m_{K^*}^2+im_{K^*}\Gamma_{K^*}} \frac{i2m_i}{(P-q-k)^2-m_i^2+i\epsilon}, \quad (16)$$

where we have defined $C_\Sigma = \frac{\sqrt{6}}{3}t_{K^{*0}\Sigma^+\to\phi p}$ and $C_\Lambda = -t_{K^{*+}\Lambda\to\phi p}$, and $t_{K^{*0}\Sigma^+\to\phi p}$ and $t_{K^{*+}\Lambda\to\phi p}$ are T-matrix elements for the re-scattering processes. As discussed before, the $K^*\Sigma^*$ mass threshold is close to the mass of Λ_c^+ and the range of the ϕp invariant mass for the decay of interest is very small, which allows us to make non-relativistic approximation for all the involved baryons and vector mesons. Therefore, we take into account only S-wave interaction for the re-scattering.

In Ref. [54], the interaction of $K^*\Sigma$, $K^*\Lambda$ and ϕp in S-wave is studied in the framework of the local hidden gauge formalism using a coupled-channels unitary approach. In the Isospin $I = 1/2$ sector, one pole around 1980 MeV is found, which couples strongly to the channels of $K^*\Sigma$, $K^*\Lambda$, and ϕN. In Fig. 9 we show the results of $|t_{K^{*+}\Lambda\to\phi p}|^2$ (solid cure) and $|t_{K^{*0}\Sigma^+\to\phi p}|^2$ (dashed curve) in the model of Ref. [54].

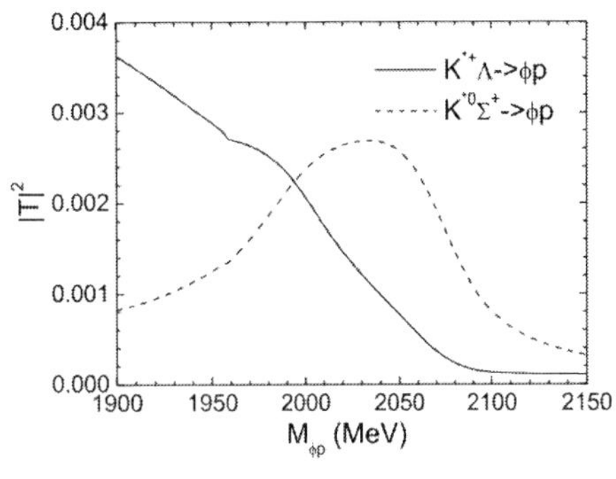

Figure 9. The square of the transition T-matrix elements for the $K^{*+}\Lambda \to \phi p$ and $K^{*+}\Sigma^{*0} \to \phi p$ as a function of the ϕp invariant mass $M_{\phi p}$ in the model of Ref. [54].

After performing the contour integration over the temporal component q^0 in Eq. (17), in the same way as shown in Refs. [55, 56], and including the finite widths of the Σ^* and K^* resonances, we get

$$t = -\frac{g_{\Lambda_c\Sigma^*K^*}gm_{\Sigma^*}}{m_\pi}\vec{\epsilon}_\phi\cdot\vec{k}\,t_T, \quad (17)$$

$$
\begin{aligned}
t_T &= \sum_{i=\Sigma,\Lambda} C_i m_i \int \frac{d^3q}{(2\pi)^3} \frac{1}{\omega_{K^*} E_{\Sigma^*} E_i} \times \\
&\frac{1}{k^0 - E_i - E_{\Sigma^*} + i\Gamma_{\Sigma^*}/2} \frac{1}{P^0 + \omega_{K^*} + E_i - k^0} \\
&\frac{1}{P^0 - \omega_{K^*} - E_i - k^0 + i\Gamma_{K^*}/2} \\
&\frac{\left[P^0\omega_{K^*} + k^0 E_i - (\omega_{K^*+E_i})(\omega_{K^*} + E_i + E_{\Sigma^*})\right]}{P^0 - E_{\Sigma^*} - \omega_{K^*} + i\Gamma_{\Sigma^*}/2}, \\
&= \frac{\sqrt{6} m_{\Sigma^+}}{3} t_{K^{*0}\Sigma^+ \to \phi p} t_T^{\Sigma} - m_\Lambda t_{K^{*+}\Lambda \to \phi p} t_T^{\Lambda}, \qquad (18)
\end{aligned}
$$

where $P^0 = M_{\Lambda_c^+}$, and

$$
\begin{aligned}
\omega_{K^*} &= \sqrt{m_{K^*}^2 + |\vec{q}|^2}, \quad E_{\Sigma^*} = \sqrt{m_{\Sigma^*}^2 + |\vec{q}|^2}, \\
k^0 &= \sqrt{m_{\pi^0}^2 + |\vec{k}|^2} = \frac{M_{\Lambda_c^+}^2 + m_{\pi^0}^2 - M_{\phi p}^2}{2M_{\Lambda_c^+}}, \\
E_i &= \sqrt{m_i^2 + |\vec{q} + \vec{k}|^2} \text{ with } i = \Sigma \text{ or } \Lambda.
\end{aligned}
$$

With the masses of the initial state Λ_c^+, the neutral pion in the final state and two of the intermediate states, for example the K^* and Λ, it is found that the region for the $\Sigma^*(1385)$ mass will produce a triangle singularity at the physical boundary [57, 58]. However, because the S-wave vertices attached to the Λ_c initial state and the ϕp final state do not introduce any momentum dependence into the loop amplitude, and the P-wave pionic vertices results in a factor of the pion momentum, the above loop integral is ultraviolet convergent.

In Fig. 10 we plot the $|t_T^{\Lambda}|^2$ of the triangle loop defined in Eq. (18) for the Λ-exchange. It can be observed that $|t_T^{\Lambda}|^2$ has a peak around 2045 MeV, which is due to the triangle singularity [29].

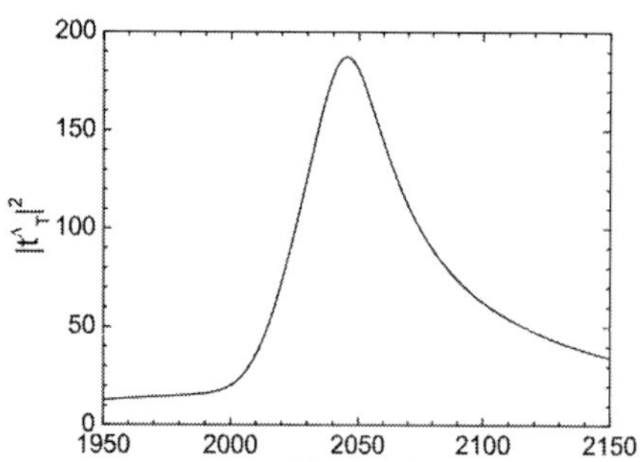

Figure 10. The square of the triangle amplitude t_T^{Λ} of the process in Fig. 7(B) for the Λ-exchange as a function of the invariant mass of ϕp.

The ϕp invariant mass mass distribution for the $\Lambda_c^+ \to \pi^0 \phi p$ decay then reads [40]

$$
\frac{d\Gamma}{dM_{\phi p}} = \frac{m_p m_{\Sigma^*}^2 g_{\Lambda_c \Sigma^* K^*}^2 g^2}{16\pi^3 M_{\Lambda_c^+} m_\pi^2} |\vec{k}|^3 |\vec{p}_\phi| |t_T|^2, \qquad (19)
$$

where $\vec{k}$ is the π^0 momentum in the rest frame of the Λ_c^+, and $\vec{p}_\phi$ is the ϕ momentum in the center-of-mass frame of the ϕp system. Furthermore, we take also the approximation

$$
\sum |\vec{\epsilon}_\phi \cdot \vec{k}|^2 \approx |k|^2, \qquad (20)
$$

where the sum runs over the polarization of ϕ meson.

Here we present the numerical results for the ϕp invariant mass distribution, $\frac{d\Gamma}{dM_{\phi p}}$, for the $\Lambda_c^+ \to \pi^0 \phi p$ decay with three models: Model I represents the calculation of the triangle diagrams in Fig. 7 with the re-scattering T-matrix elements $t_{K^{*0}\Sigma^+ \to \phi p}$ and $t_{K^{*+}\Lambda \to \phi p}$ taken from Ref. [54], which includes the contribution of a P_s state with a mass of about 1980 MeV; Model II is different from Model I by taking the $t_{K^{*0}\Sigma^+ \to \phi p}$ and $t_{K^{*+}\Lambda \to \phi p}$ as constants, and it thus represents the case without any P_s resonance; Model III represents the results obtained from the pure phase space without any special dynamic, which we take the total decay amplitude $t_{\rm III} = c_2$. For the Model II, because of the very small phase space and the closeness of the mass thresholds, all of the involved hadrons, K^*, Σ/Λ, ϕ and proton, can be treated non-relativistically. Hence, one may construct a non-relativistic effective field theory describing the interaction between vector mesons and baryons with the leading order defined by a few constant contact terms, then the total decay amplitude $t_{\rm II}$ can be written as

$$
\begin{aligned}
t_{\rm II} &= t, \text{ but with } t_{K^{*+}\Lambda \to \phi p} = \frac{\sqrt{6}}{2} t_{K^{*0}\Sigma^+ \to \phi p} \\
&= c_1 \frac{\sqrt{6}}{2} \frac{E_{K^*} + E_\phi}{4 f_\pi^2}, \qquad (21)
\end{aligned}
$$

where t is the amplitude shown in Eq. (17), and c_1 is constant. Besides, E_{K^*} and E_ϕ are the energies of K^{*+} and ϕ meson in the ϕp rest frame. In this work we take $E_{K^*} = 892$ MeV and $E_\phi = 1043$ MeV, which are obtained with the mass threshold of $K^{*+}\Lambda$. It is worth to mention that $t_{K^{*+}\Lambda \to \phi p}$ and $t_{K^{*0}\Sigma^+ \to \phi p}$ are calculated in the extreme non-relativistic limit from the vector mesons, while the method should works well in the present case which focuses on the immediate vicinity of the thresholds. There have been discussions about the validity of such an approximation in the case of vector-vector meson scatterings in the far-below threshold region, see, e.g., Refs. [59–62].

In Fig. 11, the solid, dashed, and dotted curves represent the results of Model I, II, and III, respectively. The parameter c_1 of Model II has been adjusted to the strength of the experimental data reported by the Belle Collaboration [30] at its peak around $M_{\phi p} = 2020$ MeV. The results of Model I and III are normalized such as to have the same integrated partial width as Model II. The result of Model II clearly shows a peak structure around 2.02 GeV. The origin of this peak is the triangle diagrams, in particular the Λ-exchange in Fig. 7 (B) which has a triangle singularity close to the $K^*\Lambda$ threshold ($\simeq$ 2.01 GeV). The width of this peak is comparable with the width of the K^*, which is about 50 MeV. This is a quite natural consequence as the ϕp invariant mass is the same as the $K^*\Lambda$ invariant mass so that its distribution inherits the width of the K^*. The presence of a P_s state with a mass close to the lower end

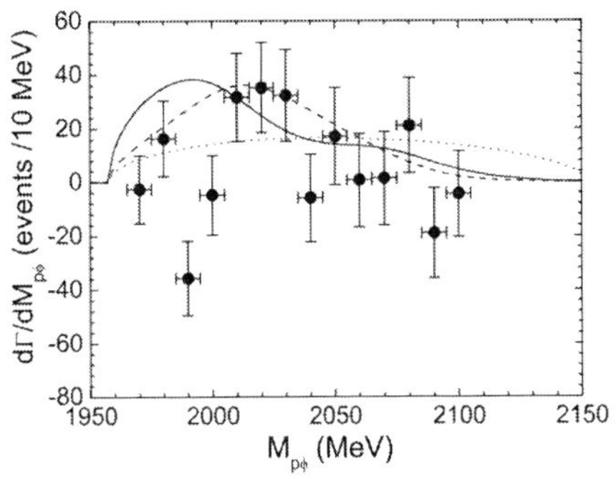

Figure 11. Invariant ϕp mass distribution of the $\Lambda_c^+ \to \pi^0 p\phi$ decay. The experimental data are taken from Ref. [30].

of the $M_{\phi p}$ distribution results in a near-threshold enhancement. Identifying such an enhancement in experiments is difficult as it requires the data to have a high statistics. In particular, it becomes much more difficult if the P_s mass is close to the $K^*\Lambda$ threshold because of the presence of kinematic singularities there. In any case, the phase space shown as Model III is very different from both Model I and Model II. Despite the low statistics of the current Belle data, the curve of Model II, whose shape is completely fixed, has a remarkable agreement with the data. In particular, the data seem to indeed show a bump structure around the $K^*\Lambda$ threshold. More data are welcome to clarify the situation.

4 Summary

We have studied the role of $N^*(1535)$ and possible ϕp state in the decays of $\Lambda_c^+ \to \bar{K}^0 N^*(1535) \to \bar{K}^0 \eta p$ and $\Lambda_c^+ \to \pi^0 \phi p$, respectively. For the $\Lambda_c^+ \to \bar{K}^0 \eta p$ decay, we first employed the molecular picture where the $N^*(1535)$ is dynamically generated from the meson-baryon interaction. In such a scenario, the weak interaction part is dominated by the charm quark decay process: $c(ud) \to (s+u+\bar{d})(ud)$, while the hadronization part takes place by the uud cluster picking up a $q\bar{q}$ pair from the vacuum and hadronizes into a meson-baryon pair, while the $s\bar{d}$ pair from the weak decay turns into a $\bar{K}^0$. The following final state interactions of the meson-baryon pairs are described in the chiral unitary model that dynamically generates the $N^*(1535)$ resonance in the $I = 1/2$ sector. Second, we take a Breit-Wigner formula to describe the distribution of the $N^*(1535)$ in the effective Lagrangian model. The above two descriptions for the $N^*(1535)$ resonance give different invariant ηp mass distributions. Future experimental measurements of the invariant ηp mass distribution studied in the present work will be very helpful to test our model calculations and constrain the properties of the $N^*(1535)$ resonance. For example, a corresponding experimental measurement could in principle be done at BESIII [63] and Belle.

For the $\Lambda_c^+ \to \pi^0 \phi p$ decay, which was proposed to be a channel to search for the hidden-strange ϕp states, we take a mechanism that the Λ_c^+ decays into the K^* and $\Sigma^*(1385)$, then the $\Sigma^*(1385)$ subsequently decays into the $\pi^0\Sigma/\Lambda$), and the K^* then interacts with the Σ/Λ) to produce the ϕp in the final state. In the re-scattering of $K^*\Sigma/\Lambda \to \phi p$, we consider cases with and without a ϕp state. For the case with the ϕp state, we take $t_{K^{*0}\Sigma^+ \to \phi p}$ and $t_{K^{*+}\Lambda \to \phi p}$ as used in Refs. [54] which produces a resonance at around 1980 MeV. While for the case without the ϕp state, both $t_{K^{*0}\Sigma^+ \to \phi p}$ and $t_{K^{*+}\Lambda \to \phi p}$ are constant. It is shown that the triangle singularities can produce a peak at around 2.02 GeV with a width similar to that of the K^*. The theoretical results of the ϕp invariant mass distribution agrees with the existing Belle data [30]. If there is a P_s state, it could distort the ϕp invariant mass distribution. However, it is difficult to be identified in the decay $\Lambda_c^+ \to \pi^0 \phi p$ because of its small phase space and the presence of triangle singularities. We look forward to more precise data from the BESIII, Belle-II and LHCb experiments in the future, which will be decisive to illuminate the role played by triangle singularities in this decay.

Acknowledgments

This work is partly supported by the National Natural Science Foundation of China under Grant Nos. 11735003, 11975041, 11747601, 11835015, 11961141012, 11475227, and the Youth Innovation Promotion Association CAS (2016367). It is also supported in part by the National Natural Science Foundation of China (NSFC) and the Deutsche Forschungsgemeinschaft (DFG) through the funds provided to the Sino-German Collaborative Research Center "Symmetries and the Emergence of Structure in QCD" (NSFC Grant No. 11621131001, DFG Grant No. TRR110), by the Chinese Academy of Sciences (CAS) under Grants No. QYZDB-SSW-SYS013 and XDPB09, and by the CAS Center for Excellence in Particle Physics (CCEPP).

References

[1] V. Crede and C. A. Meyer, Prog. Part. Nucl. Phys. **63**, 74 (2009).
[2] H. X. Chen, W. Chen, X. Liu and S. L. Zhu, Phys. Rept. **639**, 1 (2016).
[3] F. K. Guo, C. Hanhart, U. G. Meißner, Q. Wang, Q. Zhao and B. S. Zou, Rev. Mod. Phys. **90**, 015004 (2018).
[4] K. Miyahara, T. Hyodo and E. Oset, Phys. Rev. C **92**, 055204 (2015).
[5] T. Hyodo and M. Oka, Phys. Rev. C **84**, 035201 (2011).
[6] J. J. Xie and L. S. Geng, Eur. Phys. J. C **76**, 496 (2016).
[7] J. J. Xie and L. S. Geng, Phys. Rev. D **95**, 074024 (2017).
[8] M. Ablikim *et al.* [BESIII Collaboration], Phys. Rev. D **99**, 032010 (2019).
[9] C. D. Lü, W. Wang and F. S. Yu, Phys. Rev. D **93**, 056008 (2016).

[10] R. Aaij *et al.* [LHCb Collaboration], Phys. Rev. Lett. **115**, 072001 (2015).
[11] R. Aaij *et al.* [LHCb Collaboration], Phys. Rev. Lett. **122**, 222001 (2019).
[12] R. F. Lebed, Phys. Rev. D **92**, 114030 (2015).
[13] E. Klempt and A. Zaitsev, Phys. Rept. **454**, 1 (2007).
[14] V. Crede and W. Roberts, Rept. Prog. Phys. **76**, 076301 (2013).
[15] B. C. Liu and B. S. Zou, Phys. Rev. Lett. **96**, 042002 (2006).
[16] C. Helminen and D. O. Riska, Nucl. Phys. A **699**, 624 (2002).
[17] B. S. Zou, Eur. Phys. J. A **35**, 325 (2008).
[18] C. S. An and B. S. Zou, Eur. Phys. J. A **39**, 195 (2009).
[19] B. S. Zou, Nucl. Phys. A **835**, 199 (2010).
[20] B. C. Liu and B. S. Zou, Phys. Rev. Lett. **98**, 039102 (2007).
[21] A. Sibirtsev, J. Haidenbauer and U.-G. Meissner, Phys. Rev. Lett. **98**, 039101 (2007).
[22] H. Gao, H. Huang, T. Liu, J. Ping, F. Wang and Z. Zhao, Phys. Rev. C **95**, 055202 (2017).
[23] H. Gao, T. S. H. Lee and V. Marinov, Phys. Rev. C **63**, 022201 (2001).
[24] F. Huang, Z.-Y. Zhang and Y.-W. Yu, Phys. Rev. C **73**, 025207 (2006).
[25] T. Mibe *et al.* [LEPS Collaboration], Phys. Rev. Lett. **95**, 182001 (2005).
[26] K. Mizutani *et al.* [LEPS Collaboration], Phys. Rev. C **96**, 062201 (2017).
[27] A. Kiswandhi, J. J. Xie and S. N. Yang, Phys. Lett. B **691**, 214 (2010).
[28] J. J. Xie and L. S. Geng, Phys. Rev. D **96**, 054009 (2017).
[29] J. J. Xie and F. K. Guo, Phys. Lett. B **774**, 108 (2017).
[30] B. Pal *et al.* [Belle Collaboration], Phys. Rev. D **96**, 051102 (2017).
[31] K. Miyahara, T. Hyodo, M. Oka, J. Nieves and E. Oset, Phys. Rev. C **95**, 035212 (2017).
[32] L. Roca, M. Mai, E. Oset and U. G. Meißner, Eur. Phys. J. C **75**, 218 (2015).
[33] A. Feijoo, V. K. Magas, A. Ramos and E. Oset, Phys. Rev. D **92**, 076015 (2015).
[34] T. Inoue, E. Oset and M. J. Vicente Vacas, Phys. Rev. C **65**, 035204 (2002).
[35] M. Doring, E. Oset and D. Strottman, Phys. Rev. C **73**, 045209 (2006).
[36] J. A. Oller and E. Oset, Nucl. Phys. A **629**, 739 (1998).
[37] H. Y. Cheng and B. Tseng, Phys. Rev. D **46**, 1042 (1992) Erratum: [Phys. Rev. D **55**, 1697 (1997)].
[38] C. D. Lü, Y. M. Wang, H. Zou, A. Ali and G. Kramer, Phys. Rev. D **80**, 034011 (2009).
[39] J. J. Xie, B. S. Zou and H. C. Chiang, Phys. Rev. C **77**, 015206 (2008).
[40] M. Tanabashi *et al.* [Particle Data Group], Phys. Rev. D **98**, 030001 (2018).
[41] J. J. Wu, S. Dulat and B. S. Zou, Phys. Rev. C **81**, 045210 (2010).
[42] J. J. Xie, B. C. Liu and C. S. An, Phys. Rev. C **88**, 015203 (2013).
[43] Q. F. Lü, R. Wang, J. J. Xie, X. R. Chen and D. M. Li, Phys. Rev. C **91**, 035204 (2015).
[44] L. S. Geng, E. Oset, B. S. Zou and M. Döring, Phys. Rev. C **79**, 025203 (2009).
[45] M. Döring, E. Oset and B. S. Zou, Phys. Rev. C **78**, 025207 (2008).
[46] D. Jido, M. Doering and E. Oset, Phys. Rev. C **77**, 065207 (2008).
[47] T. Hyodo, D. Jido and A. Hosaka, Phys. Rev. C **78**, 025203 (2008).
[48] E. J. Garzon and E. Oset, Phys. Rev. C **91**, 025201 (2015).
[49] J. Nieves and E. Ruiz Arriola, Phys. Rev. D **64**, 116008 (2001).
[50] P. C. Bruns, M. Mai and U. G. Meissner, Phys. Lett. B **697**, 254 (2011).
[51] R. Pavao, S. Sakai and E. Oset, Phys. Rev. C **98**, 015201 (2018).
[52] L.-L. Chau, H.-Y. Cheng and B. Tseng, Phys. Rev. D **54**, 2132 (1996).
[53] H.-Y. Cheng and C.-W. Chiang, Phys. Rev. D **81**, 074031 (2010).
[54] E. Oset and A. Ramos, Eur. Phys. J. A **44**, 445 (2010).
[55] F. Aceti, J. M. Dias and E. Oset, Eur. Phys. J. A **51**, 48 (2015).
[56] M. Bayar, F. Aceti, F.-K. Guo and E. Oset, Phys. Rev. D **94**, 074039 (2016).
[57] F.-K. Guo, U.-G. Meißner, W. Wang and Z. Yang, Phys. Rev. D **92**, 071502 (2015).
[58] X.-H. Liu, M. Oka and Q. Zhao, Phys. Lett. B **753**, 297 (2016).
[59] D. Gülmez, U.-G. Meißner and J. A. Oller, Eur. Phys. J. C **77**, 460 (2017).
[60] L. S. Geng, R. Molina and E. Oset, Chin. Phys. C **41**, 124101 (2017).
[61] M. L. Du, D. Gülmez, F. K. Guo, U.-G. Meißner and Q. Wang, Eur. Phys. J. C **78**, 988 (2018).
[62] R. Molina, L. S. Geng and E. Oset, PTEP **2019**, 103B05 (2019).
[63] M. Ablikim *et al.* [BESIII Collaboration], Phys. Rev. Lett. **116**, 052001 (2016).

Partial Wave Analysis of HADES Data for Two-Pion Production in Pion-Nucleon Reactions

I. Ciepał[1,*]
for the HADES collaboration

[1]Institute of Nuclear Physics PAS, Kraków, Poland

Abstract. Pion beams are important tools to study features of baryonic resonances and their decay channels because resonances are produced in s-channel at fixed mass equal to $\sqrt{s}$. In 2014 the HADES collaboration has performed systematic measurements in the second resonance region scanning around the mass of the N*(1520). Pion beams at four momenta (0.654, 0.686, 0.738 and 0.787 GeV/c) impinging on polyethylene and carbon targets have been used. With the use of the multichannel Partial Wave Analysis (PWA) the role of the N*(1520) resonance in conjunction with the intermediate ρ-meson production has been studied. Preliminary results at 0.654 GeV/c on exclusive production of two pions ($n\pi^+\pi^-$ and $p\pi^-\pi^0$) in the final state are presented.

1 Motivation

HADES has proposed an experimental program [1, 2] focused on combined measurements of hadron and dielectron production in pion induced reactions [3–7]. It offers a great opportunity to unambiguously fix the description of baryonic resonances and their coupling to the light vector mesons ρ and ω. The combined analysis of hadronic and electromagnetic channels is crucial for validation of the Vector Meson Dominance model and will help to understand the ρ-meson production mechanism [5]. For example, studies of two pion $n\pi^+\pi^-$ and $p\pi^-\pi^0$ channels are important to obtain the branching ratio of resonances in the ρN decay channel, that is one of the main goals of the HADES experiment. On the other hand, investigations of the dielectron channel like $\pi^- p \to ne^+e^-$, which has been measured in the same experiment allows one to understand the role of the ρ meson in electromagnetic transitions in the time-like region [3–5, 8]. Therefore, to understand resonance production mechanisms a systematic energy scan and high precision data are needed.

In 2014 a large dataset of $\pi^- p$ scattering has been obtained at the four pion beam momenta 0.654, 0.686, 0.738 and 0.787 GeV/c [6, 7]. The data have been included into the multichannel PWA developed by the Bonn-Gatchina group [9]. Separations of cross sections of the two-pion final states ($\pi^+\pi^-$ and $\pi^\circ\pi^-$) into dominant channels have been performed and compared to other data. In particular the ρ-N coupling is investigated in context of electromagnetic transition form-factors.

*e-mail: izabela.ciepal@ifj.edu.pl

2 Experiment with pion beams

The High Acceptance Di-Electron Spectrometer (HADES) [10], installed at GSI Helmholtzzentrum/FAIR in Darmstadt, was designed to measure dielectrons in the 1-3.5 AGeV energy range. Its purpose is to measure and reconstruct products of heavy ion, proton and pion collisions up to a few AGeV. It allows to study both hadron and rare dilepton production in $N + N$, $\pi + N$, $p + A$, $A + A$ collisions. It provides high acceptance (18°-85°) in polar and almost full azimuthal angles, good charged particle identification (p/K/π/e) and good mass resolution of 2-3% for e+e- pairs in the light vector meson mass range.

The HADES setup together with a secondary pion beam is a very unique apparatus in the world which allows for the first time combined measurement of dielctron and hadron production. The secondary pion beam is produced in reaction of ^{14}N on ^{9}Be target with an intensity of 10^6 pions/spill. The dedicated tracking system CERBEROS [1, 11] allows for a reconstruction of the momentum of each pion beam particle with a resolution of 0.3%.
The experiment has been performed with the $(C_2H_4)_n$ (polyethylene) target and separately with the pure carbon target in order to subtract the carbon contribution from the polyethylene data. The subtraction procedure involves first establishing the ratio between yields collected with the polyethylene and carbon targets based on the $\pi^- p$ missing mass squared spectrum. The procedure takes into account also simulated spectra for all channels of interest: $\pi^- p$ elastic scattering, $n\pi^+\pi^-$ and $p\pi^-\pi^0$. A global χ^2 minimization is performed based on weighting factors assigned to each of the components which are free parameters in the fit. The fitting procedure was iterated by changing also the tracking resolution and beam

momentum value until minimum in the χ^2 was obtained. In the next step, events of the same kinematics from the polyethylene target and carbon target were grouped into bins of similar missing mass values. Then, for each carbon-polyethylene track a minimization function is calculated as a sum of squared difference between a given observable (momentum, polar and azimuthal angles, invariant masses and the angles defined in the helicity and Gottfried-Jackson frames) obtained with polyethylene and carbon targets, respectively. This method delivers pure pion-proton reaction candidates. For details see Ref. [7].

The data normalization is based on the $\pi^- p$ elastic scattering. The acceptance and efficiency corrected data in the $\theta_{\pi^-}^{CM}$ range of 60° – 110° are normalized to the distribution from the SAID database [12] of the known cross section. For details see Ref. [6].

3 Partial wave analysis

The partial wave analysis of the HADES data was performed in cooperation with the Bonn-Gatchina group [13]. This approach [13–15] was developed for the combined analysis of the pion nucleon scattering and photoproduction and is based on the so-called isobar model, introduced already in [16]. It is a two step process with the intermediate formation of meson-baryon two-particle states like $\pi\Delta$, ρ(I=1)N, σ(I=0)N. In the next step the resonances: ρ, σ and Δ decay into $\pi\pi$ and $N\pi$ final states, respectively. In this framework the world data on two-pion production reactions measured in pion-nucleon and photon-nucleon collisions [17–21] are taken into account together with the HADES data in an event-by-event maximum-likelihood method allowing for the extraction of contributions from different partial waves. In the energy range corresponding to the HADES data the largest contributions to the two-pion final states are: $\Delta(1232)\pi$, $N(1440)\pi$, $N(938)\sigma$ and $N(938)\rho$. The amplitudes of $N(1535)\pi$ and $N(1520)\pi$ are also included, but the resulting contributions are negligible.

4 Differential distributions of $n\pi^+\pi^-$ and $p\pi^-\pi^0$

A three-body decay phase space can be represented with two-dimensional Dalitz plots (the invariant masses squared of two of the three possible particle pairs) or alternatively with angular distribution in the helicity frame, which describe the angular dependence of the decay. The helicity frame is the rest frame of two final-state particles and the helicity angle is the angle between one of the particles with respect to the third one boosted to the two-particle rest frame. The helicity angles are also sensitive to the production of resonances. It should be stressed that the helicty projections have full coverage in the HADES acceptance after applying the acceptance corrections procedure described in more details in Refs. [6, 7].

In Fig. 1 the Dalitz plots for 0.654 GeV/c pion beam momentum are presented for the data (top panel) and PWA solutions for the two contributions: for $\Delta\pi$ (middle panel) and $N\rho$ (bottom panel). The enhancements corresponding to the Δ and ρ resonances visible in the data are nicely explained with the PWA plots.

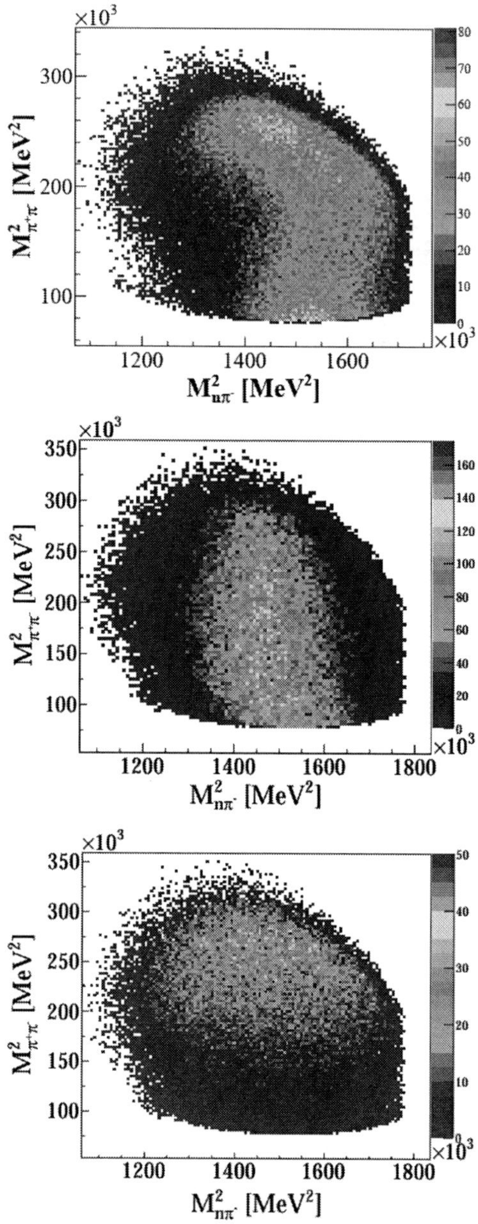

Figure 1. (Color online) Dalitz plots of $\pi^- p \to n\pi^+\pi^-$ for data (top panel) and for the channels $\Delta\pi$ (middle panel) and $N\rho$ (bottom panel) obtained from PWA at 0.654 GeV/c pion beam momentum.

Two examples of PWA decomposition for the angular distributions in different helicity frames are presented in Fig. 2. The total cross section (black line) is decom-

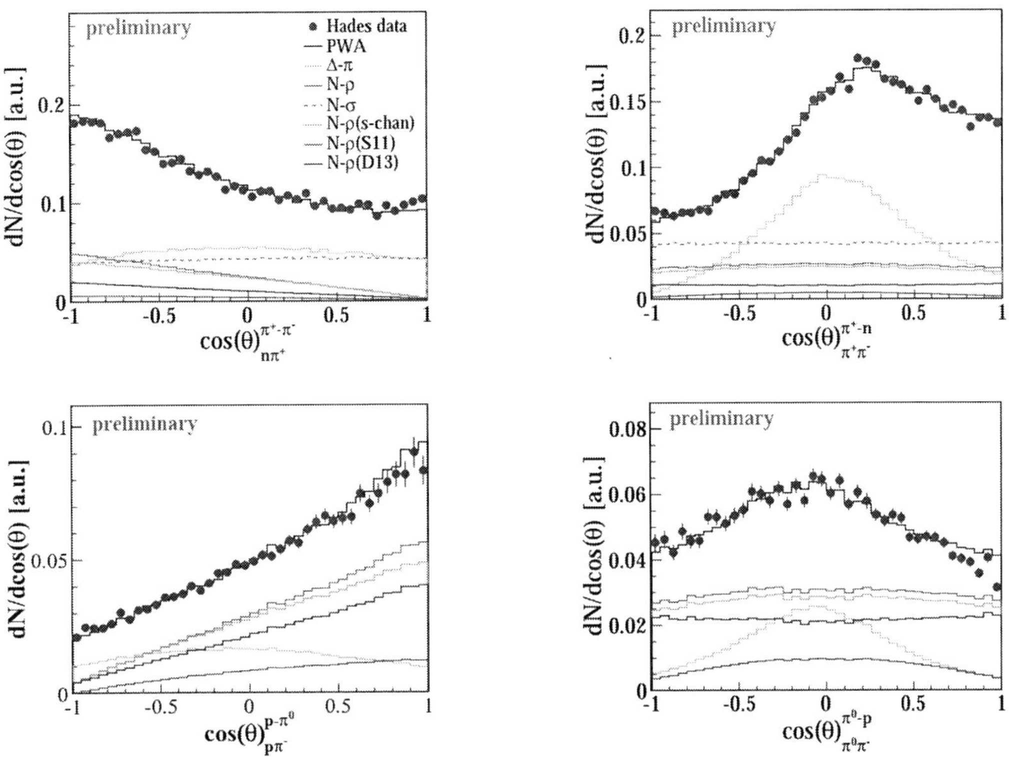

Figure 2. (Color online) Differential distributions of pions in the nucleon-pion (left column) and nucleons in the pion-pion (right column) helicity frames for the $\pi^- p \to n\pi^+\pi^-$ (upper row) and $\pi^- p \to p\pi^0\pi^-$ (lower row) reaction channels obtained at 0.654 GeV/c pion beam momentum. Colour lines represent various final state contributions (indicated in the legend).

posed into final state contributions like $\Delta\pi$, $N\rho$, $N\sigma$. In the notation of the helicity angles (e.g. $\cos\theta^{\pi^+-\pi^-}_{n\pi^+}$), the subscript refers to the choice of two-particle rest frame ($n\pi^+$) and the exponent to the chocie of particles for the calculation of angle (π^+ with respect to π^-). In the case of the pion-pion ($\pi^+\pi^-$, $\pi^0\pi^-$) helicity frame the Δ resonance is clearly visible as in the corresponding Dalitz plot in Fig. 1 (middle panel). The other components $N\sigma$ and $N\rho$ are rather uniform and in the $\pi^0\pi^-$ channel the $N\sigma$ is completely missing. In the case of the nucleon-pion ($n\pi^+$, $p\pi^-$) helicity frame the effect of the ρ meson is seen as an enhancement of large opening angles between the pion and the nucleon, what corresponds to the Dalitz plot in Fig. 1 (bottom panel). As a complement, also the $N\rho$ contributions originating from the s-channel, $S11$ and $D13$ waves are drawn allowing to study the ρ production mechanism. The $N\rho$ component is more dominant in the $p\pi^-\pi^0$ channel and ρ is mostly produced in the resonant s-channel.

5 Summary and outlook

The HADES data greatly improve the world database of two-pion production in pion-induced experiments in the energy range around the mass of the *N*(1520)* resonance. The analysis (in progress) will allow to obtain the branching ratio of *N*(1520)* resonance in the ρN decay channel and other two-pion production modes. The production of the ρ-meson is of utmost importance for the dielectron channel $\pi^- p \to ne^+e^-$, which has been measured in the same experiment [5, 8]. The e^+e^- exclusive analysis allows for the direct investigation of the *N*(1520)* resonance Dalitz decay and coupling to the ρN channel. Yet another information on the electromagnetic structure of the baryon transition can be extracted from the angular distributions of the dielectrons via fit of the spin-density matrix elements [22], which allows to parametrize the differential cross sections in a model independent way. This can be compared to the similar analysis in two-pion channels obtained from PWA.

Studies of distributions in the Gottfried-Jackson frame and analysis of data collected for other beam momenta are on-going. The ultimate goal is to extract excitation functions for the resonance contributions and branching ratios for the $N\rho$ channel.

The pion beam program will be continued in 2021 once the SIS18 will have been upgraded to serve as an injector for the FAIR facility. Recently, the HADES setup has been equipped with a large area electromagnetic calorimeter (ECAL) enabling measurements of real photons and neutral mesons via their photonic decays. The reconstruction of neutral mesons is of large importance for the complete understanding of π^--N reaction dynamics and for the complete partial wave analysis of $\pi^- p$ reactions.

6 Acknowledgements

The collaboration gratefully acknowledges the support by SIP JUC Cracow (Poland), 2017/26/M/ST2/00600; TU Darmstadt (Germany), VH-NG-823; GU Frankfurt (Germany), BMBF:05P15RFFCA, HIC for FAIR, ExtreMe Matter Institute EMMI; TU München, Garching (Germany), MLL München, DFG EClust 153, BmBF 05P15WOFCA; JLU Giessen (Germany), BMBF:05P12RGGHM; NRNU MEPhI Moscow Excellence project 02.a03.21.0005, MSERF 3.3380.2017/4.; 6 IPN, IN2P3/CNRS (France); NPI CAS Rez (Czech Republic), MSMT OP VVV CZ.02.1.01/0.0/0.0/16 013/0001677, MSMT LM2015049.

References

[1] J. Adamczewski-Musch *et al.* (HADES Collab.), Eur. Phys. J. **A 53**, 188 (2017)

[2] Proposal for experiments at SIS18 during FAIR Phase0, Properties of hadron resonances and baryon rich matter, (HADES Collab.), GSI 2017

[3] F. Scozzi, EPJ Web of Conf. **130**, 07021 (2016)

[4] B. Ramstein, EPJ Web Conf. **199**, 01008 (2018)

[5] B. Ramstein, "Studying time-like electromagnetic baryonic transitions with HADES in pion induced reactions", NSTAR2019

[6] I. Ciepał, EPJ Web Conf. **199**, 01024 (2018)

[7] W. Przygoda, EPJ Web of Conf. **130**, 01021 (2016)

[8] P. Salabura, "Exploring time like transitions in pp, $\pi^- p$ heavy ion reactions with HADES", NSTAR2019

[9] A. V. Sarantsev, JPS Conf. Proc. **10**, 010005 (2016)

[10] G. Agakichiev *et al.* (HADES Collab.), Eur. Phys. J. **A 41**, 243 (2009)

[11] R. Lalik *et al.*, 2013 IEEE Nuclear Science Symposium and Medical Imaging Conference (2013)

[12] CNS Data Analysis Center, SAID Home Page, http://gwdac.phys.gwu.edu (2018)

[13] pwa.hiskp.uni-bonn.de

[14] A. Anisovich, E. Klempt, A. Sarantsev and U. Thoma, Eur. Phys. J. **A 24**, 111 (2005)

[15] A. Anisovich *et al.*, Phys. Rev. **D 84**, 076001 (2011)

[16] D. M. Manley *et al.*, Phys. Rev. **D 30**, 904 (1984)

[17] S. Prakhov *et al.* (Crystal Ball Collab.), Phys. Rev. **C 69**, 045202 (2004)

[18] V. L. Kashevarov *et al.* ((Crystal Ball at MAMI, TAPS, and A2 Collab.), Phys. Rev. **C 85**, 064610 (2012)

[19] J. Ahrens *et al.* (GDH and A2 Collab.), Phys. Lett. **B 624**, 173 (2005)

[20] U. Thoma *et al.* (CB-ELSA Collab.), Phys. Lett. **B 659**, 87 (2008)

[21] M. Battaglieri *et al.* (CLAS Collab.), Phys. Rev. **D 80**, 072005 (2009)

[22] E. Speranza, M. Zetenyi, B. Friman, Phys. Lett. **B 764**, 282 (2017)

EPJ Web of Conferences **241**, 03002 (2020)
NSTAR 2019 http://doi.org/10.1051/epjconf/202024103002

The Λ(1405) with one pole

Eberhard Klempt[1,*]

[1]HISKP, University of Bonn, Nussalee 14-16, 53115 Bonn

Abstract. The $\Lambda(1405)1/2^-$ resonance is discussed and a fit of the Bonn-Gatchina (BnGa) partial-wave-analysis group is presented to data relevant for this resonance. In the $\Sigma\pi - N\bar{K}$ coupled channel system, no Σ pole and only one Λ pole in the mass range from the $\Sigma\pi$ threshold to 1500 MeV is required to get a good description of the data.

1 Origin of resonances in strong interactions

The strong interactions support a large variety of bound states and resonances. Different ranges of the interactions lead to different types of states. The J/ψ meson is a prototype of $c\bar{c}$ state bound in a Cornell-type potential. In contrast, the $\chi_{c1}(3872)$ has at least a strong $D^{*0}\bar{D}^0$ molecular component bound by pion exchange. For some resonances, the interpretation is controversial: the $N(1535)1/2^-$ is predicted in quark models as qqq resonance but it can as well be understood as a $\Lambda\bar{K}$–$\Sigma\bar{K}$ molecule.

Here, we study the pole content of $\Sigma\pi$ and $N\bar{K}$ S-wave interactions in the mass region below 1.5 GeV. In the quark model, one SU(3) singlet resonance is expected which is identified with $\Lambda(1405)1/2^-$ called Λ(1405) here. However, in the chiral unitary approach, two negative-parity Λ resonances and one or two Σ resonances are found.

*e-mail: Klempt@hiskp.uni-bonn.de

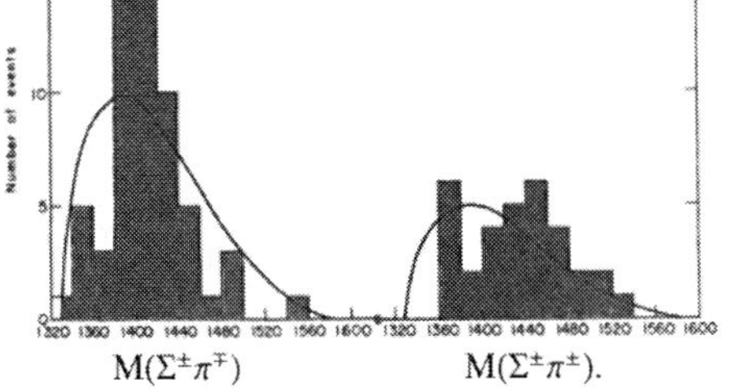

Figure 1. (color online)The neutral and like-sign M($\Sigma\pi$ invariant mass distributions from the reactions $K^-p \to \Sigma^+\pi^-\pi^-\pi^+$ and $K^-p \to \Sigma^-\pi^+\pi^+\pi^-$ [1].

2 History

In 1961, the Nobel-prize winner Luis Alvarez and his group [1] analyzed bubble chamber data on the reactions $K^-p \to \Sigma^+\pi^-\pi^-\pi^+$ and $K^-p \to \Sigma^-\pi^+\pi^+\pi^-$. The neutral $M_{\Sigma^\pm\pi^\mp}$ invariant mass distribution (with two entries per event) show a clear peak at about 1400 MeV which is absent in the like sign $M_{\Sigma^\pm\pi^\pm}$ invariant mass distribution (see Fig. 1). The Λ(1405) was discovered! A few years later, a measurement has been made of the relative signs of the resonant $K^-p \to Y^* \to \Sigma\pi$ reaction amplitudes with Y^* = Σ(1385), Λ(1405), or Λ(1520). It was shown that Λ(1405) and Λ(1520) are to be described as predominantly SU(3) singlets [2].

In 1966, Richard H. Dalitz and collaborators showed that the large K^-p scattering leads to a virtual bound state at 1410 MeV [3]. Since then, the Λ(1405) is *the* paradigma for a dynamically generated state.

In 1995, Kaiser, Siegel and Weise used a chiral Lagrangian to construct a potential for S-wave meson-baryon scattering [4]. The approach reproduced well the properties of Λ(1405).

So far, there is not necessarily a conflict between the interpretation of Λ(1405) as qqq resonance or as dynamically generated K^-p bound state. Like $N(1535)1/2^-$, Λ(1405) can be both. It could have, e.g., a qqq core coupling strongly to $\bar{K}N$ and $\Sigma\pi$ and developing a mesonic cloud with a three-quark core.

In 2000, this compatibility ceased to be valid when Oller and Meißner studied the S wave kaonnucleon interactions exploiting an Effective Field Theory (EFT) [5]. Dispersion relations were used to perform the necessary resummation of the lowest order relativistic chiral Lagrangian. They found that the low-mass region houses three negative-parity resonances, one SU(3) singlet and one SU(3) octet Λ and one Σ resonance. The spectrum was interpreted in [6], the interpretation is shown in Fig. 2. The spectrum originates from the interactions of octet baryons and octet mesons. The findings of Refs. [5, 6] have been substantiated by better calculations and various groups over the years, see. [7].

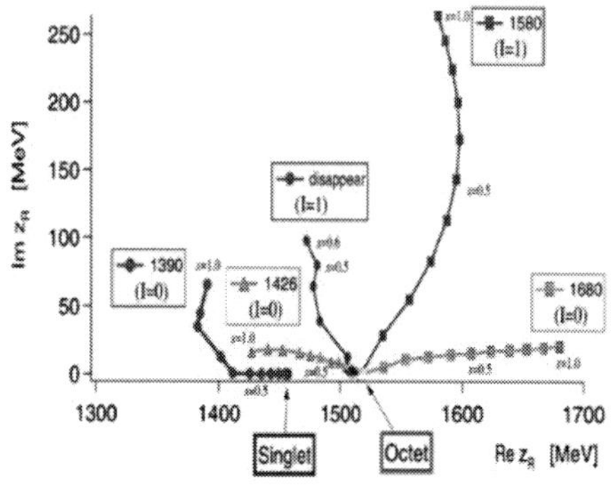

Figure 2. (color online) The $\bar{K}N$–$\Sigma\pi$ coupled-channel dynamics leads to two pairs of Λ and Σ SU(3) octets and one Λ SU(3) singlet [6].

The three Λ^* poles in Fig. 2 are combinations of the singlet state (at about 1390 MeV) and the two octet states (at about 1426 and 1680 MeV). The isovector states were found to be sensitive on the details of the coupled channel approach. The findings were confirmed in a number of further studies. They were reviewed by Maxim Mai at this conference.

In this contribution, I present an analysis of the BnGa group who fit the most relevant data on the $\Sigma\pi - \bar{K}N$ coupled channel system [8] and showed that just one resonance, $\Lambda(1405)$, is required to get a reasonable description of the data.

3 Bonn-Gatchina analysis

3.1 Bonn-Gatchina analysis

In the analysis, we use a modified K-matrix approach. Instead of the K-matrix

$$\hat{A} = \hat{K}\left(I - i\hat{\rho}\hat{K}\right)^{-1}$$

we use

$$A = K\left(I - Re\,B\hat{K} - i\hat{\rho}\hat{K}\right)^{-1} \qquad \text{with}$$

$$\int \frac{ds'}{\pi}\frac{A_{aj}(s,s')}{s'-s-i\epsilon}\rho_j(s')K_{jb}(s',s)A_{aj}(s,s)\times$$
$$Re\,B(s)K_{jb}(s,s) + iA_{aj}(s,s)\rho_j(s)K_{jb}(s)$$

$$\text{where} \qquad Re\,B(s) = ⨍\frac{ds'}{\pi}\frac{\rho_j(s')}{s'-s}.$$

Here, $⨍$ is the principle-value integral. This approach provides a correct continuation of the amplitude below thresholds.

3.2 Data used

We use the CLAS data on the three charge states $\gamma p \to K^+(\Sigma^\pm\pi^\mp)$ [9] and Crystal Ball (BNL) data on $K^-p \to \pi^0\pi^0\Lambda$ and $K^-p \to \pi^0\pi^0\Sigma^0$ at Kaon momenta in the 514 MeV/c to 750 MeV/c MeV momentum range [10]. A subset of the data is shown in Figs. 3 and 4. These two data sets [9, 10] were fitted event-by-event with a maximum likelihood fit. The data on $\gamma p \to K^+(\Sigma^0\pi^0)$ [9], shown in Fig. 3(right), were not included in the fit: these data were predicted demonstrating that $\Lambda(1405)$ and $\Sigma(1385)$ – which both contribute to the data shown in Fig. 3(left) and Fig. 3(center) – are well identified in the fit. Further included are bubble chamber data on $K^-p \to \pi^-\pi^+\pi^\pm\Sigma^\mp$ [11] (see Fig. 4, right), the differential cross sections for $K^-p \to K^-p$ and K^0n from [12] in the 1464 and 1548 MeV mass range (see Fig. 5, and total cross sections for $K^-p \to K^-p, K^0n, \Sigma^+\pi^-, \Sigma^0\pi^0, \Sigma^-\pi^+, \Lambda\pi^0$ [13–16] (see Fig. 6). The fit is constrained by data on the K^-p system at rest: by the ratios of K^-p capture rates $\gamma = \Gamma_{K^-p\to\pi^+\Sigma^-}/\Gamma_{K^-p\to\pi^-\Sigma^+}$, $R_n = \Gamma_{K^-p\to\pi^0\Lambda}/\Gamma_{K^-p\to\text{neutral}}$, and $R_c = \Gamma_{K^-p\to\pi^\pm\Sigma^\pm}/\Gamma_{K^-p\to\text{inelastic}}$ [17, 18], and the recent experimental results on the energy shift and width of kaonic hydrogen atoms $\Delta E - i\Gamma/2$ which constrain the K^-p S-wave scattering length [19, 20]. The experimental values are compared to our fit results in Table 1.

Table 1. Experimental results for the quantities listed in Eqs. (1) and the fit result.

	Data	Fit
$\Delta E - i\Gamma/2$	283±42-i(271±55)	310±15-i(332±15) eV
γ	2.38±0.04	2.40±0.02
R_n	0.189±0.015	0.209±0.008
R_c	0.664±0.011	0.671±0.010

Our fit to the elastic differential and the total cross sections is significantly lower than the data. This is a compromise with the data on the strong-interaction width Γ of the K^-p atom: if the elastic cross section is larger, the width increases as well. Fits based on the chiral unitary approach do better here even though at the expense of a larger number of parameters.

3.3 Parameters

Most relevant here are the $K^-p - \Sigma\pi$ isoscalar and isovector S-wave scattering amplitudes. The amplitudes are described by five-channel amplitudes with decays to $\Sigma^+\pi^-$, $\Sigma^0\pi^0$, $\Sigma^-\pi^+$, pK^- and nK^0. The mass differences between the three $\Sigma\pi$ and the two $N\bar{K}$ channels are taken into account, the relative couplings are fixed by Clebsch-Gordan coefficients. For the isoscalar amplitude, we us a one-pole parameterization. It depends on two real coupling constants ($g_{\pi\Sigma}$ and g_{KN}) and a bare mass value M. These parameters are defined in the fit. For the isovector channel, we use a two-pole amplitude. Each pole is characterized by the bare mass and three real coupling constants for decays into $\Sigma\pi$, $N\bar{K}$ and $\Lambda\pi$. Thus, this channel needs 8

 http://doi.org/10.1051/epjconf/202024103002

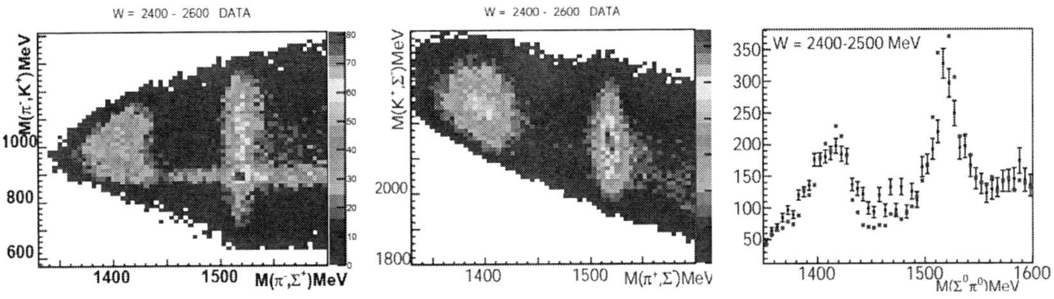

Figure 3. (color online) $\gamma p \to K^+\pi^\mp\Sigma^\pm$ and for $2400 < W = M_{\gamma p} < 2600$,MeV: $M(\pi^-K^+)$ versus $M(\pi^-\Sigma^+)$ (left) and $M(K^+\Sigma^-)$ versus $M(\pi^+\Sigma^-)$ (center) two-dimensional mass distributions for reconstructed data without acceptance correction. These data were fitted event-by-event in a likelihood fit. Right: $\Sigma^0\pi^0$ invariant mass distribution for $\gamma p \to K^+(\Sigma^0\pi^0)$ and for $2400 < W = M_{\gamma p} < 2500$,MeV. The data are not included in the fit, the points without errors are predictions.

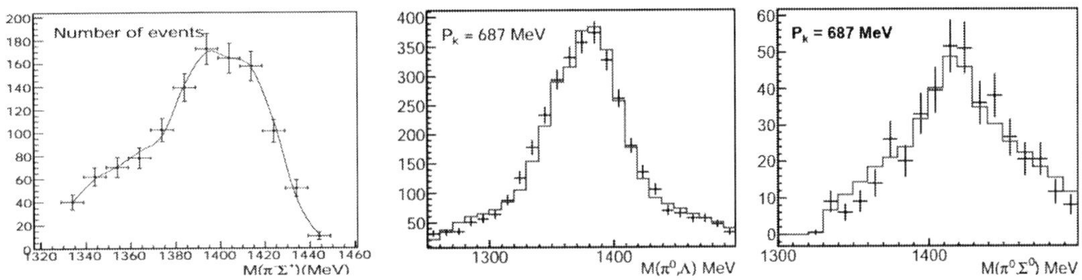

Figure 4. (color online) Left: $\Sigma^+\pi^-$ mass projection from the reaction $K^-p \to \pi^-\pi^+\pi^\mp\Sigma^\pm$ for events with $M_{\pi^+\pi^\mp\Sigma^\pm}$ compatible with $\Sigma(1670)3/2^-$ [11]. The Crystal Ball data on $K^-p \to \pi^0\pi^0\Lambda$ (left) and $K^-p \to \pi^0\pi^0\Sigma^0$ (right) [10].

parameters. Thus we have 11 parameters to describe the isoscalar and the isovector S-wave scattering amplitudes. For the fit to elastic scattering and charge exchange, we allow for a small P-wave contribution. The P-wave is described by two subthreshold vector poles – one isoscalar, one isovector pole – with twoo coupling constants.

The production mechanism of the reaction $\gamma p \to K^+\Sigma\pi$ is not known. We describe the initial state by two Breit-Wigner amplitudes in each partial wave for $J^P = 1/2^\pm, 3/2^\pm, 5/2^\pm$, and $7/2^\pm$ (8 resonances with 14 helicity amplitudes) and $2\times4+6\times8$ free coupling constants for their decays into $\Sigma(1385)\pi$, $\Lambda(1405)\pi$, $\Lambda(1520)\pi$, or NK^*. Thus 78 resonances govern the photoproduction dynamics. The properties of $\Sigma(1385)$, $\Lambda(1405)$, $\Lambda(1520)$, and K^* are taken from the RPP [21].

The bubble chamber data on $K^-p \to \pi^-\pi^+\pi^\pm\Sigma^\mp$ [11] are fitted with the assumption that the $\Lambda(1405)$ stems from $\Sigma^+(1670)3/2^-$ hyperons decaying in a cascade to $\Lambda(1405)\pi^+ \to (\Sigma^\mp\pi^\pm)\pi^+$. Parameters are the $\Sigma(1670)3/2^-$ production strength and decay probability. Contributions from $\Sigma^0(1385) \to \Sigma^\mp\pi^\pm$ with a $\approx 25\%$ fraction improve the fit.

4 Results

For the low-mass isovector S-wave, we use two K-matrix poles. One pole is found far below the $\Sigma\pi$ threshold; we assume the pole describes the effects of left-hand cuts. One pole is above the fitted region; we assign it to the lowest-mass $\Sigma(1620)1/2^-$ resonance. The low-mass isoscalar S-wave is described by one pole at about 1400 MeV. If we introduce a second pole, this pole moves either to a mass below 1 GeV, or a narrow pole with $\Gamma \approx 1$ MeV evolves with a mass coinciding with the $N\bar{K}$ threshold. We disregard this solution. Thus we conclude that the low-energy $\Sigma\pi$–$N\bar{K}$ S-wave interactions can be described by one isoscalar pole. Its mass and width are determined to

$$M_{\rm pole} = 1422 \pm 3\,{\rm MeV} \qquad \Gamma_{\rm pole} = 42 \pm 6\,{\rm MeV}$$

No further pole is required in the fit in the mass range from threshold to 1.5 GeV, neither a second pole in the Λ sector nor any pole in the Σ sector. A pole search in the complex plane identified one pole as well.

5 Discussion

In the quark model, the $\Lambda(1405)$ is a (dominantly) SU(3) singlet baryon. Experimentally, the singlet nature can be deduced from the sign of the imaginary part of the F amplitude at the resonance position, see Fig. 99.1 in [21]. In analyses based on EFT's, the narrow state at 1420 MeV is a (dominantly) octet state. This is shown in Fig. 2, and it follows from Figure 7 taken from Ref. [26]. The Figure shows the real and imaginary part of the $K^-p \to \Lambda(1405) \to \Sigma\pi$ transition amplitude. At the $\Lambda(1405)$ mass,

 http://doi.org/10.1051/epjconf/202024103002

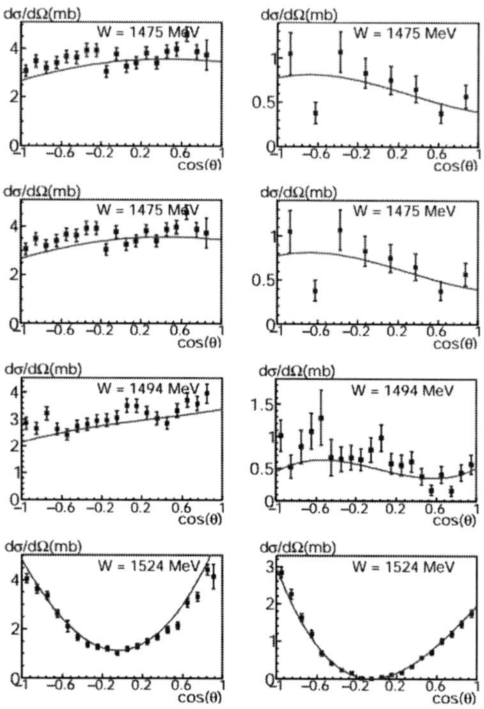

Figure 5. Selected differential cross sections for $K^-p \to K^-p$ (left) and $K^-p \to \bar{K}^0 n$ (right) from [12]. Data were taken between 1464 and 1548 MeV invariant mass and reported in 25 bins. Shown are three bins. The fit is given by the line.

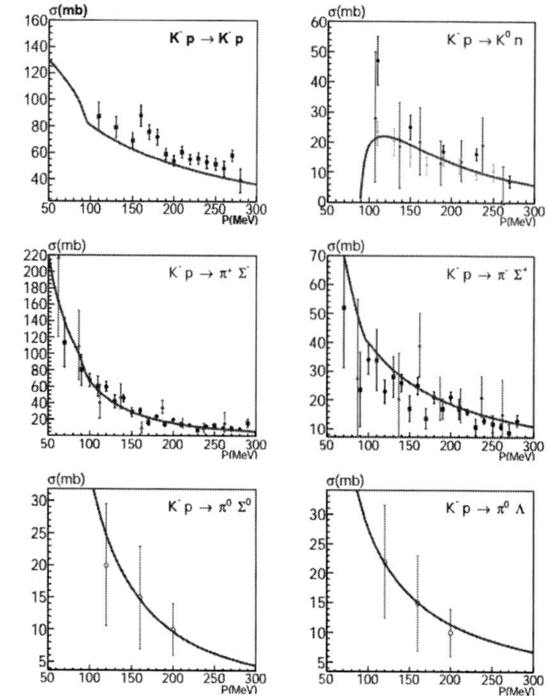

Figure 6. The total cross sections for K^-p induced reactions: $K^-p \to K^-p$, $K^-p \to \bar{K}^0 n$, $K^-p \to \pi^0\Lambda$, $K^-p \to \pi^+\Sigma^-$, $K^-p \to \pi^0\Sigma^0$, $K^-p \to \pi^-\Sigma^+$ [13–16]. The single pole $\Lambda(1405)$ fit is given as the red curve.

the amplitude is proportional to $-i$; this identifies the resonance is a SU(3)-octet resonance. The decisive difference between conventional interpretations of the $\Lambda(1405)$ and EFT-based analyses is the question: Is the $\Lambda(1405)$ dominantly a SU(3)-singlet or a SU(3)-octet state?

The $\Lambda(1405)$ is, of course, not the only Λ excitation. The spectrum of low-mass Λ and Σ resonances is known since long; recently, it has been determined again and many "candidate" states were disregarded, a few new resonances were discovered [22, 23]. The low-mass spectrum is compared to nucleon excitations in Table 2.

The $\Lambda(1405)$ can be identified as (dominantly) SU(3)-singlet state and as the expected spin partner of $\Lambda(1520)$. Its mass is rather low, this is a well-known problem of quark models. It may be noted that a simple mass formula – using the masses of nucleon, Δ and Ω and the Regge slope as only input variables – predicts 1460 MeV for the two singlet states [24]. In the (dominantly) SU(3) octet, there are the ground states and radial excitations with $(N, L^P, S) = (2, 0^+, 1/2)$ from the 56 and 70-plet, the negative-parity states $(N, L^P, S) = (1, 1^-, 1/2)$ and $(1, 1^-, 3/2)$, and the positive-parity states with $(N, L^P, S) = (2, 2^+, 1/2)$. The Λ and Σ resonances are 100 to 150 MeV above the corespponding nucleon states. A further spin doublet of Σ states with $J^P = 1/2^-$ is expected (and seen in [22, 23]) that would correspond to the $\Delta(1620)1/2^-$ and $\Delta(1700)3/2^-$ doublet. All these states were found, only

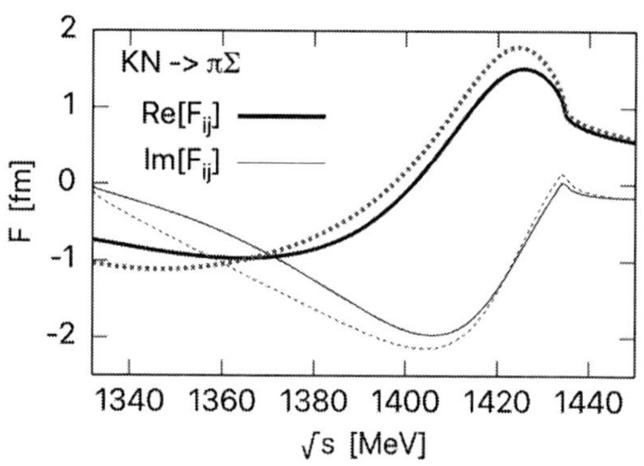

Figure 7. Real and imaginary part of the $K^-p \to \Lambda(1405) \to \Sigma\pi$ transition amplitude F [26]. At Re[F] = 0, Im[F] is negative, suggesting that the dominant part of the amplitude at 1410 MeV is a SU(3) octet.

the $J^P = 3/2^-$ from the spin triplet is again missing. The qualitative agreement – at least in the number and mass ordering of the states – of the observed states with those reported in [25] supports the quark-model interpretation of the hyperon spectrum.

One might think that the quark-model states are what they are, and $\Lambda(1350)$ comes in addition as a dynamically generated resonance. But the problem is not the Λ(1350) but the Λ(1405): in the quark model, Λ(1405) in a SU(3) singlet resonance, in EFT analyses, it is an octet. As octet resonance, it does not fit into the spectrum of Λ resonances. A possible escape would be to assume that all states in Table 3 derived from EFT's are generated in addition to the quark-model states. In this case, Λ(1405) and Λ(1670) should need to have a double-pole structure; with one Λ(1405) being a SU(3) octet, the other one a SU(3) singlet. We remind of the fact that there was the hope to find a double-pole structure for $N(1535)$ but there has never been any evidence for two close-by $N(1535)$ poles. For the Λ(1405) we do not find any hint for a double-pole structure.

Can we understand how two low-mass Λ resonances with $J^P = 1/2^-$ can appear in the analysis? We first assume that Λ(1405) is a SU(3)-singlet state. If the fit forces it to be an octet, the $K^-p \to \Lambda(1405) \to \Sigma\pi$ transition amplitude has the wrong sign; this could be compensated by adding a further SU(3) singlet resonance at about the same mass. Technically, we can thus understand why fits finding Λ(1405) as SU(3) octet state need at least one additonal resonance. If we assume, instead, the Λ(1405) to be a SU(3) octet state, there is no reason why it should show up as SU(3) singlet in fits with a single pole. The singlet nature of the Λ(1405) is decive for the interpretation of the low-lying Λ resonances.

The analyses based on EFT's find two Λ^* resonances with $J^P = 1/2^-$ in the region below 1500 MeV. Why are there two resonances in those analyses and why do we find only one? The main difference may originate from the different interpretations of resonances and interactions. In chiral perturbation theory, resonances are generated dynamically from attractive interactions. In the expansion

$$8 \otimes 8 = 1 \oplus 8_1 \oplus 8_2 \oplus 10 \oplus 27,$$

the same multiplets are expected as in the quark model:

$$3 \otimes 3 \otimes 3 = 1 \oplus 8_1 \oplus 8_2 \oplus 10,$$

except for the 27-plet which is absent in the quark model. The latter plet is supposed not to lead to attractive interactions. Thus in both scenarios, there is one singlet and two octet-states with $J^P = 1/2^-$. (In the quark model, this is a simplified view; both octets are combined with spin states to form the total-quark-spin 1/2 and 3/2.) These three expected Λ states, one mainly singlet and two mainly octet states, are assigned to different physical states (Table 3).

In our approach, these three states generate the attractive interactions: Each Λ pole leads to an attractive interaction in the $\bar{K}N$ and in the $\Sigma\pi$ channel. In EFT's, there are primarily the interactions; the $\bar{K}N$ and $\Sigma\pi$ interactions are attractive and generate the poles. Our result yields a different answer to the chicken-and-egg problem: what is first, the interaction or the pole? In our approach to the Λ(1405), the pole creates the largest part of the interactions and the pole is not generated by the interaction.

Table 2. The Λ excitation spectrum in comparison with the Σ and N excitation spectra. The $\Lambda(1850)3/2^-$ and $\Sigma(1760)3/2^-$ are missing, the masses are estimates.

8		$\Lambda(1890)3/2^+$ $\Sigma(1940)3/2^+$ $N(1720)3/2^+$	$\Lambda(1820)5/2^+$ $\Sigma(1915)5/2^+$ $N(1680)5/2^+$
8	$\Lambda(1810)1/2^+$ $\Sigma(1880)1/2^+$ $N(1710)1/2^+$		
8	$\Lambda(1800)1/2^-$ $\Sigma(1750)1/2^-$ $N(1650)1/2^-$	$\Lambda(1850)3/2^-$ $\Sigma(1760)3/2^-$ $N(1700)3/2^-$	$\Lambda(1830)5/2^-$ $\Sigma(1775)5/2^-$ $N(1675)5/2^-$
1	$\Lambda(1710)1/2^+$		
8	$\Lambda(1670)1/2^-$ $\Sigma(1620)1/2^-$ $N(1535)1/2^-$	$\Lambda(1690)3/2^-$ $\Sigma(1670)3/2^-$ $N(1520)3/2^-$	
8	$\Lambda(1600)1/2^+$ $\Sigma(1660)1/2^+$ $N(1440)1/2^+$		
1	$\Lambda(1405)1/2^-$	$\Lambda(1520)3/2^-$	
8	Λ(1116) Σ(1193) $N(940)$		

6 Outlook

What can be done to identify the SU(3) structure of Λ(1405) unambiguously? This may be achieved by a study of the reaction

$$J/\psi \to \quad \overline{\Lambda} \quad (\Sigma^{\pm}\pi^{\mp}) \qquad + c.c.$$

which is seen with a branching ratio of $(0.83\pm0.07)\cdot 10^{-3}$. The J/ψ is of course a SU(3) singlet, the Λ a SU(3) octet. Hence the $\Sigma\pi$ system forms a SU(3) octet. For zero relative orbital angular momentum in the final state, the $\Sigma\pi$ system must have the quantum numbers $J^P = 1/2^-$ or $3/2^-$. The $\overline{\Lambda}\Lambda(1405)$, $\overline{\Lambda}\Lambda(1520)$, and $\overline{\Lambda}\Lambda(1670)$ final states can be reached only by the SU(3) octet components of the excited Λ. In EFT's, the Λ(1405) is an octet state, hence the branching ratio for $J/\psi \to \overline{\Lambda}\Lambda(1405)$ should be much larger than the one for $J/\psi \to \overline{\Lambda}\Lambda(1520)$ and larger (due to the larger phase space) than the one for $J/\psi \to \overline{\Lambda}\Lambda(1670)$. In the quark model, only the latter branching ratio should be sizable and the two former branching ratios should be seen with a similar (small) rate.

I would like to thank the organizers of NSTAR2019 for the kind invitation and A.V. Anisovich, V.A. Nikonov, A.V. Sarantsev, and U. Thoma for a long and fruitful collaboration.

Note added in proof: After the conference, a second fit was made in which the presence of two Λ resonances with $J^P = 1/2^-$ was assumed. A fit with free mass and width gave a pole at the K^-p threshld and a very narrow width. When the width was fixed to 180 MeV, a shallow minimum at a mass of 1387 ± 3 MeV was observed. The description of the data was similar in quality.

References

[1] M. H. Alston, L. W. Alvarez, P. Eberhard, M. L. Good, W. Graziano, H. K. Ticho and S. G. Wojcicki, Phys. Rev. Lett. **6**, 698 (1961).
[2] R. D. Tripp, R. O. Bangerter, A. Barbaro-Galtieri and T. S. Mast, Phys. Rev. Lett. **21**, 1721 (1968).
[3] R. H. Dalitz, T. C. Wong and G. Rajasekaran, Phys. Rev. **153**, 1617 (1967).
[4] N. Kaiser, P. B. Siegel and W. Weise, Nucl. Phys. A **594**, 325 (1995).
[5] J. A. Oller and U.-G. Meißner, Phys. Lett. B **500**, 263 (2001).
[6] D. Jido, J. A. Oller, E. Oset, A. Ramos and U.-G. Meißner, Nucl. Phys. A **725**, 181 (2003).
[7] A. Cieply, M. Mai, U.-G. Meißner and J. Smejkal, Nucl. Phys. A **954**, 17 (2016).
[8] A. V. Anisovich, A. V. Sarantsev, V. A. Nikonov, V. Burkert, R. A. Schumacher, U. Thoma and E. Klempt, arXiv:1905.05456 [nucl-ex].
[9] K. Moriya *et al.* [CLAS Collaboration], Phys. Rev. C **87**, 035206 (2013).
[10] S. Prakhov *et al.* [Crystall Ball Collaboration], Phys. Rev. C **70**, 034605 (2004).
[11] R. J. Hemingway, Nucl. Phys. B **253**, 742 (1985).
[12] T. S. Mast, M. Alston-Garnjost, R. O. Bangerter, A. S. Barbaro-Galtieri, F. T. Solmitz and R. D. Tripp, Phys. Rev. D **14**, 13 (1976).
[13] W. E. Humphrey and R. R. Ross, Phys. Rev. **127**, 1305 (1962).
[14] M. B. Watson, M. Ferro-Luzzi and R. D. Tripp, Phys. Rev. **131**, 2248 (1963).
[15] M. Sakitt, T. B. Day, R. G. Glasser, N. Seeman, J. H. Friedman, W. E. Humphrey and R. R. Ross, Phys. Rev. **139**, B719 (1965).
[16] J. Ciborowski *et al.*, J. Phys. G **8**, 13 (1982).
[17] D. N. Tovee *et al.*, Nucl. Phys. B **33**, 493 (1971).
[18] R. J. Nowak *et al.*, Nucl. Phys. B **139**, 61 (1978).
[19] M. Bazzi *et al.* [SIDDHARTA Collaboration], Phys. Lett. B **704**, 113 (2011).
[20] M. Bazzi *et al.*, Nucl. Phys. A **881**, 88 (2012).
[21] M. Tanabashi *et al.* [Particle Data Group], Phys. Rev. D **98**, no. 3, 030001 (2018).
[22] M. Matveev, A. V. Sarantsev, V. A. Nikonov, A. V. Anisovich, U. Thoma and E. Klempt, arXiv:1907.03645 [nucl-ex], accepted for publication in EPJA.
[23] A. V. Sarantsev, M. Matveev, V. A. Nikonov, A. V. Anisovich, U. Thoma and E. Klempt, arXiv:1907.13387 [nucl-ex], accepted for publication in EPJA.
[24] E. Klempt, Phys. Rev. C **66**, 058201 (2002).
[25] U. Löring, B. C. Metsch and H. R. Petry, Eur. Phys. J. A **10**, 447 (2001).
[26] K. Miyahara, T. Hyodo and W. Weise, Phys. Rev. C **98**, no. 2, 025201 (2018).

Table 3. The three lowest Λ states with $J^P = 1/2^-$.

SU(3)	EFT		QM	
1	$\Lambda(1350)$		$\Lambda(1405)$	
8	$\Lambda(1405)$	$\Sigma(1401)$	$\Lambda(1670)$	$\Sigma(1620)$
8	$\Lambda(1670)$	$\Sigma(1488)$	$\Lambda(1800)$	$\Sigma(1750)$

Two Pion Photo- and Electroproduction with CLAS

Victor I. Mokeev for the CLAS Collaboration[1,*]

[1]Jefferson Laboratory, 12000 Jefferson Ave., Newport News VA, 23606, USA

Abstract. Exclusive $\pi^+\pi^-p$ photo- and electroproduction data from CLAS have considerably extended the information on the spectrum and structure of nucleon resonances. The data from the $\pi^+\pi^-p$ and $N\pi$ channels have provided results on the electrocouplings of most resonances in the mass region up to 1.8 GeV and at photon virtualities up to 5.0 GeV^2. The recent CLAS data on $\pi^+\pi^-p$ photoproduction have improved knowledge on the photocouplings of nucleon resonances in the mass range of 1.6 GeV < M_{N^*} < 2.0 GeV and on their decays to the $\pi\Delta$ and ρp final hadron states. Analyses of the combined $\pi^+\pi^-p$ photo- and electroproduction data have revealed evidence for the candidate-state $N'(1720)3/2^+$. The new results on the nucleon resonance spectrum, electroexcitation amplitudes from analysis of the CLAS $\pi^+\pi^-p$ photo- and electroproduction data, and their impact on the exploration of strong QCD are presented.

1 Introduction

Studies of exclusive $\pi^+\pi^-p$ photo- and electroproduction off protons represent an effective tool for the exploration of the spectrum and structure of excited nucleon states (N^*). This exclusive channel is sensitive to the contributions of most N^*, which allows for the determination of the nucleon resonance photo- and electrocouplings ($\gamma_{r,v}pN^*$) [1] at different photon virtualities [1]. This provides insight into the structure of most nucleon resonances with their distinctively different structural features. The exploration of different resonance electrocouplings has revealed excited nucleon state structure as a complex interplay between the inner core of three dressed quarks and an external meson-baryon cloud [3–7]. The continuum QCD analyses of the ground-state nucleon elastic form factors and $\gamma_v pN^*$ electrocouplings of different N^* starting from the QCD Lagrangian already shed light on the strong QCD dynamics underlying the generation of the dominant part of hadron mass [4, 8–10]. Essential progress has been achieved in the developments of the quark models for the description of the structure of many excited nucleon states [6, 11–16]. Extension of the results on the $\gamma_{r,v}pN^*$ electrocouplings toward larger photon virtualities and for the majority of excited nucleon states will offer a unique opportunity to explore many facets of strong QCD in the generation of different nucleon resonances.

The recent CLAS data on exclusive $\pi^+\pi^-p$ photoproduction [17] allow us to determine the photocouplings of almost all well- established resonances in the mass range from 1.6 GeV to 2.0 GeV for the first time from this exclusive channel. Analysis of the recent CLAS data [18, 19] on $\pi^+\pi^-p$ electroproduction at 1.4 GeV < W < 2.0 GeV and 2.0 GeV^2 < Q^2 < 5.0 GeV^2 demonstrated promising prospects for the extraction of the $\gamma_{r,v}pN^*$ electrocouplings of most resonances in the above mentioned kinematics area. The combined studies of $\pi^+\pi^-p$ photo- and electroproduction have revealed evidence for a new candidate-state $N'(1720)3/2^+$ [1, 2]. Further development of the Jlab-Moscow State University (JM) reaction model for the extraction of the $\gamma_{r,v}pN^*$ photo-/electrocouplings from the $\pi^+\pi^-p$ exclusive channel based on analysis of the recent CLAS data [18, 19], the new results on the $\gamma_{r,v}pN^*$ photo-/electrocouplings and their impact on the exploration of strong QCD and the observation of the new $N'(1720)3/2^+$ state will be presented in this proceeding.

*e-mail: mokeev@jlab.org

[1]$\gamma_{r,v}$ stand for the real and the virtual photons, respectively

2 Experimental Studies of $\pi^+\pi^-p$ Photo-/Electroproduction and the Reaction Model

The combination of the CEBAF CW beam and the CLAS detector with almost 4π acceptance provides a unique opportunity for the exploration of the $\gamma_{r,v}p \rightarrow \pi^+\pi^-p'$ reactions with almost complete coverage of the final state phase space. This is of particular importance for the extraction of the photo- and $\gamma_v pN^*$ electrocouplings, as well as their hadronic decay widths into the $\pi\Delta$ and ρp final states. Table 1 summarizes the available CLAS experimental data on exclusive $\pi^+\pi^-p$ photo- and electroproduction. The new data for $\pi^+\pi^-p$ photoproduction off protons [17] were obtained with the CLAS detector for the first time in terms of nine independent 1-fold differential and fully integrated $\pi^+\pi^-p$ cross sections in the center of mass energies W from 1.6 GeV to 2.0 GeV. Overall, ≈400 million $\pi^+\pi^-p$ events were selected, exceeding the statistics previously collected in this channel by a factor of ~ 50. The $\pi^+\pi^-p$ electroproduction data from CLAS [18, 19] have extended the information

Table 1. The kinematic areas covered in the CLAS measurements of exclusive $\pi^+\pi^-p$ photo- and electroproduction. Refs [17–19, 22, 23, 25] include nine differential and fully integrated cross sections. Ref [19] additionally includes the structure functions.

W-ranges, GeV	Q^2-ranges, GeV^2	Refs.
1.6-2.0	0.	[17]
1.3-1.57	0.2-0.6	[22]
1.4-2.1	0.5-1.5	[23]
1.3-1.83	0.4-1.0	[25]
1.4-2.0	2.0-5.0	[18]
1.4-2.0	2.0-5.0	[19]

Table 2. The values of $\chi^2/d.p.$ from the comparison between the nine one-fold differential $\gamma_v p \rightarrow \pi^+\pi^-p$ cross sections measured and computed within the updated JM model.

Q^2 bins GeV^2	2.0-2.4	2.4-3.0	3.0-3.5	3.5-4.2	4.2-5.0
$\chi^2/d.p.$	2.68	2.93	2.53	2.30	2.10

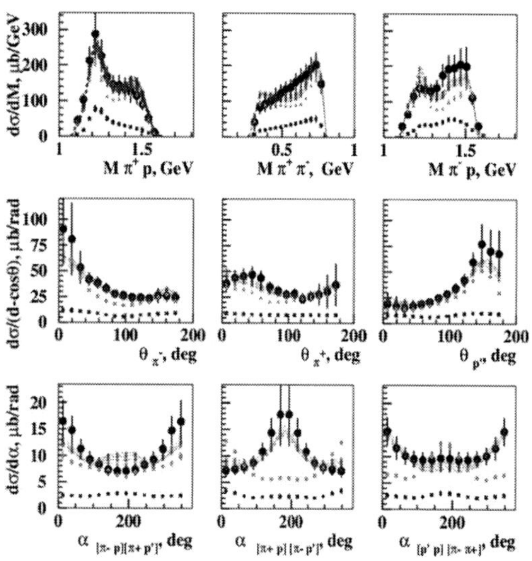

Figure 1. (Color online) Representative example of 1-fold differential cross sections (points with error bars) and the resonant/non-resonant contributions (blue/green bars) from the fits (red curves) of the CLAS $\pi^+\pi^-p$ photoproduction data at W from 1.73 GeV to 1.75 GeV.

on the nine independent single-differential and fully integrated cross sections binned in W and Q^2 in the mass range W < 2.0 GeV towards large photon virtualities of 2.0 GeV^2 < Q^2 < 5.0 GeV^2. The first results on the structure functions R_{TT} and R_{TL} for the $\gamma_v p \rightarrow \pi^+\pi^-p'$ channel have become available [19]. The measured observables can be found in the CLAS Physics Data Base [26].

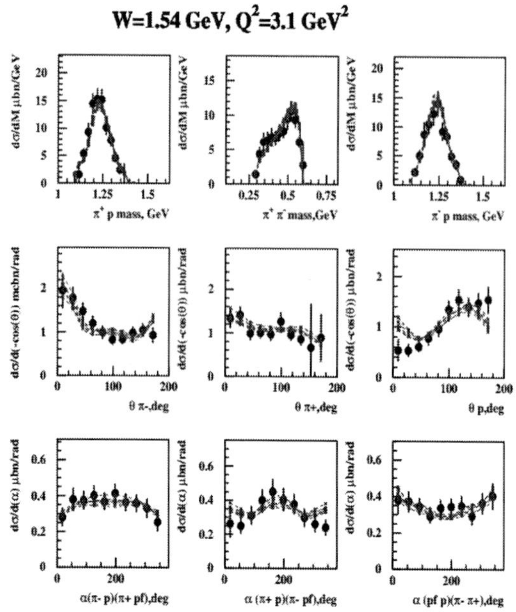

Figure 2. (Color online) Data for $\pi^+\pi^-p$ electroproduction [19] (black filled circles with error bars) and differential cross sections computed (red lines) within the updated JM model (see Section 2) at 1.53 GeV < W < 1.55 GeV and 3.0 GeV^2 < Q^2 < 3.5 GeV^2.

The recent CLAS data on the $\pi^+\pi^-p$ photo-/electroproduction cross sections and the first results on the exclusive structure functions for this channel [17–19] offer a promising prospect for the evaluation of the photocouplings for the resonances with masses from 1.6 GeV to 2.0 GeV and $\gamma_v pN^*$ electrocouplings of most excited nucleon states in the mass range up to 2.0 GeV and photon virtualities Q^2 from 2.0 GeV^2 to 5.0 GeV^2. In preparation for the extraction of the $\gamma_v pN^*$ electrocouplings from these data, the JM reaction model [9, 20, 21], which was successfully used for the extraction of $\gamma_v pN^*$ electrocouplings from the $\pi^+\pi^-p$ electroproduction data at 0.2 GeV^2 < Q^2 < 1.5 GeV^2 [9, 21], was updated. The major focus of the recent CLAS data analysis is to establish all relevant mechanisms contributing to $\pi^+\pi^-p$ photoproduction in the kinematics area of 1.6 GeV < W < 2.0 GeV and in the kinematics area of 1.4 GeV < W < 2.0 GeV and 2.0 GeV^2 < Q^2 < 5.0 GeV^2 in electroproduction from their manifestations in the kinematic dependencies of the measured differential cross sections and eventually to isolate the resonant contributions. The full $\gamma_{r,v}p \rightarrow \pi^+\pi^-p'$ amplitudes are described in the JM model as a superposition of the $\pi^-\Delta^{++}$, $\pi^+\Delta^0$, ρp, $\pi^+D^0_{13}(1520)$, and $\pi^+F^0_{15}(1685)$ sub-channels with subsequent decays of the unstable hadrons to the final $\pi^+\pi^-p$ state and direct 2π production mechanisms, where the final $\pi^+\pi^-p$ state comes about without going through the intermediate process of forming unstable hadron states. The JM model incorporates contributions from all well-

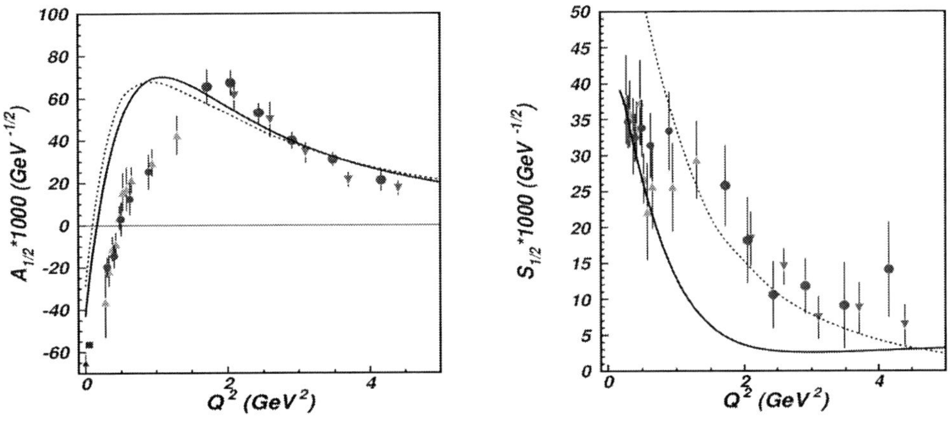

Figure 3. (Color online) CLAS results on the $N(1440)1/2^+$ electrocouplings from analysis of the recent data on $\pi^+\pi^-p$ electroproduction [18, 19] (magenta filled triangles), from the analyses of $N\pi$ electroproduction [32] (red filled circles), and from the previous CLAS data on $\pi^+\pi^-p$ electroproduction [9, 21] (green filled circles). The results at the photon point are taken from the PDG [29] (blue filled triangle) and from analysis of the CLAS $N\pi$ photoproduction data [33] (blue filled square) The $N(1440)1/2^+$ electrocouplings computed within the DSE [10] are shown by the solid lines. The results from the novel light front quark model [6, 11] are shown by the dashed lines.

Table 3. Resonance photocouplings determined from analysis of the $\pi^+\pi^-p$ photoproduction data from this work in comparison with the previous results from the PDG average [29] and from multichannel analysis [31].

Resonances	$A_{1/2}\times10^3$ from $\pi^+\pi^-p$ GeV$^{-1/2}$	$A_{1/2}\times10^3$ PDG ranges GeV$^{-1/2}$	$A_{1/2}\times10^3$ multichannel analysis [31] GeV$^{-1/2}$	$A_{3/2}\times10^3$ from $\pi^+\pi^-p$ GeV$^{-1/2}$	$A_{3/2}\times10^3$ PDG ranges GeV$^{-1/2}$	$A_{3/2}\times10^3$ multichannel analysis [31] GeV$^{-1/2}$
$\Delta(1620)1/2^-$	29.0±6.2	30 – 60	55±7			
$N(1650)1/2^-$	60.5±7.7	35 – 55	32±6			
$N(1680)5/2^+$	−27.8±3.6	−18 – −5	−15±2	128±11	130 – 140	136±5
$N(1720)3/2^+$	80.9±11.5	80 – 120	115±45	−34.0±7.6	−48 – 135	135±40
$\Delta(1700)3/2^-$	87.2±18.9	100 – 160	165±20	87.2±16.4	90 – 170	170±25
$\Delta(1905)5/2^+$	19.0±7.6	17 – 27	25±5	−43.2±17.3	−55 – −35	−50±5
$\Delta(1950)7/2^+$	−69.8±14.1	−75 – −65	−67±5	−118.1±19.3	−100 – −80	−94±4

established N^* states of four-star PDG rating [29] with the observed decays to the $\pi\Delta$ and ρp final hadron states. For the resonant amplitudes, a unitarized Breit-Wigner ansatz that accounts for the transition between the same and different resonances in the dressed resonant propagator was employed, which makes the resonant amplitudes consistent with restrictions imposed by a general unitarity condition [21]. Analysis of the $\gamma_{r,v}p \rightarrow \pi^+\pi^-p$ differential cross sections in the previously unexplored kinematic areas of 1.6 GeV < W < 2.0 GeV (photoproduction) and of 1.4 GeV < W < 2.0 GeV and 2.0 GeV^2 < Q^2 < 5.0 GeV^2 (electroproduction) reveals no evidence for the contribution of mechanisms other than those the employed in the previous JM model version [9, 17, 20, 21]. However, we modified the parameterization of the extra-contact terms in the $\pi\Delta$ sub-channels and also updated the ansatz for the description of the non-resonant contributions in the ρp and $\pi^+F^0_{15}(1685)$ sub-channels. After these modifications, a good description of all available $\gamma_{r,v}p \rightarrow \pi^+\pi^-p$ differential and integrated cross sections was achieved in the entire kinematics area covered by the recent CLAS measurements [17–19] and Table 1. The quality of the data description can be seen from the values of $\chi^2/d.p.$ listed in Table 2 and from the range of $\chi^2/d.p.$ < 1.31 for the computed differential cross sections selected in the photoproduction data fit [17]. A representative example of the differential cross section description together with the resonant and non-resonant contributions inferred from the CLAS $\pi^+\pi^-p$ photoproduction data [17] is shown Fig. 1. An example of the electroproduction data description is shown in Fig. 2. Credible isolation of the resonant contri-

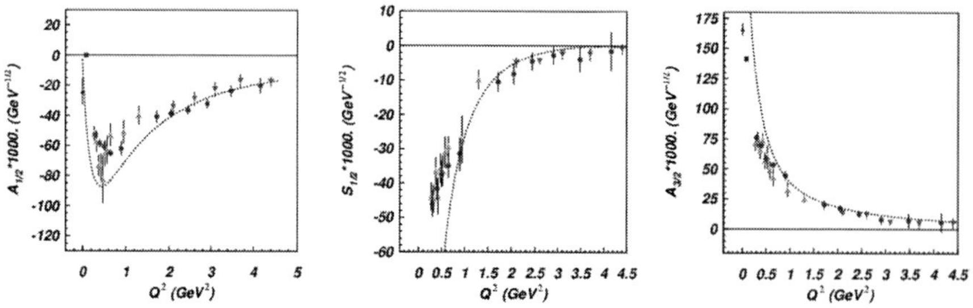

Figure 4. (Color online) CLAS results on the $N(1520)3/2^-$ electrocouplings from analysis of the recent data on $\pi^+\pi^-p$ electroproduction [18, 19] (magenta filled triangles), from the analyses of $N\pi$ electroproduction [32] (red filled circles), and from the previous CLAS data on $\pi^+\pi^-p$ electroproduction [9, 21] (green filled circles). The results at the photon point are taken from PDG [29] (blue filled triangles) and from analysis of the CLAS $N\pi$ photoproduction data [33] (blue filled squares) The results from the novel light front quark model [6, 11] are shown by the dashed lines.

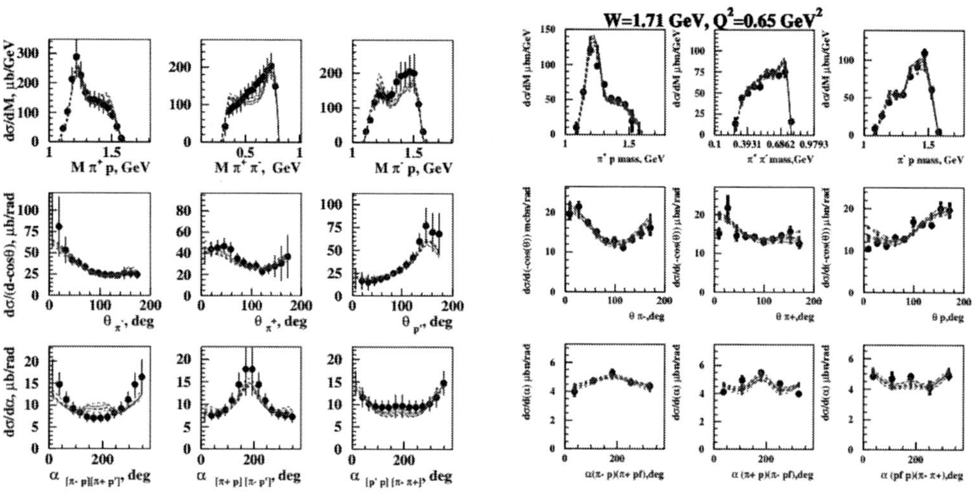

Figure 5. (Color online) Description (red curves) of the CLAS $\pi^+\pi^-p$ photoproduction data (black filled circles with error bars) at W = 1.74 GeV and Q^2= 0. GeV^2 [17] (left) and electroproduction data at W = 1.74 GeV and Q^2= 0.65 GeV^2 [23] (right) achieved in the fit within the JM model after implementation of the candidate $N'(1720)3/2^+$ state.

butions achieved in the data fit is vital for the extraction of the resonance photo-/electrocouplings.

3 Extending Knowledge on the Resonance Spectrum/Structure from CLAS $\pi^+\pi^-p$ Photo-/Electroproduction Data

The CLAS data for the $\pi^+\pi^-p$ photoproduction channel play a critical role in the extraction of the photocouplings of higher-lying nucleon excited states (M > 1.60 GeV), which decay preferentially to the $\pi\pi N$ final states, e.g. $\Delta(1620)1/2^-$, $\Delta(1700)3/2^-$, $N(1720)3/2^+$ [29, 40], and the $N'(1720)3/2^+$ candidate state [2]. The $\pi^+\pi^-p$ channel has the largest cross section among the studied $\pi\pi N$ photoproduction channels [30]. The resonance photocouplings extracted from the CLAS data are listed in Table 3 and compared with the resonance photocoupling ranges in the PDG2018 [29] and the results of the multichannel analysis [31]. There is good agreement in the magnitude and sign of the photocouplings between our results and the photocoupling ranges in the PDG listings. On the other hand, for several resonances, the photocouplings determined from the multichannel analysis are different from ours. Implementation of our $\pi^+\pi^-p$ photoproduction data into the global multichannel analyses will essentially improve our knowledge on the photocouplings and hadronic

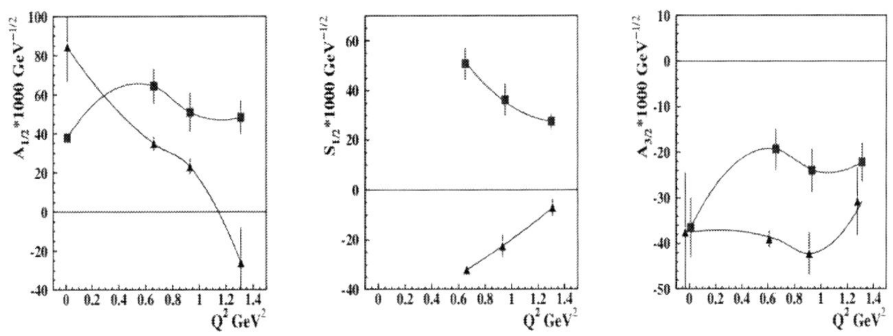

Figure 6. (Color online) Photo-/electrocouplings of the conventional $N(1720)3/2^+$ (filled triangles) and candidate $N'(1720)3/2^+$ (filled squares) states determined from the CLAS data on $\pi^+\pi^-p$ photo- and electroproduction [17, 23] within the JM model.

decay parameters of the resonances in the mass range of $W > 1.6$ GeV.

The preliminary results on the $\gamma_v pN^*$ electrocouplings for the $N(1440)1/2^+$ and $N(1520)3/2^-$ resonances have become available at 2.0 GeV^2 < Q^2 < 5.0 GeV^2 from the $\pi^+\pi^-p$ electroproduction channel for the first time. The electrocouplings were determined from a fit of the recent CLAS $\pi^+\pi^-p$ electroproduction data [18, 19] at 1.4 GeV < W < 1.55 GeV and 2.0 GeV^2 < Q^2 < 5.0 GeV^2. Representative example of the computed cross sections selected in the data fit with $\chi^2/d.p. < 1.46$ are shown in Fig. 2. The electrocouplings of the $N(1440)1/2^+$ and $N(1520)3/2^-$ resonances determined at 2.0 GeV^2 < Q^2 < 5.0 GeV^2 from the CLAS $\pi^+\pi^-p$ data [19] are shown in Fig. 3 and Fig. 4 along with the previously available results from CLAS [9, 21, 32] and theoretical expectations [6, 10, 11]. Consistent results for the $\gamma_v pN^*$ electrocouplings of the $N(1440)1/2^+$ and $N(1520)3/2^-$ resonances in a wide Q^2-range from 0.2 GeV^2 to 5.0 GeV^2, which have been determined in independent analyses of the dominant meson electroproduction channels πN and $\pi^+\pi^-p$, shown in Fig. 3 and Fig. 4, validate the credible extraction of these quantities in a nearly model independent way. This success also demonstrates the capabilty of the JM model for the credible extraction of the N^* parameters from independent studies of $\pi^+\pi^-p$ electroproduction.

Full information on the electrocouplings of most excited nucleon states extracted from the CLAS data can be found in Ref. [34] and in the web pages of the Refs. [27, 28].

The electrocouplings of the $N(1440)1/2^+$ resonance were evaluated starting from the QCD Lagrangian within the continuum QCD Dyson-Schwinger Equation (DSE) approach [10]. The DSE predictions are shown in Fig. 3 by the solid lines. The DSE evaluations are applicable at photon virtualities where the inner quark core becomes the major contributor to the resonance structure at Q^2 above 1-2 GeV^2. In this range of photon virtualities, they offer a good description of the experimental results on the $N(1440)1/2^+$ electrocouplings, which was

Table 4. Masses and the total and partial hadronic decay widths to the $\pi\Delta$ and ρp final hadron states of the convention $N(1720)3/2^+$ and candidate $N'(1720)3/2^+$ resonances obtained from the CLAS $\pi^+\pi^-p$ data [17, 23] fit within the JM model.

Resonance	$N'(1720)3/2^+$	$N(1720)3/2^+$
Mass, GeV	1.715-1.720	1.743-1.753
Total width, MeV	120 ± 6	112 ± 8
Branching fraction ($\pi\Delta$), %	46-64	38-55
Branching fraction (ρp), %	3-13	23-49

achieved with *exactly the same* dressed quark mass function as the one employed in the previous evaluations of the electromagnetic elastic nucleon form factors and the magnetic $p \to \Delta(1232)3/2^+$ form factor [8]. Furthermore, the dressed quark mass function employed in the evaluations of the aforementioned electroexcitation amplitudes was computed starting from the QCD Lagrangian [36]. This success conclusively demonstrates the capability of gaining insight into the strong QCD dynamics underlying the dominant part of hadron mass generation from the combined experimental results on the nucleon elastic form factors and the $\gamma_v pN^*$ electrocouplings analyzed within continuum QCD approaches. In the near term future the $\Delta(1600)3/2^+$ electrocouplings will become available from the CLAS $\pi^+\pi^-p$ electroproduction data [18, 19]. They will be confronted with the parameter-free prediction from the continuum QCD [37] obtained with the same dressed quark mass function as employed in the DSE computations of the nucleon elastic form factors and the $\Delta(1232)3/2^+$ and $N(1440)1/2^+$ electrocouplings. If universality of the dressed quark mass function is confirmed, this will validate the credible access to the hadron mass generation dynamics from the data on nucleon elastic form factors and the electrocouplings of different excited nucleon states.

Table 5. $N(1720)3/2^+$ hadronic decays determined from the independent fits to the data on charged double pion photo- [17] and electroproduction [23] off protons accounting only for contributions from conventional resonances.

Resonance state	N^* total width MeV	Branching fraction for decays to $\pi\Delta$	Branching fraction for decays to ρN
$N(1720)3/2^+$			
electroproduction	126.0 ± 14.0	64% - 100%	< 5%
photoproduction	160.0 ± 65.0	14% - 60%	19% - 69%

The novel light front quark model that incorporates the parameterized momentum-dependent quark mass with parameters adjusted to the data on the Q^2-evolution of the nucleon elastic form factors [6, 11] was developed for the description of the $\gamma_v pN^*$ electrocouplings. The model also accounts for the contributions from the meson-baryon cloud. It was found that the implementation of the momentum-dependent dressed quark mass is absolutely needed in order to reproduce the behavior of the nucleon elastic form factors at $Q^2 > 2.0$ GeV^2. A successful description of the electrocouplings of many resonances in the mass range up to 1.7 GeV was achieved at $Q^2 > 2.0$ GeV^2 with the same momentum- dependent quark mass as that used for the successful description of the nucleon elastic form factors. The description of the $N(1440)1/2^+$ and $N(1520)1/2^+$ electrocouplings is shown in Figs. 3 and 4 by the dashed lines. This success within this conceptually different phenomenological framework compared to the continuum QCD approaches emphasizes the relevance of the dressed quark with running mass as an active component in the structure of the ground and excited nucleon states.

The studies of the combined $\pi^+\pi^-p$ photo- and electroproduction data with CLAS have revealed evidence for the existence of the new candidate state $N'(1720)3/2^+$ [2]. The contribution from the $N'(1720)3/2^+$ has been observed, in addition to several other new baryon states, from a global multi-channel analyses of the exclusive meson photoproduction data with decisive impact from CLAS data on KY photoproduction [3, 38]. For the first time signals from the $N'(1720)3/2^+$ candidate state have been observed in analyses of the CLAS $\gamma_v p \rightarrow \pi^+\pi^-p$ electroproduction data [23], which shows a pronounced structure in the W-dependence of the fully integrated cross sections at $W \approx 1.7$ GeV and in all Q^2 bins covered by the measurements.

Further studies of the CLAS $\gamma_{r,v}p \rightarrow \pi^+\pi^-p$ photo- and electroproduction cross sections [17, 23] have been carried out within the framework of the current version of the JM model outlined in Sec. 2. All nine single-differential cross sections were included in the data fits carried out independently for the photo- and electroproduction data. In the fits we simultaneously varied the electrocouplings and decay widths to the $\pi\Delta$ and ρN final hadron states for all resonances that contribute to the region of the structure at $W \approx 1.7$ GeV. We also simultaneously varied the non-resonant parameters of the JM model.

The hadronic decay widths of each resonance remain the same in all Q^2-bins in the electroproduction data fit, while for the photoproduction data, the resonance hadronic couplings were varied independently. We found that for the description of all six angular distributions in the bins of W and Q^2 at $W \approx 1.7$ GeV, an essential contribution from the resonances with spin-parity J^π=3/2$^+$ is needed. Therefore, we compare the two possibilities for description of the $\pi^+\pi^-p$ photo- and electroproduction data at $W \approx 1.7$ GeV: either by considering the contribution from the conventional $N(1720)3/2^+$ resonance or by implementing in addition a contribution from the new $N'(1720)3/2^+$ state with parameters fit to the CLAS data. Both methods provided a reasonable description of the photoproduction data with $\chi^2/d.p. < 1.31$ (1.6 GeV < W < 2.0 GeV) accounting for the contribution from the statistical and dependent from the final hadron state kinematic variable systematic data uncertainties, as well as of the electroproduction data with $\chi^2/d.p. < 3.0$ (1.65 GeV < W < 1.8 GeV) accounting for the statistical data uncertainties only. Representative examples of the description of the photo- and electroproduction differential cross sections are shown in Fig. 5. The results in Fig. 5 were obtained accounting for the contributions from the $N'(1720)3/2^+$ candidate-state, but the fit quality remains almost the same if $N'(1720)3/2^+$ candidate-state is taken out and the photo- /electrocouplings of the conventional $N(1720)3/2^+$ resonance are fit to the data without the contribution from the new candidate state.

On the other hand, as is shown in Table 5, the decay widths of the conventional $N(1720)3/2^+$ resonance to the $\pi\Delta$ and ρp final hadron states inferred from independent analyses of $\pi^+\pi^-p$ photo- and electroproduction depend considerably on the presence or absence of the $N'(1720)3/2^+$ candidate-state. In the case where the contributions from only conventional resonances are taken into account, the $N(1720)3/2^+$ decays into the ρN final state inferred from the data fit on charged double pion photo- and electroproduction off protons differ by more than a factor of four. Since the resonance hadronic decay widths should be Q^2-independent, this makes it impossible to describe both the $\pi^+\pi^-p$ photo- and electroproduction cross sections when only contributions from conventional resonances are taken into account.

By implementing a new $N'(1720)3/2^+$ baryon state with the parameters inferred from the CLAS $\pi^+\pi^-p$ data fit a successful description of all nine single-differential $\gamma_{r,v}p \rightarrow \pi^+\pi^-p$ photo- and electroproduction cross sections has been achieved. Furthermore, the hadronic decay widths of all resonances in the third resonance region as inferred from the fits at different Q^2 remain Q^2-

independent in the entire range of photon virtualities up to 1.5 GeV^2 that is covered by the CLAS measurements [17, 23]. This is a strong evidence for the existence of the new $N'(1720)3/2^+$ baryon state. Indeed, if the implementation of a new baryon state would just effectively describe the non-resonant contributions, then it would not be possible to reproduce the CLAS $\pi^+\pi^-p$ data in a wide Q^2-range with Q^2-independent decay widths of such a fake resonance because of the pronounced evolution of the $\pi^+\pi^-p$ non-resonant parts with Q^2. The successful description of the CLAS $\pi^+\pi^-p$ photo- and electroproduction data for 0. GeV^2 < Q^2 < 1.5 GeV^2 with Q^2-independent decay widths of both the $N'(1720)3/2^+$ and $N(1720)3/2^+$ resonances into $\pi\Delta$ and ρp final hadron states, validates the existence of the new baryon state $N'(1720)3/2^+$ in a nearly model- independent way.

The results on the Q^2-evolution of the $N'(1720)3/2^+$ and $N(1720)3/2^+$ $\gamma_v pN^*$ electrocouplings determined from the CLAS data fit are shown in Fig. 6. Their rather different Q^2-evolution emphasizes the contributions from both the $N'(1720)3/2^+$ and $N(1720)3/2^+$ excited nucleon states into the $\pi^+\pi^-p$ photo- and electroproduction in the third resonance region. The masses and the total and partial hadronic decay widths of the $N'(1720)3/2^+$ and $N(1720)3/2^+$ resonances are listed in Table 4. The conventional $N(1720)3/2^+$ and candidate $N'(1720)3/2^+$ states have the same spin-parities. Their masses are slightly diferenrt and the total decay widths are almost the same within the data uncertainties. However, the different patterns for their decays to the $\pi\Delta$ and ρp final hadron states and different Q^2-evolution of their electrocouplings allow us to identify these two different resonances. So far, the $N'(1720)3/2^+$ is the only "missing" baryon state for which the experimental results on the Q^2-evolution of the electroexcitation amplitudes have become available. These results offer insight into the structure of the so-called "missing" baryon states, paving a way towards understanding the peculiar features of these resonances which have made them so elusive from the observation in experiments for such a long time. In the studies of the conventional and exotic baryons within the chiral quark-soliton model [39], the new baryon state $N(1726)3/2^+$ was predicted as a member of the 27-plet. The mass and spin-parity of this predicted state are consistent with the observed $N'(1720)3/2^+$ candidate-resonance. When the predictions from this or other quark models on the $N'(1720)3/2^+$ electrocouplings will become available, they will shed light on the nature of the new baryon state observed in the CLAS $\pi^+\pi^-p$ photo- and electroproduction data. This success also demonstrates the importance of combined studies of exclusive photo- and electroproduction, both for the validation of new baryon states and for the further search for new excited nucleon states.

4 Conclusion and Outlook

Studies of exclusive $\pi^+\pi^-p$ photo- and electroproduction with CLAS have extended the information on the nucleon resonance photo- and electroexcitation amplitudes. The photocouplings of almost all nucleon resonances in the mass range from 1.6 GeV to 2.0 GeV have become available for the first time from the studies of $\pi^+\pi^-p$ photoproduction. The entries on the photocouplings and $\pi\Delta$ and ρp hadronic decays of most resonances in the mass range from 1.6 GeV to 2.0 GeV have been updated in the PDG by results from CLAS [40].

Studies of the combined $\pi^+\pi^-p$ photo- and electroproduction data have provided evidence for the candidate-state $N'(1720)3/2^+$. The contribution from this state is needed in order to describe both photo- and electroproduction data with Q^2-independent masses and total and partial decay widths of the $N(1720)3/2^+$-conventional and $N'(1720)3/2^+$ candidate resonances to the $\pi\Delta$ and ρp final hadron states. The $N'(1720)3/2^+$ is the only "missing" baryon state observed so far for which the information on the Q^2-evolution of the electroexcitation amplitudes has become available. These results allow us to gain insight into the structure of "missing" resonances and eventually to understand why these states have remain elusive from the observation for such a long time. Predictions on the Q^2-evolution of the "missing" resonance electrocouplings from different models is the urgently needed next step towards approaching this challenging objective.

The $\gamma_v pN^*$ electrocouplings have become available from the CLAS $\pi^+\pi^-p$ electroproduction data at 0.2 GeV^2 < Q^2 < 5.0 GeV^2 for the resonances in the mass range up to 1.6 GeV and at 0.5 GeV^2 < Q^2 < 1.5 GeV^2 for most resonances in the mass range from 1.6 GeV to 1.8 GeV. Consistent results on nucleon resonance electrocouplings from independent studies of the $N\pi$ and $\pi^+\pi^-p$ electroproduction channels with different non-resonant contributions validate the credible extraction of these quantities. Physics analyses of the CLAS results on the $\gamma_v pN^*$ electrocouplings within the framework of the continuum QCD approach [4, 8, 10] and the novel relativistic quark model [6] have revealed the dressed quarks with momentum-dependent masses as active degrees of freedom in N^* structure. The CLAS results on the $\gamma_v pN^*$ electrocouplings of different resonances are of particular importance in order to gain insight into the strong QCD dynamics underlying the generation of the dominant part of hadron mass encoded in the momentum dependence of running dressed quark mass. Consistent results on the dressed quark mass function from independent studies of excited nucleon states of different structure validate credible access to this key ingredient of strong QCD.

In the future, the electrocouplings of most excited nucleon states in the mass range up to 2.0 GeV will become available at 2.0 GeV^2 < Q^2 < 5.0 GeV^2. They will be obtained from the recent CLAS $\pi^+\pi^-p$ electroproduction data [18, 19] analyzed within the framework of the updated JM model outlined in Section 2. This essential increase of the information on the $\gamma_v pN^*$ electrocouplings will open up new avenues for the exploration of many facets of strong QCD in generation of the excited nucleon spectrum. The insight into the structure of most N^* will allow us to address the challenging open problems of the Standard Model in the strong QCD regime, including: a) insight into the dressed quark mass generation in different resonances in connection with hadron mass generation; b) studies of the dressed-quark-gluon vertex and diquark correlation complexity in the structure of orbital excited resonances; c) insight into dynamical chiral symmetry breaking from the data on electrocouplings of the resonances parity-partners; d) search for new states of baryon matter and exploration of their structure. Furthermore, the information on electrocouplings of most excited nucleon states makes it possible to determine the resonance contributions in inclusive electron scattering [34, 35] and gain insight into the behavior of parton distributions in the ground state nucleons at close to unity values of Bjorken variable x_B bridging the efforts on the nucleon structure studies in the resonance and deep inelastic scattering regions.

This work was supported by the U.S. Department of Energy, Office of Science, Office of Nuclear Physics under contract DE-AC05-06OR23177

References

[1] V.I. Mokeev *CLAS Collaboration*, Few Body Syst. 46 **59**, (2018).

[2] V.I. Mokeev *et al.*, Phys. Lett. B 135457 **805**, (2020).

[3] V.D. Burkert, *The 12th International Workshop on the Physics of Excited Nucleons - N*2019*, (2019). See these proceedings.

[4] C.D. Roberts, *The 12th International Workshop on the Physics of Excited Nucleons - N*2019*, (2019). See these proceedings.

[5] V.D. Burkert and C.D. Roberts, Rev. Mod. Phys. 011003 **91**, (2019).

[6] I.G. Aznauryan and V.D. Burkert, Few Body Syst. 98 **59**, (2018).

[7] H. Kamano, Few Body Syst. 24 **59**, (2018).

[8] J. Segovia, Few Body Syst. 1185 **55**, (2014).

[9] V.I. Mokeev *et al.*, Phys. Rev. 025206 **C 93**, (2016).

[10] J. Segovia *et al.*, Phys. Rev. Lett. 171801 **115**, (2015).

[11] I.G. Aznauryan and V.D. Burkert, Phys. Rev. 055202 **C 85**, (2012).

[12] G. Ramalho, Few Body Syst. 92 **59**, (2018).

[13] G. Ramalho, *The 12th International Workshop on the Physics of Excited Nucleons - N*2019*, (2019). See these proceedings.

[14] I. Obukhovsky *et al.*, Phys. Rev. 014032 **D 89**, (2014).

[15] M.M. Giannini and E. Santopinto, Chin. J. Phys. 020301 **53**, (2015).

[16] E. Santopinto, *The 12th International Workshop on the Physics of Excited Nucleons - N*2019*, (2019). See these proceedings.

[17] E. Golovatch *et al., CLAS Collaboration*, Phys. Lett. B 371 **788**, (2019).

[18] E.L. Isupov *et al., CLAS Collaboration*, Phys. Rev. 025209 **C 96**, (2017).

[19] A. Trivedi, Few Body Syst. 45 **60**, (2019).

[20] V.I. Mokeev *et al.*, Phys. Rev. 045212 **C 80**, (2009).

[21] V.I. Mokeev *et al., CLAS Collaboration*, Phys. Rev. 035203 **C 86**, (2012).

[22] G.V. Fedotov *et al., CLAS Collaboration*, Phys. Rev. 015204 **C 79**, (2009).

[23] M. Ripani *et al., CLAS Collaboration*, Phys. Rev. Lett. 022002 **91**, (2003).

[24] M. Ripani *et al.*, Nucl. Phys. A 220 **673**, (2000).

[25] G.V. Fedotov *et al., CLAS Collaboration*, Phys. Rev. 025203 **C 98**, (2018).

[26] CLAS physics database, http://clasweb.jlab.org/physicsdb.

[27] Nucleon Resonance Photo-/Electrocouplings Determined from Analyses of Experimental Data on Exclusive Meson Electroproduction off Protons ,https://userweb.jlab.org/~mokeev/resonance_electrocouplings/.

[28] Fits of the resonances electrocouplings. ,https://userweb.jlab.org/~isupov/couplings/.

[29] M. Tanabashi *et al., Particle Data Group*, Phys. Rev. 025203 **D 98**, (2018).

[30] U. Thoma, Int. J. Mod. Phys. A 280 **20**, (2019).

[31] V. Sokhoyan *et al.*, Eur. Phys. J. A 95 **51**, (2015).

[32] I.G. Aznauryan *et al., CLAS Collaboration*, Phys. Rev. 055203 **C 80**, (2009).

[33] M. Dugger *et al. (CLAS Collaboration)*, Phys. Rev. C **79**, 065206 (2009).

[34] A.N. Hiller Blin *et al.*, Phys. Rev. 035201 **C 100**, (2019).

[35] A.N. Hiller Blin, *et al., The 12th International Workshop on the Physics of Excited Nucleons - N*2019*, (2019). See these proceedings.

[36] D.Binosi *et al.*, Phys. Lett. B 183 **742**, (2015).

[37] Ya Lu *et al.*, Phys. Rev. 034001 **D 100**, (2019). .

[38] A.V. Anisovich *et al.*, Phys. Rev. Lett. 062004 **119**, (2017).

[39] G-S.Yang and H-C. Kim, arXiv:1809.07489[hep-ph].

http://doi.org/10.1051/epjconf/202024103003

[40] The Review of Particle Physics (2019),
http://pdg.lbl.gov/.

Study of $\bar{p}p \to \overline{\Xi}^{+}\Lambda K^{-}$ with the $\overline{\text{P}}$ANDA Detector

Jennifer Pütz[1,*], *Albrecht* Gillitzer[1], *James* Ritman[1], and *Tobias* Stockmanns[1] on behalf of the PANDA Collaboration
[1]Forschungszentrum Jülich, Wilhelm-Johnen-Straße, 52428 Jülich, Germany

Abstract. For a deeper insight into the mechanisms of non-perturbative QCD it is essential to understand the excitation pattern of baryons. Up to now only the nucleon excitation spectrum has been subject to systematic experimental studies, while very little is known on excited states of double or triple strange baryons.
In studies of antiproton-proton collisions the $\overline{\text{P}}$ANDA experiment is well-suited for a comprehensive baryon spectroscopy program in the multi-strange sector. A large fraction of the inelastic $\bar{p}p$ cross section is associated to final states with a baryon-antibaryon pair together with additional mesons, giving access to excited states both in the baryon and the antibaryon channel.
For final states containing a $\overline{\Xi}^{+}$ Ξ^{-} pair, cross sections up to μb are expected, corresponding to production rates of $\sim 10^{6}$/d at a luminosity $L = 10^{31}\,\text{cm}^{-2}\,\text{s}^{-1}$. This study focuses on excited Ξ^{-} states decaying into ΛK^{-}. A strategy to reconstruct the reaction $\bar{p}p \to \overline{\Xi}^{+}$ Ξ^{*} and its charge conjugate channel with the $\overline{\text{P}}$ANDA detector will be presented.

1 Motivation

In QCD the elementary particles of the strong interaction are quarks and gluons, whereas the minimum quark content of hadrons observable in nature is either three quarks (baryons) or a quark-antiquark pair (mesons). How these composite systems emerge is far less well understood.
To understand the inner structure of baryons and mesons it is necessary to study the excitation pattern of these particles. In baryon spectroscopy one of the main goals is to answer the following questions: Which degrees of freedom are relevant for the excitation mode of baryons? Can they be described by a three-quark or a quark-diquark structure? How important is the dynamics in baryon-meson systems for the generation of resonant baryon states?
Up to now an intense worldwide effort is ongoing to study the nucleon and Δ excitation spectrum with photo-induced reactions [1–3] while much less information is available on the excitation spectrum of double and triple strange baryons Ξ and Ω. Experimental data on Ξ resonances are coming almost entirely from bubble chamber experiments and not much information was added to PDG [4] since 1988. The study of Ξ baryons will provide independent information on the excitation pattern and allows to verify if the picture deduced from the excited nucleon and Δ states is correct and has a general validity for baryons in the light quark sector. $\bar{p}p$ collisions provide an ideal entrance channel for baryon spectroscopy in the multi-strange sector since the large cross section allows to collect high statistics data in a reasonable time frame. With the $\overline{\text{P}}$ANDA detector it will be possible to access the excited states of $\overline{\Xi}^{+}$ and Ξ^{-} directly.

*e-mail: j.puetz@fz-juelich.de

2 The $\overline{\text{P}}$ANDA Experiment

The $\overline{\text{P}}$ANDA experiment is one of the four major projects of FAIR. It has a wide physics program which includes baryon and meson spectroscopy, studies of QCD dynamics, nucleon structure and electromagnetic processes as well as the study of hadrons in nuclei and hypernuclear physics [5].
The $\overline{\text{P}}$ANDA detector consists of two parts, the *Target Spectrometer* which has a cylindrical geometry and surrounds the interaction point, and the *Forward Spectrometer* for the reconstruction of high-energy, forward-boosted particles. The modular and flexible design allows the $\overline{\text{P}}$ANDA detector to provide almost 4π acceptance for both charged and neutral particles.
One of the properties of hyperons is their long lifetime leading to their typical long decay length and the characteristic topology of the associated particle tracks in the detector. For the reconstruction of these particles a central tracking system is used which consists of the Micro Vertex Detector (MVD), the Straw Tube Tracker (STT) and the Gas Electron Multiplier (GEM) stations. The MVD is the innermost part of the tracking system enclosing the interaction point and is a silicon based high precision vertex detector. The STT consists of gas filled drift tubes arranged in a cylindrical geometry around the beam axis enclosing the interaction point and the MVD. Along the beam line the STT is followed by the GEM stations which are also gas based detectors with planar geometry. The GEM stations are important to reconstruct the trajectories of particles emitted in the forward direction within the Target Spectrometer. The tracking system used in the "Forward Spectrometer" is called Forward Tracking Stations (FTS) and consists of three pairs of tracking stations positioned

along the beam line. It is based on self-supporting straw tubes which are also used for the STT. The FTS measures the deflection of the particles within a dipole field. Therefore, it is planned to place the first station in front, the second pair within and the third behind the dipole magnet. The different tracking systems will provide a high tracking resolution. In addition, the information from these detectors is combined with other components of $\overline{\text{P}}$ANDA to provide a good particle identification.

3 Study of $\bar{p}p \to \bar{\Xi}^+ \Lambda K^-$ and $\bar{p}p \to \Xi^- \bar{\Lambda} K^+$

3.1 Event Generation

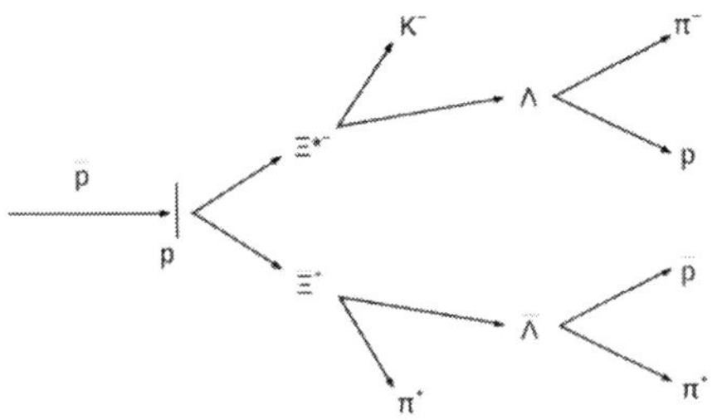

Figure 1. Illustration of the reaction including the decay chain to the detected final state.

For this study 20 million signal events are generated with the event generator EvtGen [6]. The events are generated according to the decay tree shown in Figure 1. Here, the chosen momentum of the antiproton beam is 4.6 GeV/c which corresponds to a center-of-mass energy of $\sqrt{s}$ = 3.25 GeV and which is about 320 MeV above the production threshold of $\bar{\Xi}^+ \Lambda K^-$ or $\Xi^- \bar{\Lambda} K^+$. The given phase space allows the population of the resonant states $\Xi(1690)^-$ and $\Xi(1820)^-$ as well as $\bar{\Xi}(1690)^+$ and $\bar{\Xi}(1820)^+$. Table 1 lists the properties of the chosen resonances according to [4]. The fraction between resonant and non-resonant signal which is used as input in EvtGen is shown in Table 2. This assumed ratio is based on the measured total production cross section of both the excited states and ground states of single strange hyperons, see [7].

Table 1. Implemented mass and width of the resonances.

State	Mass [MeV/c²]	Γ [MeV/c²]
$\Xi(1690)^-$	1690	30
$\Xi(1820)^-$	1823	24

Since there are neither experimental data nor theoretical predictions for the reaction $\bar{p}p \to \bar{\Xi}^+ \Xi^{*-}$, the phase space model (PHSP), generating an isotropic angular distribution, is chosen for each decay mode.

EvtGen does not take into account the curved trajectories of $\bar{\Xi}^+$ and Ξ^- in the magnetic field of the solenoid as well as the interaction of both particles with the detector volume. For this reason the propagation of $\bar{\Xi}^+$ and Ξ^- is managed by the transport code GEANT4 [8]. As opposed to $\bar{\Xi}^+$ and Ξ^-, $\bar{\Lambda}$ and Λ have various decay modes. Here, the focus is on $\Lambda \to p + \pi^-$ and $\bar{\Lambda} \to \bar{p} + \pi^+$ which have a branching ratio of BR($\Lambda \to p + \pi^-$) = 0.634[4] each. To avoid the generation of unwanted final states, the branching ratio of $\bar{\Lambda}$ and Λ for this decay mode is set to 100 % in GEANT4, respectively. This implies that the final result has to be scaled by the corrected branching ratios.

Table 2. Decay modes and ratios used as input for the event generation.

Fraction	Channel	Channel (c.c.)
0.4	$\bar{\Xi}^+ \Xi(1690)^-$	$\Xi^- \bar{\Xi}(1690)^+$
0.4	$\bar{\Xi}^+ \Xi(1820)^-$	$\Xi^- \bar{\Xi}(1820)^+$
0.2	$\bar{\Xi}^+ \Lambda K^-$	$\Xi^- \bar{\Lambda} K^+$

3.2 Reconstruction

The detector parts described in Section 2 measure the momentum of the final state particles, i.e. p, $\bar{p}$, K^-, K^+, π^-, and π^+. Therefore, the reconstruction procedure starts with those final state particles and proceeds backwards in the decay chain shown in Figure 1.

The standard tracking algorithms used in PandaRoot [9] – the software framework of $\overline{\text{P}}$ANDA– assume a particle's origin at the interaction point. As already mentioned in Section 2, hyperons have a long decay length and thus decay several centimeters away from the interaction point. This implies, that a reconstruction of the final state particles is not possible with the standard tracking algorithms. Since no realistic algorithm for secondary particles is implemented in PandaRoot, yet, an ideal pattern recognition algorithm is used for the reconstruction of the final state particles. This means, that the hits caused by the charged particles are grouped based on the information of the generated particles. As a consequence, particles with only one hit in any of the tracking detectors will be reconstructed. In order to obtain realistic reconstruction efficiencies, a track filter is used selecting only those candidates which cause at least four hits in either the MVD, STT or GEM detector. This selection criterion is motivated by the trajectory of a charged particle in a homogenous magnetic field. The projection of the trajectory onto, e.g. the x-y-plane (assuming that the particle moves in z-direction), is a circle which is defined by three points in the plane. A fourth hit will confirm the track hypothesis.

Two different selection strategies for composite state particles and the full tree reconstruction are chosen to analyze 10 million events each. The first strategy is a sequential fit procedure in which the candidate selection is based on several cuts on fitted values. In the second case, as the name full decay tree fit implies, the candidate selection is done on the fully reconstructed event. Here, the results of the selection based on the full decay tree fit are presented.

In this analysis no PID information contained in the generated MC event sample is used for reconstructing the final state particles. This implies that for a given charge sign each of the corresponding candidates lists, for example π^+, K^+ and p, is filled with the same candidates based on all tracks for this charge sign found within the event. The candidate lists then only differ in the mass value assigned to their members based on the respective hypothesis. Since the event selection is completely done on the last stage of the reconstruction, a pre-selection of the events, which requires that the event is reconstructable, in general, is done. Reconstructable means that per event at least three candidates for each charge sign have to be available, otherwise the event is rejected. Table 3 shows the reconstruction efficiency for

Table 3. Reconstruction efficiencies for the final state particles from $\bar{p}p \rightarrow \overline{\Xi}^+ \Lambda K^-$ (left) and $\bar{p}p \rightarrow \Xi^- \bar{\Lambda} K^+$ (right). The statistical error is $\sigma \approx 0.1\,\%$.

type	rec. eff.[%]	type	rec. eff.[%]
π^-	71.2	π^+	70.6
π^+ ($\bar{\Lambda}$)	68.6	π^- (Λ)	68.3
π^+ ($\overline{\Xi}^+$)	73.7	π^- (Ξ^-)	73.1
K^- (res.)	84.9	K^+ (res.)	86.7
K^- (cont.)	85.1	K^+ (cont.)	86.9
p	88.7	p	86.2
$\bar{p}$	82.3	$\bar{p}$	83.4

the reconstructed final state particles for $\bar{p}p \rightarrow \overline{\Xi}^+ \Lambda K^-$ and $\bar{p}p \rightarrow \Xi^- \bar{\Lambda} K^+$. Here, the reconstruction efficiency is defined as the number of reconstructed true candidates divided by the number of generated candidates. The results show a lower reconstruction efficiency for π^- and π^+ with respect to K^-, K^+, p and $\bar{p}$. There are two reasons for the reduced reconstruction efficiency of the pions. Due to the lifetime of the hyperons, π^- and π^+ are produced several centimeters away from the interaction point. Therefore, there are either only a few hits left in the MVD, or none at all. In addition, the track fitting algorithm which uses a Kalman Filter [10] has difficulties to handle low-momentum particles which are produced several centimeters away from the interaction point. In the studied channel, π^- and π^+ have a total momentum up to 800 MeV/c resulting in a loss of reconstructed candidates. After the candidate lists for the final state candidates have been filled, the composite state candidates are combined. To reduce the number of wrongly combined candidates having a mass much higher than the nominal mass of the corresponding hyperon, a mass window cut of ±15 MeV around the nominal hyperon mass is applied. Subsequently, the candidate is passed to the next stage of the reconstruction.

If all particles are reconstructed and combined to a candidate for the full tree, a fit on the full decay tree is applied. For the fit a recursive fitting routine called DecayTreeFitter [11] is used which applies vertex fits as well as kinematic fits with mass constrains for the hyperon masses and a momentum constraint to the initial four-momentum vector of the $\bar{p}p$ system to each candidate. The final selection of the candidate is then done by a cut on the fit probability shown in Figure 2. Candidates

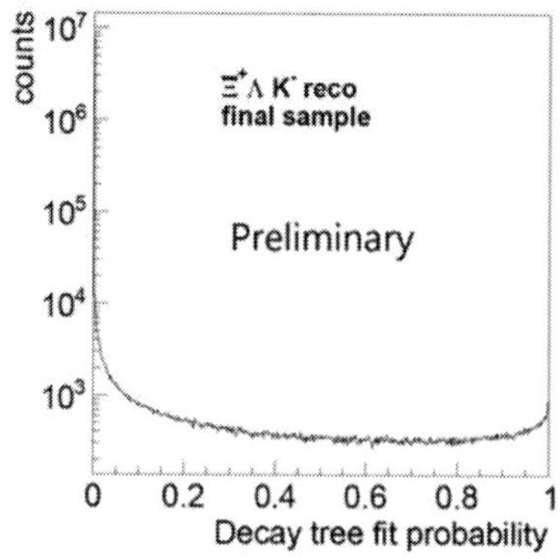

Figure 2. Decay tree fit probability.

with a fit probability less than 0.01% are rejected. In addition, only candidates for which the fit converged are chosen as final selected candidates. The reconstruction efficiencies achieved for the full reconstruction of the decay tree are 5.4 % for $\overline{\Xi}^+$ Λ K^- and 5.5 % for Ξ^- $\bar{\Lambda}$ K^+. Both samples have a purity of about 98 %. In addition, the fraction of the decay modes is determined to ~ 38 % for $\overline{\Xi}^+$ $\Xi(1690)^-$, ~ 42 % for $\overline{\Xi}^+$ $\Xi(1820)^-$, and ~ 20 % for $\bar{p}p \rightarrow \overline{\Xi}^+ \Lambda K^-$, which is consistent with the input shown in Table 2. Figure 3 shows the Dalitz

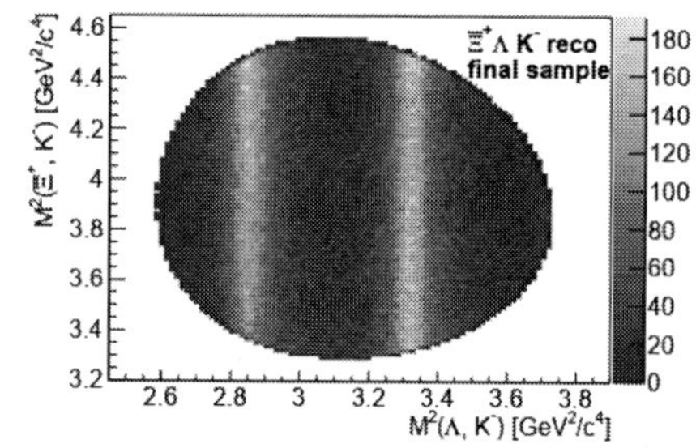

Figure 3. Dalitz plot for the final selected $\overline{\Xi}^+$ Λ K^- candidates. Adapted from [12].

plot for the final reconstructed candidates. Here, the squared mass of the Λ K^- system is plotted against the squared mass of the $\overline{\Xi}^+$ K^- system. Both resonances are clearly observable as vertical bands in the Dalitz plot, the $\Xi(1690)^-$ in the squared mass range between 2.8 GeV^2/c^4 and 2.95 GeV^2/c^4 and the $\Xi(1820)^-$ in the region between 3.25 GeV^2/c^4 and 3.4 GeV^2/c^4. The Dalitz plot ratio (Fig. 4) between the MC truth partners of the final reconstructed candidates and the generated candidates appears to be almost flat within the statistical errors which implies that the detector acceptance is structureless for this channel.

Small deviations are visible in the corresponding ratios of the mass distributions for the Λ K^- and $\overline{\Xi}^+$ K^- systems. The ratio for the Λ K^- system (Figure 5) shows a decrease of about 20 % towards lower masses. The reason for this

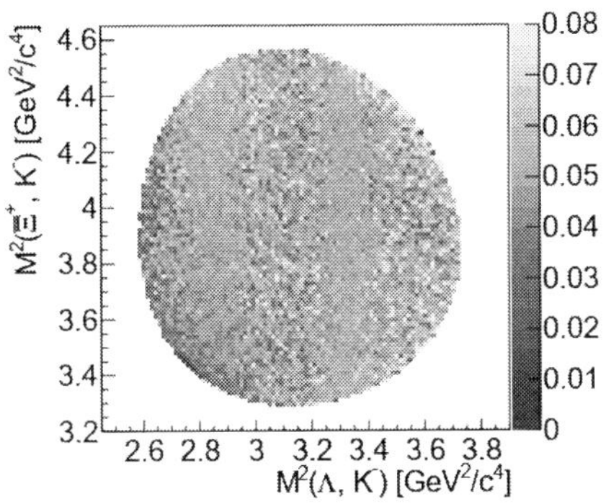

Figure 4. Ratio of the MC truth partner of the final selected and the generated sample. Adapted from [12]

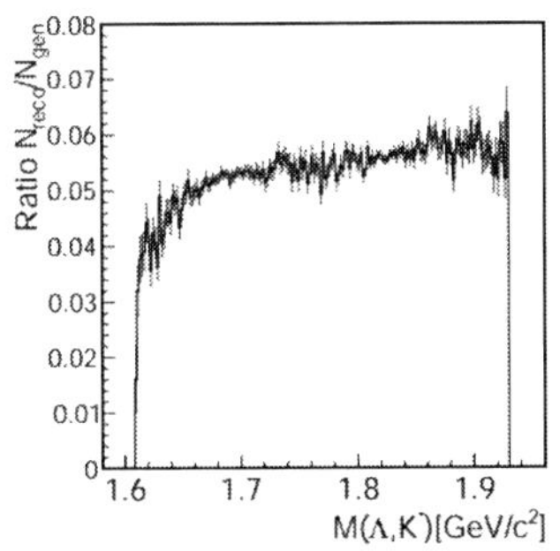

Figure 5. Ratio of the generated and the MC truth partner of the reconstructed Λ K^- mass distribution. Adapted from [12]

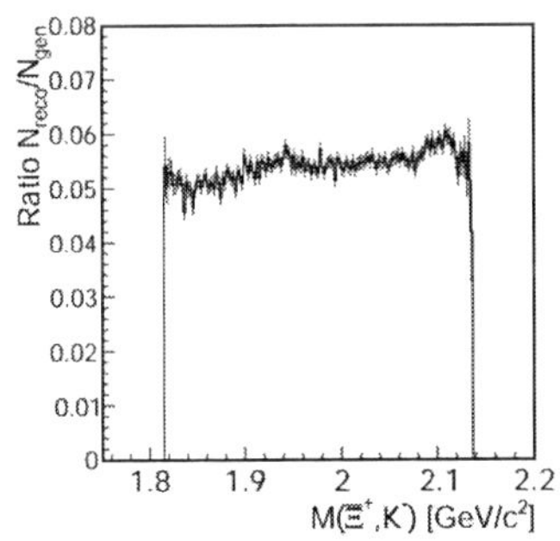

Figure 6. Ratio of the generated and the MC truth partner of the reconstructed $\bar{\Xi}^+$ K^- mass distribution. Adapted from [12].

behavior has to be explored in more detail based on the angular and momentum distributions of all final state particles. Figure 6 shows the corresponding ratio for the mass distribution for $\bar{\Xi}^+$ K^- which is within the statistical fluctuation almost flat and varies within ±9 %.

To determine the mass and width of the Ξ resonances, two Voigt functions and an additional polynomial function

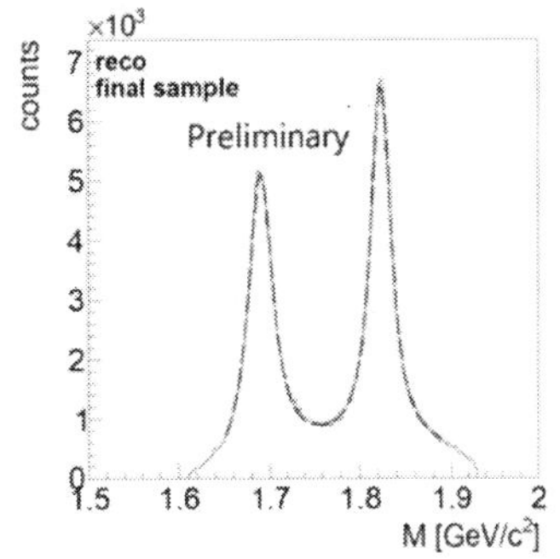

Figure 7. Mass distribution of the final selected Λ K^- candidates.

are fitted to the mass distribution of the Λ K^- system, shown in Figure 7. Table 4 summarizes the fit results for the Ξ^* and $\bar{\Xi}^*$, respectively. The error on the fit values is dominated by systematic uncertainties and is estimated by varying the fit region, systematically. The instrumental width was determined separately by fitting a double Gaussian function to the deviation of the reconstructed and the generated mass distribution. The achieved mass and width for $\Xi(1690)^-$ and $\bar{\Xi}(1690)^+$ are in good agreement with the values listed in Table 1. In case of $\Xi(1820)^-$ and $\bar{\Xi}(1820)^+$, the fitted mass values are in good agreement with the input values, while the achieved width differs from the input values by $1.8\,\sigma$ for $\Xi(1820)^-$ and by $2.3\,\sigma$ for $\bar{\Xi}(1820)^+$, respectively.

Table 4. Fit results for Ξ^* and $\bar{\Xi}^*$

State	Mean [MeV/c²]	Γ [MeV/c²]
$\Xi(1690)^-$	1690.0 ± 0.1	30.1 ± 0.6
$\Xi(1820)^-$	1823.0 ± 0.1	22.9 ± 0.4
$\bar{\Xi}(1690)^+$	1690.2 ± 0.1	30.2 ± 0.6
$\bar{\Xi}(1820)^+$	1823.1 ± 0.1	22.7 ± 0.4

4 Background Studies

Beside the investigation of the signal, a study of about 100 million background events at the identical $\bar{p}$ momentum of 4.6 GeV/c are generated. The event generation was performed with the Dual Parton Model (DPM) [13] event generator, modeling the inelastic hadronic background in $\bar{p}p$ collisions. In the analysis these background events were subject to the same selection strategies as used for the signal sample. For both strategies no event out of 100 million generated events was accepted. Assuming that this non-observation of background corresponds to 2.3 events at 90 % confidence level, a lower limit for the signal-to-background ratio can be calculated. The signal-to-background ratio is calculated by

$$\frac{S}{B} = \frac{\sigma_{sig} \cdot \epsilon_{sig} \cdot b_{sig}}{\sigma_{bg} \cdot \epsilon_{bg}}, \tag{1}$$

where σ_{sig} is the signal and σ_{bg} the background cross section, b_{sig} is the total branching fraction of the final selected

signal evens and ϵ_{sig} and ϵ_{bg} the reconstruction efficiencies for signal and background, respectively. For the so far unmeasured signal section $\sigma_{sig} = 1\,\mu b$ is assumed. This assumption is based on the measured total production cross sections of $\Xi^- \,\overline{\Xi}^+$ production [7] and on the production of excited states and ground states of single strange hyperons [14] in $\bar{p}p$ collisions. Furthermore, the inelastic cross section for $\bar{p}p$ at a beam momentum of 4.6 GeV/c is $\sigma_{bg} = 50\,mb$ [4].
As already discussed in Section 3.1, the branching ratio for Λ into p and π^- as well as for $\bar{\Lambda}$ into $\bar{p}$ and π^+ was set to 100 %. This is corrected in the following by taking the correct branching ratio $b_{sig} = b_\Lambda^2 = 0.483$ for the final selected signal into account. The reconstruction efficiency achieved with the full decay tree fit is $\epsilon_{sig} = 5.4\,\%$ for $\overline{\Xi}^+ \,\Lambda\, K^-$ and $\epsilon_{sig} = 5.5\,\%$ for $\Xi^- \,\bar{\Lambda}\, K^+$. The reconstruction efficiency for the background events at 90 % confidence level is $\epsilon_{bg} = 2.3 \cdot 10^{-8}$. The resulting lower limits for the signal-to-background ratio are listed in Table 5.

Table 5. Signal-to-Background ratio.

	S/B
$\overline{\Xi}^+ \,\Lambda\, K^-$	> 19.1
$\Xi^- \,\bar{\Lambda}\, K^+$	> 19.5

5 Summary

The study of in total 20 million signal events for the process $\bar{p}p \to \overline{\Xi}^+ \,\Xi^*$ where $\Xi^* \to \Lambda\, K^-$ as well as for the charge conjugate channel is presented. Here, for the signal, two Ξ resonances $\Xi(1690)^-$ and $\Xi(1820)^-$ as well as their charge conjugate were chosen. For this analysis no Monte Carlo information is used to assign the particle type to the reconstructed candidates. With the full decay tree fit a reconstruction efficiency of 5.4 % was achieved for $\overline{\Xi}^+ \,\Lambda\, K^-$ and 5.5 % for $\Xi^- \,\bar{\Lambda}\, K^+$, and the purity for both samples is about 98 %. The reconstructed fraction of the resonant and non-resonant contribution is consistent with the input. Assuming the luminosity to be $L = 10^{31}cm^{-2}s^{-1}$, the reconstruction efficiency corresponds to at least 18,000 fully reconstructed events per day.
The comparison between the reconstructed values for the masses and widths of the Ξ resonances with the input, shows that the masses for both resonances are in good agreement with the input. In case of the $\Xi(1690)^-$, also the width is in good agreement with the input while the widths for the $\Xi(1820)^-$ and its charge conjugate partner differ by about two standard deviations from the input value.
About 100 million events generated with the DPM generator are studied with the same analysis strategies as applied to the signal events. This study of background events shows that the hadronic background is sufficiently suppressed since no event out of 100 million generated events was accepted. The usage of open PID does not reduce the purity and the signal-to-background ratio significantly.

6 Outlook

This first study, although still based on some simplifications, shows already the promising perspective of investigating Ξ resonances with $\overline{\text{P}}$ANDA. One of these simplifications is the usage of an ideal tracking algorithm, as long as a realistic tracking algorithm is not yet available. As soon as a tracking algorithm for secondary particles will be available this study should be redone.
Another point is that in the analysis with the sequential fit procedure ideal particle identification (PID) was used which leads to the fact, that the results cannot be directly compared. Therefore, the analysis should be redone with open PID.
It is already mentioned, that the angular distribution of the reaction $\bar{p}p \to \overline{\Xi}^+ \,\Xi^*$ is unknown. Therefore, an isotropic angular distribution is used for the event generation. This simplification should be replaced by a more realistic approach based on future theoretical studies.
As discussed in Section 1 the information on Ξ resonances is poor. In case of both $\Xi(1690)^-$ and $\Xi(1820)^-$, only tentative assignments of spin and parity are available. In order to explore the feasibility of determining these quantum numbers, a partial wave analysis of the reaction $\bar{p}p \to \overline{\Xi}^+ \,\Lambda\, K^-$ and $\bar{p}p \to \Xi^- \,\bar{\Lambda}\, K^+$ is required. First studies in this respect are already ongoing.

References

[1] The Electron Stretcher Accelerator (ELSA), https://www-elsa.physik.uni-bonn.de/elsa-facility_en.html
[2] The Mainz Microtron (MAMI), https://www.blogs.uni-mainz.de/fb08-nuclear-physics/accelerators-mami-mesa/the-mainz-microtron
[3] Thomas Jefferson National Accelerator Facility (JLab), https://www.jlab.org
[4] M. Tanabashi, *et al.* (Particle Data Group), Phys. Rev. **D 98**, 030001 (2018)
[5] W. Erni, *et al.*, arXiv preprint, arXiv:0903.3905 (2009)
[6] D. J. Lange, Nucl. Instr. Meth. A **462**, 152 (2001)
[7] V. Flaminio, *et al.*, CERN-HERA-**84-01** (1984)
[8] S. Agostinelli, *et al.*, Nucl. Instr. and Meth. in Phys. Res. A **506**, 250–303 (2003)
[9] S. Spataro, Journal of Physics: Conference Series **331**, 032031 (2011)
[10] A. Fontana *et al.*, Journal of Physics: Conference Series **119**, 032018 (2008)
[11] W. D. Hulsbergen, Nucl. Instr. and Meth. in Phys. Res. A **552**, 566–575 (2005)
[12] J. Pütz, PhD Thesis, Rheinische Friedrich-Wilhelms-Universität Bonn (2020)
[13] A. Capella, NATO ASI Series B: Physics **333**, 287–310 (1994)
[14] B. Musgrave *et al.*, Il Nuovo Cimento **35**, 735 (1965)

EPJ Web of Conferences **241**, 03005 (2020) http://doi.org/10.1051/epjconf/202024103005
NSTAR 2019

From Experimental Data to Pole Parameters in a Model Independent Way
(Fixed-t Single-Energy Partial Wave Analysis (SE PWA) and Laurent + Pietarinen Model (L+P) for Pole Extraction)

Alfred Švarc[1,2,*]

[1]Rudjer Bošković Institute, Bijenička cesta 54, P.O. Box 180, 10002 Zagreb, Croatia
[2]TESLA BIOTECH d.o.o., Mandlova 7, 10000 Zagreb, Croatia

Abstract. As it has been shown in the invited talk at the NSTAR-2017 conference [1] it is in principle possible to extract pole parameters directly from experiment with minimal model dependence, and in this contribution one way to achieve it in practice by fixed-t analyticity has been demonstrated. Namely, unconstrained partial-wave amplitudes obtained at discrete energies from fits even to complete sets of independent observables which are required to uniquely reconstruct reaction amplitudes do not vary smoothly with energy, and are in principle non-unique. We have demonstrated that this behavior can be ascribed to the continuum ambiguity. We have applied continuum ambiguity invariance to pseudo-scalar meson photoproduction and showed that for a complete set of pseudo-data the non-uniqueness effect can be removed through a phase rotation generating "up-to-a-phase" unique set of single energy partial wave amplitudes. Now we show that for real data this method does not work, but another approach - fixed-t analyticity solves the problem. We present the final results for the η and π^0 -photoproduction [2, 3]. Extracting pole positions from partial wave amplitudes is the next step, and we summarize the essence of the new, Laurent + Pietarinen expansion method applicable for continuous and discrete data. It is based on applying the Laurent decomposition of partial wave amplitude, and expanding the non-resonant background into a power series of a conformal-mapping-generated, quickly converging power series obtaining the simplest analytic function with well-defined partial wave analytic properties which fits the input. Unifying both methods in succession, one constructs a model independent procedure to extract pole parameters directly from experimental data without referring to any theoretical model.

1 Introduction

Partial-wave analysis, a textbook method to identify resonances and attribute them quantum numbers, is a standard procedure used to analyze a wide class of experimental data (see for instance ref. [4]). In the quest for a unique set of partial-wave amplitudes, experimental programs worldwide have attempted to measure complete sets of observables needed to perform the unambiguous reaction-amplitude reconstruction. Programmatic studies of photoproduction experiments at Jefferson Lab [5], Mainz [6] and Bonn [7] are now producing the data required to do complete experiments, in terms of either helicity amplitudes or multipoles, motivating a reexamination of the ambiguities associated with multipole analyses.

It is well known that single-channel physical observables remain invariant with respect to a general energy- and angle-dependent phase rotation which turns out to introduce the non-uniqueness into the amplitude reconstruction. This invariance is the so-called continuum ambiguity, and has been extensively analyzed in mid-70s through the mid-80s [8–10]. The interconnection between continuum ambiguity and partial wave analysis has never been extensively discussed as most studies were made in the context of πN elastic scattering where the optical theorem and an application of elastic unitarity practically eliminate the continuum ambiguity as a source of non-uniqueness. In the inelastic domain, up to now, the main attention has been paid only to handling angle independent phase rotations on the level of partial waves where the analytic structure remains untouched [11–13].

It is a general knowledge that as a direct consequence of continuum ambiguity the unconstrained partial-wave amplitudes obtained at discrete energies even from fits to complete sets of eight independent observables [14] which are required to uniquely reconstruct reaction amplitudes do not vary smoothly with energy, and are in principle non-unique [15]. In [1] it has been shown that for a complete set of pseudo-data this problem can be easily solved by simple phase rotations. However, in this paper we show that such an approach is not valid for the real data, and we propose another method based on imposing fixed-t analyticity which solves the problem.

Extracting poles from partial waves is yet another issue. Up to now, there was no reliable way to extract pole parameters from SE partial waves, but a new and simple single-channel method (Laurent + Pietarinen expansion) applicable for continuous and discrete partial-wave data has been recently developed [16–18]. It is based on applying the Laurent decomposition of partial wave amplitude, and expanding the non-resonant background into a power series of a conformal-mapping, quickly converging power series obtaining the simplest analytic function with well-defined partial wave analytic properties which fits the input. The method is particularly useful to analyse partial wave data obtained directly from experiment because it works with minimal theoretical bias since it avoids con-

*e-mail: svarc@irb.hr

structing and solving elaborate theoretical models, and fitting the final parameters to the input, which is the only alternative standard procedure employed nowadays. The generalization of this method to multi- channel case is also developed and presented.

Unifying both methods in succession, one constructs a model independent procedure to extract pole parameters directly from experimental data without referring to any theoretical model.

2 Unconstrained SE partial wave analysis (SE-PWA)

In ref. [1] it has been in details shown that unconstrained SE-PWA for η-photoproduction is discontinuous in energy, but for the convenience of the reader let us repeat the essence.

When we perform an unconstrained SE-PWA of a complete set of pseudo-data generated from the ETA-MAID15a model [19] and put initial values to 10 % randomized ETA-MAID15a model solutions, helicity amplitudes at fixed energy and fixed angle which we obtain are shown in Fig. 1.

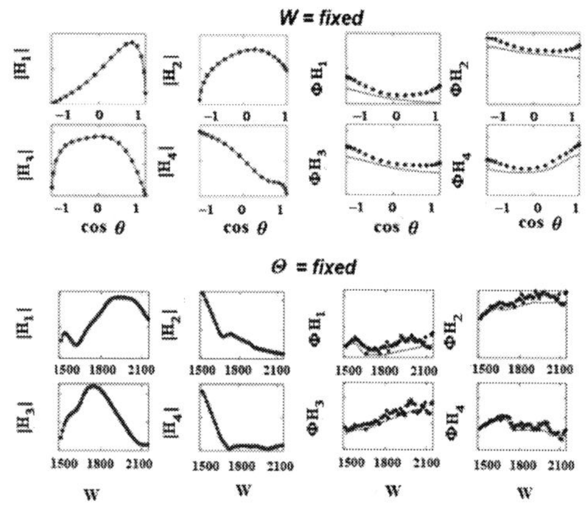

Figure 1: Absolute values and phases of helicity amplitudes at fixed energy and fixed angle obtained in unconstrained SE-PWA for a complete set of MAID pseudo-data [19]. Full lines are predictions of ETA-MAID15a model, and full dots are the result of unconstrained SE-PWA. The absolute scale is unimportant to see the effect.

We see that absolute values are, as expected, always reproduced, while phase is not; instead it is different and continuous at the fixed energy, but different and discontinuous at fixed angle. In both cases the reason is that no connection between phases at neighbouring energies is imposed. However, for the fixed energy, analyticity introduced through partial wave decomposition guarantees continuity, but for fixed angle no conditions whatsoever are imposed, so the result is even discontinuous.

In ref. [1] we have explicitly shown that the main reason for the problem is the unspecified and free phase at neighbouring energies, and that rotating that phase at each energy to a continuous phase solves the problem. We introduce the following angle-dependent phase rotation simultaneously for all four helicity amplitudes:

$$\tilde{H}_k^{SE}(W,\theta) = H_k^{SE}(W,\theta)\cdot e^{i\Phi^{15a}_{H_N}(W,\theta) - i\Phi^{SE}_{H_N}(W,\theta)}$$
$$k = 1,\ldots,4 \qquad (1)$$

where H_N is any out of four helicity amplitudes, $\Phi^{SE}_{H_N}(W,\theta)$ is the single-energy phase and $\Phi^{15a}_{H_N}(W,\theta)$ is the continuous phase, in this case the phase of the generating solution ETA-MAID15a.

However, this method simply does not work for the real data. If we take the data base from ref. [2], and make the phase rotation with H_1 defined by Eq. (1) we obtain:

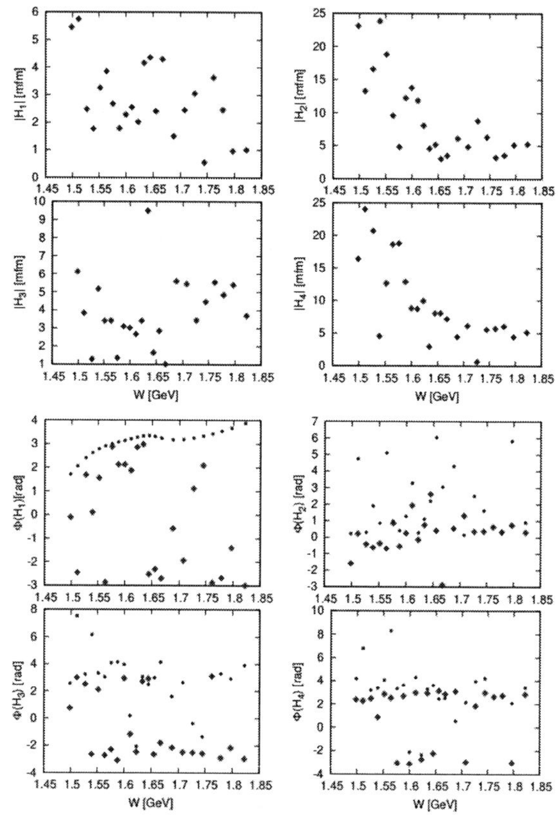

Figure 2: (Color online) Helicity amplitudes at fixed angle for the real data of ref.[2] unconstrained, before the rotation (red symbols) and after the phase rotation with H_1 defined by Eq. (1) (blue symbols).

We see that the only smooth function is the phase of helicity amplitude H_1 after the rotation. Absolute values are identical before and after the rotation as they should be, but they are discontinuous as are the phases of helicity amplitudes H_2 - H_4. The reasons for that are numerous: the used data base is certainly incomplete, but errors in some observables are also rather big, and some data also might be mutually incompatible (not proven yet). So, the method used for pseudo-data for real data does not work.

3 Fixed-t analyticity

Fixed-t analyticity, the method developed by G. Hoehler in Karlsruhe [20], developed for η photoproduction in ref. [2], and with great success applied to π-nucleon scattering in ref. [3], can easily be applied for imposing the continuity of phase. It is based on the idea that imposing analyticity in t automatically imposes continuity in both, energy and angle as t is in general a function of energy and angle. This is illustrated in Fig. 3 where Mandelstam (W,t) graph is shown for η-photoproduction. W is center of mass energy and t is standard Mandelstam variable for η photoproduction defined in ref. [2].

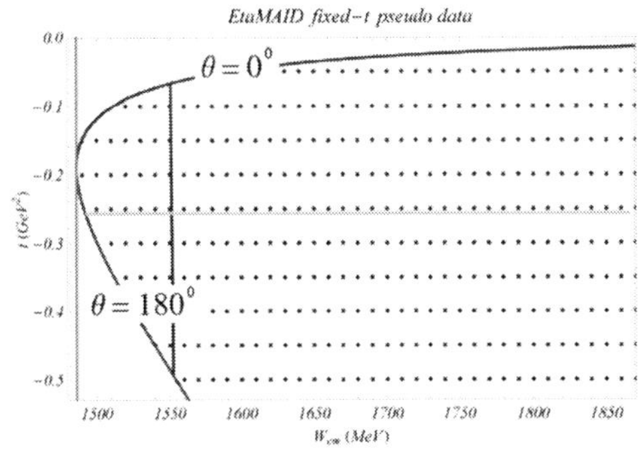

Figure 3: (Color online) Mandelstam (W,t) graph is shown for η-photoproduction.

Fixed-W PWA is performed by using the data along the red vertical line, and there is no connection between phases at nearby energies, and as it was demonstrated it causes the discontinuity in energy variation. However, fixed-t PWA is performed along the yellow horizontal line, and analyticity in t automatically imposes continuity on both, angle and energy. So nearby energies are treated simultaneously, and continuity in energy is automatically achieved. Therefore, a two step procedure was invented [2, 20]:

Step 1:

Obtaining a fit of an analytic function to all experimental data at fixed–t which automatically creates continuity in energy and angle.

However, after *Step 1* we are still model dependent and are still at the level of reaction amplitudes and DO NOT have partial waves yet! So, we have to introduce *Step 2*.

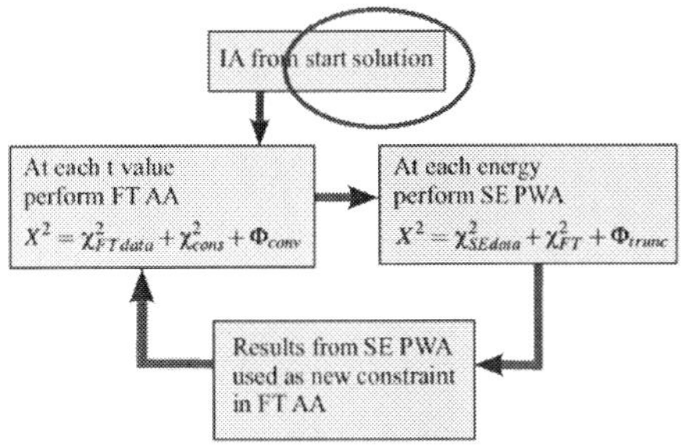

Figure 4: Iterative scheme for the fixed-t minimization. Circle designates the only step of the process which depends on some theoretical assumptions as ETA-MAID15a theoretical parameters are taken as start values for the first minimization. Figure is taken over from ref. [2].

Step 2:

Create partial waves through a constrained SE PWA using continuity in energy and angle obtained in Step 1 as a penalty function.

In more details, we have to perform the following actions:

Step 1:

- We create the data base in the form $O_i = f(W_{fixed}, t)$ from the existing experimental data base $O_i = f(W_{fixed}, \Theta)$.
- Instead of modeling reaction amplitudes as function of W and t we expand them using Pietarinen variable (Pietarinen expansion [16]).
- We fit the artificial data base $O_i = f(W_{fixed}, t)$ with these expansions (each reaction amplitude) by fitting coefficients of the Pietarinen expansion, and starting with coefficients of the Pietarinen expansion of the initial solution.
- In this way and obtain a result which automatically analytic in energy and angle, and optimally describes the input.

Step 2:

We want to obtain continuous SE PWA. This means we have to ensure that neighboring points in energy belong to the same solution (solution having the same phase)! We do it via PENALTY FUNCTION TECHNIQUE.

$$X^2 = \chi^2_{\text{SEdata}} + \chi^2_{\text{FT}} \quad (2)$$

where χ^2_{FT} is a constraining term given by

$$\chi^2_{\text{FT}} = q_{\text{cons}} \sum_{k=1}^{4} \sum_{i=1}^{N^C} \left[\frac{\text{Re } H_k(E,\theta_i)^{\text{fit}} - \text{Re } H_k(E,\theta_i)^{\text{FT}}}{\varepsilon^{\text{Re}}_{k,i}} \right]^2 + q_{\text{cons}} \sum_{k=1}^{4} \sum_{i=1}^{N^C} \left[\frac{\text{Im } H_k(E,\theta_i)^{\text{fit}} - \text{Im } H_k(E,\theta_i)^{\text{FT}}}{\varepsilon^{\text{Im}}_{k,i}} \right]^2$$

where FT subscript indicates that that the corresponding amplitude is a solution of a fixed-t Pietarinen fit, and q_{cons} is a tunable penalty-factor.

The whole process is implemented using the iteration scheme depicted in Fig. 4.

So, the process effectively goes like this:

First we perform the minimization in t. To do so, we create the artificial data base at fixed-t. We do it by SPLINE3 fit of existing experimental data to obtain values of observables at the values of angle which is not actually measured, but which is needed for the chosen and fixed value of t at various energies. Then we identify the fitted value with the value of that observable at chosen t. This, somewhat artificial data base, is then fitted with a specific form of reaction amplitudes. In our method we use invariant amplitudes, which are calculated from standard helicity amplitudes. To impose analyticity in t, we use the conformal mapping generated Pietarinen expansion [2]. Namely, to claim that the real and imaginary part of a function belong to the same analytic function, one has to know the value of the function on the complete real axes in order to be able to calculate the dispersion integral. However, this is never the case for the measured data. So, we are using the following trick: when we fit the data with a Pietarinen expansion of reaction amplitudes, then real and imaginary part by construction satisfy dispersion relations. So, when our fitting parameters are taken to be Pietarinen coefficients, the analyticity is automatically imposed. However, determining initial values of Pietarinen coefficients for reaction amplitudes is non-trivial, and this is the only model-dependent step of our procedure. We take reaction amplitudes from a certain theoretical model (MAID-[21], Bonn-Gatchina-[22],Juelich-Bonn-[23],...), and expand them into Pietarinen series. The obtained coefficients are the starting values of our minimization. We perform the minimization and obtain a set of reaction amplitudes which at the same time fit the data, and are analytic in t, hence continuous in W and Θ. The next step is fixed-W PWA constrained with the analytic and continuous reaction amplitudes in t, coming from the Step 1. Constraining is defined by set of equations (2) via penalty function technique. The idea is to perform a fit to the data, and obtain a good representation of the data, but at the same time require that the helicity amplitudes be as close to the analytic amplitudes obtained in Step 1 as possible. In that way we get a *reasonable* representation of data, and *approximate* analyticity. However, we do not expect that neither the ideal agreement with the data, nor exact analyticity are achieved immediately. Therefore, we introduce iterations. We return the obtained set of helicity amplitudes into Step 1 as initial amplitudes, we fit these amplitudes with Pietarinen expansion, and repeat the Step 1, but this time with amplitudes which are closer to the fitted data set, than energy dependent (ED) model amplitudes of first step are. Then we perform the fit, and go into Step 2 with the obtained result. We expect that in the second step, the obtained result will be closer to the data, and satisfy the analyticity properties better than in first step. And we repeat the procedure as long as we do not get a stable result. At the end of our procedure we obtain the amplitudes (or partial waves) which are as close to the data as possible, and and the same time almost analytic with the analyticity imposed in Step 1. However, let us observe that there is still is some model dependence is Step 1 as we have prescribed a definite, very simple, form of analyticity by introducing Pietarinen expansion. And with this procedure we create a final set of partial waves. We give the final set of partial waves obtained with this procedure for η-photoproduction in Fig. 5 (see ref. [2]).

In this result we see four set of partial waves obtained by using two theoretical models to generate initial values: MAID-[21] and Bonn-Gatchina-[22]. Observe that we have two groups of solutions which are obviously different. The reason is that the overall phase of partial wave decomposition for the used models is notably different, and it influences partial waves. For details see ref. [1]. The main message of that paper is that if we want to compare partial waves, we have to match the overall phase, and that has to be done correctly, on the level of amplitudes and not on the level of partial waves. Therefore, we have "rotated" the solution corresponding to the Bonn-Gatchina initial phase to the MAID phase, and solution corresponding to the MAID initial phase to Bonn-Gatchina phase. So, we have four solutions: 1. MAID as initial model with the MAID phase (black squares); 2) MAID as initial model rotated to Bonn-Gatchina phase (green inverted triangles); 3) Bonn-Gatchina as initial model at the Bonn-Gatchina phase (blue triangles); and 4) Bonn-Gatchina as initial model rotated to MAID phase (red circles). We see that in most of partial waves we

see a good agreement between MAID and Bonn-Gatchina results at the same phase, with the exception of one: there is a notable discrepancy in M1- multipole. So, we need new measurements to resolve the uncertainty. This, as seen in Fig. 6 at the randomly chosen and representative energy of W=1600 MeV, shows that presently used data set is incomplete, and that we need more measurements to clarify the difference. We see that in mea-

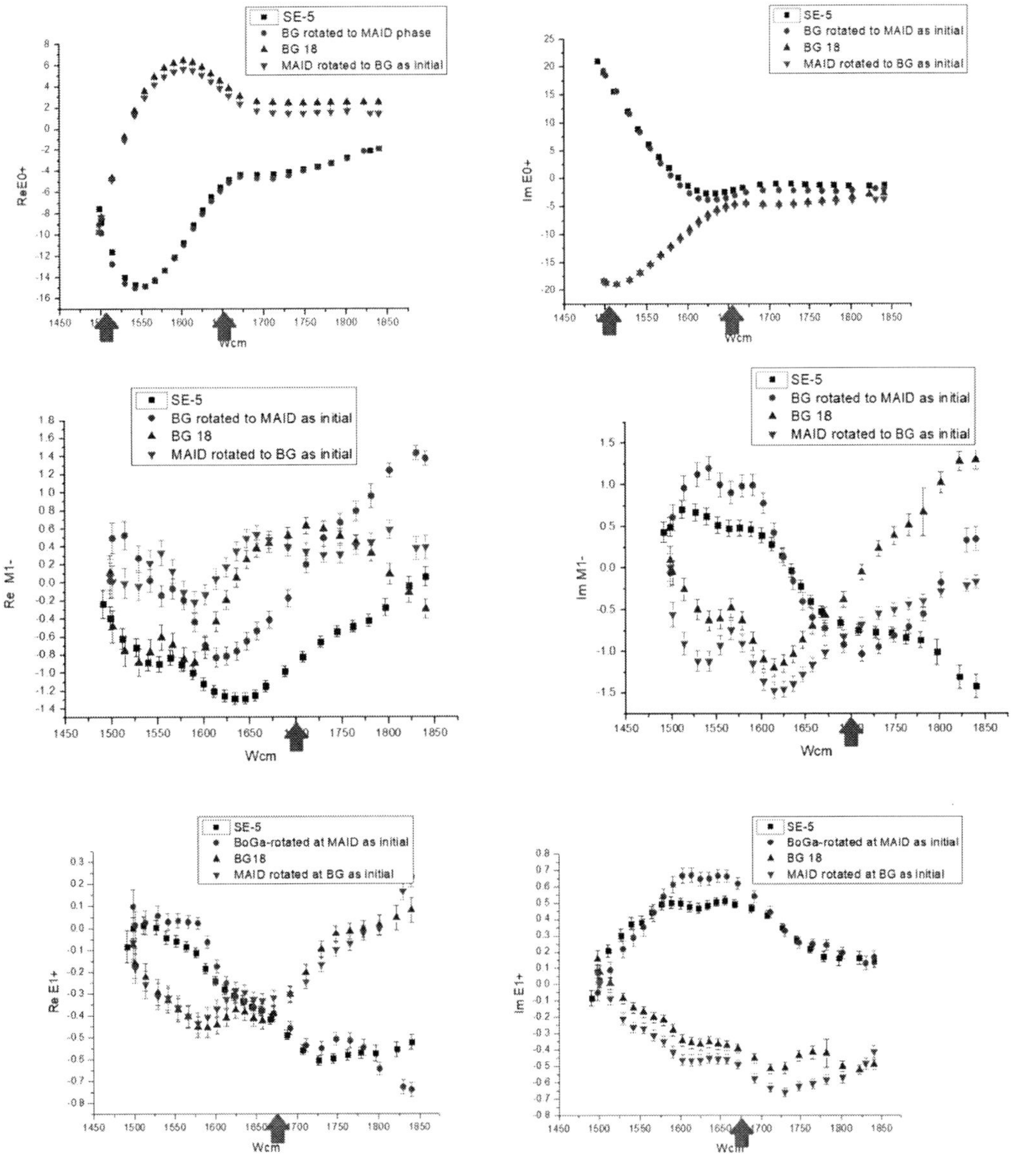

Figure 5: (Color online) Final set of η-photoproduction partial wave amplitudes obtained in ref. [2] for the lowest multipoles.

sured observables $d\sigma/d\Omega$, F, T and Σ all four SE solutions agree, while the difference in M1- multipole is notably seen in observables P, G, H and O_z. Therefore, new measurements are needed.

The situation with overall phase ambiguity is very different for π^0 photoproduction. In η photoproduction the overall phase is free, and fixed-t analyticity fixes the phase of SE solution to always different phases determined by the phase of initial solution. So, all four generated SE solutions look notably different. For pion photoproduction, the phase of all four initial solutions, which is determining the phase of the final SE solution, is fixed by the Watson's theorem [24], and the whole discrepancy can be solely attributed to the data uncertainty. Let us briefly elaborate on that. Since the electromagnetic interaction is in pion photoproduction treated to the first order in the coupling constant, the complexity of the structure functions is completely due to the pion-nucleon interaction. Therefore, each partial wave component F_i^{α} of the structure functions may be written in the form

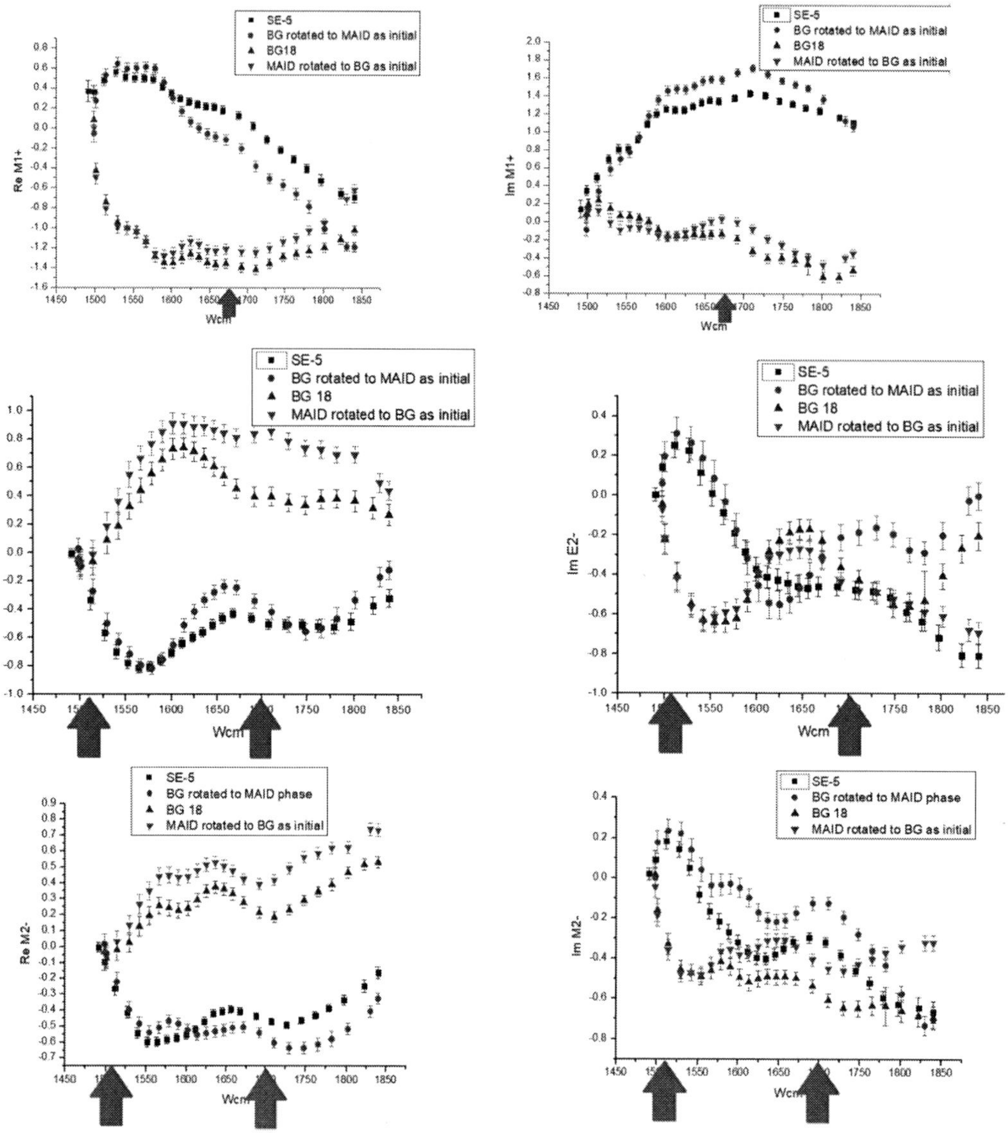

Figure 6: (Color online) Final set of η-photoproduction partial wave amplitudes obtained in ref. [2] for higher multipoles.

$F_i^{\alpha} = e^{i\delta_{\alpha}} R_i^{\alpha}$, where R_i^{α} is a real function of the kinematical values and δ_{α} is the phase shift for elastic πN scattering in that particular channel α. Therefore, the phase of pion photoproduction is at lower energies determined by the phase of pion elastic scattering, and that is the condition which is obeyed by all ED solutions. Hence, this unique phase is in all four cases transferred to all four presented SE solutions, and they all must have very similar phase. Therefore, the strong similarity of not only dominant, but of all partial waves in pion photoproduction at lower energies strongly indicate that the data base is here almost complete, but only almost. Some differences among all four solutions, however, indicate that some improvements, as we shall demonstrate later, are still possible.

Therefore, in π^0 photoproduction there is no ambiguity in the overall phase at energies lower than 1650 MeV. In Fig. 7 we show a set of fixed-t solutions for pion photoproduction when MAID-[21], Bonn-Gatchina-[22], Jülich-Bonn-[23] and SAID-[25] are taken as initial solutions in the first step of iterative procedure.

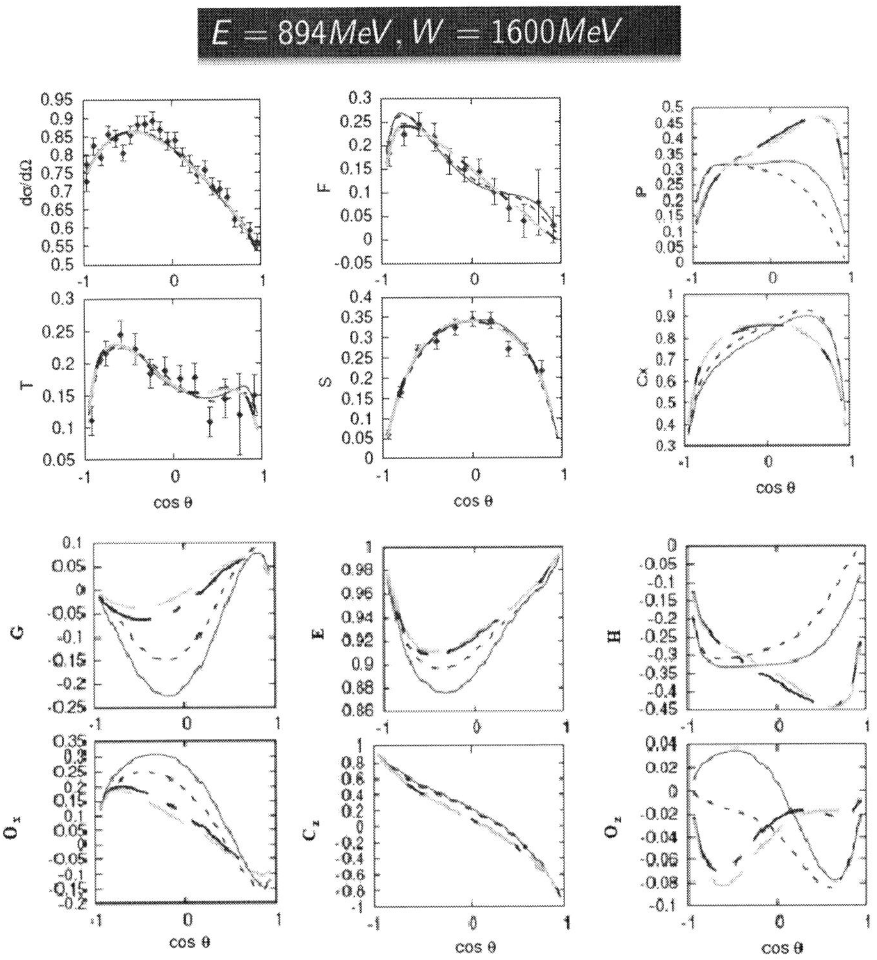

Figure 7: (Color online) The comparison of four obtained SE solutions with measured and unmeasured observables. The colors of the curves shown here correspond to the colors of the single-energy multipoles shown in Figs 5 and 6.

4 Laurent + Pietarinen expansion

The driving concept behind the Laurent-Pietarinen (L+P) expansion was the aim to replace an elaborate theoretical model by a local power-series representation of partial wave amplitudes [16]. The complexity of a partial-wave analysis model is thus replaced by much simpler model-independent expansion which just exploits analyticity. The L+P approach separates pole and regular part in the form of a Laurent expansion, and instead of modeling the regular part in some physical model it uses the conformal mapping to expand it into a rapidly converging power series with well defined analytic properties. So, the method replaces the regular part calculated in a model by the simplest analytic function which has correct analytic properties of the analyzed partial wave (multipole), and fits the data. In such an approach the model dependence is minimized, and is reduced to the choice of the number and location of branch-points used in the L+P expansion.

The L+P expansion is based on the Pietarinen expansion used in some former papers in the analysis of pion-nucleon scattering data [27–30], but for the L+P model the Pietarinen expansion is applied in a different manner. It exploits the Mittag-Leffler expansion[1] of partial wave amplitudes near the real energy axis, representing the regular, but unknown, background term by a conformal-mapping-generated, rapidly converging power

[1] Mittag-Leffler expansion [31]. This expansion is the generalization of a Laurent expansion to a more-than-one pole situation. For simplicity, we will simply refer to this as a Laurent expansion.

 http://doi.org/10.1051/epjconf/202024103005

Table 1: Formulae defining the Laurent+Pietarinen (L+P) expansion.

$$T^a(W) = \sum_{j=1}^{N_{pole}} \frac{x_j^a + \imath\, y_j^a}{W_j - W} + \sum_{k=0}^{K^a} c_k^a\, X^a(W)^k + \sum_{l=0}^{L^a} d_l^a\, Y^a(W)^l + \sum_{m=0}^{M^a} e_m^a\, Z^a(W)^m$$

$$X^a(W) = \frac{\alpha^a - \sqrt{x_P^a - W}}{\alpha^a + \sqrt{x_P^a - W}}; \quad Y^a(W) = \frac{\beta^a - \sqrt{x_Q^a - W}}{\beta^a + \sqrt{x_Q^a - W}}; \quad Z^a(W) = \frac{\gamma^a - \sqrt{x_R^a - W}}{\gamma^a + \sqrt{x_R^a - W}}$$

$$D_{dp}^a = \frac{1}{2\,N_W^a - N_{par}^a} \sum_{i=1}^{N_W^a} \left\{ \left[\frac{\mathrm{Re}\, T^a(W^{(i)}) - \mathrm{Re}\, T^{a,exp}(W^{(i)})}{Err_{i,a}^{\mathrm{Re}}} \right]^2 + \left[\frac{\mathrm{Im}\, T^a(W^{(i)}) - \mathrm{Im}\, T^{a,exp}(W^{(i)})}{Err_{i,a}^{\mathrm{Im}}} \right]^2 \right\} + \mathcal{P}^a$$

$$\mathcal{P}^a = \lambda_c^a \sum_{k=1}^{K^a} (c_k^a)^2\, k^3 + \lambda_d^a \sum_{l=1}^{L^a} (d_l^a)^2\, l^3 + \lambda_e^a \sum_{m=1}^{M^a} (e_m^a)^2\, m^3 \qquad D_{dp} = \sum_a^{all} D_{dp}^a$$

a ..... channel index N_{pole} ..... number of poles $W_j, W \in \mathbb{C}$

$x_i^a,\, y_i^a,\, c_k^a,\, d_l^a,\, e_m^a,\, \alpha^a,\, \beta^a,\, \gamma^a ... \in \mathbb{R}$

$K^a,\, L^a,\, M^a ... \in \mathbb{N}$ number of Pietarinen coefficients in channel a.

D_{dp}^a ..... discrepancy function in channel a N_W^a ..... number of energies in channel a

N_{par}^a ..... number of fitting parameters in channel a $\mathcal{P}^a$ ..... Pietarinen penalty function

$\lambda_c^a,\, \lambda_d^a,\, \lambda_e^a$ ..... Pietarinen weighting factors $x_P^a,\, x_Q^a,\, x_R^a \in \mathbb{R}$ (or $\in \mathbb{C}$).

$Err_{i,a}^{\mathrm{Re,\,Im}}$ ..... minimization error of real and imaginary part respectively.

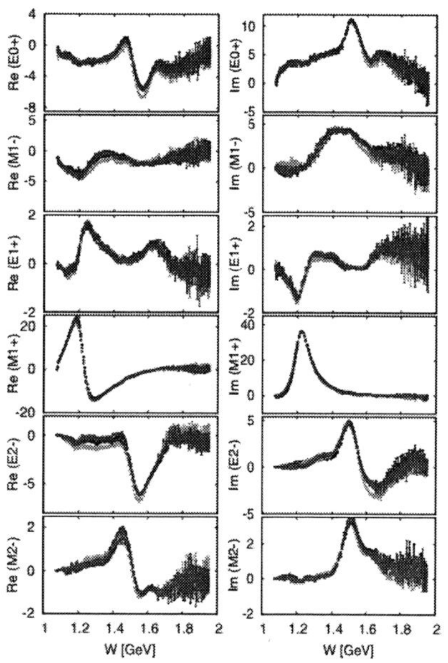

Figure 8: (Color online) SE solutions for the few lowest partial waves obtained using different models as initial solutions (BnGa (black),JuBo (blue), SAID (red) and MAID (green)). Multipoles are in units of mfm. Figure is taken over from ref. [3].

series called a Pietarinen expansion[2]. The method was used successfully in several few-body reactions [17, 18, 32], and recently generalized to the multi-channel case [33]. The formulae used in the L+P approach are collected in Table 1.

In the fits, the regular background part is represented by three Pietarinen expansion series, all free parameters are fitted. The first Pietarinen expansion with branch-point x_P is restricted to an unphysical energy range and represents all left-hand cut contributions. The next two Pietarinen expansions describe the background in the physical range with branch-points x_Q and x_R respecting the analytic properties of the analyzed partial wave. The second branch-point is mostly fixed to the elastic channel branch-point, the third one is either fixed to the dominant channel threshold, or left free. Thus, only rather general physical assumptions about the analytic properties are made like the number of poles and the number and the position of branch-points, and the simplest analytic function with a set of poles and branch-points is constructed. The method is applicable to both, theoretical and experimental input, and represents the first reliable procedure to extract pole positions from experimental data, with minimal model bias.

The generalization of the L+P method to a multi-channel L+P method is performed in the following way: i) separate Laurent expansions are made for each channel; ii) pole positions are fixed for all channels, iii) residua and Pietarinen coefficients are varied freely; iv) branch-points are chosen as for the single-channel model; v) the single-channel discrepancy function D_{dp}^a (see Eq. (5) in ref. [32]) which quantifies the deviation of the fitted function from the input is generalized to a multi-channel quantity D_{dp} by summing up all single-channel contributions, and vi) the minimization is performed for all channels in order to obtain the final solution.

[2] A conformal mapping expansion of this particular type was introduced by Ciulli and Fisher [27, 28], was described in detail and used in pion-nucleon scattering by Esco Pietarinen [29, 30]. The procedure was denoted as a Pietarinen expansion by G. Höhler in [20].

5 Conclusions

Unconstrained SE PWA is discontinuous because the continuity of continuum ambiguity phase at neighbouring energies is not ensured. One of the ways to achieve the continuity in energy is imposing fixed-t analyticity. We present four sets of solutions for η-photoproduction, and four sets of solutions for π^0-photoproduction are presented. Importance of Watson theorem for fixing the phase is stressed for π^0-photoproduction. New methods of imposing the continuity in phase with the final result of getting a continuous PW solution are under development.

Acknowledgment

This work was supported by the Deutsche Forschungsgemeinschaft (SFB 1044).

References

[1] A. Švarc, Y. Wunderlich, H. Osmanović, M. Hadžimehmedović, R. Omerović, J. Stahov, V. Kashevarov, K. Nikonov, M. Ostrick, L. Tiator, and R. Workman, Few-Body Syst (2018) 59:96.

[2] H. Osmanović, M. Hadžimehmedović, R. Omerović, J. Stahov, V. Kashevarov, K. Nikonov, M. Ostrick, L. Tiator, and A. Švarc, Phys. Rev. **C 97**, 015207 (2018).

[3] A. Švarc, H. Osmanović, M. Hadžimehmedović, R. Omerović, J. Stahov, M. Gorchteyn, and V. Kashevarov, K. Nikonov, M. Ostrick, L. Tiator, nucl-th-arXiv:1908.05167, Phys. Rev. **C 100**, 055203 (2019).

[4] A.D. Martin and T.D. Spearman: Elementary Particle Theory, North-Holland Publishing Company, Amsterdam 1970.

[5] https://www.jlab.org/physics

[6] https://www.blogs.uni-mainz.de/fb08-nuclear-physics/accelerators-mami-mesa/the-mainz-microtron/

[7] https://www-elsa.physik.uni-bonn.de/index.html

[8] D. Atkinson, P.W. Johnson and R.L. Warnock, Commun. mat. Phys. **33** (1973) 221.

[9] J.E. Bowcock and H. Burkhard, Rep. Prog. Phys. **38** (1975) 1099.

[10] D. Atkinson and I.S. Stefanescu, Commun. Math. Phys. **101**, 291 (1985).

[11] A.V. Anisovich, R. Beck, E. Klempt, V.A. Nikonov, A.V. Sarantsev, U. Thoma, Eur. Phys. J. **A48**, 15 (2012).

[12] L. Tiator, D. Drechsel, S. S. Kamalov and M. Vanderhaeghen, Eur. Phys. J. ST **198**, 141 (2011).

[13] A. M. Sandorfi, S. Hoblit, H. Kamano, and T.-S. H. Lee, J. Phys. G: Nucl. Part. Phys. **38** (2011) 053001.

[14] W. T. Chiang and F. Tabakin, Phys. Rev. C **55**, 2054 (1997).

[15] A. Švarc, Y. Wunderlich, H. Osmanović, M. Hadžimehmedović, R. Omerović, J. Stahov, V. Kashevarov, K. Nikonov, M. Ostrich, L. Tiator, R. Workman, Phys. Rev. **C 97**, 054611 (2018).

[16] A. Švarc, M. Hadzimehmedovic, H. Osmanovic, J. Stahov, L. Tiator, and R. L. Workman, Phys, Rev. **C88**, 035206 (2013).

[17] A. Švarc, M. Hadzimehmedovic, R. Omerovic, H. Osmanovic, and J. Stahov, Phys, Rev. **C89**, 0452205 (2014).

[18] A. Švarc, M. Hadzimehmedovic, H. Osmanovic, J. Stahov, L. Tiator, and R. L. Workman, Phys, Rev. **C89**, 65208 (2014).

[19] V. L. Kashevarov, L. Tiator, M. Ostrick, Bled Workshops Phys., **16**, 9 (2015).

[20] G. Höhler, *Pion Nucleon Scattering*, Part 2, Landolt-Bornstein: Elastic and Charge Exchange Scattering of Elementary Particles, Vol. 9b (Springer-Verlag, Berlin, 1983).

[21] L. Tiator, M. Gorchteyn, V. L. Kashevarov, K. Nikonov, M. Ostrick, M. Hadžimehmedović, R. Omerović, H. Osmanović, J. Stahov, and A. Švarc v, arXiv:1807.04525v1 [nucl-th], and references therein.

[22] A. V. Anisovich *et al.*, Phys. Rev. C **96**, 055202 (2017), and references therein.

[23] D. Rönchen, M. Döring, and U.-G. Meißner, arXiv:1801.10458v1 [nucl-th], and references therein.

[24] K. M. Watson, Phys. Rev. **95**, 228 (1954).

[25] For SAID solutions GE09 and E429 see Ref. [26].

[26] E. F. McNicoll *et al.* [Crystal Ball at MAMI Collaboration], "Study of the $\gamma p- > \eta p$ reaction with the Crystal Ball detector at the Mainz Microtron(MAMI-C)," Phys. Rev. C **82**, 035208 (2010). Erratum: [Phys. Rev. C **84**, 029901 (2011)]

[27] S. Ciulli and J. Fischer in Nucl. Phys. **24**, 465 (1961).

[28] I. Ciulli, S. Ciulli, and J. Fisher, Nuovo Cimento **23**, 1129 (1962).

[29] E. Pietarinen, Nuovo Cimento Soc. Ital. Fis. **12A**, 522 (1972).

[30] E. Pietarinen, Nucl. Phys. **B107**, 21 (1976).

[31] Michiel Hazewinkel: *Encyclopaedia of Mathematics*, Vol.6, Springer, 31. 8. 1990, pg.251.

[32] A. Švarc, M. Hadžimehmedović, H. Osmanović, J. Stahov, and R. L. Workman, Phys. Rev. **C91**, 015207 (2015).

[33] A. Švarc, M. Hadžimehmedović, H. Osmanović, J. Stahov, L. Tiator, R. L. Workman, Phys. Lett. **B755** (2016) 452–455.

Determining dominant partial waves in photoproduction via moment analysis

Yannick Wunderlich[1,*]

[1]Universität Bonn, Helmholtz-Institut für Strahlen- und Kernphysik, 53115 Bonn, Germany

Abstract. Important insights into the excitation spectra of baryons are provided by measurements of polarization observables in reactions that involve particles with spin. The photoproduction of a single pseudoscalar meson constitutes an example-reaction that has been under intense investigation recently. We present the basic method of moment-analysis for pseudoscalar meson photoproduction, in which just the angular distributions are analyzed. Using this method, the total angular momentum quantum number of the dominant partial waves contributing in the data can be extracted quickly. Furthermore, the Legendre-coefficients extracted from the angular distributions show interesting composition-patterns in terms of multipoles and allow for instructive comparisons to models. In this contribution, recent results for moment analyses of polarization data for the photoproduction of pions and eta-mesons are shown.

In baryon spectroscopy, the extraction of resonance parameters from scattering data represents the most important central problem. A standard-approach to the solution of this problem, which has taken center stage over the last 50 years, is given by fits of so-called energy-dependent (ED) partial-wave analysis (PWA-) models. Well-known examples for such approaches are the Bonn-Gatchina (BnGa-) model [1], the Jülich-Bonn (JüBo-) model [2, 3], the MAID-analysis [4] and the SAID-PWA [5]. A comprehensive overview of these examples is given in reference [6]. In each case, a reaction-theory is constructed, which satisfies the well-established theoretical S-Matrix principles [7] (analyticity, unitarity and crossing) with varying degrees of rigor. Within the context of such a model, resonances are extracted as poles on the second Riemann-sheet of the scattering amplitude.

Another approach to analyze scattering observables is represented by single-energy (SE) fits, or truncated partial-wave analyses (TPWAs). The simplest possible example is given by $2 \rightarrow 2$-scattering of spinless particles. The only possible observable, i.e. the differential cross section σ_0, is defined in terms of the amplitude as $\sigma_0 = |A(W, \theta)|^2$. The infinite partial-wave series for the amplitude reads

$$A(W, \theta) = \sum_{\ell=0}^{\infty} (2\ell + 1) A_\ell P_\ell(\cos\theta). \qquad (1)$$

In case one truncates this series at some maximal angular momentum quantum number ℓ_{max}, one obtains the following expansion for the cross section (see reference [8])

$$\sigma_0(W, \theta) = \frac{q}{k} \sum_{n=0}^{2\ell_{\text{max}}} a_n^{\sigma_0}(W) P_n(\cos\theta), \qquad (2)$$

$$a_n^{\sigma_0}(W) = \sum_{\ell,k=0}^{\ell_{\text{max}}} A_\ell^*(W) C_{\ell k}^n A_k(W). \qquad (3)$$

Here, for the scalar example, the coupling-coefficients $C_{\ell k}^n$ in the bilinear equation (3) are well-known and given by [8, 9]

$$C_{\ell k}^n = \langle \ell, 0; k, 0 | n, 0 \rangle^2 \frac{(2\ell + 1)(2k + 1)}{(2n + 1)}. \qquad (4)$$

Here, the $\langle \ell, m; \ell', m' | n, M \rangle$ are the usual Clebsch-Gordan coefficients.

The central problem in a SE fit is then to *solve* for the real- and imaginary parts of generally phase-constrained[1] partial waves. Thus, one first extracts the Legendre-moments using equation (2). Then, one solves the bilinear equations (3) and it is this second step in which multiple discrete ambiguities [9, 10] can, and probably will, occur. Such ambiguities require additional theoretical input on some subset of the partial waves in order to be resolved. However, one could also remain, in a first instance, with the moment-analysis using (2) and try to use it in order to learn as much about the present dataset as possible.

This latter idea can be generalized to more complicated reactions involving particles with spin without a lot of additional effort. The photoproduction of a single pseudoscalar meson is one example for such a reaction. We write the process in the most general form as $\gamma N \longrightarrow \varphi B$, where φ is a pseudoscalar meson and B a recoil-baryon. The following formalism can thus be applied to the most commonly met example of pion-photoproduction $\varphi B = \pi N$, but also other reactions such as eta-photoproduction $\varphi B = \eta N$ or the production of kaons $\varphi B = K\Lambda, \ldots$ may be studied.

The photoproduction reaction is described using 4

*e-mail: wunderlich@hiskp.uni-bonn.de

[1]Some kind of constraint has to be introduced on the overall phase of the partial waves due to the bilinear nature of the equations (3).

Type	$\check{\Omega}^{\alpha}$	β_{α}	γ_{α}	Type	$\check{\Omega}^{\alpha}$	β_{α}	γ_{α}
	σ_0	0	0		$\check{O}_{x'}$	1	0
$\mathcal{S}$	$\check{\Sigma}$	2	−2	$\mathcal{BR}$	$\check{O}_{z'}$	2	−1
	$\check{T}$	1	−1		$\check{C}_{x'}$	1	0
	$\check{P}$	1	−1		$\check{C}_{z'}$	0	+1
	$\check{E}$	0	0		$\check{T}_{x'}$	2	−1
$\mathcal{BT}$	$\check{G}$	2	−2	$\mathcal{TR}$	$\check{T}_{z'}$	1	0
	$\check{H}$	1	−1		$\check{L}_{x'}$	1	0
	$\check{F}$	1	−1		$\check{L}_{z'}$	0	+1

Table 1: We list here the parameters needed to evaluate the angular parametrizations for the 16 polarization observables of pseudoscalar meson photoproduction given in equations (5) and (6). Such a Table was first published by Tiator [15] (see also references [9, 14]).

complex spin-amplitudes, for instance CGLN-amplitudes $\{F_i(W,\theta), i = 1,\ldots,4\}$ [11], accompanied by 16 polarization observables [12, 13]. The observables are defined in terms of the amplitudes as bilinear hermitean forms and they divide into the subsets of group $\mathcal{S}$ observables $\{\sigma_0, \Sigma, T, P\}$ and furthermore three classes of beam-target ($\mathcal{BT}$), beam-recoil ($\mathcal{BR}$) and target-recoil ($\mathcal{TR}$) observables with four quantities each. The CGLN-amplitudes can be expanded into a well-known partial-wave series using electric and magnetic multipoles $\{E_{\ell\pm}, M_{\ell_\pm}\}$ [11, 12]. In case one truncates the multipole-series at some $\ell_{\max}$, the observables acquire the following standard-form [9, 15], which is tantamount to the equations (2,3) in the scalar case:

$$\check{\Omega}^{\alpha}(W,\theta) = \frac{q}{k} \sum_{n=\beta_{\alpha}}^{2\ell_{\max}+\beta_{\alpha}+\gamma_{\alpha}} \left(a_{\ell_{\max}}\right)_n^{\check{\Omega}^{\alpha}}(W)\, P_n^{\beta_{\alpha}}(\cos\theta), \quad (5)$$

$$\left(a_{\ell_{\max}}\right)_k^{\check{\Omega}^{\alpha}}(W) = \left\langle \mathcal{M}_{\ell_{\max}}(W) \right| C_k^{\check{\Omega}^{\alpha}} \left| \mathcal{M}_{\ell_{\max}}(W) \right\rangle . \quad (6)$$

where we have adopted a notation by Chiang and Tabakin [13] to denote the observables $\check{\Omega}^{\alpha}$ and the index α runs as $\alpha = 1,\ldots,16$. The $P_n^{\beta_{\alpha}}(\cos\theta)$ are associated Legendre polynomials. The multipoles which are present in a certain truncation-order define the vector

$$\left|\mathcal{M}_{\ell_{\max}}\right\rangle = \left[E_{0+}, E_{1+}, M_{1+}, M_{1-}, E_{2+}, E_{2-}, \ldots, M_{\ell_{\max}-}\right]^T . \quad (7)$$

A set of parameters which defines the photoproduction moment-expansion for all 16 observables is given in Table 1 (cf. reference [14]). Tiator [15] first published similarly formalized expansions for an expansion into powers of $\cos(\theta)$.

The matrices defining the coefficients (6) as bilinear forms are hermitean and have dimensions $(4\ell_{\max}) \times (4\ell_{\max})$, for $\ell_{\max} \geq 1$. We have to report that during the course of the thesis [9], no closed expression for these matrices has been found, which would be analogous to the result (4) from the scalar reactions. However, they can be calculated numerically for each relevant truncation order $\ell_{\max}$. The results for group $\mathcal{S}$ and $\mathcal{BT}$ observables in the order $\ell_{\max} = 5$ have been collected in the appendices of [9]. As an example, the matrix $C_0^{\check{E}}$ is shown in Table 2 for the truncation order $\ell_{\max} = 2$.

It is a fact that once the matrices $C_k^{\check{\Omega}^{\alpha}}$ are evaluated, one observes that in all cases the matrixes decompose into *blocks* of interference-terms among multipoles of certain definite orbital angular momentum quantum number ℓ. Therefore, we introduce a short-notation [14] for these interference-blocks, using the spectroscopic notation of $S, P, D, \ldots$ for the angular momenta $\ell = 0, 1, 2, \ldots$. For example, a contribution coming from an interference-block

$$\sum_{\mathcal{M},\mathcal{M}'=\{E,M\}} \sum_{p,p'=\{\pm\}} c_{p,p'}^{\mathcal{M},\mathcal{M}'} \mathcal{M}_{0p}^{*} \mathcal{M}'_{1p'} . \quad (8)$$

between S- and P-wave multipoles would be just written as $\langle S, P \rangle$. Each Legendre-moment generally contains multiple such blocks and therefore, several of the short-forms for interference-blocks are added together in order to describe the partial wave decomposition of a particular moment (see Table 2).

The question whether or not specific minimal subsets of the bilinear forms (6) allow for an unambiguous amplitude extraction in so-called *complete experiments* has been addressed in the classic work by Omelaenko [16] and in the more recent references [9, 17].

The formalism outline above allows for moment-analyses in pseudoscalar meson photoproduction. The following two main aspects are important:

(I) $\ell_{\max}$-analysis: The parametrization (5) of the angular distribution is fitted for different $\ell_{\max}$. The χ^2/ndf is compared for different fits, ascending from the lowest possible order. In case the goodness of fit is unsatisfactory, one has to increase $\ell_{\max}$. Once a good fit is obtained, the resulting $\ell_{\max}$ gives, in a lot of cases, already quite a good estimate for the maximal angular momentum detectable in the data. Plots of χ^2/ndf vs. energy show 'bumps' whenever new important contributions from higher angular momenta enter the data (cf. Figures 3,4, top).

(II) Model-comparisons: The fitted Legendre-moments $\left(a_{\ell_{\max}}\right)_k^{\check{\Omega}^{\alpha}}$ can be compared to the right hand side of equation (6), i.e. to the definitions of the moments as bilinear forms in terms of multipoles. The latter can be evaluated using multipoles $\mathcal{M}_{\ell}$ stemming from an ED model. These comparisons can be performed switching on/off

$$C_0^{\check{E}} = \left[\begin{array}{c|ccc|cccc} 1 & 0 & 0 & 0 & 0 & 0 & 0 & 0 \\ \hline 0 & 3 & 3 & 0 & 0 & 0 & 0 & 0 \\ 0 & 3 & -1 & 0 & 0 & 0 & 0 & 0 \\ 0 & 0 & 0 & 1 & 0 & 0 & 0 & 0 \\ \hline 0 & 0 & 0 & 0 & 6 & 0 & 12 & 0 \\ 0 & 0 & 0 & 0 & 0 & -1 & 0 & -3 \\ 0 & 0 & 0 & 0 & 12 & 0 & -3 & 0 \\ 0 & 0 & 0 & 0 & 0 & -3 & 0 & 3 \end{array}\right] \equiv \langle S,S\rangle + \langle P,P\rangle + \langle D,D\rangle$$

Table 2: The matrix $C_0^{\check{E}}$ which defines the coefficient $(a_2)_0^{\check{E}}$ for an expansion of $\check{E}$ up to $\ell_{\max} = 2$ is shown. Every matrix element defines a particular multipole-interference term. As a generic feature, the matrices $C_n^{\check{\Omega}^\alpha}$ decompose into blocks of interference terms between multipoles with definite orbital angular momentum quantum number ℓ (cf. equations (6) and (7)). Therefore, a short-notation for interference-terms is introduced, using the spectroscopic notation of $S, P, D, \ldots$ for $\ell = 0, 1, 2, \ldots$.

certain model partial waves. In this way, sometimes one obtains valuable information on which partial wave interferences are important. Furthermore, one can try to interpret the results of such comparisons in view of important physical effects visible in the data.

We commence with the discussion of results obtained from new data for π^0- and η-photoproduction [18–20].
In case of π^0-photoproduction, we consider a dataset for the observable $\check{E}$ recently measured by the A2-collaboration at MAMI, and which has been first published in the PhD-thesis [18]. The formula (5) reads, when adapted to the observable $\check{E}$, as follows (see also Table 1)

$$\check{E}(W,\theta) = \frac{q}{k} \sum_{n=0}^{2\ell_{\max}} (a_{\ell_{\max}})_n^{\check{E}}(W)\, P_n(\cos\theta). \quad (9)$$

This expression has been fitted to the data for different $\ell_{\max}$ and an order of $\ell_{\max} = 4$ was sufficient to yield a satisfactory fit over the whole measured energy-region. The plot of χ^2/ndf vs. energy, as well as several exemplary angular distributions, are shown in Figure 3.
Furthermore, we show plots of the two selected Legendre-moments $a_0^{\check{E}}$ and $a_2^{\check{E}}$ in Figure 1 (see also the thesis [18]). In these plots, the energy-region has been zoomed in, in order to facilitate a more detailed view of the fit-results for the moments. Furthermore, a model-comparison is shown for the Bonn-Gatchina solution BnGa 2017_02, which has been used to re-evaluate both of the moments using different truncation orders. We see that for $a_0^{\check{E}}$ below the $p\eta$-threshold, i.e. for roughly $W < 1500$ MeV, BnGa S- and P-waves are largely sufficient to describe the fit-results for the moment. Very close to the $p\eta$-threshold, a small correction due to D-waves is needed. Then, for all energies above the $p\eta$-threshold, the BnGa F-waves are required to correct the continuous curve such that it coincides with the fitted values for the Legendre-moment. One feature of the shown result is striking: exactly at the $p\eta$-threshold, the fitted values show a pronounces, sudden change in direction, or 'cusp'. Furthermore, analytic S-Matrix theory requires amplitudes to have branch-point singularities precisely at the energies corresponding to the opening of thresholds [7, 21]. Such singularities lead to cusp-like behaviours and they are basically a direct consequence of the combination of the principles of analyticity and unitarity. We conclude that this particular polarization measurement is precise enough to show such sophisticated theoretical effects directly in the data, or more precisely in the extracted Legendre-moments. Furthermore, we see in the comparison-plot that for $a_0^{\check{E}}$, the cusp-effect enters due to a $\langle D, D\rangle$ interference-term.
The same effect can be seen in the plot for $a_2^{\check{E}}$ on the right of Figure 1. The extracted values for the Legendre-moment show a pronounced cusp right at the $p\eta$-threshold. Furthermore, the cusp enters into the BnGa model-curves once the D-waves are included. Here, this cusp-effect is due to the $\langle S, D\rangle$- and $\langle D, D\rangle$ interference-blocks. However, contrary to the results for the moment $a_0^{\check{E}}$, (small) F-wave corrections are important already below the $p\eta$-threshold.
For the process of η-photoproduction, we consider a dataset for the beam-asymmetry $\check{\Sigma}$, which has been recently measured by the CBELSA/TAPS-collaboration [18–20]. When using the general expression (5) for the observable $\check{\Sigma}$, one obtains the following angular parametrization

$$\check{\Sigma}(W,\theta) = \frac{q}{k} \sum_{n=2}^{2\ell_{\max}} (a_{\ell_{\max}})_n^{\check{\Sigma}}(W)\, P_n^2(\cos\theta). \quad (10)$$

The resulting plots of χ^2/ndf vs. energy, as well as some fitted angular distributions, can be seen in Figure 4. Over the whole fitted energy-region, again a truncation-order of $\ell_{\max} = 4$ is found to yield a satisfactory fit quality.

As an example for an extracted Legendre-moment, we consider here the quantity $a_4^{\check{\Sigma}}$. A plot of the extracted values, as well as a comparison to the Bonn-Gatchina solution BnGa 2014_02, are shown in Figure 2a. The fitted Legendre-moment, which is consistent with zero in the lower energy region, shows a pronounced rise, which looks like a cusp-like structure, right at the $p\eta'$-threshold. Furthermore, the solution BnGa 2014_02 cannot describe this structure. As can be seen in some exemplary angular distributions shown in Figure 2a, the rise of the moment $a_4^{\check{\Sigma}}$ to values significantly larger than zero is in correspondence with the appearance of a backward-peak in the

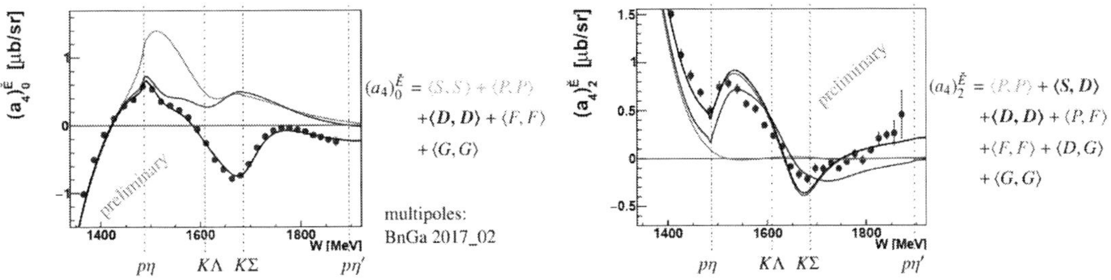

Figure 1: Shown are the two Legendre-moments $a_0^{\check{E}}$ (left) and $a_2^{\check{E}}$ (right), extracted from a recent dataset for the observable $\check{E}$ measured in the reaction $\gamma p \to \pi^0 p$ by the A2-collaboration at MAMI [18]. The coefficients have been extracted using equation (9) in the truncation-order $\ell_{\max} = 4$. The fitted values for the moments are shown as blue dots. Continuous curves show the respective Legendre-moment, evaluated using multipoles from the recent Bonn-Gatchina solution BnGa 2017_02, up to the finite truncation orders: $\ell_{\max} = 1$ (green curve), $\ell_{\max} = 2$ (blue), $\ell_{\max} = 3$ (red) and $\ell_{\max} = 4$ (black). The composition of the respective Legendre moment as a bilinear form in the multipoles is written in the short-notation on the right of the respective plot. Threshold-energies of important photoproduction final-states are illustrated by dash-dotted vertical lines. (color online)

measured angular distribution for $\check{\Sigma}$. This backward-peak needs the modulation $a_4^{\check{\Sigma}} P_4^2(\cos\theta)$ in order to be described. This will be discussed in more detail in upcoming publications (for instance [19]).

For a comparison of the extracted values for the moment $a_4^{\check{\Sigma}}$ to a more recent BnGa-solution, see Figure 2b. There, it is seen that the more recent solution BnGa 2017_02 can describe the cusp a lot better and that furthermore, the cusp-structure enters the model-curve once the G-waves are included, via an $\langle S, G\rangle$ interference-block. For a more involved discussion of the physical meaning of this cusp-effect in η-photoproduction data, see upcoming publications [19].

We conclude that in a new era of polarization-measurements, the precision of the data has become good enough to be directly sensitive to singularities in the amplitudes as dictated by S-matrix theory. This puts new increased demands to the construction and fit of ED models. Furthermore, moment-analysis is the ideal method to make sensitivities to such singularities visible.

http://doi.org/10.1051/epjconf/202024103006

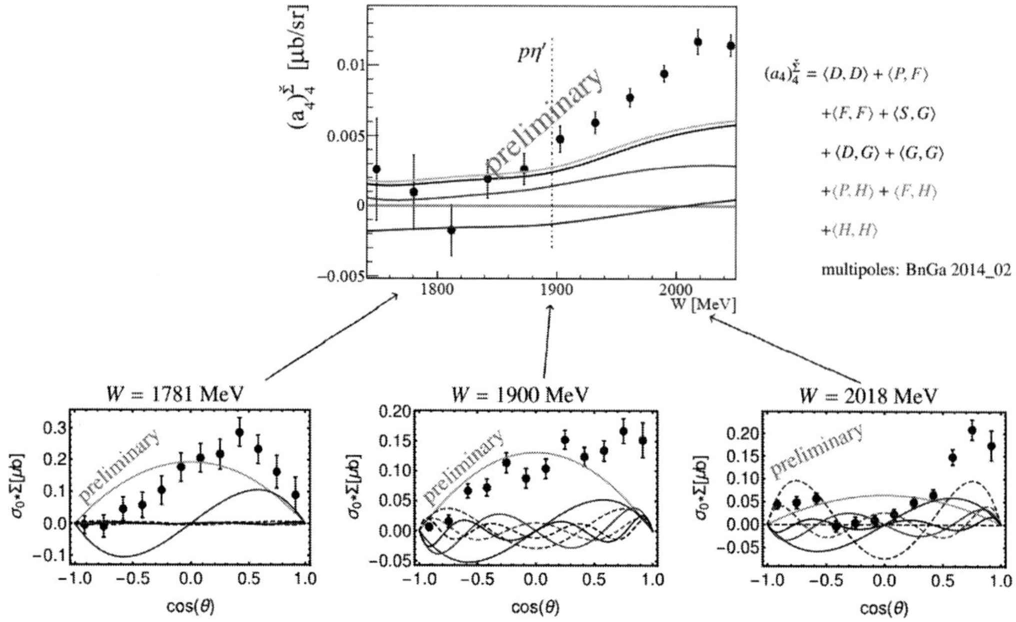

(a) The Legendre-moments $a_4^{\check{\Sigma}}$, extracted from a recent dataset for the observable $\check{\Sigma}$ measured in the reaction $\gamma p \rightarrow \eta p$ by the CBELSA/TAPS-collaboration [18–20] is shown at the top. The coefficient has been extracted using equation (10) in the truncation-order $\ell_{\max} = 4$. The fitted values are shown as black dots. Continuous curves show the Legendre-moment, evaluated using multipoles from the Bonn-Gatchina solution BnGa 2014_02, up to the finite truncation orders: $\ell_{\max} = 1$ (green curve), $\ell_{\max} = 2$ (blue), $\ell_{\max} = 3$ (red), $\ell_{\max} = 4$ (black) and $\ell_{\max} = 5$ (cyan). The composition of the Legendre moment as a bilinear form in the multipoles is written in the short-notation on the right of the plot. The threshold-energy of the $p\eta'$ final state is illustrated by the dash-dotted vertical line. Furthermore, three angular distributions of the profile function $\check{E} = \sigma_0 E$ are shown at the bottom of the figure. For the three exemplary energies, the different contributions from individual Legendre-moments to the fitted function are shown as modulations. In particular, these are the contributions (cf. equation (10)): $a_2^{\check{\Sigma}} P_2^2(\cos\theta)$ (green solid line), $a_3^{\check{\Sigma}} P_3^2(\cos\theta)$ (blue solid line), $a_4^{\check{\Sigma}} P_4^2(\cos\theta)$ (blue dashed line), $a_5^{\check{\Sigma}} P_5^2(\cos\theta)$ (red solid line), $a_6^{\check{\Sigma}} P_6^2(\cos\theta)$ (red dashed line), $a_7^{\check{\Sigma}} P_7^2(\cos\theta)$ (black solid line), $a_8^{\check{\Sigma}} P_8^2(\cos\theta)$ (black dashed line). Arrows indicate the corresponding energy-bins in the plot of $a_4^{\check{\Sigma}}$. (color online)

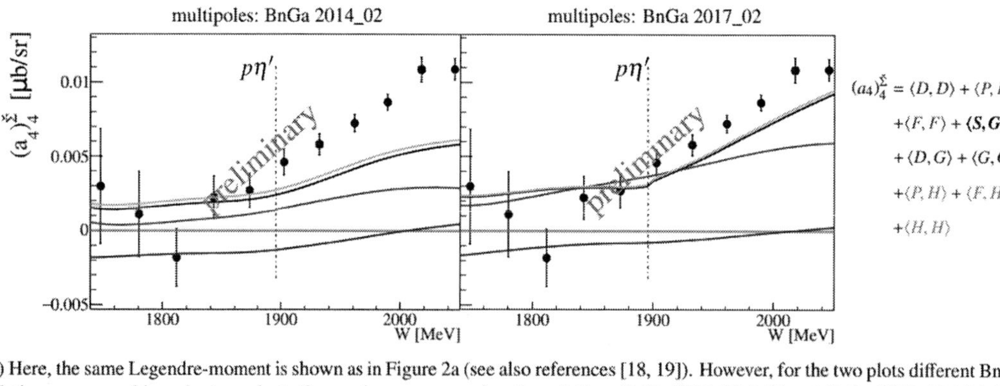

(b) Here, the same Legendre-moment is shown as in Figure 2a (see also references [18, 19]). However, for the two plots different BnGa-solutions were used in order to evaluate the continuous curves, i.e. the solutions BnGa 2014_02 (left) and BnGa 2017_02 (right). In particular, in case of BnGa 2017_02, one can see the $p\eta'$-cusp in the continuous curves once the G-waves are included. The composition of the Legendre-moment in terms of the short-notations for interference terms is shown on the right of the plots. The threshold-energy of the $p\eta'$ final state is shown as a dash-dotted vertical line. (color online)

Figure 2: Plots showing the Legendre-moment $a_4^{\check{\Sigma}}$, extracted from η-photoproduction data [18–20].

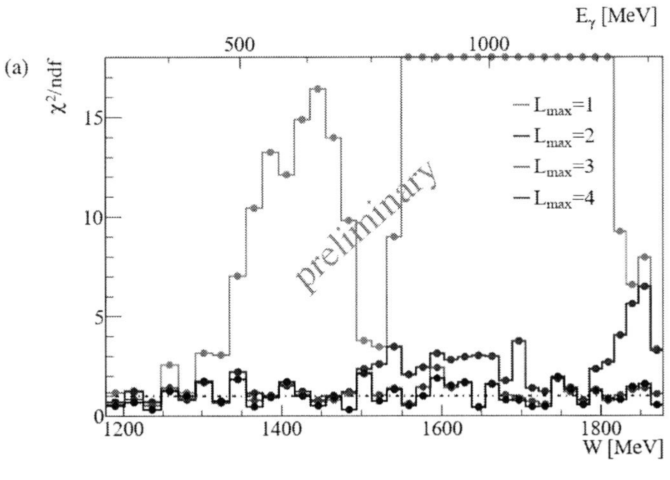

Figure 3: The double-polarization observable $\check{E}$ data from the A2-collaboration at MAMI [18], with only statistical error, was fitted using associated Legendre polynomials according to the Legendre moment expansions (5,9) and truncating the expansion at $\ell_{max} = 1, \ldots, 4$. (a) The resulting χ^2/ndf values for fits with different orders ℓ_{max} as a function of the center of mass energy W and photon LAB-energy E_γ are shown. (b) 12 out of 38 selected angular distributions of $\check{\Sigma}$ (black points) are plotted together with fit-curves for different ℓ_{max} (solid lines), starting at W= 1484 MeV up to 1855 MeV. (color online)

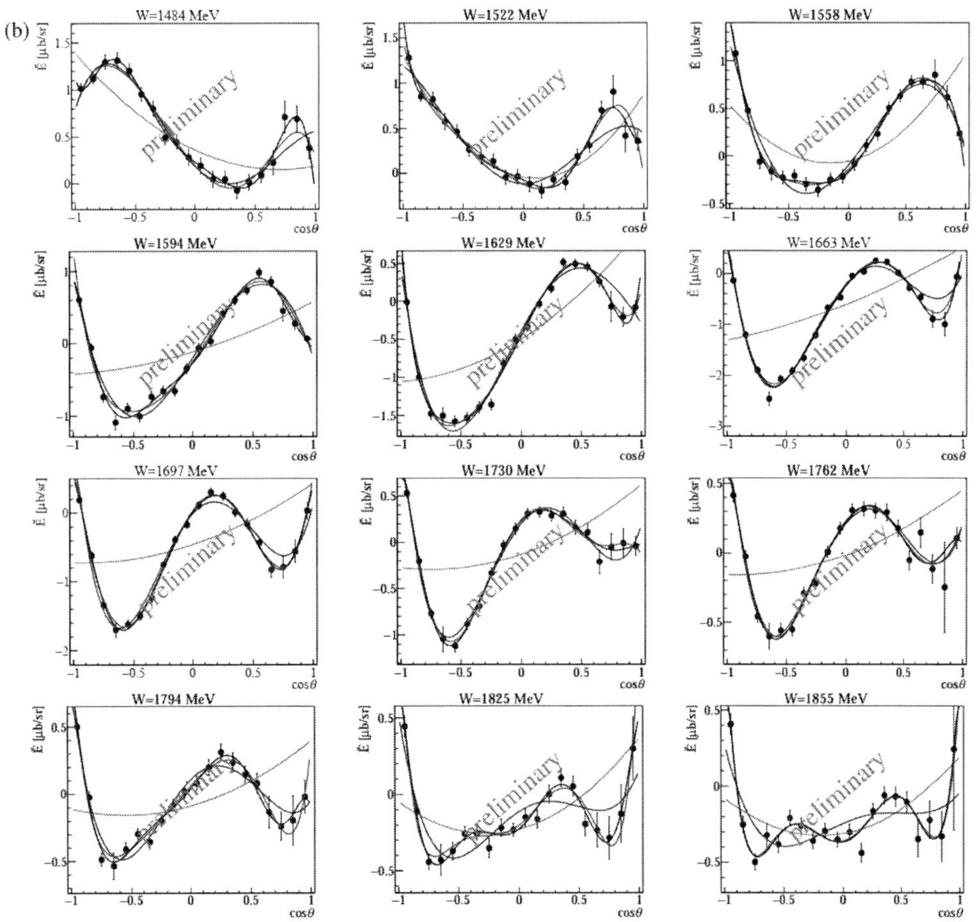

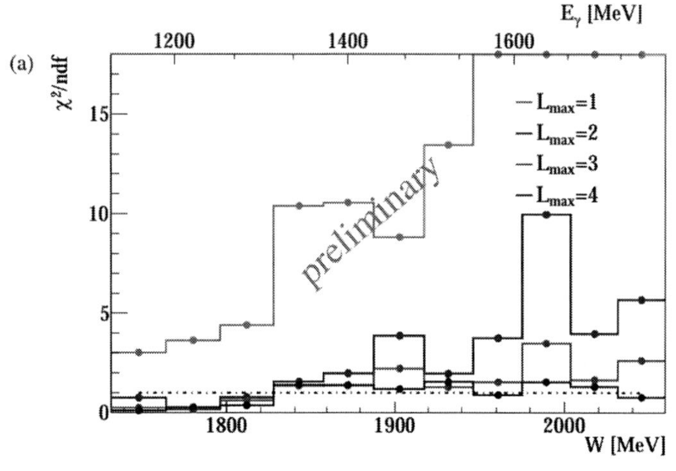

Figure 4: The beam-asymmetry $\check{\Sigma}$ data from the CBELSA/TAPS-collaboration [18, 19], with only statistical error, was fitted using the expansion into Legendre moments given in equations (5,10) and truncating the expansion at $\ell_{\max} = 1, \ldots, 4$. (a) The resulting χ^2/ndf values for fits with different orders $\ell_{\max}$ as a function of the center of mass energy W and photon LAB-energy E_γ are shown. (b) 6 out of 11 selected angular distributions of $\check{\Sigma}$ (black points) are plotted together with curves corresponding to fits with different $\ell_{\max}$ (solid lines), starting at W= 1748 MeV up to 2045 MeV. (color online)

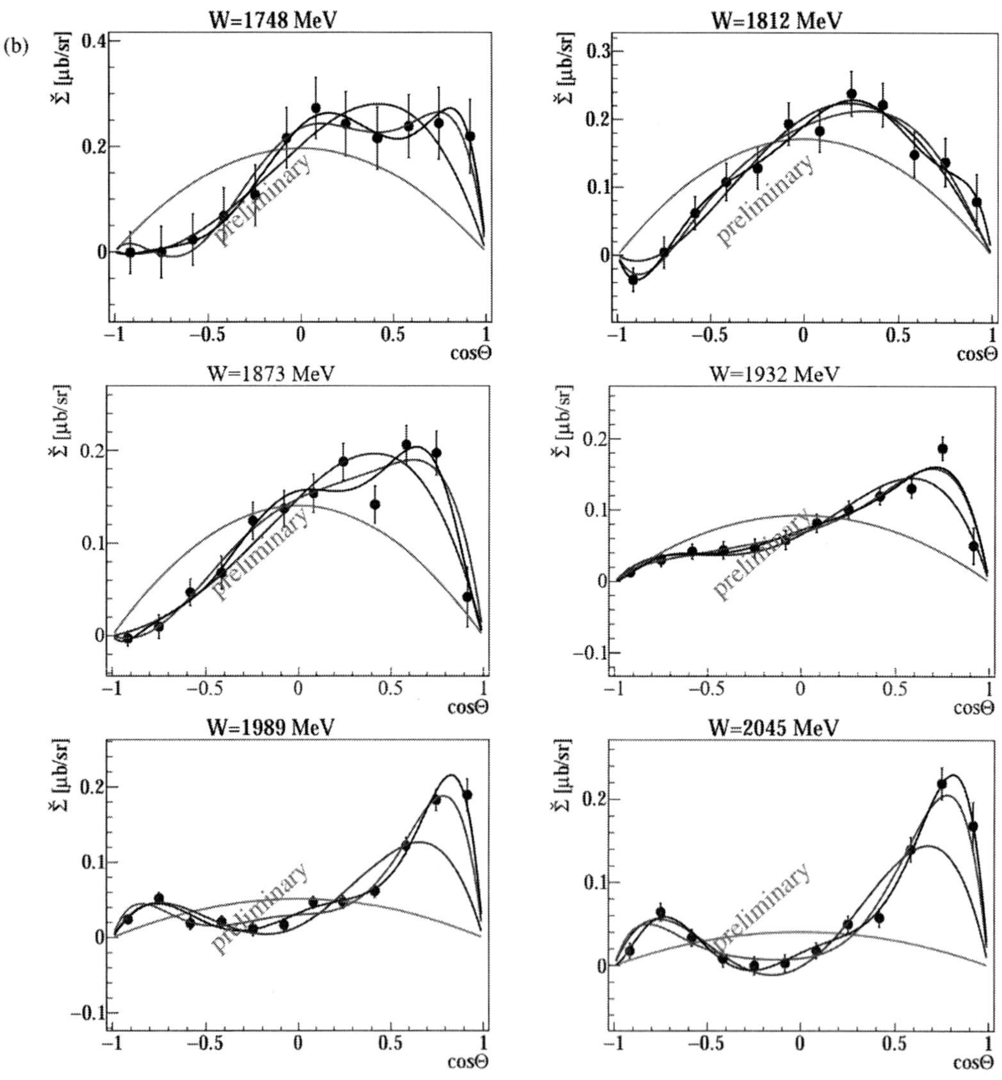

References

[1] A. V. Anisovich, R. Beck, E. Klempt, V. A. Nikonov, A. V. Sarantsev and U. Thoma, Eur. Phys. J. A **48**, 15 (2012).

[2] D. Ronchen *et al.*, Eur. Phys. J. A **49**, 44 (2013) doi:10.1140/epja/i2013-13044-5 [arXiv:1211.6998 [nucl-th]].

[3] M. Doring, C. Hanhart, F. Huang, S. Krewald and U.-G. Meißner, Phys. Lett. B **681**, 26 (2009) doi:10.1016/j.physletb.2009.09.052 [arXiv:0903.1781 [nucl-th]].

[4] D. Drechsel, S. S. Kamalov and L. Tiator, Eur. Phys. J. A **34**, 69 (2007).

[5] R. A. Arndt, W. J. Briscoe, I. I. Strakovsky and R. L. Workman, Phys. Rev. C **74**, 045205 (2006).

[6] A. V. Anisovich *et al.*, Eur. Phys. J. A **52**, no. 9, 284 (2016) doi:10.1140/epja/i2016-16284-9 [arXiv:1604.05704 [nucl-th]].

[7] R. J. Eden, P. V. Landshoff, D. I. Olive and J. C. Polkinghorne, "The analytic S-matrix", Cambridge University Press, (1966).

[8] J. E. Bowcock and H. Burkhardt, Rep. Prog. Phys. **38**, 1099 (1975).

[9] Y. Wunderlich, "The complete experiment problem of pseudoscalar meson photoproduction in a truncated partial wave analysis", url: http://hss.ulb.uni-bonn.de/2019/5353/5353.htm, PhD-thesis, University of Bonn (2019).

[10] A. Gersten, Nucl. Phys. B **12**, p. 537 (1969).

[11] G. F. Chew, M. L. Goldberger, F. E. Low, and Y. Nambu, Phys. Rev. **106**, 1345 (1957).

[12] A. M. Sandorfi, S. Hoblit, H. Kamano, T. -S. H. Lee, J. Phys. G **38**, 053001 (2011).

[13] W.-T. Chiang and F. Tabakin, Phys. Rev. C **55**, 2054 (1997).

[14] Y. Wunderlich, F. Afzal, A. Thiel and R. Beck, Eur. Phys. J. A **53**, no. 5, 86 (2017).

[15] L. Tiator, AIP Conf. Proc. **1432**, 162 (2012).

[16] A. S. Omelaenko, Sov. J. Nucl. Phys. **34**, 406 (1981).

[17] Y. Wunderlich, R. Beck and L. Tiator, Phys. Rev. C **85**, 055203 (2014).

[18] F. Afzal, "Measurement of the beam and helicity asymmetries in the reactions $\gamma p \to p\pi^0$ and $\gamma p \to p\eta$", url: http://hss.ulb.uni-bonn.de/2019/5551/5551.htm, PhD-thesis, University of Bonn (2019).

[19] F. Afzal *et al.*, "Precise beam asymmetry Σ data for $\gamma p \to \eta p$ in the $p\eta'$ threshold region", in preparation (2019).

[20] F. Afzal [CBELSA/TAPS Collaboration], "Recent results from the CBELSA/TAPS experiment at ELSA", contribution to *NSTAR2019* (2019).

[21] S. Ceci, M. Doring, C. Hanhart, S. Krewald, U.-G. Meissner and A. Svarc, Phys. Rev. C **84**, 015205 (2011).

AUTHOR INDEX

AUTHOR INDEX

AUTHOR INDEX

EDP Sciences – Web of Conferences
17, Avenue du Hoggar
Parc d'Activité de Courtabouf
BP 112
F-91944 Les Ulis Cedex A
France

ISBN 978-1-7138-1919-6

CURRAN ASSOCIATES INC.
proceedings.com